ELECTRICAL CORONAS

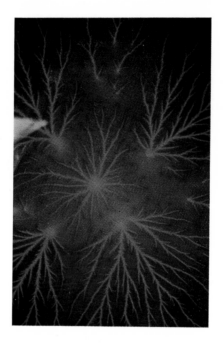

Color autograph of corona streamers taken with a
film transverse to the corona gap axis using a positive point with impulse potential made at 2 cm
from the anode point in the point-to-plane gap.

LEONARD B. LOEB

ELECTRICAL CORONAS

Their Basic
Physical Mechanisms

1965
UNIVERSITY OF CALIFORNIA PRESS
BERKELEY AND LOS ANGELES

PHYSICS

University of California Press
Berkeley and Los Angeles, California
Cambridge University Press
London, England
Library of Congress Catalog Card Number: 64-18642
© 1965 by The Regents of the University of California
Printed in the United States of America

ELECTRICAL CORONAS

*This book is gratefully dedicated to those
who made it possible:*

MY GRADUATE STUDENTS,
FOR WORK OVER THE YEARS,
and the OFFICE OF NAVAL RESEARCH
AND THE RESEARCH CORPORATION OF NEW YORK,
FOR FINANCIAL SUPPORT OF THE STUDENTS

Preface

It might seem that such a relatively inconspicuous form of electrical discharge would hardly merit a book of the length of the present volume. It is indeed true that most of the manifestations, extending as they do from minute, inconspicuous glow discharges in air, ranging from millimeters in length and some tenths of a millimeter in breadth to tenuous glowing spheres some tens of centimeters in diameter in the form of St. Elmo's fire, are neither awe-inspiring nor dramatic in appearance. Many of them cannot be seen except in a well-darkened room. Yet these same obscure discharge forms represent unique, highly interesting, and even dramatic sequences of events. Thus, for example, the Trichel pulse corona in air at 760 mm rises from a state of no current to a current of 10^{-2} amp and a current density of 100 amp/cm^2 in two one-hundred millionths of a second and dies off in some three one-hundred millionths of a second more.

Industrially these seemingly unimpressive phenomena are of great importance and use. They are responsible for a considerable power loss in high tension lines and increase costs by imposing the use of more massive metals than would otherwise be indicated. They lead to insulator deterioration and consequent costly flashover with power blackout. They are responsible for electrical noise, often called static on the radio, especially where insulators are exposed to moist air and humidity. Without corona discharge no dust would be removed by electrical precipitation, and air pollution control would be more difficult. Electrically coded and transmitted messages can be printed out as letters on paper tape at the rate of 10,000 letters per second by its means. It is of use in certain electrochemical processes. The Geiger counter is a corona discharge device. Coronas are being used extensively now in dry-process ore beneficiation, thus permitting the utilization of ores that are otherwise of too low grade to work economically. Such coronas are also being used effectively in the deëmulsification of

crude oil–brine mixtures. They are also used in the discharge of precipitation statically electrified aircraft in flight. They are used now and can
become an efficient means of discharging statically electrified surfaces of
wool and paper in the wool and papermaking industries. They may become
still more important in static elimination with the increasing use of plastics
in industry. These are just a few of the industrial applications and incon-
veniences associated with the corona discharge.

Probably far more important in the contributions of the study of coronas
to our civilization are the end they serve in making us acquainted with the
processes of electrical breakdown in gases. It was the positive point, or
wire, electrical corona that led physicists to the discovery of the vital
third type of secondary ionization processes leading to electrical breakdown.
This was the important role of photoionization in the gas. The reason why
corona study has done perhaps as much, or more, than the study of any
other discharge modes in furthering our knowledge of the basic processes
and how they act stems from the fact that the breakdowns associated with
them are localized in the isolated high electrical fields of one polarity.
Thus the other electrode, since it is either so distant or has so low a surface
field that it contributes little to the breakdown process, allows a more
detailed analysis of what takes place at electrodes of different sign before
breakdown. These isolated, highly asymmetrical fields, which are associated
with and cause the appearance of coronas, have the advantage over all
other geometrical arrangements in that the potential and current range
between the initial breakdown at the electrode surface and the complete
breakdown of the gap to a transient arc via the electrical spark is in most
cases remarkably large. This gives one the opportunity to study quite in
detail the various steps and mechanisms in such breakdown processes.
Under other conditions these phenomena are such complicated functions
of the interplay of processes active at both electrodes and on such short
time scales that they preclude proper analysis. It is for this reason that the
study of these phenomena has proven so immensely rewarding.

Initially, corona discharges were studied largely by engineers and occasionally by physicists whose interest was aroused by some curious aspect
or other of their behavior. However, neither the physical mechanisms
themselves, nor the value which these have in the delineation of breakdown
processes, could have been known to the earlier investigators until the
proper physical techniques were at hand.

Progress began in the late 1930's with the availability of single-sweep
cathode ray oscilloscopes of time resolution of 10^{-5} second or better. Since
1950, with the technical developments of the post-World War II period,
with time resolution progressively improving from 10^{-7} to a few times
10^{-9} second, with photomultipliers, with square wave high potential pulses
of short duration, and finally with the improvement in gas purification

and vacuum techniques, the whole field has been thrown open to investigation. Hence since about 1937 an extensive and progressively more accurate fund of information has been accumulated. This has been published in a large variety of journals of a world-wide scope in the intervening years. Progress has been especially rapid since the late 1950's, in consequence of improved techniques, so that today as this book goes to press about 20 per cent of the information is as yet not fully published in the journals.

In 1936 the author's attention was called to certain discrepancies between observed corona behavior and the classical Townsend theory of gaseous breakdown. This discrepancy launched the author and his group of graduate students into an extensive and prolonged study of the coronas and their associated phenomena. Through encouragement by funds allotted to support the doctoral candidates engaged in these studies, first by the Research Corporation and later by both the Research Corporation and the Office of Naval Research, the progress made in the pioneering studies by this group, amplified by many brilliant contributions from research laboratories throughout the world, required that this material be systematically brought together in a book.

In 1953 work was begun on a preliminary assembly of the material in the form of two chapters of an intended book on discharge through gases. The publication of volume XXII of the *Encyclopedia of Physics*, which appeared in 1956, led the author to abandon completion of the half-finished book on the electrical breakdown of gases, since the chapters written or in preparation for the book had been adequately covered by the *Encyclopedia*. Following this decision the completion of the investigations of static electrification in the author's laboratory in 1956 necessitated the publication of a book covering these studies, which had also been supported by funds from the Office of Naval Research. This book, entitled *Static Electrification*, was published in 1958. The one aspect of the phenomena of electrical breakdown of gases not included in volume XXII of the *Encyclopedia* was that of the coronas. Stimulated by the rapid and exciting advances in this field beginning around 1960, the author was encouraged to begin anew the task of gathering together the progress in this field, starting from the collection of the material compiled in 1953. Writing was started in 1961.

This book has therefore been written to gather under one cover the cogent but widely scattered material on the subject—so to speak to separate the wheat from the chaff, but above all insofar as possible to analyze the phenomena in terms of the basic processes active and to the end that they may serve as a diagnostic for other discharge studies, as well as assisting the engineer utilizing these phenomena to achieve his objectives more easily. This book does not and is not intended to portray the applied aspects of the phenomena nor to suggest such applications. A study of the properties of these discharges unquestionably will indicate many new and useful applications to inventive minds.

The first chapter serves to introduce this subject, to give essential definitions, and to present the various gap forms in use, which at the same time are of such a nature as to permit some knowledge of the fields across the gaps before space charge distortion intervenes. The second chapter gives a condensed summary of the essential basic processes of gaseous electronics involved and derives the quantitative theories governing those of the manifestations which are readily derived and will frequently be invoked in what follows. The subject is then developed in terms of essentially three main divisions: (1) highly asymmetrical gaps in which virtually all the breakdown processes initiate from the highly stressed electrode with minimum, or negligible, contribution from processes at the other electrode; (2) asymmetrical gaps in which the other electrode plays a minor role in the breakdown process; and (3) symmetrical gaps with highly asymmetrical fields at both electrodes. In general the asymmetrical gap cases, 1 and 2, are divided into separate sections dealing with the highly stressed electrode acting either as anode or cathode. Under these headings two types of gases will be treated: those that let the electrons exist as free electrons and those in which electrons attach to make negative ions. Under these general headings as well, certain separate topics will be treated in appropriate sections at the proper place. The highly stressed anode is discussed in chapter 3, which is in turn divided into two chief sections, one on attaching gases where negative ions play a role, and the other on gases in which electrons remain free.

Chapter 4 deals with the highly stressed cathode and is again divided into electron attaching and free electron gases. The data on the free electron gases, in contrast to the electron attaching gases and in general work with relatively pure and very pure gases, is unfortunately all too meager. This stems from the relatively late introduction of techniques of achieving outgassed surfaces, high vacua, and gas electrophoretic cleansing in a manner amenable to general laboratory practice. Chapters 5 and 6 deal with corona phenomena in asymmetrical gaps in which the gap length, or geometrical form, is such as to involve some participation of the less highly stressed electrode. The ideal and most frequently used arrangement here is the coaxial cylindrical geometry, although what is there observed applies nearly equally well to shorter gaps of other forms. Chapter 5 covers the highly stressed anode, and chapter 6 the highly stressed cathode, again subdivided into free electron and electron attaching gases. The concluding section of chapter 6 appropriately treats the subject of relative breakdown thresholds as a function of gas species and pressure. It serves admirably to illustrate how sensitive such thresholds are to various influences and how dangerous it is to generalize and predict behavior in any particular instance without a very careful study.

Chapter 7 is relatively brief. It applies the principles developed in the previous chapters to the analysis of non-uniform fields but symmetrical electrode forms. It touches on the various breakdown mechanisms that can occur insofar as knowledge permits, both for impulse and static breakdowns. It deals with the rectifying properties of asymmetrical gaps and concludes with a discussion of the influence of external light on the breakdown of such gaps, with analysis of actions such as are analogous to the notorious Joshi effect treated, however, for gas systems where the basic processes can be analyzed.

Of special interest are the following features of the book. Chapter 3 devotes some space to a most important discovery by Hermstein of a new form of glow discharge about a highly stressed anode, in which a negative ion space charge cloud creates a high anode potential drop akin to the cathode fall produced by the positive ion in the common cathode glow discharge. This Hermstein glow discharge, discovered in 1960 and published in a relatively obscure journal, explains the extensive potential range between corona threshold and arc-over with highly stressed anode in electron attaching gases. If its presence is ensured, the sparking potential of some asymmetrical gaps can be raised by many kilovolts. The same chapter summarizes recent developments in the study of the streamer mechanism of the electrical spark by the research group of Raether in Hamburg. It presents in detail the recent most startling revelations of the streamer mechanism deriving from the Lichtenberg figure technique, using impulse potentials begun by Nasser in 1959 in West Berlin and later carried on in the author's laboratory. Much of this work, including that on negative streamers, is still in the process of publication. The recent discoveries by Waidmann, of the influence of the absorption coefficients of various gases on the propagation of the breakdown streamers, appear in chapter 3, and the recent discoveries of Tamura on the same subject using point-to-plane corona, including some as yet unpublished data, are featured in chapter 2. Chapter 5 details the very revealing studies concerning the progress of the breakdown in coaxial cylindrical geometry in free electron gases, made possible by the α particle triggering techniques of Lauer, Colli and Facchini, and Huber with anode wire. This work quite logically leads into an extensive study of the physical mechanisms of the Geiger counter, with theoretical interpretation devoid of the complication of counting instrumentation. It includes a recent study of the propagation of the pulse as detected by photomultiplier, as well as some discussion of the halogen-inert gas counters.

A late insertion in this book is a very important section (chapter 4, section A) on the work of Miyoshi, not as yet published in English, on the negative point-to-plane corona in air over an extensive range of pressures and gap geometry. This leads to some very interesting generalizations concerning breakdown modes in such gaps. The section in chapter 6 dealing with the question of relative sparking potential thresholds with highly stressed anode and cathode is most significant in the diagnosis of such problems. Chapter 4 also includes a section on the electrical wind. Finally, the section of chapter 7 dealing with a critique in terms of known mechanisms of the peculiar influence of light on threshold potentials, of breakdown in cells with metal or glass electrodes, and the rectifying action of such cells is the first all-around critique of the notorious Joshi effect. This analysis leads to an indication of the direction in which investigations should be carried out in order to delineate the physics of such light action. Recent sensational progress dating from the Sixth International Conference on Ionization Phenomena in Gases in Paris in 1963 and other important advances since the text has gone to press are included in the Appendixes.

It is the hope of the author that this compilation will prove of interest and value to those investigators who desire a more profound understanding of the physics of electrical breakdown processes as well as to the industrial physicist and engineer whose tasks take him into the nature of corona discharges and gaseous breakdown studies.

Contents

1

Introduction:
Definitions, Gap Forms,
Basic Mechanisms, and
Fundamental Relations

A. INTRODUCTION

1. GENERAL CONSIDERATIONS TYPIFYING CORONA BREAKDOWN

In principle these discharges do not differ from the discharges with uniform electrical fields, since the same basic mechanisms, a first coefficient and a second coefficient such as are involved in the uniform field breakdowns, must occur in the same generic relationships. They do, however, differ in that with non-uniform fields or asymmetrical electrodes breakdown initiates in the high fields now concentrated at one or both of the electrodes. The weak field between the electrodes, or on one of them in asymmetrical gaps, however, precludes a complete *breakdown of the whole gap to an arc at or near the threshold despite discharges at one or both of the electrodes.* Thus the breakdown will occur in a sequence of steps over a considerable range of potentials from the first manifestation to the complete spark breakdown. The action of the low field region, aside from yielding fields too low for breakdown streamer propagation at the lower thresholds, is involved with the action of space charges which in some cases facilitate the breakdown procedure and in others impede it. It thus happens that the non-uniform field geometries lead to a stepwise breakdown spreading out in voltage and sometimes in time as a sequence of events which in the uniform field

1

breakdown takes place nearly in one act with a time scale in microseconds or less. In consequence such discharges have been studied for the purpose of following breakdown mechanisms under various conditions as well as in their own right.

While any form of non-uniform field gap might be studied, the phenomena of interest are more clearly delineated if the electrodes represent the extremes of divergence in shape, such as a point-to-infinite-plane gap or coaxial cylinders with the inner cylinder of much smaller radius than the outer cylinder. These have the property of associating the breakdown sequences with *the one electrode* of small radius or high curvature. Hence the anode breakdowns may successively be studied more or less *independently of the action of the other or large electrode*. Other geometrically asymmetrical systems, such as confocal paraboloids,[1.1] or hyperboloids of revolution[1.2] to plane, and concentric spherical or sphere-to-plane[1.3] systems, have been studied. Probably the one most used because of its convenience has been the point-to-plane employing a hemispherically capped cylinder for the point. The other frequently studied system is the coaxial cylindrical system.[1.4] The coaxial cylinder, confocal paraboloids, and the hyperboloid or sphere to infinite plane geometries have the advantage of permitting calculation of the axial electrical fields preceding the first breakdown threshold. After the first notable threshold is reached the fields, except for coaxial cylinders, are largely incalculable because of the presence of space charge distortion. If ion mobilities are known, the coaxial cylindrical geometry permits a fairly good first-order approximation of the fields at steady state for most of the gap at lower currents and adequate pressures. The field along the axis of symmetry for the hemispherically capped cylindrical point-to-plane gap before breakdown can be computed to about one per cent for certain fixed ratios of gap length to point diameter, as shown in the work of E. E. Dodd,[1.5] who carried out the study for one ratio. The tedious computation can only be carried out for one ratio of point diameter to gap length at a time. Dodd's analysis of axial fields before breakdown may be carried out for other ratios if desired. The point-to-plane and coaxial cylindrical systems are easy to perfect and adjust, which makes them particularly suited to study. The fields for confocal paraboloids, sphere-to-plane, concentric spheres, and hyperboloid-to-plane are given by exact analytical expressions. There are differences in the appearance of the phenomena between the point and coaxial cylindrical configurations. The point type geometry has one very small localized region at which the discharge acts, making it convenient to observe and study, since the charge is concentrated in a local area in two dimensions. The intense convergence of the field also enhances certain effects relative to the coaxial cylinders. With the coaxial cylinders the discharge can start at any point along the central wire and spread radially and laterally. It differs also in that it, as well as the confocal and concentric sphere geometries, conserves photons by multiple reflections from the outer cylinder while these are generally lost by scattering in point-to-plane geometry if the gap is not relatively short. That is, such geometries usually include more influence of the larger electrode. Such confined geometries also conserve products of chemical reaction, while ventilation is far better in the point-to-plane type of gap. Thus, these factors lead to somewhat different manifestations in the two types of gaps. In fact, the isolated electrodes yield the more truly characteristic coronas, while the confined geometries often yield pseudo coronas.

2. DIFFICULTIES INHERENT IN STUDIES BEFORE THE FAST CATHODE RAY OSCILLOSCOPE

There was, in the past, much confusion and misunderstanding about the breakdown mechanisms active and the nature of the various thresholds observed, which are now well recognized and understood. In part, the confusion came from different interpretations and nomenclature applied to the thresholds for self-sustaining discharge in a gap with non-uniform field. Some researchers defined the threshold noted as the appearance of a *measurable* current, some in terms of *visible* glows or manifestations, and still others, disregarding the thresholds for minor currents, based their definitions primarily on the more complete breakdown of the gap, that is, on the appearance of a transient arc or a stable arc or glow through the medium of a spark. Further difficulty was encountered from lack of uniformity in geometries used. Much of the earlier work was done with larger conductors and with alternating potential because of the industrial importance of the partial breakdowns and sparkovers for points and wires. Alternating potentials were also resorted to, since the higher potentials required far longer gaps and since large conductors with adequately steady potentials were hard to achieve. Again, much work has been done with non-uniform field gaps of less asymmetrical type, such as point-to-point breakdown. Some difficulties stemmed from unsuspected disturbing geometrical factors. For example, while fields about two opposing plane, flat-ended, long cylindrical rods are symmetrical, they are not symmetrical if one rod is materially shorter than the other and rests on a large plane electrode. Also, asymmetries relative to grounded insulators that acquire surface charges and exposure to surfaces with floating potentials caused trouble. All these studies yielded their own characteristic sequence of mechanisms which added further confusion.

Most of the work preceded the development of the cathode ray oscillograph, with its time resolution in transient studies of 10^{-5} sec or less. Thus, much early effort was restricted to current and potential characteristic studies, which alone are not too significant. Much of the work was done without noting the important changes in form, color, or spectra of the discharges. Many such details yield valuable clues as to the character of the mechanisms active. Perhaps the greatest advances were made in the analysis of the lower pressure coaxial cylindrical coronas through the studies of J. S. Townsend[1.6] and his pupils, and later through the investigations associated with the positive wire corona which became the basis of the Geiger counter after 1928.[1.7]

3. GENERAL PROPERTIES AND BEHAVIOR OF CORONAS; THRESHOLDS AND DEFINITIONS

In general, it was recognized that, as potential was raised in such gaps, some sort of a breakdown appeared at the highly stressed electrode with relatively low currents. These currents, lying in the order of a microampere at relatively low potentials, frequently gave unsteady or fluctuating currents near threshold. At other times such difficulties were not noticed. For short potential ranges above thresholds, currents increase linearly with potential. With increase in potential the discharge also generally became more stable and steady. At this stage, depending on pressure and the nature of the gas, luminosity of some form or other appeared near the highly stressed electrode. Thereafter current increased somewhat faster than linearly, approaching a parabolical increase with potential. Depending on the gap, the current and potential could be increased by considerable margins until a point was reached yielding a *complete breakdown* of the gap to a transient arc through the medium of a *spark* with brilliantly luminous channel at the higher pressures. The breakdown to a spark usually occurred at voltages from two to six times that for onset of the first glow and current. As gaseous purity[1.8] was adequately increased in later years, with the absence of negative ion-forming gases, it became clear that in very pure inert gases, as well as H_2 and N_2, with clean cathodes the highly stressed negative electrode corona did not occur in the absence of external current-limiting resistors, but the gap broke down to a transient arc via a spark. Even in very pure inert gases the potential range between positive corona threshold and spark was much reduced. Again, very confusingly, the breakdown *thresholds* in the same gas were at times observed to be lower for the highly stressed negative electrodes; under other conditions, such as different pressures, they were lower for the highly stressed positive electrode. This situation was somewhat puzzling as interpreted in terms of Townsend's basic breakdown theory,[1.6] especially with the confusion existing as to the significance of the thresholds. The potential differences required to initiate the pre-sparking currents and luminosities were always far below the breakdown for a uniform field gap of the same length. In most cases, even the sparking potential for the non-uniform field gap was markedly lower than for the same length uniform field gap. The potential for the *spark transition* for non-uniform gaps, with the possible exception of extremely pure inert gases and H_2 and N_2, is universally lower for the highly stressed positive electrode than for the negative, unless the pressure is sufficiently low to preclude a filamentary spark.

Except at relatively low pressures, the luminous manifestations at the highly stressed electrode near the threshold for the low currents take on various characteristic shapes, such as glows, multiple spots, haloes, coronas, brushes, streamers, etc. In consequence, these luminous manifestations gave to the phenomena the general name *coronas*. It comes from the French word *couronne*, literally crown, which typifies one of the various forms observed. *This expression, corona, will be used to describe the general class of luminous phenomena appearing associated with the current jump to some microamperes at the highly stressed electrode preceding the ultimate spark breakdown of the gap.* Where observed, the sudden current jump, usually just preceding the initial appearance of the corona and the associated value of the potential, will be designated as the *corona threshold*. The threshold for the appearance of a corona form may be further classified in terms of the characteristic phenomenon or mechanisms associated with it, such as the *burst pulse threshold*, the *streamer threshold*, the *Trichel pulse threshold*, or the *glow discharge threshold*. The current at many such thresholds is pulsating or intermittent in nature. Depending on the geometry and the spectroscopic nature of the gas, the intermittent or pulsed thresholds may not show luminosity in all cases. If the potential is raised on the order of some hundreds of volts above *threshold*, the frequencies of the intermittent pulses become so great that they merge to a nearly steady but slightly fluctuating current. Transition from intermittent to the steady state is sometimes sharp and is described as the *onset of steady corona*. Above the onset of steady corona there will be a limited region, in which current increases nearly proportional to potential increase. This is called the *Ohm's law regime*. After this the current increases more rapidly than potential, that is, parabolically, eventually leading to a complete *spark breakdown*, which will be so designated.

Actually the great variety of phenomena with varied geometry, over an extended range of pressures and gases with varying degrees of purity and condition of electrode surfaces, is such that the superficial observation of thresholds or onsets of the past years do not present any coherent or intelligible assembly of data. In fact, analyses with modern facilities reveal in many instances self-sustaining discharges of low-order setting in well below the microampere current jump and luminosity previously designated as thresholds.[1.9] These occur frequently with negative electrodes of small radii. With points and wires of limited area, these currents are in a range of values that are not readily measured with common instruments such as microammeters. Thus, both they and their significance have been lost to the investigators until quite recently. Ignorance of their existence has been the source of added confusion in interpretation. They, in common with the pre-breakdown currents observed in uniform field gaps *preceding* spark breakdown, very near the sparking threshold, act to *initiate or interfere* with breakdown by means of "conditioning" the cathodes, soiling the gas or building up space charges.[1.10] Similar effects are produced if field intensification of the triggering electron current reaches too high a value.[1.10] Since most metal electrodes are covered with films of oxide or contamination, either introduced or from standing even in fairly good vacuum, secondary emission coefficients will differ from those on the same metal after clean-up or modification by impact of positive ions during the discharge. Thus, initial breakdown potentials may differ from the sustaining potential or that of the freshly discharge-conditioned breakdown potential.

Such changes in onset and offset potentials lead to a sort of threshold and *current hysteresis loop* which will be so designated.[1.9] Such action causes fluctuation in the visual form of the discharge between a diffuse glow over the surface and a spot discharge with or without change of current.

4. Initiating or Triggering Electrons

It is perhaps essential at this point to introduce one aspect of corona studies common also to uniform gap breakdown, namely, the question of initiating or, as we will now term them, *triggering*, electrons. Above threshold or onset, self-sustaining discharges maintain themselves, but it is impossible to start them without initial or triggering electrons in the gap to start the primary multiplication.[1.11] While much study has been devoted to the statistical time lags of sparks in uniform gaps caused by triggering, few such studies[1.11] have been made with coronas.[1] Had they been made, the importance of triggering electrons to reduce statistical time lags would have been manifest. As the later studies indicated that the thresholds of the sustained discharges frequently are preceded by statistically distributed intermittent or pulsed current fluctuations, it became of importance to ensure adequate initiating electrons to trigger the discharge and thus facilitate ascertaining the correct threshold values.[1.12] With uniform field gaps, except for very short gaps, this initial triggering presented no serious

---

[1] Another time lag involved in all breakdown phenomena is the formative time lag, that is, the time between a triggering electronic event and the achievement of the discharge. The latter is significantly characteristic of the breakdown mechanism unless influenced by conditioning predischarges, in which case it becomes meaningless.

problem for breakdown under steady potentials where time lags were not sought for. This is not the case either with time-dependent and alternating potential studies in uniform gaps or, in particular, with the corona gaps. In principle, if electrons are liberated from the cathode in a gap with a highly stressed positive electrode and a distant but extensive cathode, or if volume ionization is used, there is not too much difficulty in providing enough triggering electrons at 760 mm pressure. Natural ionizing events, cosmic rays and so on, create 20 ion-electron pairs/cm^3 per second, which is often adequate, since these accumulate in time. All electrons from the volume will be drawn either as electrons or negative ions into the positive point. There in the high field electrons will detach from negative ions and produce avalanches of suitable size. Conservation of such charges is still better with the positive wire in coaxial cylindrical systems. Thus, at most, a bit of thorianite ore placed a few centimeters from the gap yields field intensified ion currents at potentials below threshold of the order of 10^{-13} amp or better at atmospheric pressure. As pressures are reduced, illumination of the outer cathode with ultraviolet light of about 2537 Å from a quartz Hg arc may be needed. The intensity should be kept low since the field intensified currents, if too large, may create troublesome space charges.

In contrast, the situation with a highly stressed negative electrode is quite different. Electrons move outward from the very limited high field region at the wire or point surface. With small radii or high curvature, this field is very close to the wire and the volume is excessively small. Electrometers cannot measure the pre-threshold current even if thorianite ore is used. In this case, either the radioactivity should be on the point or, better, the wire should be illuminated by ultraviolet light. This will furnish triggering electrons. The light used must be of adequate intensity because of the small surface area of the point or wire. While the need for such triggering had been recognized, beginning with the early studies of Kip,[1.12] the successful achievement of it was another matter. Neither Kip, Hudson, or English[1.12] was able to get reproducible and consistent triggering. The logical procedure using ultraviolet illumination appeared to give capricious results, sometimes *lowering*, at other times *raising* appearance thresholds. If properly used the light should yield consistent and slightly lowered threshold potentials. Therefore, recourse was had to the dusts of MgO, Al$_2$O$_3$, or similar highly insulating substances on the points. These provide triggering electrons by either field emission or the Malter[1.13] effect once an initial breakdown charges them with positive ions. Such triggering is not reliable, as the dust wears off on running a discharge and often gives spurious and uncontrollably irregular effects. H. W. Bandel[1.14] has clarified the difficulties of Kip, Hudson, and English. In all but Hudson's work unfiltered room air was used for the convenience of ultraviolet illumination. Under those circumstances dust was frequently present on the needle.

Thus, under the combined action of strong ultraviolet illumination and dust, the triggering was too great and pre-discharge space charge altered the thresholds. With carefully adjusted intensity of ultraviolet light and no dust, the threshold was observed to be slightly lowered and consistent values were obtained. Care must be used with some metals, employing such illumination in air, as the O_3 created oxidizes the metal in time. Actually, the only studies of consequence on time lags in corona were the more recent ones of M. Menes[1.11] (in the laboratory of L. H. Fisher), B. Gaenger, and possibly some work of R. Strigel.[1.15] These were largely formative time lags for positive points which should and do fall in the 10^{-7} sec time range. Data on negative points are mostly meaningless because of a conditioning and hysteresis action.[1.9,1.10]

5. Visual Appearances of Coronas

A few general comments might be made about the visual appearances. With highly positively stressed electrodes, perhaps three rather characteristic forms are noted. In the intermittent regime *just above threshold*, the most highly stressed area of an electrode will be covered by a seemingly tightly adhering velvety glow as burst pulses appear. This is particularly notable and becomes intense and thin at higher pressures, especially above onset of steady corona. With coaxial cylinders it will appear as spots at onset. As current increases with potential increase, the glow will spread. In the intermittent regime, shortly above threshold, the glow will extend out into the gap a millimeter or two when the pre-onset streamers appear. From points, the glow may have a paintbrush-like appearance with a short, intensely bright stem and a flare below. These will appear at the point tip or in spots on a wire. The forms which at high pressures project into the gap, including the paintbrush-like manifestations, are due to the appearance of pre-onset streamers, and the range is a rough indication of streamer advance in that gap. The tightly adhering glow of the quiescent steady corona, beyond the pre-onset streamer potential, is a form of a newly interpreted glow discharge with negative space charge, characteristic of electron-attracting gases such as air, which suppresses the streamers.[1.16] Generally speaking, with or without negative ions the glow will adhere closely to the anode near threshold. As potential increases, it will project out into the gap, especially where the glows are restricted to patches. Reduction in pressure causes an increase in thickness or extension of all glows into the gap roughly in inverse proportion to the pressure.

At very high potentials approaching spark breakdown, especially in gaps in which point radius is small compared to gap length, a narrow bundle of luminosity composed of many superposed filamentary streamers extends well into and across the gap along the axis. These streamers are a manifestation of incipient breakdown streamers. They do not always appear preceding breakdown with larger electrodes and short gaps. Reduction in pressure will broaden and diffuse all such luminosities in three dimensions because of augmented diffusion.

For the highly stressed negative electrodes at threshold and above breakdown, the luminosity occurs in limited spots in air and in electron-attaching gases. For a point of small to moderate radius, there will at onset be one spot. This discharge is an intermittent localized glow discharge of some 0.02 cm diameter near the cathode surface and extends into the gap perhaps a millimeter or two. Decreasing pressures reveal it to be complex, having a Crookes dark space, a constricted negative glow, a Faraday dark space, and a flaring positive column. This gives it a shaving brush-like appearance. The shape is dictated by the relative potential gradients along the discharge axis and in the space outside the discharge. In free electron gases at threshold, the discharges may appear first as a faint glow extending into the gap which contracts to a spot and then reverts to the initial form, oscillating between the two forms. As potentials and current increase these glows settle down at points to yield the same general configuration as for air but are more diffuse and extend farther into the gap. In some gases the spots will be cusped with the base on the electrode. As potentials increase, the spot at first increases in intensity, but presently a second spot appears, then a third. On a wire the spots multiply along the wire. In some cases they are in ceaseless motion, in others they are quiescent at one point. Both positive and negative glow patches or spots appear to, and actually do, repel each other. On larger smooth surfaces, the quiescent spots, owing to mutual repulsion and field gradients, form regular patterns. Reduction in pressure, as indicated, extends the discharges in three dimensions, roughly inversely proportional to pressure. Near breakdown in air and possibly in other gases, though not readily observed, the negative glow increases in extent, and at lower pressures the Crookes dark space and a faint general glow extending to the negative electrode are all that is seen near the point where pulses cease. Near breakdown, a very sharp luminous spike will usually appear along the axis of the spot in the fan, and with increase in potential it will extend backward toward the negative glow. At a point where this spike approaches the cathode, the spark occurs. In air in the region just referred to, the intermittent pulses characteristic of the threshold and above have ceased. When a spark-suppressing gas like Freon is mixed with air, the discharge takes on a fantastic number of curious and elaborate forms which are not understood.

The spectral behavior of breakdowns in air is of some significance because it reveals important discharge characteristics. Possibly similar inferences could be drawn from single lines or bands in other gases. Low electrical fields yield ionization and excitation that favor the arc spectrum of N_2 and nitric oxide which yields a reddish purple glow. This is characteristic of the fan in the negative point discharge and the outer areas of the paintbrush in the positive streamer region. High electrical fields produce highly excited N_2 molecules, the spectra of which are represented largely by the second positive group bands, with just a few second negative bands shown. These lie largely in the green to violet end of the spectrum and yield an intense electrical blue light. They have recently been photographed on color film in air. This color will be noted in the closely adhering glow of the steady positive corona, in the filamentary breakdown streamer channels, and in the axis of the streamers as well as in spark channels. The streamer tips have very high fields. The negative glow of the Trichel pulse corona in air is also quite white or bluish, as fields here are indeed high. Thus, the color in discharge manifestations in air is a guide to field conditions. In other gases, the colors follow spectral distribution but do not clearly designate high and low field regions except possibly as they affect single lines or bands. In some gases, notably in O_2 and CO, spectra are of such low intensity in the visible as to delay the appearance of luminosity until thresholds have been exceeded by a considerable margin.

B. AXIAL FIELD CALCULATIONS FOR VARIOUS GAP FORMS

One may now turn to the fields existing for the various gap forms. In most cases only fields along the axes are given as they are the most intense. They are also more simply expressed in algebraic language and are easy to compute.

1. COAXIAL CYLINDRICAL CONDENSER[1,4]

The capacity C for a condenser of length l with radii of the inner and outer cylinders a and b respectively is given by

$$(1.1) \qquad C = \frac{lD}{2 \log b/a}.$$

Usually it is more convenient to express the capacity and so on, per unit length. Here D is the dielectric constant of the medium. Since we deal with gases, its value will be taken as 1.000 for practical purposes. The potential V_r at any point r cm distant from the axis is

$$(1.2) \qquad V_r = \int_b^a X \, dr = \int_b^a -\frac{2q}{r} \, dr = 2q \log b/r.$$

If q is the charge per unit length of condenser, $q = CV$ with $C = \dfrac{1}{2 \log b/a}$.

Thus

(1.3)
$$V_r = V \frac{\log b/r}{\log b/a}.$$

The field X at r cm from the axis is then

(1.4)
$$X_r = V/(r \log b/a).$$

2. CONFOCAL PARABOLOIDS

The field at any point along the axis of a gap between confocal paraboloids is given by

(1.5)
$$X_x = \frac{V}{\log (f/F)} \frac{1}{x + f}.$$

Here V is the applied potential, f the focal length of the smaller paraboloid (the point), and F is that of the larger paraboloid. Referred to the focus, the field X at r cm from that point is

(1.6)
$$X_r = \frac{V}{\log f/F} \frac{1}{r}.$$

3. HYPERBOLOID OF REVOLUTION OPPOSITE A PLANE

Bennett[1,2] gives an expression for a field at the tip of the hyperboloid in terms of the radius of curvature R of the tip of the point and the distance b of the tip from the plane as

(1.7)
$$X_p = \frac{2V}{R \log (4b/R)}.$$

The field X_x distant x but near the point (that is, in the critical region for breakdown) is given as

(1.8)
$$X_x = \frac{2V}{R \log (4b/R)} \frac{R}{R + x}.$$

4. SPHERE TO PLANE AND CONCENTRIC SPHERICAL GAPS[1,3]

The field strength X_r at any distance r from the center for inner and outer concentric spheres of radii a and b at an applied potential V is

(1.9)
$$X_r = V_r \frac{ab}{r^2(b-a)},$$

and at the surface of the smaller sphere where r = a,

(1.10)
$$X_a = V_a \frac{b}{a(b-a)}.$$

For a relatively distant plane with b $\sim$ 10a, the potential for the field in the critical region of ionization distant r from the center near the surface of radius a may safely be set as

(1.11)
$$X_r = \frac{V}{r}.$$

5. FIELD ALONG THE AXIS FOR THE HEMISPHERICALLY
CAPPED CYLINDER AGAINST AN INFINITE PLANE

This problem has been solved by E. E. Dodd.[1.5] The important parameter in such a calculation is the ratio L/a, the length of the gap L along the axis to the radius of the point a. The solution of this problem is exceedingly tedious mathematically. It was carried out numerically for one ratio of L/a. The choice of this quantity was based on the reasoning that a fairly large a allows for use of lower values of applied potential, since the high field region extends farther into the gap. However, if L is relatively small, the chief virtue of the point—its isolation from secondary action at the plane—is lost. Thus a larger a requires a larger L. Here again limitations are imposed by convenience in building the gap for controlled studies. If the plane radius R equals L, the gap length, the infinite plane condition is approximated to within experimental uncertainty. Thus, since the glass walls of the enclosing envelope must be distant from the plane, the envelope diameter must be 24 cm with an R of 8 cm. For a gas of relatively high breakdown threshold, such as air at 760 mm pressure, the value of the radius a in centimeters was fixed as 0.025 < a < 0.05. This led to the choice of L/a for Dodd's computation of L/a = 160. It was thus, for practical reasons, termed the "standard" point-to-plane geometry. When Bandel later undertook a complete study of corona phenomena in clean, dry, filtered air over a large range of conditions of pressure and gap lengths with "standard" geometry, this choice proved to have one disadvantage. The larger radius point yielded conveniently low starting potentials. For the positive point, it proved to be so efficient that near the transition point from corona to a transient arc in a streamer spark, the potential range at which the pre-breakdown streamers appeared for convenient gaps requiring points of larger a was so narrow that these interesting phenomena could not be advantageously studied.

TABLE 1.1

x in Units of ρ	x/a	X_x in Terms of $V_{0/a}$	Calculated V/V_0	Trough Model V/V_0
160	0.0	0.6053	1.00	1.000
159.9	0.1	0.5009	0.9450	0.933
159.8	0.2	0.4224	0.8990	0.897
159.7	0.3	0.3597	—	—
159.6	0.4	0.3142	—	—
159.5	0.5	0.2760	0.7968	0.824
159.4	0.6	0.2449	—	—
159.2	0.8	0.1976	—	—
159.0	1.0	0.1639	0.6910	0.725
158.8	1.2	0.1389	—	—
158.6	1.4	0.1197	—	—
157.4	1.6	0.1048	—	—
157.2	1.8	0.09275	0.5931	0.631
157.6	2.4	0.06806	—	—
157.0	3.0	0.05306	0.5098	0.556
156.0	4.0	0.03822	—	—
155.0	5.0	0.02957	0.4318	0.483

Dodd's potential distribution is given in table 1.1. The quantity ρ is the ratio of the position x in question as measured from the plane to gap length L; x is the distance from the point tip of radius a. Thus ρ has a value of 160.00 at a distance of L cm from the tip of the point to the plane or at the distance x = 0 from the point tip. A value of ρ = 155.0 represents the value at 0.97L from the plane and represents a distance x = 5a from the tip. At this distance, the electrostatic field already has become so small as to preclude further ionization by collision in laboratory gaps. For purposes of application to calculation, the values of ρ are more conveniently expressed in terms of the ratio of the distance from the point tip x relative to tip radius a, the limits used in calculating the Townsend integral for avalanche size. The electrostatic field intensity X_x at x is then given in units of the applied point potential V_0 divided by point radius a; that is, V_0/a. The corresponding potential V is described relative to V_0. For comparison the values derived through measurement of equipotentials in a large three-dimensional electrolytic tank are given. The calculated values are accurate to $\pm 1\%$. Further accuracy of calculation is not warranted because of the deviations from infinite plane geometry that are due to the limited plane radius. The trough measurements are low in the high field region and high in the lower field region. This is largely owing to the difficulty of placing the probes accurately in regions of small x/a, since even in the square tank a is reduced by practical considerations limiting L.

REFERENCES

1.1 L. B. Loeb, J. H. Parker, E. E. Dodd, and W. N. English, Rev. Sci. Instr. *21*, 47 (1950).

1.2 L. B. Loeb, A. F. Kip, G. G. Hudson, and W. H. Bennett, Phys. Rev. *60*, 715 (1941).

1.3 W. Hermstein, Arch. Elektrotech. *45*, 214 (1960).

1.4 L. B. Loeb, *Fundamentals of Electricity and Magnetism*, 3d ed. (John Wiley, New York, 1947, and Dover Publications, New York, 1961), p. 217.

1.5 L. B. Loeb, J. H. Parker, E. E. Dodd, and W. N. English, Rev. Sci. Instr. *21*, 42 (1950).

1.6 L. B. Loeb, *Fundamental Processes of Electrical Discharge in Gases* (John Wiley, New York, 1939), pp. 486–494. J. S. Townsend, Phil. Mag. *28*, 83 (1914).

1.7 S. Werner, Z. Physik *90*, 356 (1934). S. Werner, Z. Physik *92*, 705 (1934).

1.8 G. L. Weissler, Phys. Rev. *63*, 96 (1943). C. G. Miller and L. B. Loeb, J. Appl. Phys. *22*, 614 (1951).

1.9 C. G. Miller and L. B. Loeb, J. Appl. Phys. *22*, 494, 614 (1951). W. N. English and L. B. Loeb, J. Appl. Phys. *20*, 710 (1949).

1.10 Kip, A. F., Phys. Rev. *54*, 141 (1938). W. N. English and L. B. Loeb, J. Appl. Phys. *20*, 707 ff. (1949). L. B. Loeb, A. F. Kip, G. G. Hudson, and W. H. Bennett, Phys. Rev. *60*, 717 (1941). These papers deal with triggering, ultraviolet light, and so on. For conditioning and soiling actions, see both references of 1.9.

1.11 M. Menes and L. H. Fisher, Phys. Rev. *94*, 1 (1954).

1.12 See first three references 1.10.

1.13 L. Malter, Phys. Rev. *49*, 879 (1936). H. Paetow, Z. Physik *111*, 770 (1939). Also H. Jacobs, Phys. Rev. *84*, 877 (1951). H. Jacobs, J. Freely, and F. A. Brand, Phys. Rev. *88*, 492 (1952). D. Dobischek, H. Jacobs, and J. Freely, Phys. Rev. *91*, 800 (1953).

1.14 H. W. Bandel, Phys. Rev. *84*, 95 (1951).

1.15 R. Strigel, Wiss. Veroffent. a.d. Siemens Konzern *11*, 52 (1932); *15*, 1 (1936). B. Gaenger, Z. Elektrotechnik *39*, 516 (1949).

1.16 W. Hermstein, Arch. Elektrotech. *45*, 209, 279 (1960).

2

Fundamental Definitions and Relations Pertinent to Electrical Breakdown

It is important next to introduce the fundamental Townsend and later concepts and definitions bearing on breakdown thresholds and currents.

A. FUNDAMENTALS OF ELECTRICAL CARRIERS IN GASES[2.1]

An electron placed in a uniform electrical field of strength $X = V/d$, where V is potential applied and d is the interplanar distance, begins to ionize atoms or molecules of the gas once it has an energy derived from the field exceeding a value E_i, the ionizing potential of the atomic gas species present. Symbolically this action is represented by an equation,

$$A + e \rightarrow A^+ + 2e.$$

Values of such potentials or electron energies are expressed in electron volts (ev), that is, the energy gained by an electron in falling through one volt potential difference, which will be used throughout. Tables 2.1 and 2.2 give a list of values of these for commonly used gases. The chance of ionization on impact relative to a non-ionizing impact is zero at threshold and is roughly given by $P(i) = \beta(E - E_i)$ reaching about 0.1 at $E = 3E_i$.

Between a lower energy E_e and E_i, the electrons excite the gas atoms to higher energy states (represented symbolically by A^*)

$$A + e + KE \rightarrow A^* + e$$

from which the atom recovers in from 10^{-10} to 10^{-7} sec by radiating light

15

TABLE 2.1

MOLECULAR CRITICAL POTENTIALS, IN ELECTRON VOLTS

Gas	First Electronic Excitation Potential	Energy of Dissociation	Ionization Potential
H_2	11.47	4.477	15.422
O_2	1.635	5.115	12.2
N_2	5.23	9.762	15.576
O_3	—	6.17	—
NO	5.38	6.507	9.25
CO	6.04	11.111	14.00
CO_2	10.0	5.46	13.7
H_2O	7.6	9.511	12.6
Cl_2	2.27	2.481	13.2
Br_2	1.71	1.97	13.3
I_2	1.472	1.542	9.0
N_2O	—	—	11.0
OH	4.06	4.45	12.9
HCl	9.62	4.40	—
HBr	—	3.72	—
He_2	—	—	4.251
SO_2	—	—	12.1
Cl^-	—	—	3.78
Br^-	—	—	3.52
S^-	—	—	3.6

quanta or photons. These photons have a frequency ν given by $E_e = h\nu$. For certain atomic or molecular species (that is, inert gas atoms, group II metals, and similar molecular configurations such as N_2), the lifetime may be that of a metastable state (represented by A^m) which can extend to seconds. These are low electronic levels for which transition to the ground state is forbidden by selection rules. At energies above E_i, electrons can excite and ionize or just excite states of energy which emit photons of $h\nu > E_i$. If $E_i > h\nu$, photons can be absorbed by other atoms and reëmitted. For certain states such as the lowest excitation potential in certain atomic species, the probability of such action is large and the photons are absorbed and reëmitted thousands of times, so that it takes a long time for them to escape from the site of their creation through the gas and reach the walls ($\sim 10^{-4}$ sec in some cases). These radiations are said to be imprisoned. Most radiations are only moderately absorbed and quickly (with the speed of light) reach the walls and electrodes. If $E_e = h\nu$ exceeds the work function of the metals upon which light impinges, there is a good chance, γ_p, ($10^{-5} < \gamma_p < 10^{-1}$), that they will eject electrons. If these electrons escape into the gas there is a chance χ of their diffusing back to the electrode, a chance which decreases as the ratio of field strength X to pressure p, X/p, increases. The loss of emitted electrons through back diffusion in the

TABLE 2.2

ATOMIC CRITICAL POTENTIALS OF CERTAIN ELEMENTS, IN ELECTRON VOLTS
(Data largely from C. Moore, Atomic Energy Levels U.S.
Bureau of Standards Circular 467, vols. I and II [1949, 1952])

Element	First Excitation Potential	Metastable Levels				Ionization Potential
		$3P_1$	$3P_2$	$3P_0$	$1S_0$	
H	10.16	—	—	—	—	13.595
	$3S_1$	$3P_1$	$3P_2$	$3P_0$	$1S_0$	
He	19.81	20.96	20.96	20.96	20.62	24.580
Ne		16.53	16.62	16.72	—	21.559
A		11.62	11.53	11.72	—	15.755
Kr		9.98	9.82	10.51	—	13.996
Xe		8.39	8.28	9.4	—	12.127
Mg		2.71	2.715	2.705	—	7.644
Cu		1.882	1.895	1.879	—	6.111
Ba		1.115	1.138	1.185	—	5.210
Hg		4.89	5.47	4.67	—	10.434
. . . .						. . .
Li	1.845	—	—	—	—	5.390
Na	2.11	—	—	—	—	5.138
K	1.61	—	—	—	—	4.339
Rb	1.56	—	—	—	—	4.176
Cs	1.38	—	—	—	—	3.893
O	1.967, 9.15	—	—	—	—	13.614
N	2.382, 3.576, 10.33	—	—	—	—	14.54
C	12.63	—	—	—	—	11.264
F	12.71	—	—	—	—	17.418
Cl	0.109, 8.825	—	—	—	—	13.01
Br	—	—	—	—	—	11.84
I	2.34	—	—	—	—	10.44

gas can have values χ of the order of $0.01 < \chi < 1$. If $E_e = h\nu$ exceeds E_i for its own atomic species or for some impurity or for other atomic species present, there is a good chance (f_2) that the gas atoms or molecules will be *photoelectrically ionized*. Contrary to belief, this action has relatively high probabilities, f_2, $10^{-5} < f_2 < 10^{-1}$ or atomic cross sections (10^{-20} to 10^{-16} cm²). This is particularly so in air or in the presence of impurities Λ_i where $E_{iA i}$ is near but below the value of the $h\nu$ emitted. Thus, photons produced in some gases can even photoionize their own species. Photons are absorbed in a gas species by ionization and scattering, through absorption and reëmission such that the intensity I_0 in progressing through x cm of a given gas declines to I according to the well-known attenuation law $I = I_0 e^{-\mu x}$

where μ is called the absorption coefficient. The atomic scattering or absorbing cross section is σ, with $\sigma N_1 p = \mu$, N_1 as the atomic density at 1mm, and p the gas pressure. The quantity $1/\mu$ is called the mean free path for absorption, or for photoexcitation, depending on the specific nature of the acts causing it. For photoionization, $1/\mu$ is the *photoionizing free path*.

The electrons created by ionization in the ionizing field are free to pick up energy and ionize or excite on their own account. Ionization leads to an exponential increase of electrons as they advance in a field. Under some circumstances these electrons can strike certain molecules such as O_2 with sufficient energy so as to attach to them to form O^- negative ions, dissociating the O_2 molecule and leaving a free O atom. This is known as *dissociative attachment*. The energy of attachment in O^- is uncertain and may be 1.456 ev or 2.0 ev. Since the energy to dissociate O_2 into two O atoms is 5.08 ev, this attachment sets in at an energy of 3.624 or 3.08 ev. Cross sections vary with energy but lie around 2×10^{-19} cm^2. Less energetic electrons may attach slowly to O_2 molecules and give O_2^- ions in triple impacts with O_2. These ions have a binding energy of 0.4 ev. Electrons spontaneously dissociate Cl_2 or SF_6 and give Cl^- ions or SF_6^- or other ions. In HCl^- it takes a few tenths of a volt of energy to attach to form Cl^-. It has recently been observed that at an $X/p \sim 20$ the O^- begin to interact with O_2 molecules to form O_2^-. The O_2^- ions begin to break up at $X/p = 90$, while dissociative attachment also sets in at around $X/p \sim 7$.[1] The data for cross sections have not been corrected for detachment loss.

If not drawn to the electrodes as current or engaged in ionizing, the electrons will diffuse to the walls or out of the observing space. Diffusion increases with electron velocity and in inverse proportion to the pressure. It superposes itself in three dimensions on directed field motions. If electron energies are high in the gas near an attracting positive electrode, they will not at once enter that electrode. Instead, they will diffuse in the field, ionizing, exciting, or expending their excess energy in collisions with atoms until they are absorbed by the electrode.

The positive ions, once created, move toward the negative electrode with a *drift velocity* v characteristic of their nature and of the gas in which they find themselves. At lower ratios of field strength to pressure, the velocity v is proportional to the field X and to $1/p$, or $1/\rho$, where ρ is the gas density; that is, it varies as X/p. The ionic velocity in unit field and at normal air density (0° and 760 mm pressure) (N.T.P.) is called the reduced mobility, and values cited are usually expressed in cm/sec per volt/cm, or better, cm^2/volt sec. The drift velocity v divided by the field strength X is called the *mobility*, designated by k or by some authors as μ. Usually at fields

---

[1] Recent studies by Frommhold [2.37] indicate that perhaps the O^- may lose its electron directly more readily than suspected above an $X/p \approx 20$. The whole question of the behavior of negative ions is still in a state of confusion.

$$\mu = \frac{v}{E}$$

where much ionization by collision occurs, the ionic drift velocity is no longer proportional to X/p but increases as about $\sqrt{X/p}$ and the reduced mobility decreases in proportion to $\sqrt{p/X}$. This transition occurs when the random energy of the ions derived from the field surpasses the thermal energy of agitation, i.e., $\frac{1}{2}Mv^2 > \frac{3}{2}kT$, with k the gas constant per molecule.

The electrons likewise have drift velocities in the field. Here, owing to their low mass and many relatively *elastic* impacts with atoms (except when exciting and ionizing), the energy greatly exceeds thermal energy. Thus, they acquire distributions about a mean energy value which depends on X/p and lies much above that for thermal agitation $\frac{3}{2}kT$ at T° absolute. Under certain circumstances, and at number densities exceeding 10^8 per cm^3, the energy distribution among the electrons is Maxwellian. Otherwise the distribution takes on quite a diversity of forms depending on how the electron free path, between collisions with atoms, varies with energy. Important for ionization and excitation is the form of the energy distribution curve under conditions of X/p in the high-energy portion exceeding values of E_e and E_i. The efficiency of ionization at different electron energies also affects the ionization by collision. It rises linearly from E_i to about $3E_i$, where it lies around 10%. While individual electrons, as they progress in a field, are constantly changing their energies by gain from the field and loss by collisions, collisions are so frequent that the electrons as a group move across finite distances with a mean drift velocity v in the field direction. This varies in proportion to $\sqrt{X/p}$ at ionizing energies and well below, as is the case for ions at high X/p values. Electron drift velocities bear no direct relation to the ionic drift velocities in a given gas. In general, they are on the order of 200 or more times higher than those for the ions, the value being relatively higher the heavier the ions. With these high drift velocities the electrons are subject to correspondingly high rates of diffusion. A very fundamental relation exists between mobility μ and diffusion coefficient D, which reads $\mu/D = e/kT$, in which e is the ionic or electronic charge and kT is the energy of thermal agitation at the effective absolute temperature T of the carriers. In high fields at which the $\sqrt{X/p}$ law for ionic drift velocities applies, the effective ionic temperature T exceeds the gas temperature by small factors for ions. For electrons in virtue of elastic impacts $T = \eta T$ where T is the gas temperature and η can be of the order of 100 or more. The quantity η has been evaluated by Townsend and his pupils and may be found in tables.[2.1]

B. BASIC INTERACTION OF IONS, METASTABLES, AND ATOMS

The nature and behavior of the excited, metastable states and positive ions are appearing today to be much more complicated than were originally believed under classical atomic theory. It must be realized that even at 1 mm pressure at 0° C the normal gaseous ions make 10^6 or more collisions per second while electrons make on the order of 10^9 collisions per second at thermal energies. As the effective temperature T in a field increases, these values are exceeded, especially at higher pressures. It must be realized that, while normal atoms and molecules as observed in gases are chemically relatively inert and not very reactive, this is no longer true for such configurations when they are in excited electronic states E_e or are ionized. The normal dynamic stability ensured in the unexcited state is disturbed, and the excited and ionized configurations are exceedingly reactive. Thus even the inert gases form compounds like A_2^+, A_2^*, He_2^+, $NeHe^+$, and AH^+.[2]

The following reactions may be listed in symbolic form:

a. If E_m, E_e, or E_i of the atom or molecular species A (in the respective form of A^m, A^* or A^+) is equal to or greater than E_i for a species B or molecule BB, then:

(1) $A^m + B \rightarrow A + B^+ + e$ ionization by metastables,

(2) $A^* + B \rightarrow A + B^+ + e$ ionization by excited states,

(3) $A^+ + B \rightarrow A + B^+$ charge exchange, and

(4) $KE + A_1^+ + A_2 \rightarrow (A_1 + KE) + A_2^+$ charge exchange with the neutralized atom A_1 keeping its kinetic energy while A_2 gets the charge. This is a most likely reaction in many species where A_1 and A_2 are of the same species. Here cross sections may be such that every other impact leads to charge exchange.

b. If E_m or E_e for atom A is $< E_i$ for atom B or molecule BB but in excess of E_e or E_D for dissociation of BB, then

(5) A^m or $A^* + B \rightarrow A + B^*$,

(6) A^m or $A^* + BB \rightarrow A + 2B + KE$.

If $2E_m$ for A is greater than E_i for A,

(7) $A^m + A^m \rightarrow A^+ + A + e + KE$. Ionization by metastable interaction thus causes ionization to appear long after excitation, since A^m are generated in large numbers and in a clean gas diffuse slowly.

If certain higher states of E_e or A together with heat of formation of a molecule A_2 exceed E_i for A_2, then

(8) $A^* + A \rightarrow A_2^+ + e$. This reaction has a high probability. Where in low-energy electron swarms more A^* than A^+ are created, this type of reaction is responsible for much of the ionization produced. For example,

---

[2] Most such configurations are only stable in the ionized or excited states. One set of compounds of Xe, that is, XeF_4, are stable under normal conditions. They derive from the extreme polarizability of the large Xe atom in the higher-order effective fields of tetrahedrally symmetrical halogen or oxygen atoms or molecules. They do not occur for the lighter inert gas atoms.

in A gas at an $X/p = 30$ at pressures greater than 10 mm, this accounts for 70% of the ions. These excited states of argon have a lifetime of $\approx 10^{-7}$ sec. Again there is another reaction

(9) $A^+ + 2A \rightarrow A_2^+ + A$. This changes atomic ions to molecular ions. It increases as p^2 and at tens of millimeter pressure in milliseconds accounts for complete change of A^+ to A_2^+. This is a form of clustering or complexion formation. Another very important type of reaction occurs in A and probably other inert gases.

(10a) $A^m + 2A \rightarrow A_2^* + A$.

(10b) $A_2^* \rightarrow A + A + h\nu$. This slow two-step reaction proportional to p^2 yields excited states and photons of A_2^* which are not absorbed by the atomic gas A but can liberate electrons from electrodes, and is an important mechanism in A gas above 100 mm pressure. Another reaction

(11a) $A^+ + 2B \rightarrow A^+B + B$ represents complex ion formation and may lead to

(11b) $A^+B + B \rightarrow AB_2^+$, which represents further complex ion formation or clustering. It can be produced by attraction of a small ion like Li^+ for a strong electrically polarizable molecule like Xe or a naturally dipolar molecule such as H_2O.

Some forms of complex ion formation are quite stable, and often surprising in nature. With 1% H_2 in A at a few millimeters pressure most ions are AH^+. Ions $NeHe^+$ have been found predominating in mixtures, while at higher pressures in He, Ne, and A the molecular ions predominate after some milliseconds. The normal ion in O_2 is O_2^+, which is retarded by frequent charge exchange and loss of momentum in the field direction by collisions with O_2 molecules. The same appears true of O_2^- in O_2, and of A^+ and He^+ in A and He, respectively. In H_2 gas the ion is predominantly H_3^+, which is retarded by frequent proton exchange with H_2 in impacts. Nitrogen gas has four ions, N^+, N_2^+, N_3^+, and N_4^+, the latter predominating at low X/p values and higher pressures. Hg vapor present to 10^{-3} mm (its vapor pressure at room temperature) will in a few milliseconds convert most ions of E_i greater than 10.34 volts to Hg^+ ions by charge exchange.

(12) Ions with $E_i \geq 2\phi e$, where ϕe is the work function for a metal surface, will on impact with the surface yield secondary electrons. The probability that impact of these ions will liberate electrons on impact is designated by γ_i. γ_i for singly charged ions ranges between 0.25 and 0.01 or less depending on the ions, ϕe, the pressure of gas, and X/p. If $E_i < 2\phi e$, but if the ions have sufficient kinetic energy (> 150 volts), they will also liberate secondary electrons. The value of γ_i is 10^{-6} at 150 volts and may reach 10^{-2} at 1000 volts. Emission of secondary electrons by this kinetic process may be quite large on dirty or negative ion-coated metal surfaces. Metastable atoms if $E_m > \phi e$ will also liberate secondary electrons on

reaching a metal surface. This process, called secondary electron liberation by ion or metastable impact on the cathode, plays an important role in gaseous breakdown. γ_i varies rather widely with surface conditions. It is high for gassy but not oxidized surfaces. Increase in ϕe reduces γ_i. In contrast, γ_p, depending on photoelectric effects, is not quite so sensitive to change in surface conditions and varies differently from γ_i with changes. Oxide layers generally reduce γ_i but do not alter γ_p as radically. Glass has a $\gamma_p \sim 10^{-3}$, water has a very low γ_p at wave lengths above 1000 Å, while ice has a γ_p of the same order as that of glass. γ_i is not known for other than metals, but values doubtless exist for other substances. Both increase as X/p increases. Ions like H_3^+, Hg^+, and similar configurations have low values of γ_i, while He^+ and Ne^+ and, even more, He^{++} and Ne^{++} are very effective.

The bombardment of cathodes by positive ions in many discharges blasts atoms and surface layers off the surfaces by a process called sputtering. This alters γ_i and γ_p in time. Chemically active gases will tend to reform surface coatings. At a pressure of 10^{-10} mm, monolayers of ambient gas form on surfaces in 3.6×10^4 sec. At 10^{-7} mm, monolayers will form in the order of seconds. Sputtering rate depends on ion species, potential, pressure, and lattice constants. It is lower than adsorption time for lattice atoms but may clean gas films faster than they can form. Thus γ_i especially is subject to alteration by predischarges and changes in time. γ_i may also vary with current density when a competing chemical change is active. Deposits of carbon on electrodes are particularly effective in lowering γ_i.

C. IONIZATION BY COLLISION, ELECTRON AVALANCHES, AND SECONDARY ACTIONS

It is now essential to consider what happens when electrons are liberated from the cathode in a uniform field in a gap. The electrons, after some small distance x_0, gain their equilibrium energy distribution in the field in the gas. As they proceed toward the anode along x, the direction of the applied field X at a pressure p, they will ionize. After an electron ionizes there are two electrons that start with nearly zero energy and quickly gain energy enough to ionize anew. There are thus four new electrons after the second ionizing event; these again ionize and create eight. Thus the one initial electron, starting from the cathode, on advancing x cm in the field direction, will have created $n = e^{\alpha x}$ electrons. On arrival at the anode where $x = d$, there will be $n_d = e^{\alpha d}$ new electrons. Since these move with a high drift velocity, they will reach the anode, leaving their positive ion companions virtually at the place where they were created. The positive ions then gradually drift toward the cathode. The quantity α, called the *first Townsend* coefficient, or *coefficient for ionization by electron impact*, represents the number of new electrons created per centimeter path in the field direction. Its reciprocal $1/\alpha$ represents the *average ionizing free path* or distance between ionizing events. Thus an initial electron current $i_0 = n_0 e$ will have generated a current of $i = ne$ electrons on arrival at the anode, where $i = i_0 e^{\alpha d}$. One electron and its progeny of $e^{\alpha d}$ electrons on arrival at the anode constitute an *electron avalanche*. The quantities α are found to obey a relation $\alpha/p = f(X/p)$ characteristic of each gas. The $\alpha/p - X/p$ curve rises asymptotically from zero, increases rapidly, goes through a point of inflection at a roughly linear section, and then slowly levels off at high values of X/p. Ionization sets in so gradually that there is no sharply defined threshold value. The form of this curve depends on the tail of the energy distribution curve for the electrons, on the probability of ionization, and, inversely, on electron drift velocity. As these vary unpredictably with X/p, no single analytical function can describe the $\alpha/p - X/p$ curve over its whole length.

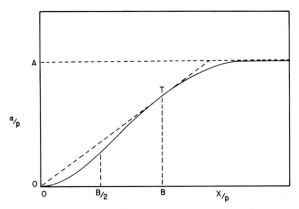

FIG. 2.1. A typical $\alpha/p - f(X/p)$ curve. The points B/2 and B refer to a constant in the exponent of Townsend's approximate relation (equation 2.1) giving the point of inflection and tangent to the curve from the origin.

Figure 2.1 shows a typical curve of α/p as a function of X/p drawn to an extended scale of abscissae. Beyond the apparently flat maximum, most observers note a decline. Since this region is in the Morton–Johnson regime, where α/p is both a $f(X/p)$ and distance x that the electrons have traveled, as well as in one where experimental errors are large, too much reliance cannot be placed on this feature. It is not, in consequence, shown. The *rise* can accurately be fitted to an exponential of the form $\alpha/p = A_1 e^{B_1 X/p}$. The next section can be fitted by a relation $\alpha/p = A_2(X/p - B_2)^2$. There is a linear section at the point of inflection. The upper portion is given by $(\alpha/p + B_3)^2 = A_3 X/p$.

Townsend, in an early theory based on a naïve concept, derived an equation which yields a curve of the general form depicted. That equation has since been correctly derived on *assumptions applying to certain limited regions of the curve.* It has the form

$$(2.1) \qquad\qquad \alpha/p = Ae^{-Bp/X}.$$

This form can obviously be fitted to any section of the curve with some accuracy, but *never fits the whole* observed curve for any one set of values of A and B. In figure 2.1, representing such a function approximately, it is seen that A represents the asymptotic limit of the peak of the curve. The point of inflection occurs at $B/2$, and the point B marks the tangent to the curve drawn from the origin. The relation is often used in approximate calculations such as in Geiger counter theory. If the whole range of α/p values are involved, it is incorrectly applied. Otherwise, in a limited range with A and B chosen to fit that region, it is as good as the empirical relations indicated above. It should be noted that, by expanding or contracting the scale of abscissae relative to that of the ordinates, the observed curve can be so distorted as to give illusory agreement over a large section with any reasonable curves. This in principle is the same as fitting sections by different functional forms, and the agreement is illusory.

For A gas there is an empirical variant of Townsend's relation used by Ward, of the form $\alpha/p = Ae^{-(Bp/X)^2}$, which apparently accurately covers the observed range of values. This is a pure coincidence, as there is no theoretical justification for this form. The points $B/2$ and B of the Townsend expression, representing the point of inflection and point of tangency, respectively, in figure 2.1, are very important relative to the observed curves. The point of inflection marks the region of X/p values in which charge accumulation in a uniform field gap ceases to aid and begins to hinder breakdown to a spark. The point of tangency represents the beginning of the Morton–Johnson regime, at which α/p is no longer a function of X/p alone.

There is another quantity often used in connection with α/p. If α/p is divided by X/p one has a new ratio $\eta = \alpha/X$. This represents the number of new ions/centimeter path per unit field strength. That is, η represents the number of ions created per volt of potential, as field strength times distance equals the potential drop. The quantity η is the ionizing efficiency function or, as it is sometimes called, the Stoletow–Philips function. $1/\eta$ or X/α represents the number of volts that are required to create an ion pair. If $\eta = \alpha/X$ is plotted as a function of X/p it rises at first rapidly, reaches a peak, and declines equally rapidly, as shown in figure 2.2. In the Townsend equation for α/p, the peak is reached when $X/p = B$. The sig-

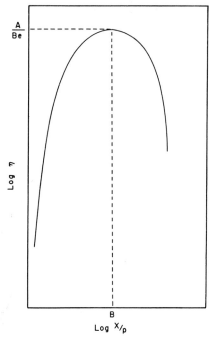

Fig. 2.2. The Stoletow-Philips function $\eta = \alpha/X$ plotted to a log-log scale as a f(X/p). Note that the maximum efficiency occurs at X/p = B. Beyond that point electrons gain energy from the field faster than they lose it by collisions. Thus as electrons advance along x in the field, energy goes up, efficiency decreases, and η varies as X/p and x.

nificance of this is that, as X/p first increases, ionization becomes more efficient, because less electron energy goes to elastic and excitation losses and more goes to ionization. However, as the electrons get much more than about $2E_i$ of energy, they cease to ionize efficiently and the ions gain energy faster than they can dissipate it in ionization. Thus after this peak, the electrons gain energy as x increases at constant X/p, and as they gain energy, η decreases. Thus they gain energy more and more rapidly the longer they move in the field. This excess energy goes to waste at the anode unless the electrons enter a field free space and can expend it in ionization. From this point on, α/p ceases to be a unique function of X/p but depends on X/p and on the distance x traversed along the field. Measurements and use of measured values of α/p beyond this point are meaningless and yield erroneous values.

The uniform field values of α/p given in tables are averages obtained from statistically determined values of individual avalanches. They also represent maximum values, since collection time in the static measuring technique is great. In the first place, since ionization is a chance event, individual sequences of ionizing events in $e^{\alpha d}$ may encounter bad fortune and αd will fall well below its average value. Other sequences will be fortunate and αd will exceed the average value. These fluctuations cannot be ignored where a few single events are considered. Common observations of $e^{\alpha d}$ cover ranges from 10^4 to 10^8 electrons, so that αd covers between nine and eighteen successive random events. Thus successive fluctuations of αd by a factor of three in such sequences are not rare. Again it is assumed that all the electrons in the avalanche are created by electron impact of 10^{-12} sec or less and thus that progress across the gap proceeds as an exponentially increasing electron swarm that arrives about simultaneously at the anode. This assumption can be quite in error. As noted in A, in ranges of X/p which cover many breakdown conditions, over 75% of the ions are created from A^* states by impact with neutral atoms to give A_2^+ and e. This occurs within a time 10^{-7} sec. Thus there will be some delay and diffusion in time and space of ions created when crossing times of the avalanche are of the same or shorter times. And thus, while there are tables of measured values of α/p as a function of X/p, the values given have use and significance only up to the peak of the α/X curves, or to the point of tangency for a line drawn from the origin to the highest point in the $\alpha/p - X/p$ curve. For individual avalanches, actual values will differ widely from the mean. The values of α/p, where temporal scales are short—for example, where crossing times are less than 10^{-7} sec—may be less. At lower pressures—for example, less than about 10 mm Hg pressure—α/p in gases like A and He will be likely to vary with p as well as X/p. It must be noted that attachment-forming negative ions will reduce the amplification indicated here, and that where negative ion formation is likely, special corrections are needed. Again, current values of these may be in error since detachment has not been allowed for.

D. FIELD INTENSIFIED CURRENTS
AND BREAKDOWN THRESHOLDS

Again regarding, for simplicity, the parallel plane uniform field gap, assume that n_0 electrons are liberated from the cathode per second by some outside source giving a current $i_0 = n_0 e$. At the anode $n = n_0 e^{\alpha d}$ electrons and $n_0(e^{\alpha d}-1)$ ions will have been left along the way. Half the ions will be within $1/\alpha$ cm from the anode, and three-fourths of them at $2/\alpha$ from the anode. In the process of creating ions, impacts at energies above E_e will create a comparable number w of excited states per centimeter advance among the atoms. If X/p is low, the fraction $f = w/\alpha$ of excited states of significance relative to the ions will be larger than unity. If X/p is very high, f will be comparable to but less than 1, decreasing as X/p increases. However, many higher excited states may be created. Some of these, through actions such as reaction equations 1, 2, and 10a of the preceding section B, will produce molecules with their own species or else atomic ions of impurity. The others will radiate photons of energy $h\nu$, which, because of a fractional geometrical factor of $g \lesssim 0.5$ or less, will be directed back to the cathode. Absorption of these photons en route will reduce their number by a factor $e^{-\mu x}$, depending on the distance from x at which they were created in the gas to the cathode. Arrived there, a fraction θ will have sufficiently large values of $h\nu$ to succeed in liberating further electrons, of which a fraction χ will escape back diffusion and head for the anode. Thus roughly $nfge^{-\mu d}\theta\chi = n\gamma_p$ will liberate electrons from the cathode. These will proceed across the gap and in a time $\tau_e = d/v_e$ sec will reach the anode as the second *photoelectrically produced generation*. Here v_e is the electron drift velocity. Thus with sufficient precision one may write that, after the first avalanche arrives, a second-generation avalanche of size $n\gamma_p e^{\alpha d}$ will arrive at the anode.

Once ionization by collision begins with *no* appreciable photon generation, the n_0 triggering electrons liberated by some external agency will yield a *field intensified current* given by

(2.2)
$$n\epsilon = n_0\epsilon e^{\alpha d} = i = i_0 e^{\alpha d}.$$

Since it has been indicated that $\eta = \alpha/X$ and η is likewise a function of X/p, one may replace αd by $d = \eta V$ in a uniform field gap since $X = V/d$, whence $d = V/X$. Thus the Townsend relation can be written directly in terms of the potential V as

(2.3)
$$i = i_0 e^{\eta V}.$$

As X/p increases, enough photons are created by electron impact and get back to the cathode to add an observable γ_p action. Then, more precisely, the current will be

(2.4)
$$i = \frac{i_0 e^{\alpha x}}{1 - \dfrac{\theta w g}{(\alpha - \mu)}\left[e^{(\alpha-\mu)x} - 1\right]}.$$

If μ can be neglected, as it often can be, and $x = d$, then

(2.5) $$i = \frac{i_0 e^{\alpha d}}{1 - \theta f g(e^{\alpha d} - 1)} = \frac{i_0 e^{\alpha d}}{1 - \gamma_p(e^{\alpha d} - 1)} = \frac{i_0 e^{\eta V}}{1 - \gamma_p(e^{\eta V} - 1)}.$$

These currents are *field intensified ionization currents*. In many cases where breakdown occurs, $e^{\eta V} = e^{\alpha d} \gg 1$ so that the field intensified current reads

(2.6)
$$i = \frac{i_0 e^{\alpha d}}{1 - \gamma_p e^{\alpha d}} = \frac{i_0 e^{\eta V}}{1 - \gamma_p e^{\eta V}}.$$

This last equation has an interesting implication. If $\gamma_p e^{\eta V} = \gamma_p e^{\alpha d} \geq 1$, then i is *independent of i_0 and a self-sustaining current exists.* For, if $\gamma_p e^{\alpha d}$ is unity, it means that each avalanche will on the average sustain itself independently of i_0. If $\gamma_p e^{\eta V} = \gamma_p e^{\alpha d} > 1$, the current will augment until dissipative factors such as diffusive loss, recombination of carriers, and space change effects reduce the rate of increase to zero. This could imply a very large increase of current. Thus

$$(2.7) \qquad \gamma_p e^{\eta V} = \gamma_p e^{\alpha d} = 1$$

sets the threshold for a self-sustaining discharge or *breakdown* and the value of $V = V_t$, the *Townsend breakdown* voltage.

In addition to this action it must be recalled that the positive ions move back to the anode with a drift velocity $v_i = k_0(X/p)$. The half of the ions created within $1/\alpha$ cm of the anode reach the cathode in a time t_i, roughly $t_i = d/v_i$. Arriving, they liberate a fraction θ_i of electrons, of which a fraction χ overcomes back diffusion such that a fraction $\theta_i \chi = \gamma_i$ of the n ions from the first avalanche leaves the cathode and produces $\gamma_i e^{\alpha d}$ electrons and ions in the *second generation.* Thus, there is also a γ_i-produced secondary ionization which creates a field intensified current

$$(2.8) \quad i = \frac{i_0 e^{\alpha d}}{1 - \gamma_i(e^{\alpha d} - 1)} \quad \text{or alternately} \quad i = \frac{i_0 e^{\eta V}}{1 - \gamma_i(e^{\eta V} - 1)}$$

where V is the potential. Again, neglecting 1 relative to $e^{\alpha d}$, one may set the condition for a self-sustaining discharge due to *ionization by ion impact on the cathode* as

$$(2.9) \qquad \gamma_i e^{\alpha d} = \gamma_i e^{\eta V} = 1.$$

Since γ_i depends on the energy of the ions and since χ and other factors depend on X/p, γ_i and γ_p in general are also functions of X/p, but γ_i and γ_p will vary in a different fashion with X/p. Since γ_i and γ_p act independently, one may lump them additively into a single term $\gamma = (\gamma_p + \gamma_i)$ within the accuracy of most measurements, provided these involve time scales in which with $t_i \gg t_e$ both processes are active. Diffusion of metastable atoms to the cathode produces a third coefficient, γ_m, which has a still longer time scale and is operative only in very pure inert gases. With the advent of fast oscilloscopes, so that growth of current could be analyzed on time scales of less than t_i and comparable with t_e, it became clear that the γ measured with Townsend's static method were composite. This circumstance is important, for if a breakdown process over gaps of one centimeter occurs on a time scale of 10^{-6} sec only a γ_p can be active. However, if the time scale exceeds 10^{-5} sec γ_i may be involved, and if it is 10^{-3} or 10^{-4} sec one may also expect metastable action γ_m to be involved. The photons active in γ_p must usually be of a kind that do not experience delay and dissipation by resonance absorption. In some such cases there may be a delay in creating the states to produce γ_p. Thus in the inert gases, the photons yielding γ_p, notably in A, are created by the reactions 10 of the preceding section B. These are pressure-dependent and suffer a delay in the order of 3 μsec in their generation from A^m in collision with two A atoms. This action sets the threshold for a γ_p breakdown at near atmospheric pressures. At some tens of millimeter pressure, breakdown proceeds largely by the γ_i mechanism, since triple impacts of A^m with 2 A atoms becomes infrequent. It appears that in N_2 and in air, the secondary mechanisms which involve γ_p may suffer a delay per generation of the order of $3t_e$. Considering a γ from whatever cause, Townsend set as the threshold for a breakdown the condition that $\gamma e^{\alpha d} = 1$, or alternately

$$(2.10) \qquad\qquad \gamma e^{\eta V t} = 1.$$

Here γ can be γ_p alone, γ_i alone, $\gamma_p + \gamma_i$, etc. If $\gamma e^{\alpha d} > 1$, then, as noted, the self-sustaining current augments.

It must next be recalled that γ is a pure probability for which an average value is obtained by observing some 10^7 or so simultaneous avalanches. Again $e^{\alpha d}$ is governed by chance, since the probability of ionization in a given impact is a factor. Actually, since, in most discharges such as are dealt with here, $9 < \alpha d < 18$, the probable fluctuations of αd can be between ± 3 and ± 4 in single avalanches. This can produce a large variation in $e^{\alpha d}$. It was shown by Wijsman[2.2] at the author's suggestion that if one set $M = \gamma e^{\alpha d} = \gamma e^{\eta V}$ then the breakdown probability P_0 is

$$(2.11) \qquad\qquad P_0 = 1 - 1/M.$$

This makes the breakdown probability 0 at threshold, for any subaverage avalanche can break off the sequence. As the amplification factor M given by $\gamma e^{\alpha d} = n$ exceeds unity, the chance of breakdown increases. But in any event sustained current is not always ensured, since sooner or later an unfortunate avalanche sequence can interrupt the discharge. It is noted that P_0 becomes 0.9 only when $M = 10$. However, other factors hitherto neglected influence this purely statistical consequence. In a gap which has once broken down there are always delayed carriers, for example, electrons or negative ions, that yield electrons in high fields. With any prolonged sequence of avalanches the regularity of successive ionizing generations must be disorganized by the all-pervasive action of electron diffusion in space and time. Thus, a later *re-ignition* is ensured even when an interruption occurs. With many initial triggering electrons the chance of continuity is much increased.[2.3] It is, however, clear that this more or less guarantees a *fluctuating current regime* at threshold, though the range of the fluctuating regime in M values and especially in potential may be small.

There are, however, other agencies that enter into the growth of a discharge. To appreciate these a word must be said concerning the growth of a discharge, either from a single avalanche and from a triggering electron current i_0 from the cathode. Bandel[2.4] observed these growths oscilloscopically for discharges starting with an i_0 in air, using uniform field geometry in the author's laboratory. Later observations were made by Kluckow in Raether's laboratory, for air, N_2, and H_2, for both single avalanches and currents i_0, and most recently by Schlumbohm. The theory was developed initially by Steenbeck and later by Bartolomeyczk and others.[2.5] Legler,[2.6] Frommhold, and others in Raether's laboratory have done beautiful work in extending Wijsman's statistical analyses of avalanches to actual breakdown sequences. The avalanche advances in a uniform field at practically constant velocity, for it is only for very large avalanches at high X/p under special conditions that the avalanche space charge retards and reduces ionization.[2.7] Therefore basic temporal growth of the discharge can be assumed to follow an exponential increase in time as well as in x. The growth will then turn out to be exponential about a characteristic time constant τ which is determined by the participation of a γ_p or a γ_i and thus about the characteristic τ_e or τ_i between successive generations. The condition for growth including γ_p and γ_i as derived by Bartolomeyczk, in simplified form, is

$$(2.12) \qquad 1 = \gamma_i e^{\left(\alpha - \frac{1}{\tau v_i}\right) d} + \frac{g\theta w}{\alpha - \mu} e^{\left(\alpha - \mu - \frac{1}{\tau v_e}\right) d}.$$

A much more accurate series of calculations geared to various experimental situations have been set up by Bandel, Davidson, Auer, Miyoshi, and Ward.[2.5] Analyses pertinent to the studies of the Raether school have been set up by Legler.[2.6]

E. EVALUATION OF THE TOWNSEND INTEGRAL, SPACE CHARGES AND POLARITY, AND THE INFLUENCE OF SPACE CHARGES ON BREAKDOWN

Most of the analyses above omitted the influence of space charges. These are, in principle, the really important agencies which intervene at or near thresholds of self-sustaining currents. They may lead to interruption and intermittence of a notable sort for considerable potential ranges near threshold, or else they will increase the certainty of continuation of avalanches and even lead to a complete breakdown via a spark to a transient arc. The influence of space charges depends on geometrical field considerations in relation to the variation of α/p as a function on X/p.

1. Lowered Potential in Uniform
   Field Geometry

The uniform field represents a unique case where the positive space charge at the anode, once it builds up to a certain density, at which M exceeds unity by a certain value, leads to accelerated growth and a spark breakdown. This was anticipated first in theory by Steenbeck and independently discovered by Varney, White, Loeb, and Posin as a consequence of experimental data.[2.8] It was later confirmed by Crowe, Bragg and Thomas,[2.8] and has been more completely developed by Ward for A.[2.8] It was directly observed by Bandel and later by Kluckow.[2.9]

The accurate formal abstract analysis can be visualized physically quite simply. Assume such a gap with metal electrodes. When the ionization at the anode has built up sufficiently rapidly, especially if a γ_p is active, the electrons created in the last $1/\alpha$ cm are present in large quantities with positive ions. Electron entrance into the anode is delayed until the electrons have lost energy by collision. The space charge of the positive ions also retards electron removal. Thus, in principle, there is a highly conducting plasma out to about $1/\alpha$ to $1/2\alpha$ from the anode after the first few avalanche sequences. This virtually extends the conducting anode out into the gap by $1/\alpha$ cm, projecting its potential V to a point $d - 1/\alpha$ from the cathode. This increases the field from $X = V/d$ to $X' = V/(d - 1/\alpha)$ but shortens the gap length from d to $d' = d - 1/\alpha$. Now the avalanche size is given by $e^{\alpha d}$. To simplify reasoning let $\alpha/p = f(X/p) = D(X/p)^q$, which can, if q is properly chosen, represent its behavior as earlier indicated. This makes

$$e^{\alpha d} = e^{pD(X/p)^q d} \quad \text{and}$$
$$e^{\alpha' d'} = e^{pD(X'/p)^q d'}$$

if p is held constant. Now for uniform fields $X = V/d$ and $X' = V/d'$. Thus

$$e^{\alpha d} = e^{pD(V/pd)^q d} \quad \text{and}$$
$$e^{\alpha' d'} = e^{pD\,[V/(p[d - 1/\alpha])]^q\,(d - 1/\alpha)}.$$

If $q = 1$, then $e^{\alpha d} = e^{DV}$ and $e^{\alpha' d'} = e^{DV}$. The avalanche size is unaltered by space charge. If, however, $q > 1$ then α/p increases more rapidly than linearly with X/p, and $e^{\alpha' d'} > e^{\alpha d}$. Here the space charge increases the size of the avalanche. If q is < 1 then $e^{\alpha' d'} < e^{\alpha d}$ and the space charge diminishes the size of the avalanche. Thus one notes that for uniform fields, as α/p increases more rapidly or more slowly than X/p, the space charge enhances or diminishes avalanche size. The increase in $e^{\alpha d}$ by this action of space charges increases M for the same V, and *ionization will therefore build up autocatalytically.* At low pressures this will lead to a glow discharge when ion densities become so great that rearrangement of fields leads to the adverse space charge effects in a system of cathode and anode falls of potential. At higher pressures, as space charge builds up, avalanches will grow to a critical size leading to a new type of breakdown ending in a filamentary spark breakdown at very slightly above a Townsend threshold in air. If α/p increases more slowly than X/p, the positive ion space charge will inhibit the discharge and raise the breakdown threshold V_s above the expected V_t.

2. THE TOWNSEND INTEGRAL FOR AVALANCHES IN NON-UNIFORM FIELDS

When one goes to non-uniform fields certain modifications of the basic breakdown equations appear which are of paramount importance. Because fields are no longer uniform, as they are in equation 2.2, $e^{\alpha d}$ has its argument αd replaced by $\int_a^b \alpha dx$. This follows, since X/p is a function of X or r, along the field direction, so that X/p must be expressed as a function of x, and this in turn must be converted to giving values of α as a $f(x)$, or of r, with p assumed constant. If α/p can be expressed as a single analytical expression for X/p and if one knows the form of the field X as a $f(x)$, one can insert its analytical form into that expression and integrate. This matter will shortly be discussed more at length. Again, since the fields are high in the neighborhood of cathode or anode, or both, one of these will break down first. This breakdown will create the following situation: The ionization by impact reduces to zero in the lower field portions at distances very small compared to gap length. The upper limit b of the integral will be replaced by a quantity x_0, or r_0, with $a < x_0 \ll b$. Since the mobility of positive carriers is much less than that for electrons, the space charge distributions will differ near the negative and positive electrodes. Furthermore, there will be a unipolar space charge distribution which accumulates in the low field region and which has the same sign as the electrode at which breakdown occurs.

In order to calculate breakdown thresholds on the Townsend theory, it is therefore essential to be able to calculate the quantity $\int_a^{x_0} \alpha dx$ from the smaller electrode surface of radius to a, the point x_0 or r_0 at which ionization by collision ceases in the gap. Since no single analytical function exists for the whole range of α values, the exact but tedious expedient of plotting α/p as a function of X/p and X/p as a function of x to appropriate scales must be resorted to. Then by crossplots α/p, or $(\alpha/p)p = \alpha$ can be plotted as a function of x. Integration must then be performed graphically by planimeter or by use of one of the many devices for such procedures. In using this procedure it must also be recognized that the data for α/p beyond the point of tangency to the $\alpha/p - X/p$ curve from the origin cease to be accurate. In this region $\int \alpha dx$, as computed from the integral using observed values of α from uniform field data, may be as much as $\frac{1}{4}$ to $\frac{1}{6}$ the actual value. E. E. Dodd[2.10] developed a method for evaluating this troublesome integral in certain cases which will save considerable time. This is particularly true where it is desired to derive the variation of a threshold V_T with pressure.

Knowing the field X along the line of integration, where $X = Vg(x)$ with V the applied potential, one can change the variable of integration to the quotient X/V by the transformation $dx = h(X/V)d(X/V)$. Then if the electrode system used is such that $h(X/V)$ is homogeneous, $h(kt) = K^n h(t)$, the V can be transferred outside the integral and the integrand can be further transformed to a function of the quotient of the field divided by gas pressure p. Thus

$$(2.13) \qquad \int_a^{x_0} \alpha dx = \left(\frac{p}{V}\right)^{n+1} \int_{g(a)V/p}^{X_x} \alpha h \left(\frac{X}{p}\right) d \left(\frac{X}{p}\right).$$

However, $\alpha/p = f(X/p)$, so that

$$(2.14) \qquad \int_a^{x_0} \alpha dx = -p \left(\frac{p}{V}\right)^{n+1} \int_{X_x/p}^{g(a)V/p} f \left(\frac{X}{p}\right) h \left(\frac{X}{p}\right) d \left(\frac{X}{p}\right).$$

In this relation the integrand is independent of p and V, and thus can be set up for any gap directly if $\alpha/p = f(X/p)$ is known. The variables p and V appear only in the external factor and in the upper limit of integration at the high field electrode. The lower limit is zero or some prearranged constant at which further multiplication in the gap ceases. This enables a preview of the variation of the Townsend integral as a function of p and V. Conversely if the variation of the value of the integral, say with p, is known, the dependence of V on this parameter can be predicted.

For coaxial circular cylindrical geometry the distance x is replaced by the radii a and b and the variable r from the axis with X given by equation 1.4: $X_r = V/(r \log b/a)$. Then

$$(2.15) \qquad dr = -\frac{1}{(X/V)^2 \log (b/a)} d(X/V).$$

In this instance $n = -2$. In consequence

$$(2.16) \qquad \int_a^{r_c} \alpha dr = \frac{V}{\log (b/a)} \int_0^{\frac{V/p}{a \log b/a}} \left(\frac{X}{p}\right)^{-2} f \left(\frac{X}{p}\right) d \left(\frac{X}{p}\right).$$

Here the integrand is also independent of electrode dimensions.

From this relation the potential threshold V_s, which depends essentially on $\int_{x_0}^{a} \alpha dr$, may be used as an example for burst pulse onset. In this case, the integral has an essentially constant value since the effective γ changes relatively slowly. If p decreases then V/p in the upper limit must remain *nearly* constant so that V will also decrease with p. To compensate for this, V/p must increase. Hence V will fall more slowly than directly as p. If $f(X/p)$ is of the form $A(X^2/p)$ (a common variation in some X/p ranges), then the integral will be of the form $A'(V^2/p)$, so that as p decreases V will vary as $\sqrt{p}$. If $f(X/p)$ increases more rapidly with X/p, then V will vary more rapidly with p. If $f(X/p) = A(X/p)$, V will have the form BV log V/p and V will be relatively insensitive to changes in p. The relation $f(X/p) = A(X/p)$ represents the point of inflection in the rapidly varying $\alpha/p - X/p$ relation. Beyond this, $f(X/p)$ is progressively changing more and more slowly following perhaps fractional exponents. In this region V is independent of p, or begins to vary as some inverse power of p. In actual practice *no single functional relationship can represent* $\alpha/p = f(X/p)$ *because of the changing form of the distribution laws.*

Not far beyond the point of inflection the electrons cease to remain in equilibrium in the field; for example, X/p = 400 for air; 150 for H_2; 40 for Ne; ≈ 100 for A and Kr; and 3 for $10^{-3}\%$ A in Ne. This leads to an increase in ionization beyond that given by $e \int \alpha dr$, as Morton and Johnson have shown.[2.11] Actually the maximum ionization occurs in the region of $\alpha/p = A(X^2/p)$ or faster, and there is a fairly close agreement for an observed V_s varying as $\sqrt{p}$ in much of the counter and burst pulse corona study.

Usually the evaluation of this integral is used at the breakdown threshold, V_T, where space charges do not intervene. With impulse potentials of short duration above threshold the integral may be computed for values of V in excess of V_T. This will be of importance in the development of positive streamers. Here increase in V rapidly increases $\int \alpha dx$ to values such that the avalanche appears to increase to astronomical proportions. However, Tholl[2.38] in Raether's laboratory has recently shown that, when $e^{\alpha d}$ in uniform field geometry exceeds 10^6 electrons, α decreases. It declines to about 0.55 of its value when $e^{\alpha d}$ reaches 10^9, or $\alpha d \approx 20$. From there on the avalanche develops into a breakdown streamer. How far this applies to the non-uniform case is not established, but these facts require due caution in the use of the integral.

3. Ionization Characteristics for the
   Highly Stressed Anode and Cathode

It is now of importance to consider the curves of α as a function of X and x in a gap in order to evaluate the integrals of αdx as a function of x. This enables one to study the form of the ionization curve as a f(x) in the neighborhood of the highly stressed anode and cathode. The curves to be shown and discussed were calculated for a point-to-plane gap. The point radius a in the cases illustrated ranged around 0.019 cm or 0.025 cm in a point-to-plane gap that followed standard geometry with $L/a = 160$, and thus the gaps were 3.1 and 4 cm. The potentials used in calculations were 5000 volts, which is near the threshold for burst pulses, pre-onset streamers, or Trichel pulse coronas at 760 mm of room air.

Figure 2.3 is a composite of curves drawn for the 0.019 cm point at 5000 volts. The scale of ordinates applies to all curves if multiplied by the appropriate factor indicated on the plot. The curve for the potential as a function of x from the point, as given by Dodd's factors, is not plotted. However, X/p is plotted, and multiplication of the ordinates by 10 gives the value needed. It ranges from 200 to 40 and is well below the Morton–Johnson regime in air. The data of Masch in this range of values enables α/p to be found for the values of X/p. These yield the curve for α as a f(x) with ordinate scale values multiplied by 100. Note that α is very high. It drops to negligible values at x = 0.025 cm. That is, the area of the $\alpha - x$ curve below x = 0.02 cm, or about one point radius, is less than 1% of the $\int \alpha dx$.

The integral of αdx calculated for an *electron starting at the cathode* and progressing outward from the point is shown in the long-dashed curve to the right. Its scale factor is unity. The one for an *electron approaching the anode* is the dashed mirror image of the one moving outward, and is on the left with scale factor unity.

The area for both integrals is 12.5, independent of the sign of the charge. Hence the total number of ions produced is the same. When one calculates the values of the quantity $\exp \int_0^x \alpha dx$ yielding the number of ions produced as a function of x for the two $\int \alpha dx$–x curves, there appears a startling difference in curve shapes. While the total number of electrons and ions generated is the same, that is, $e^{12.5}$ or 2.6×10^5, the number of ions as a function of x differs widely in the two cases. This comes from the appearance of $\int_0^x \alpha dx$ in the exponent of e. Thus while electrons leaving the cathode are initially few, they multiply fairly rapidly in the high field. As the field decreases with increasing x, they multiply more slowly. However, as their number is by then very great, they continue to multiply even in very low fields. For electrons approaching the anode, the low multiplication in low fields results in few electrons until they are within x = 0.005 cm of the anode. Then, since in each $1/\alpha$ (≈ 0.0005 cm near the anode) the number of electrons doubles, the rise is very steep.

These last two dashed curves, to the left for incoming electrons to the anode and to the right for the outgoing electrons, with ordinates multiplied by 10^5, are of great significance for breakdown from anode and cathode points. The slopes of these curves (their first derivatives) are significant, as they represent the *rate of ionization* as a function of x. These rates are important, since on the avalanche time scale of $<10^{-8}$ sec, the positive ions created are practically immobile. The electron drift velocities are more than 200 times greater than those of the ions. Thus the derivatives of these curves represent the distribution of the avalanche ions remaining after the electrons have moved on in the field.

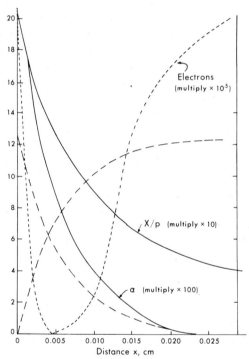

Fig. 2.3. Composite of curves at 5000 volts for a 0.019 cm radius point to a 3.1 cm distant plane. Abscissae are distances x from the point surface in centimeters, ordinates represent X/p, α, number of electrons, and ions near the point. The scale to be multiplied by the appropriate factors is designated for each curve. The values of X/p are given by the scale to the left multiplied by 10 in volts per centimeter per millimeter Hg. The curve for α as a f(x) is given by the scale to the left multiplied by 100. The values for the $\int \alpha \, dx$ for electrons moving out from the point are the *heavily* dashed curves, right; and those moving in to the point are the similar curves, left. The ordinate value is multiplied by 1 with equal maximum values of 12.5. The total electrons generated as a f(x) as the electrons move away from or toward the point are shown by the dotted lines, left and right, respectively. The ordinate scale must be multiplied by 10^5. These come from exp $\int \alpha \, dx$ at each value of x.

The curve for the slope, or the rate of ionization, that is, the number of positive ions as a function of x for electrons approaching the anode, is not shown. The curve will be closely similar to that for the number of electrons. The curve, being of the form $(\int \alpha dx) e^{\int \alpha dx}$, will, properly scaled, drop more sharply than that for $e \int \alpha dx$. This signifies that most of the positive ions are created *very close to the anode*—on the order of one or two $1/\alpha$ of the anode. They form a dense plasma of ions and electrons. The electrons in the high field would appear to enter the anode at once. Thus it might be expected that they would do so before they lost all their ionizing energy. Experimental evidence of Morton and of Johnson[2.11] indicates that this does not happen. The electrons must expend their energies gained in the high field regions in diffusive ionizing paths near the anode before they enter the anode. The number of ions collected by anodes under such conditions is observed to be within 10% as large as those created by a highly stressed cathode where electrons move outward and spend *all* their energy in ionization and excitation.

In consequence of the highly charged plasma of electrons and ions in the thin ionizing sheath near the anode surface and of the positive ion space charge after the electrons enter the anode, the field at a, the anode surface, is effectively extended outward into the gap. This is indicated by the potential diagram of figure 2.4, where the potential before a series of successive avalanches near the anode is plotted as a f(x). This curve, calculated for Trichel by M. Sitney for confocal paraboloids at threshold, shows the sequences as indicated in the legend.[2.12] Figure 2.5 represents the same picture in more detail close to the anode for the gap used in figure 2.3. The upper curve shows the sort of distortion produced. Ionization continues in the steep fall beyond the space charge.

Since too much space charge is created as the *ions* move outward into the gap, the field at the anode is reduced to a point where the avalanche sequence and discharge may terminate. Here, contrary to the case of the uniform field, irrespective of whether α/p increases faster than X/p, the space charge is inhibitory. This results from the low field regions, which delay ion movement and permit space charges in the gap to accumulate.

Under some conditions, to be mentioned later, where electrons attach to form negative ions in the gas in low field regions, as in air, a new phenomenon appears. Photoelectrons created beyond the region $x = r_0$ where ionization ceases will attach to form negative ions. These will advance into and accumulate in the region near the anode at values of x but beyond that at which $X/p < 90$, at which ions shed their electrons. They thus create a negative ion space charge sheath. This decreases the potential in that region and sharply enhances the anode field. Beyond the zone of ionization the positive ion space charge is now more bound by the negative ions, and the potential again increases. This leads to the schematically indicated poten-

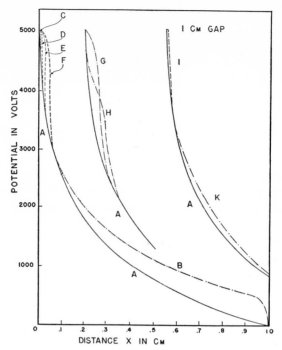

Fig. 2.4. Potential distribution and burst formation with a 0.5 mm diameter point 1μA current and 1 cm gap. Curve A is the undistorted potential in each case. Curve B shows the space charge distortion near the *cathode* due to extinct bursts. Curves C, D, E, and F show successively the approximate potentials after the first, second, fifth, and tenth avalanches in burst formation. Curves G and H show later stages of a burst where diffusion and movement in the field occur. Curves I and K show a new burst starting as the old one recedes. At higher fields this may overtake the old burst, forming a streamer.

tial diagram of figure 2.5 in the lower curve. This sheath and its consequent glow corona were recently discovered by Hermstein.[1.16] The phenomenon is termed the Hermstein negative ion sheath glow corona. The field is very high but so short that avalanches cannot increase to yield streamers. Thus the pre-onset streamer corona is interrupted by the glow. Because of the nature of the potential distribution shown, this glow spreads over the anode surface.

For electrons moving away from the cathode, the derivative of the ionization and the ionization curve itself are shown in figure 2.6. This curve was calculated under similar but slightly different conditions from those of figure 2.3. Here it is seen that the positive ions are created in a zone of maximum rate of ion production at some distance from the cathode surface. The zone serves to slow up the rate of ionization within itself but greatly

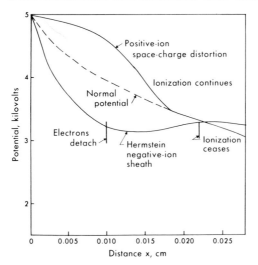

FIG. 2.5. Detailed sketch of potential near the anode as a function of distance caused by positive ions and electrons near the point in the upper curve. The undistorted potential fall in the absence of space charge is dashed. The lower curve indicates the distortion when negative ions are produced outside the ionizing zone and accumulate near the anode in the Hermstein glow regime in air. Note that the potential with positive space charge is above that for the undistorted gap. Electrons are radially drawn into the discharge from outside. With the Hermstein glow the discharge is negative relative to the outside space. Electrons move out radially, and the discharge spreads over the surface.

increases the field over an effective stretch near the cathode. On the anode side this cloud also increases the field slightly, causing ionization beyond r_0 once the cloud builds up. Of course, the ion cloud is not entirely stationary but moves toward the cathode. As it does so, it increases the cathode field and keeps the positive ion space charge cloud from becoming too large and dense. Since the distance to the cathode is small, the positive ions reach it in a high field and relatively quickly. The positive ions bombard the cathode surface, remove its oxide layers, and augment the γ_p process that started the breakdown by adding an effective γ_i to the secondary liberation. If the electrons created remain free, they move away sufficiently rapidly so as not to produce much of a space charge backfield. In fact, unless the gap is very long there is produced by virtue of this action a field increase by positive ions, cleaning of the cathode, and γ_i in high fields, resulting in a most unstable condition. There is no limit to the increase of the current if power supply is adequate until either a glow discharge or an arc of high current capacity is created. The new space charge regions connected with the higher density plasmas and charge generation under the conditions named above finally limit current. Thus, such gaps with highly stressed cathode, once threshold is reached, continue to increase in current to yield a *new stable form of discharge.* During the transition period such a condition leads to a negative current-potential characteristic. That is, current can increase even with a decrease in potential. This is often called a negative resistance characteristic. Under these conditions a stable lower current corona can be sustained only through the use of an external high resistance in the series with the power source. This negative characteristic has frequently been observed in corona studies with highly stressed cathode.

4. Negative Ion Space Charges

If electrons attach either dissociatively or directly to atoms or molecules, negative ion space charges are generated. Where attachment takes place through dissociative attachment in O_2, electron energies required are of the order of 3.1 or 3.6 volts. As this is less than the ionization potentials of 12 to 15 volts in the air gases, it occurs in the lower field edge of the positive space charge ion cloud and in the region just beyond the space charge hump in figure 2.6. In the higher field regions at $X/p > 90$, electrons detach from the ions and remain free, as seen in figure 2.5 with Hermstein's glow.

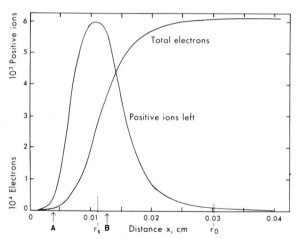

Fig. 2.6. This shows the electron multiplication for a 0.025 cm radius point at about 5000 volts when it is the cathode. The electrons increase along the curve to the right, and the scale of ordinates is multiplied by 10^4. The derivative of this curve is the peaked curve showing the distribution of the positive ions as a $f(x)$ left behind by the electrons. The scale of ordinates is to be multiplied by 10^3. This nearly immobile positive ion space charge increases the field at the cathode and moves slowly toward the cathode.

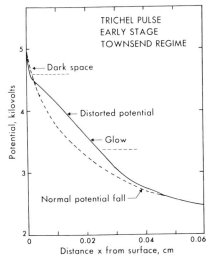

FIG. 2.7. The potential distribution resulting from the action of the space charges of many avalanches of figure 2.6 in building up highly energetic electrons that ionize out in the field and attach to form negative ions beyond this. This is the initial phase of a negative Trichel pulse breakdown in air. The space charge distorted fall is the solid line; the normal fall is dashed.

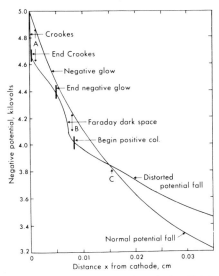

FIG. 2.8. The potential fall at the height of the Trichel pulse activity. The high-energy electrons from the space charge cloud field at the cathode produce a small glow discharge with Crookes dark space, negative glow, Faraday dark space, and positive column. The negative ion space charge exerts its influence around and beyond C. This eventually chokes off the discharge.

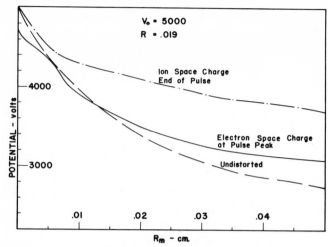

Fɪɢ. 2.9. The end of the Trichel pulse is shown in the dash-dotted curve. The earlier phase at the height of the pulse is the solid line. The undistorted field is the dashed line. The space charge must now clear to the right before a new pulse can begin.

The negative ions are soon in a field much lower than applies to the positive space charge and are, in part, held back by that charge. They thus accumulate in considerable density. Their influence is sufficient to reduce the cathode field and terminate breakdown, since the corresponding positive ions rapidly go to the cathode. Once breakdown ceases, the applied field gradients sweep away negative and positive ions so that a new breakdown can occur after a clearing time for the space charge. Thus, here again, ions of the sign opposing the stressed electrode enhance breakdown, but ions of the same sign as the electrode, collecting in the lower field region, terminate breakdown leading to fluctuating currents. Such a situation is depicted schematically in the sequence of figures of the negative Trichel pulse corona in air at the start, at its brightest, and on extinction (figs. 2.7, 2.8, and 2.9).

5. Eꜰꜰᴇᴄᴛɪᴠᴇ Cᴀʟᴄᴜʟᴀʙʟᴇ Iᴏɴ Sᴘᴀᴄᴇ Cʜᴀʀɢᴇ
   Dɪsᴛᴏʀᴛɪᴏɴ ɪɴ Cᴏᴀxɪᴀʟ Cʏʟɪɴᴅᴇʀs;
   Tʜᴇ Sᴘᴀᴄᴇ Cʜᴀʀɢᴇ Cᴏɴᴛʀᴏʟʟᴇᴅ
   Cᴜʀʀᴇɴᴛ-Pᴏᴛᴇɴᴛɪᴀʟ Cᴜʀᴠᴇ

In this case, where space charges do not actually terminate a discharge until removed, because the fields after some increase in potential above threshold are certainly adequate to remove them, they will act to limit the breakdown current, leading to a stable discharge regime with a positive resistance characteristic. Such is the case somewhat above threshold for positive corona in air, where negative ions initiate a new type of glow discharge, discovered by Hermstein,[1.16] and particularly for the positive coaxial cylindrical corona in a clean, free electron gas such as pure H_2, N_2, or the inert gases. The equation for the ion stabilized coaxial cylindrical corona discharge was early derived by Townsend[1.6] and is given here because of its exemplary value. From it certain other conclusions and properties of such corona-type breakdown will be derived.

In considering the space charge limitation of current, it is clear from the field considerations indicated that ionization with or without an ion space charge will be confined to a limited region lying axially distant between a and r_0. This distance for a wire of some 10^{-3} cm radius will not exceed a few tenths of a millimeter and even at 10 mm pressure will not exceed a few millimeters because of the rapid field attenuation, while b is usually several centimeters. When r_0 becomes appreciable, relative to b, then the law that is to be derived is no longer applicable. In coaxial cylindrical geometry, as the relations to be derived will show, the positive ion density beyond the ionizing zone is constant. This follows since, as the area of a given sheath dr thick increases with increase in r, the field strength, and in consequence the velocity of the ions is reduced in the same measure so that ion concentration remains invariant. Another peculiarity of an ionic space charge stabilized corona current is that, as potential increases, ion production and concentration increases. This rapidly absorbs the increase in potential, so that the potential drop across the ionizing zone remains constant.

Let n be the number density of ions in transit in a zone dr thick of 1 cm length at a distance r from the axis. If dq ions are created per centimeter length of anode in a time dt, these are distributed within $dr = vdt$, where v is the ionic drift velocity. The number density of ions is

(2.17)
$$n = \frac{dq}{2r\pi(dr)} = \frac{dq}{2\pi r(vdt)} = \frac{1}{2\pi rv}\frac{dq}{dt}$$

ions/cm^3 in unit length of anode. Now $v = KX_r = \dfrac{Kp_0}{p} X_r$ with (eq. 1.4) $X_r = V/(r \log b/a)$. Thus n becomes (from eq. 2.17)

(2.18)
$$n = \frac{r \log b/a(dq)}{2\pi K(p_0/p)Vr(dt)} = \frac{\log b/a}{2\pi K(p_0/p)V}\frac{dq}{dt}.$$

Here it is seen that n is independent of r but depends on $\log b/a$, dq/dt, $1/V$, p_0/p, and $1/K$, where K is the reduced mobility. If dq/dt remains constant, n decreases linearly with pressure. It varies inversely with mobility and with potential V.

For this geometry the potential at a point r from the axis is given by

$$(2.19) \qquad V_r = \int_b^r X \, dr = \int_b^r -\frac{2Q}{r} = 2Q \log b/r.$$

Here $Q = CV$, the charge per cm on the wire with the cylindrical condenser system having a capacity C. C is given (by eq. 1.1) as $1/(2 \log b/a)$ if D, the dielectric constant for the gas, is set as unity. Then (eq. 1.3) $V_r = V \log b/r/(\log b/a)$ and the field is (eq. 1.4) $X_r = V/(r \log b/a)$. The contribution of the field X per centimeter length produced by the positive space charge of density n in transit inside a distance r, now in a continuous sheath, is 4π times the number of ions present divided by the area at r. Thus

$$(2.20) \qquad \Delta X_r = \frac{4\pi n(\pi r^2)}{2\pi r} = 2\pi n r = \frac{dV}{dr}.$$

If this space charge extends sensibly from a to b, since r_0 is sufficiently close to a, the space charge potential V_n becomes

$$(2.21) \qquad V_n = 2\pi n \int_a^b r \, dr = \pi n(b^2 - a^2).$$

At low pressures the limit must be changed to r_0 instead of to a. Since the potential on the wire anode is V, the applied potential, the potential required to maintain the discharge across the ionizing zone can be found as

$$(2.22) \qquad V - V_n = V - \pi n(b^2 - a^2).$$

This fact also signifies that, in the first calculation of the field X_r driving the ions across, the acting potential is reduced by the space charge itself. Thus in place of $X_r = V/(r \log b/a)$, one should, to a second approximation at b, correctly use

$$(2.23) \qquad X_r = \frac{V - \pi n(b^2 - a^2)}{r \log b/a}.$$

Now again, since the integration above was carried out to a instead of r_{0g}, with no space charge active in the plasma of the ionizing zone, one should include r_{0g} instead of a. One must also restore the field due to the space charge inside any point r, that is, $2\pi n r$, to the field due to the charge on the wire, to get the more correct value of X to be used:

$$(2.24) \qquad X_r = \frac{V - \pi n(b^2 - a^2)}{r \log b/a} - 2\pi n r.$$

The discharge initiates and burns if at r_0 the potential is V_g, the threshold potential. Integration of X from r_{0g} to b yields the relation

$$(2.25) \qquad V_g = [V - \pi n(b^2 - a^2)] \frac{\log r_{0g}/a}{\log b/a} + \pi n(r_{0g}^2 - a^2) \quad \text{or}$$

$$(2.26) \qquad V - Vg = \pi n \left[(b^2 - a^2) - (r_{0g}^2 - a^2) \frac{\log b/a}{\log r_{0g}/a} \right].$$

Here n is constant in r. Since the current per unit length of cylinder is $i = 2\pi r n v_+$, with $v_+ = k_+ X_r = \dfrac{K_1}{p} \dfrac{V}{\log b/a}$ and with K, the mobility at 1 mm pressure, one has the value of

(2.27) $$n = \frac{ipr \log b/a}{2\pi r K_1 V} = \frac{ip \log b/a}{2\pi K_1 V}.$$

One thus has the relation for the current and potential as

(2.28) $$V(V - V_g) = \frac{\log b/api}{2K_1}\left[(b^2 - a^2) - (r_{0g}^2 - a^2)\frac{\log b/a}{\log r_{0g}/a}\right].$$

This relation is the second-order approximation and was derived by S. Werner[2.13] in connection with slow Geiger counter theory. It is still not quite accurate, since in deriving the value for n the influence of space charge has not been included.

To discuss the meaning of this relation one can bypass the more accurate expression applicable to the condition, where r_{0g} cannot be neglected. This permits its use, for general discussion, to consider gases at intermediate and atmospheric pressure by neglecting a^2 relative to b^2 and $r_{0g}^2 - a^2$ compared to b^2. This yields the current potential relations for the coaxial cylindrical geometry at steady corona as originally derived by J. S. Townsend[1.6] in 1914, the form of

(2.29) $$V(V - V_g) = \frac{ipb^2}{2K_1} \log b/a, \quad \text{or}$$

(2.30) $$i = (V - V_g)V\left[\frac{2K_1}{pb^2 \log b/a}\right].$$

Townsend and later H. F. Boulind[2.14] in his laboratory used the relation to evaluate the mobility K_1, all quantities being measurable or known. The theory failed at lower pressures and gave unreasonably high values for K, since r_0 was neglected. The equation applies generally and has been verified by S. Werner,[2.13] the author, and C. G. Miller.[2.15] W. H. Bennett and L. H. Thomas deduced a similar relation for coronas in H_2 and N_2.[2.16]

If K is constant, then i increases in proportion to $V(V - V_g)$. This is the characteristic parabolic increase in current with potential observed for cylindrical corona currents until the limits of applicability of the basic assumptions are reached. This form is quite similar to that observed for other asymmetrical gap coronas with space charge limitation, though probably varying slightly in form with different geometrical conditions. Analysis of some point-to-plane coronas using this analogy yields the correct order of magnitude for K. Near threshold where $V - V_g$ varies much more rapidly than V, the current increases as

(2.31) $$i = (V - V_g)\frac{2KV}{pb^2 \log b/a} = \frac{V - V_g}{R}.$$

with

(2.32)
$$R = \frac{pb^2 \log b/a}{2KV}.$$

This is termed the Ohm's law regime noted near threshold for self-sustaining space charge controlled coronas by numerous observers, including Kip and S. Werner. In this region V_g maintains the breakdown current and $V - V_g$ takes care of the space charge fall in potential across the low field region. An interesting confirmation of this situation came from studies of C. G. Miller[2.15] in coaxial geometry with wire negative. In this study, at low pressures in air, pure N_2, or N_2 with 1% O_2, once the potential reached an appropriate value as V_g, the system went *over to an arc with no corona*. Here the electrons do not attach to yield negative ions and the space charge limitation is absent. Placing an external limiting resistor in the circuit led to breakdown with a limited current. If the iR drop in the external circuit was deducted from the applied V, the currents rose uniformly at a constant value of $V - iR$. The same situation can be observed at pressures so low that r_0 approaches b. In any event, the potential across the ionizing zone remains at close to V_g no matter what the applied potential V is; the excess $V - V_g$ goes to an iR drop *inside or outside* the discharge as conditions determine.

6. CURRENT INCREASE AT CONSTANT
   POTENTIAL ACROSS THE IONIZING ZONE

This constancy of V_g across the narrow ionizing sheath raises a question as to how *current increases* to give the increased space charge, with V remaining at V_g across the ion-generating sheath. Consideration of this problem led the author to a qualitative picture of what was occurring.[2.17] At a steady current, assume that the potential is raised by a very slight amount, ΔV. This at once increases the amplification factor M for each electron avalanche. Thus, the number of photons creating the γ_p, or its equivalent, is at once increased. These lead to *more* and *new electron avalanches*, which originate from areas on the cathode that were not emitting avalanche successions before. Thus, new avalanches are launched and the surface density of ionizing avalanches increases. On arrival at the anode more positive ions are created. As these move out of the ionizing sheath, the space charge begins to increase above that which existed before ΔV was applied. By the time the ion increase reaches the cathode, the potential drop across the space charge has risen to $V + \Delta V - V_g$, restoring the potential across r_{0g} to V_g and placing the new potential increment across the space charge. Thus V_g remains sensibly constant but i and $V - V_g$ increase the space charge in proportion to the increase in V.

This description, published by the author, was developed mathematically by Colli, Facchini, Gatti, and Persona to a quantitative theory and proven to be correct.[2.17] The first two of these investigators had observed that coronas of this type, when studied with the oscilloscope, showed certain regular oscillations which were not understood. The development of the theory indicated that, when potential across the corona changed by ΔV, as by line fluctuations or otherwise, the initial rise of the current due to increase of ionization in the sheath tended to overproduce the positive ions. This led to an initial overshoot of current followed by an ultimate reduction in current by virtue of the excess space charge. When this was collected, the potential was again in excess across the ionizing zone, again yielding a current increase. This produced current or current and potential oscillations across the system on a time scale of ion crossing times. The difficult theoretical study, in fact, predicted the oscillations which had been observed.

F. THE EQUIPOTENTIAL LINES
AND THE PHYSICAL FORM OF
THE DISCHARGE

Having presented the current characteristics of the discharges, it is now of interest to call attention to another aspect of field conditions affecting discharges. Let it be assumed that a positive point-to-plane corona discharge is operating, creating its positive ion space charge distortions along the discharge axis. In the spaces adjoining this region, where there is no breakdown, the normal field gradient is present. This situation is depicted in figures 2.4 and 2.10. The lower solid line of figure 2.10 represents the undistorted potential across the gap before breakdown. The solid lines represent the potential at different times as a positive ion streamer propagates across the gap. This was drawn by the author on the basis of certain data given by Trichel and data on streamer channels.[2.12,2.18] Such progress of the potential streamer fronts has been photographically confirmed by E. Nasser, through the Lichtenberg figures of advancing streamers as they cross a gap, and measured under varying applied potentials.[2.19] For the present, the nature of a streamer breakdown will be accepted without discussion. It is noticed that all along the path followed by the discharge, the conducting channel pushes the potential in the breakdown channel above that existing in the gap off the axis, in which there is no discharge. It is noted that the potential of the discharge is everywhere positive to the surrounding space. This means that any electrons, such as photoelectrons, created radially about the discharge channel will be drawn into it. If radial fields are high enough, they will produce electron avalanches converging into and flowing up the discharge channel to the anode. The ionization will

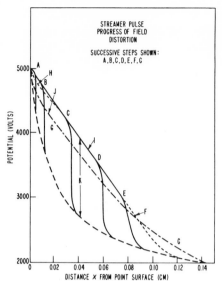

FIG. 2.10. Potential near the anode point when a streamer is created and propagates outward from this point. Dashed line is the undistorted field due to the point potential. Succession of solid and dotted lines A, B, C, D, E, and F is the streamer tip potential at various times as it advances out across the gap. The dot-dashed line G is the distortion after the streamer has reached the cathode or ceased progress. H is the start of the recovery of the field as the ions move away. I is the potential fall down the streamer channel. J is the distorted anode fall caused by the defunct streamer channel. K is the potential difference between streamer channel and undistorted gap at that point.

be most intense along the breakdown channel, and the forces are conservative, tending to keep the channel more or less confined. Equipotential surfaces could thus be roughly drawn for a given point of advance, as in figure 2.11. Note, also, that, if there were two such channels parallel to each other and in proximity to an intervening region of no discharge, such channels, having the same sign of potential because of their space charges, will *repel each other*. One channel will also reduce the field of its neighbor relative to free space.

It is now of interest to depict the situation existing with a negative point-to-plane discharge in air.[2.18] Here, because of the presence of positive and negative ions in the discharge, the potential along the channel is highly distorted. This channel potential and the undistorted potential for three phases of the discharge are shown in figures 2.7, 2.8, and 2.9, representing the initial stage, the luminous, conspicuous, and most active phase, and the final stage. Here, negative potential is plotted with positive values. The dashed lines in figures 2.7 and 2.9 are the undistorted channel. Most

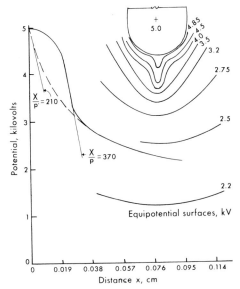

Fig. 2.11. The equipotential surfaces created about the streamer channel from a positive point depicted, as previously, to the left. The values of X/p for the undistorted field (dashed) and in front of the streamers are indicated on the figure. The accompanying equipotential surfaces through the channel and near the point are also indicated. The strong axial and radial fields can be inferred.

interesting is the fully developed discharge of figure 2.8. Here it is seen that the *outside potential* is negative to the dark space in which electrons are accelerated before reaching the highly ionized positive space charge. It also is slightly negative, or perhaps equipotential, to the highly conducting negative glow area where ionization is most active. Below this is the so-called Faraday dark space, in which the potential between the negative glow and negative space charge, due to dissociative attachment to negative ions, occurs. From the region of negative ion space charge in the positive column, the discharge potential is negative to the undistorted surroundings. This has the result that in the regions of the dark spaces, electrons are accelerated toward the axis of the discharge. Thus, particularly the Crookes and Faraday dark spaces are strongly constricted. The negative glow region is not distorted one way or the other but is confined by the constrictive effects of the dark space regions. The weakly ionizing and attaching region of the so-called positive column is, on the contrary, negative to the surrounding space. In consequence, the electrons in this region diverge and the column flares. This action nicely accounts for the constricted breakdown region along the axis in the dark spaces and glow and the shaving-brush-like flare in the faintly glowing positive column. Here again, the charges along the breakdown channels have the same sign relative to intervening regions of no discharge. Thus, two negative corona spot breakdowns will repel each other.

This description went unrecognized until W. Finkelnburg [2.20] carried out some studies in the anode and cathode regions of arc discharges by shooting arrowlike probes through different regions and recording the potential distributions oscilloscopically. With these he was able to plot the equipotential surfaces surrounding the space charges in anode and cathode regions. He observed these to be of the same form as described above. In fact, on the basis of this work on coronas, he was able to present the diagrams here indicated in terms of potential fall by means of equipotential surfaces in agreement with his arc studies. His equipotentials plotted for the negative point corona in air are shown in figure 2.12.

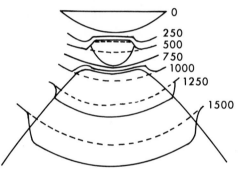

Fig. 2.12. Equipotential surfaces about the point cathode, labeled O in the Trichel pulse corona discharge at its height. This was derived by W. Finkelnburg on the basis of diagrams of the author, such as figure 2.8. This in turn was based on observations of the dimensions of glow regions by W. N. English, using a telemicroscope at lower pressures and extrapolating measurements by the principle of similitude. Here the cathode potential is zero and the solid lines represent equipotentials in volts. The dashed lines are the lines due to the undistorted field.

The field distribution for the steady positive point corona in air, where a negative ion space charge close to the anode surface creates a sharp potential fall, with the positive ion space charge and decreased fall farther out in the gap, presents an interesting picture. Here, as seen in figure 2.5, the undisturbed space adjoining the discharge at the wire surface is strongly positive relative to the region of breakdown. Thus, avalanches will naturally move outward parallel to the surface, trying to spread the discharge along the surface. This trend is augmented by photoionization in the gas, as will presently be seen. Outward in the gap, the fields reverse the discharge along the axis, being positive to the outer spaces. The discharge here attempts to contract. Thus, one could expect a *cusped* discharge with base along the wire. In supposedly pure N_2 gas, which one may suspect of having become contaminated through breakdown, Miller[1.9,2.15] observed just such cusped breakdown spots. In any event, the interrupted discharges near threshold, presumably through action of positive ions alone, were confined spots which repelled each other. As the discharge became quiescent because of negative ion formation, as Hermstein indicated, the discharge was cusped with an expanded base along the wire. Thus, discharge shapes and repulsions are governed by the potential distribution between regions of no discharge and discharge.

G. BASIC THEORY OF BREAKDOWN MODES AT ISOLATED, HIGHLY STRESSED ANODES DUE TO PHOTOELECTRIC IONIZATION IN THE GAS; BURST PULSES AND STREAMERS

1. INTRODUCTION

It is next essential to consider the nature of the breakdown in asymmetrical fields in which the large low field electrode plays the part of a current collector only. In other words, it does not furnish a secondary mechanism. As indicated in the preceding discussions, in coaxial cylinders there is a tendency for the outer cylindrical electrode, when it is the cathode, to dominate the breakdown through its function of furnishing the secondary mechanism. This appeared to be clear in Miller's[1.9,2.15] studies with coaxial cylinders, as well as those of Lauer [2.21] and Huber.[2.22]

On the other hand, studies of Townsend with coaxial cylinders had indicated some action causing breakdown on positive wires that did not depend on secondary action at the cathode. This was most certainly the case for the positive point-to-plane type of breakdown with fairly long gaps. Townsend, with his usual acuity, was led to ascribe this breakdown to some secondary action *in the gas*.[2.23] At that time (in the first two decades of the century), he believed that positive ions could ionize by impact in the gas. He, in fact, had earlier invoked such ionization to account for the secondary mechanism γ in uniform field breakdown. To this end he assigned the quantity β to be the number of ions produced by collision of the positive ion per centimeter advance in the field direction as it approached the cathode. In the case of the positive corona, β would play the same dominating role that electrons did for the negative point. It was not until many years later, during the 1930's, that many studies of ionization by positive ions ended by disproving this theory.[2.24] Direct measurements revealed that positive *ions* cannot ionize gases below some sixty ev energy, and fast *neutral atoms* created from these by charge exchange cannot do so under some 50 volts. In many gases, energies of 100 or more volts are needed. It was, then, quite well established that, because of the large momentum loss of ions by collision, there was little chance that they could acquire the 30 to 100 ev of energy needed in gases even at values of $X/p \sim 1000$.

In the meanwhile, atomic studies on photoelectric ionization of gases had been revealing that this could occur for values of $h\nu$, even below E_i, for certain gases through action on molecular or polymer species and/or through interaction of excited states. In 1934, Cravath, in the author's laboratory, and Dechéne, in France, showed, respectively, that light from coronas in air and from sparks in air could photoionize the air quite effectively.[2.25] This was also later confirmed by Raether and his school.[2.26] In the meantime, in the intensive research for creating faster Geiger counters, earlier workers in this field, notably Greiner and Stever,[2.27] respectively, proposed photoelectric ionization in the gas and proved the same to be the basic propagating mechanism in counter action. Since coaxial cylindrical geometry is used there, the mechanism was always complicated by interference by secondary effects from the negative outer cylinder. Elaborate precautions accordingly had to be taken to reduce this cathode action to zero.

Studies in the author's laboratory on the so-called burst pulse and streamer coronas in point-to-plane geometry from 1936 on led to the conviction that in air [2.28] the breakdown at the positive points was determined by photoelectric ionization in the air. This led to the consideration of the absorption coefficients of photoionizing photons and the influence of these as a secondary agency triggering breakdown at positive points. It was known from Cravath's measurements that the absorption coefficient for the ionizing corona radiation in room air at 760 mm pressure had a value of $\mu \approx 10$ cm^{-1}. Dechéne, for sparks, had observed a spectrum of wave lengths which were filtered out as gap length increased, ranging from possibly 100 cm^{-1} down to perhaps 5 cm^{-1}. Weissler[2.29] and Po Lee's later studies indicated that a continuum starting at 1000 Å and extending down to 400 Å in O_2 was involved with coefficients extending over a large range of values. Later measurements by Raether's students[2.26] indicated the very active photon to have a μ of 32.5 cm^{-1}. The later work of Przybylski showed this to be 38.5 cm^{-1}. He also discovered that one for N_2 photons in O_2 has a $\mu = 25$ cm^{-1} at 760 mm of O_2.[2.26] Cravath's value of μ for air would make the coefficient due to O_2, which is present to one part in five, come to 50 cm^{-1}, which, considering his early techniques, is close to Przybylski's value. It is clear that, while there may be a broad spectrum of values, the most active photons are those of around 30 cm^{-1} active on pure O_2 at 760 mm. Thus in air, $\mu \approx 6$ cm^{-1}. The quantity $1/\mu$ represents the average distance for photoionization in the gas. Thus, in air at N.T.P., $d_\mu \approx 0.16$ cm, and in O_2 it is ≈ 0.033 cm. There is little doubt that sources of even more active wave lengths have values of d_μ of the order of 0.01 cm and less. This circumstance will then have an important influence on the form taken by a breakdown from a highly stressed positive point.

2. Conditions Leading to Burst Pulses and Pre-Onset Streamer Pulses

Assume that in air at N.T.P., the avalanches increase in size as potential V on the positive point is raised. Eventually they will reach a magnitude such that the photoelectrons produced by electron impacts within $1/\alpha$ cm from the anode will create a sizable number of photons. These photons proceed outward and in a zone of a definite average radius create new electrons. Now at 1.6 mm outward along the axis of the point, the photoelectrons created are in a field that is too weak to cause multiplication. As they move in, however, they create new ions and electrons at the point surface. If these are sufficient in number to perpetuate the avalanche sequence, there will be a local self-sustaining corona. At threshold, however, the positive space charge, as it accumulates after several avalanche generations, will soon extinguish the corona, since the ionization required to yield Hermstein's[1.16] negative ion space charge is inadequate at this stage. However, note that the most effective ionization at 1.6 mm out from the surface will lie along and near the surface of the point if the high field area is of sufficient extent. This will permit the spread of the corona discharge laterally over the point surface. Field distortion might tend to restrict this spread, but as the positive space charge increases along the axis, quenching the discharge along the axis, the lateral spread will occur. This spread could be extensive along a wire anode where the field remains high. It is, in principle, the nature of the action of the Geiger counter. Here, for fast Geiger counters, gaseous mixtures are developed which are properly suited to give the most rapid build-up and spread, and equally rapid quenching without retriggering by actions at the cathode. Thus, a fluctuating discharge, which at threshold spreads over the available surface, is to be expected.

Next, if pure O_2 is used in place of air such that d ≈ 0.033 cm, the photoelectrons are produced very close to the ionizing zone at the surface. They will not, however, get very far away from the discharge region laterally. This, with the conservative nature of the fields, will limit ionization to the area of initial breakdown. At the self-perpetuating threshold, the discharge will originate in a limited area, and by the time the space charge has built up to quench the discharge locally, it will not have spread along the surface sufficiently to avoid the quenching action at the center. Thus, discharge will take place in local bursts that do not spread.[2.22,2.30]

Further increase in potential and suppressions of a γ action at the cathode will increase the avalanche size such that at a point on the anode surface many photons will be created. In addition, once the avalanche electrons enter the anode, the dense highly localized space charge, before it moves, will create a space charge density such that it will effectively add to the field in the gap. This presupposes either very large initial single avalanches at the anode or a convergence of many smaller avalanches at the same local area before the ion space charge can move away from the point. The space charge field will be a maximum along the point axis, and before the positive ions move, it will add to the field at the point. The increased field along the axis may then become great enough to draw in the photo-electrons produced within the 0.33 mm in the highest part of the field along the axis. Once this occurs, the discharge will propagate itself outward along the point axis as a *breakdown streamer*. Thus, a short d_μ and a high space charge field at the anode propagates the *breakdown outward along the field axis* instead of along the surface. The avalanche size needed to initiate the *breakdown* streamer process in air at N.T.P. lies in the order of 10^8 electrons. However, propagation depends as much on the space charge density as on the total number of ions, and for pre-onset streamers this may be $\approx 10^5$ at threshold.

Both these new breakdown modes at isolated highly stressed anodes, attributable to a new secondary mechanism—photoelectric ionization of the gas—lead to threshold relations resembling in general form the Townsend conditions with a γ_p and γ_1. They have, on simplifying assumptions, been deduced by the author[2.31] for the burst pulse (lateral spread) and by the author and Wijsman[2.32] for the streamer (axial propagation) thresholds. These are presented in the following sections.

3. The Burst Pulse or Pseudo Geiger Counter Pulse and Geiger Pulse Thresholds

An electron moving from r_0 to a in the field of a positive point on wire creates $\exp \int_{r_0}^{a} \frac{-}{} \alpha dr$ new electrons. Each avalanche electron will also create $f \exp \int_{r_0}^{a} \alpha dr$ photons of wave lengths short enough to photoionize atoms or molecules of the gas. Here f is defined as the ratio w/α introduced at the beginning of chapter 2. At the low fields encountered, f will probably be greater than unity and decline somewhat as X/p increases. Assume for convenience that there is just one ionizing wave length of absorption coefficient μ active. About half the photons liberated will be directed into the gas away from the point as most of the ionization is created within $1/\alpha$ cm of the surface of radius a. Thus, the geometrical survival ratio g will be set as 0.5. If $r_0 - a$ is larger compared to a, g will be greater than 0.5 and photons reflected from the metal will also contribute slightly. Of the gf exp $\int_r^a \alpha dr$ photons moving outward, a fraction β will proceed into the gas a distance in excess of r_0 and create new electrons that may ionize effectively. This fraction β is roughly given by $e^{-\bar{r}_0 \mu}$ for a hemispherical point, as shown by the author.[2.31] Here $\bar{r}_0$ is an average path length greater than r_0. The paths are oriented in all directions from the wire or point so that $\bar{r}_0$ will be greater than r_0 for all photons except those created at r_0 normal to the surface. The average of r_0 for a hemispherical point with $a > r_0$ is roughly given by $r_0 = r_0/\cos\bar{\theta}$. Here $\bar{\theta}$ is an average angle of photoemission with the normal to the point given by

$$(2.33) \qquad \bar{\theta} = \int_0^{\pi/2} \theta \sin\theta \, d\theta \Big/ \int_0^\pi \sin\theta \, d\theta \approx 1 \text{ radian.}$$

A rigorous calculation by R. J. W. Wijsman differs little numerically from the approximate value above, which makes $\bar{r}_0 = 1.86 r_0$. If f_1 is the fraction of the $fe^{-1.86\mu r_0}$ photons that ionize, the photon production by the avalanche may be set as

$$(2.34) \qquad g\beta f \exp \int_0^a \alpha dr = f_1 gf e^{-1.86\mu r_0} \exp \int_{r_0}^a \alpha dr.$$

If now

$$(2.35) \qquad gff_1 e^{-1.86\mu r_{0g}} e^{\int_0^a \alpha dr} = gff_1 e^{\int_{r_{0g}}^a \alpha (dr - 1.86\mu r_{0g})} \geq 1,$$

the discharge becomes self-sustaining at a distance $r_0 = r_{0g}$ for the breakdown threshold. As all photons absorbed may not produce ions, f_1 is the fraction of the photons created in this region which succeed in ionizing. This discharge can be intermittent because of space charge accumulation. It may also remain localized because of the large value of μ. Thus, when the potential is raised to V_g such that X_{0g}/p yields the condition expressed by the equation above, a burst pulse type of corona discharge sets in.

A study by Tamura[2.33] has extended and adapted this theory to the evaluation of the effective μ_0, and $f_1 f$ from the observation of threshold values in the gases air, O_2, and N_2 and H_2. Tamura assumes a hyperboloidal point-to-plane gap for which the essential field is

$$(1.8) \qquad X = \frac{2V}{a \log (4l/a)} \frac{a}{a + x} = X_0 \frac{a}{a + x},$$

in which a is the focal distance at the tip, l is the gap length, and x is the distance from the tip at which X is desired. Here X_0 is the field at the tip.

It is next *assumed* that the *approximate Townsend* expression for α/p as a function of X/p, to wit

(2.1) $$\alpha/p = Ae^{-BX/p} = Ae^{-(1/\epsilon)}$$

applies for the limited region of X/p ranging from $100 < X/p < 800$ volt/cm per millimeter. From curves of Masch the constants A and B are given in table 2.3.

TABLE 2.3

Gas	A, ions/cm × mm	B, volt/cm × mm
Air	15	365
N_2	12	342
O_2	8	213

The quantity $\epsilon = (X/p)/B$ is the ratio of reduced field strength to pressure. It is the quantity X/p represented in units of B at a distance x from the point tip. In terms of x then, ϵ becomes $\epsilon = \epsilon_0 a/(x + a)$, where ϵ_0 is the reduced field strength at the tip surface along the axis; that is, $\epsilon_0 = (X_0/p)/B$. The multiplication M for an avalanche starting at x is

(2.2)
$$M = \exp \int_0^x \alpha dx = \exp Ap \int_0^x e^{-Bp/X}dx$$

$$= \exp \int_0^x e^{-1/\epsilon}dx = \exp Ap \int_0^x e^{-(x+a)/a\epsilon_0}dx$$

$$= \exp Apa\epsilon_0(1 - e^{-x/a\epsilon_0})e^{-1/\epsilon_0}$$

as modified in this chapter, section E.2, and by factors above. Now consider equation 2.35, which reads

(2.35)
$$\beta fM = \oint f_1 f e^{-1.86\mu r_0} e^{\int_a^{a+r_0} \alpha dr} = 1,$$

if the origin is changed to the center of a. Here r_0 is the distance from the tip at which ionization has virtually declined to zero. $\beta = 0.5f_1e^{-1.86\mu r^0}$ is the chance that a photoionizing photon creates an avalanche beyond r_0 from the point surface. f is the ratio of photoionizing photons to electrons, μ is the absorption coefficient for photoionization, and f_1 is the chance that these photons photoionize the gas when absorbed.

Near the point M can be expressed by equation 2.2. Set $\mu = \mu_0 p$, where p is the pressure of the gas, and μ_0 is the absorption coefficient at 1 mm. This then leads to the condition for a localized burst pulse in the form

(2.36) $$0.5f_1f \exp\left[-1.86\mu_0 pr_0 aApe_0(1 - e^{-r_0/a\epsilon_0})e^{-1/\epsilon_0}\right] = 1, \quad \text{or}$$

$$\exp\left[aAp\epsilon_0(1 - e^{-r_0/a\epsilon_0})e^{-1/\epsilon_0} - 1.86r_0\mu_0 p\right] = \frac{2}{f_1f}.$$

In logarithmic form the threshold condition becomes

$$(2.37) \qquad \epsilon_0(1 - e^{-r_0/\epsilon_0 a})e^{-1/\epsilon_0} - \frac{1.86 r_0 \mu_0}{aA} = \frac{1}{aAp} \log \frac{2}{f_1 f}.$$

Set

$$(2.38) \qquad K = \frac{1.86 \mu_0}{A} \quad \text{and} \quad \rho_0 = \frac{r_0}{\epsilon_0 a};$$

the equation 2.37 assumes the form

$$(2.39) \qquad \epsilon_0\{(1 - e^{-\rho_0})e^{-1/\epsilon_0} - K\rho_0\} = \left(\frac{1}{aAp}\right) \log \frac{2}{f_1 f}.$$

If now p becomes very large, then the righthand term goes to zero. Thus K, the reduced absorption coefficient, becomes

$$(2.40) \qquad K = \frac{1 - e^{-\rho_0}}{\rho_0} e^{-1/\epsilon_{00}}.$$

As assumed, most of the ionization occurs within r_0 of the point. That is, r_0 is the smallest value of x which sensibly affects M in equation 2.2. This renders $e^{-x/a\epsilon_0}$ in equation 2.2 sensibly zero, so that $e^{-\rho} = e^{-r_0/\epsilon_0 a} \fallingdotseq 0$. The quantities $e^{-\rho}$, $1 - e^{-\rho}$, and $(1 - e^{-\rho})/\rho$ are plotted against $\rho = x/\epsilon_0 a$ in figure 2.13.

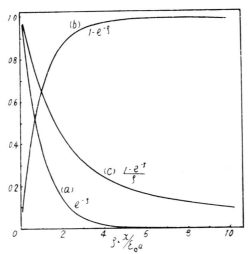

Fig. 2.13. Computed curves of Tamura for use in evaluating his functions used in equations 2.39 and 2.40. Here $e^{-\rho}$, $1 - e^{-\rho}$, and $\dfrac{1 - e^{-\rho}}{\rho}$ are plotted against $\rho = \dfrac{x}{E_0 a}$. The curves are self-explanatory.

It is seen that on the scale of plotting, $e^{-\rho}$ becomes 6.7×10^{-3} at $\rho = 5$. Thus one may reasonably set $\rho = 5 = x/\epsilon_0 a$, or set r_0 as $5\epsilon_0 a$ and $\rho_0 = 5$. This makes $1 - e^{-\rho_0} = 0.9933$ (that is, it includes 99.3% of the electrons created), so that $(1 - e^{-\rho_0})/\rho_0 = 0.1985$. Hence $K = 0.1985 e^{-(1/\epsilon_0 a)}$.

Placing equation 2.39 in equation 2.40, we find

$$(2.41) \qquad \log \frac{2}{f_1 1} = aАp\epsilon_0 (e^{-1/\epsilon_0} - e^{-1/\epsilon_{00}})(1 - e^{-\rho})$$

$$\doteqdot aАp\epsilon_0 (e^{-1/\epsilon_0} - e^{-1/\epsilon_{00}}).$$

From the starting potential V_s at different pressures, the values of $X_{s/p}$ can be inferred leading to the values of ϵ_0. These can be plotted against $1/p$, that is, reciprocals of p for the various gases. In all cases, the ϵ_0 plotted against $1/p$ fell on straight lines, which is truly remarkable in view of the approximations made. The intercepts of these on the axis of ordinates yielded values of ϵ_{00}. Then from $K = 0.1985 \exp(1/\epsilon_{00})$, the values of K may be deduced and, since $K = 1.86\mu_0/A$, the evaluation of μ_0 follows. These values are 720 cm^{-1} for air, 460 cm^{-1} for N_2, and 440 cm^{-1} for O_2 at 760 mm pressure. Plotting $\epsilon_0(e^{-1/\epsilon_0} - e^{-1/\epsilon_{00}})$ against $1/p$ should give straight lines for ff_1. The line is nearly straight for air, but for O_2 and N_2 the lines are slightly curved. Doubtless, since f could depend on ϵ_0 or ϵ_0/p, they should not be straight lines. Values for $f_1 f$ lie between 0.5 to 0.7 for air, 0.4 to 0.6 for N_2, and 0.3 to 0.5 for O_2. This theory yields values of μ which agree remarkably well with known values of absorption continua in the gases and with measured absorption coefficients for photoionization. It is to be noted that these deal primarily with the threshold of a localized burst pulse corona such as observed by Miller, Loeb, and by Huber in pure O_2. It does not apply to the photons needed for the spread of the burst pulse along a wire or over a surface. Here, as indicated below, μ_0 must be in the neighborhood of 10 cm^{-1} at 760 mm pressure for propagation.

It appears that initially Tamura had used a very fine needle. Later measurements were made with needles of different focal lengths. Here the curves for ϵ_s as a function of $1/Aap$ showed different slopes for the various points, yielding different values of ϵ_{00} for different radius points, or for different X/p regions tested. This result is not surprising. It means that as a changes, r_0 changes, and as X/p changes, r_0 changes. Thus the most effective values of μ_0 will also change. Since the photoionization is caused by wave lengths in a considerable spread of continua and since photons can come from several wave length regions in these gases, some of which may not have been too pure, this result should be expected. Thus the value of the *effective* absorption coefficient depends on X/p. It is thus essential to modify the application of the theory to determine the most efficient active coefficients.

Starting from

$$(2.39) \qquad \epsilon_0\{(1 - e^{-\rho_0})e^{-1/\epsilon_0} - K\rho_0\} = \frac{1}{aAp}\log\frac{2}{f_1 f}$$

for a gap of constant a, p, ϵ_0, the value of βfM depends on the value of the critical distance r_0. If one takes for this critical distance the *optimum* value of r, which makes the value of βfM a maximum, then the value of ρ_0 corresponding to this critical value r_0 would make the lefthand side of equation 2.39 a maximum. Thus Tamura sets a new value for K:

$$(2.42) \qquad K = e^{-\left(\frac{1}{\epsilon_0} - \rho_0\right)} \quad \text{or}$$

$$(2.43) \qquad \rho_0 = \log\frac{1}{K} - \frac{1}{\epsilon_0}, \quad \text{so that equation 2.39 becomes}$$

$$(2.44) \qquad \epsilon_0 K\{e^{\rho_0} - (1 + \rho_0)\} = \frac{1}{Aap}\log\frac{1}{0.5 f_1 f}.$$

If one assumes that only one kind of radiation is active and that the relations hold even for very high pressures, so that at the value of the reduced field strength for threshold the value of ϵ_0 would tend to a finite value ϵ_∞ as pressure increases indefinitely, then from equation 2.44

$$(2.45) \qquad \{e^{\rho_0} - (1 + \rho_0)\}_{p\to\infty} \to (\rho_0)_{p\to\infty} \to 0$$

and from equations 2.42 and 2.43 $K = e^{-1/\epsilon_{0\infty}}$ and

$$(2.46) \qquad \rho_0 = 1/\epsilon_{0\infty} - 1/\epsilon_0.$$

Equation 2.43 becomes

$$(2.47) \qquad \epsilon_0 e^{-1/\epsilon_0} - e^{-1/\epsilon_{0\infty}}\left(\epsilon_0 + \frac{\epsilon_0}{\epsilon_{0\infty}} - 1\right) = \frac{1}{Aap}\log\frac{1}{0.5 f_1 f}.$$

When $\epsilon_{0\infty}$ is very large, equation 2.47 can be approximated by

$$(2.48) \qquad \epsilon_0 - \epsilon_{0\infty} \approx \frac{\epsilon_{0\infty}^2}{Aap}\log\frac{1}{0.5 f_1 f}.$$

Then for a given point in a gap in a certain gas, if one plots ϵ_0 against $1/p$, it will approximate a straight line provided $f_1 f$ is nearly constant or varies slowly. The intersection of this straight line with the ϵ_0 axis gives $\epsilon_{0\infty}$.

In a gas in which a certain number of lines of different absorption coefficients are active and/or f varies rapidly with X/p, equation 2.47 is valid for only a given range of X/p or ϵ_0. In this event the effective value of $\epsilon_{0\infty}$ may be assigned from extrapolation of the curve by plotting ϵ_0 against $1/p$ in the X/p range to the ϵ_0 axis, and the value of the effective μ may be estimated from equation 2.44. The value of $f_1 f$ may then be derived from equation 2.47.

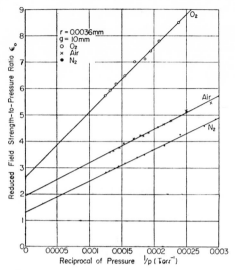

FIG. 2.14. Tamura's plots for E_0 against $1/p$ with p in millimeters for O_2, air, and N_2.

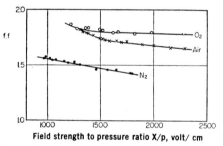

FIG. 2.15. Tamura's values of ff_1 plotted as a function of X/p in volts per centimeter per millimeter for O_2, air, and N_2.

Preliminary results obtained are shown in figures 2.14 and 2.15 or $f\epsilon_0$ plotted against $1/p$ in millimeters^{-1} and for f_1f_2 in the X/p range of 1000 to 2500 for the small point for air, O_2, and N_2. It is seen that values range between 1.5 and 2 and change relatively slowly, decreasing as X/p increases. More interesting is the fact that for O_2 and N_2 there are a series of values of μ_{760}, reduced to standard conditions as X/p varies with point or pressure. Around X/p of 10^3, values of μ_{760} range around 3700. At X/p at 300 and 100, μ_{760} lies around 100 and 5, while for X/p at 50 and 70, values are 10^{-2} and 10^{-3}. This result comes from a private communication from Tamura dated October, 1962, and the data are drawn in figures 2.16, 2.17,

2.18, and 2.19 for O_2, air, N_2, and H_2. The data in H_2 have since been published.[2.33] These data are indeed very interesting and in a measure account for the differences of values of μ_{760} observed by different workers.

Alder, Baldinger, Huber, and Metzger,[2.34] in their excellent analysis of the spread of the Geiger counter pulse down a wire, set up a condition for the threshold of a Geiger counter discharge that *spreads down the tube*. They used a primitive expression for the Townsend avalanche integral. Corrected by the proper integral, their expression becomes

$$(2.49) \qquad gf_1f_0 \frac{p_a}{p} \exp\left(\int_{r_{0g}}^a \alpha dr\right)$$

$$\left[\exp\left(-\mu_0\theta v \frac{p_i}{p}\right) + \mu_0\theta v \frac{p_i}{p} \operatorname{Ei}\left(-\mu_0\theta v \frac{p_i}{p}\right)\right] \geq 1.$$

Here, in general, the terms are analogous to those in equation 2.35, viz., gf_1. However, p_a is the partial pressure of the gas creating the ionizing photons, and p_i is the partial pressure of the gas being photoionized with p, the total gas pressure. Actually f and μ, as defined in equation 2.35, are the same as $f = f_0(p_a/p)$, and $\mu = \mu_0(p_i/p)$ in equation 2.49. This distinction arises from the usage in Geiger counter work of mixed gases with the

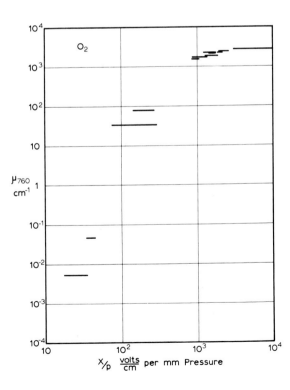

Fig. 2.16. Tamura's values for the several absorption coefficients in O_2 at different X/p values for various point radii used.

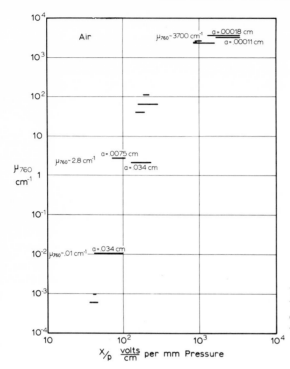

FIG. 2.17. Tamura's values for the several absorption coefficients in air at different X/p values for various point radii used.

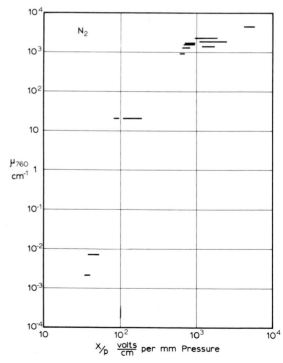

FIG. 2.18. Tamura's values for the several absorption coefficients in N_2 at different X/p values for various point radii used.

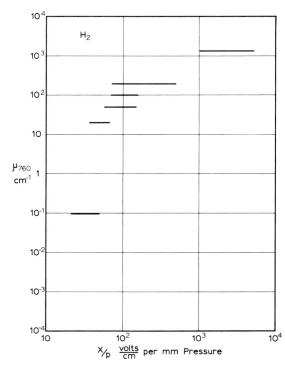

FIG. 2.19. Tamura's values for the several absorption coefficients in H_2 at different X/p values for various point radii used.

different functions indicated. In equation 2.49 Ei stands for the exponential integral, and the term under Ei represents the equivalent of the correction for the directions of avalanche generation appearing under the value $1.86r_0$ in the term $e^{-1.86\mu r_0}$ in the expression 2.34. The new term θv in equation 2.49 is the one involved in the propagation of the pulse along a wire with velocity v. $\theta v = x_0$ measures the critical distance along the wire needed to propagate the pulse with a velocity v. Here θ is the longest time needed for photon emission from the beginning of the avalanche.

The difference between the author's[2.31] expression for *stationary* threshold and that for a *propagating* threshold is that for propagation there is a critical length $x_0 = \theta v$, while all that is needed for the start of any breakdown is the critical avalanche ionizing distance in the gas.

It is of interest to regard θ. θ is composed of the time for creation of the avalanches, the time for photon travel, and the time of radiation. If $v(r)$ is the electron drift velocity, as a function of distance from the anode, then θ cannot be less than $\tau = \int_{r_{0g}}^{a} [1/v(r)]\, dr$. If the photon comes from a state of a lifetime τ_f, then θ becomes $\theta \int_{r_{0g}}^{a} dr/v(r) + \tau_f$. The photon transit time across 2 mm is negligible.

The relation $x_0/\theta = v$ only defines the distance needed to propagate a pulse with velocity v. Neither x_0 nor v is fixed to any physical process by the theory involved. In practice v can be observed and θ is definitely limited at one extreme by the processes above. However, there is never any limitation on the values of v and x_0 as long as the ratio is θ. However, there are upper and lower limits on v that are not given in the propagation equation. The upper limit of v is set by equation 2.35. This relation defines a minimum average distance r_{0g} of avalanche travel to the anode at the corona threshold V_g. It is determined by the field geometry, X as a function of r, the variation of α/p with X/p and thus with r, the partial pressures p_a and p_i, and probabilities f_0 and f_i defined in relation to equation 2.49. Below the average value of r_{0g}, a self-sustaining corona cannot start. Since the avalanche must be complete to be fully effective, θ is the limiting time. To have a discharge that may or may not spread, a limiting distance r_{0g} is needed to allow the discharge to build up at the lowest potential. As potential increases, the critical value of r_{0g} decreases. Thus $r_{0g} \geq x_0 = v\theta$, and $r_{0g}/\theta \geq x_0/\theta = v$. As x_0 is less than r_{0g} then the maximum velocity of spread v_m is set by

$$(2.50) \qquad\qquad v_m = r_{0g}/\theta.$$

This really says that the velocity of propagation can never exceed the velocity of avalanche creation and completion of photoionization.

There is also a lower limit to the velocity of spread. If x_0 becomes very small, the velocity of spread will become very slow. If the velocity of spread is so slow that the local quenching space charge builds up before the pulse has spread from the region, it cannot spread but will extinguish itself. Assume that it takes only a succession of n avalanches to quench the corona by space charge. This n may not be very large. Only ten avalanches were needed to quench Lauer's photon-conditioned γ_p pulses in H_2. The time for n successive avalanches in succession is $n\theta$. Thus if

$$(2.51) \qquad\qquad r_{0g}/v_1 = n\theta = \frac{r_{0g}\theta}{x_{01}}, \; v_1 = r_{0g}/n\theta.$$

Thus $vr_{0g}/n\theta$ sets a limitation on the velocity and on the minimum value of $x_{01} = v_1\theta = r_{0g}/n$.

These considerations apply to initiation by a single electron avalanche. If, as in Geiger counters, pulses are propagated by large localized ionizing events, velocities will increase with the size of the initiating electron cloud. This follows, since so many photons will be created that the many that range beyond $1/\mu_0$ will contribute to the spread. This means an effectively larger x_0, for x_0 depends on the distances at which photoelectrons are liberated within a time that is comparable with the formative time of the avalanche. Furthermore, avalanches now being multiplied many-fold need not be completed to propagate the pulse. Thus in principle an increased velocity, v, will depend on an increased x_0 and the decreased time of growth of avalanches sufficiently to cause propagation. The increase in the triggering event magnitude and avalanche size with potential has been observed.

4. THE STREAMER PULSE THRESHOLD

The derivation of the condition for streamer propagation is best made under simplifying assumptions.[2.32] It is advantageous to assume a uniform field geometry and a single photoionizing photon of absorption coefficient μ. It will be simpler to assume a single electron avalanche as the initiating agency, although a group of converging avalanches in asymmetrical fields is more likely, and some from α particles have been observed. Let a single electron start at the cathode and traverse the distance x_c to the anode. It creates $e^{\alpha x_c}$ electrons and positive ions. At x_c the avalanche will, by radial diffusion of the electron cloud, have expanded so that the electron cloud is confined to a roughly cylindrical volume, say $2/\alpha$ cm deep and $\bar{r}$ cm in radius. The diffusion occurs because of the high random velocity of the electrons, which have a diffusion coefficient D during the avalanche transit time t. Thus, one has $\bar{r}$ given roughly by

$$(2.52) \qquad \bar{r} = \sqrt{6Dt} = \sqrt{\frac{6Dx_c}{v_e}} = \sqrt{6\tfrac{2}{3}U_t \frac{x_c}{X}} = \sqrt{\frac{4}{3} \frac{\lambda_0{}^2 e X x_c}{p^2 v_e{}^2 m}}.$$

Here v_e is the electron drift velocity, U_t is the terminal energy of the electrons in a field X, e is the electronic charge, m is its mass, λ_0 is the average electronic free path at N.T.P., and p is the pressure. To estimate the electrical field along the axis caused by a space charge of positive ions left by the avalanche electrons absorbed by the anode, it is convenient to calculate the radial field of the electron density contained in a sphere of radius $\bar{r}$ instead of that for the cylindrical element of volume $\pi r^2(2/\alpha)$ or some other configuration. This spherical approximation, at most, differs from the more realistic space charge distribution by a factor less than 2, as ascertained by Weissler. The number N of ions in this volume at x_c is determined by the density of ions in this region multiplied by $(4/3)(\pi\bar{r}^3)$. From equation 2.2, $dn/dx = \alpha e^{\alpha x}dx$, the rate of ionization. The density of the ions is given by the number of positive ions created in a length dx at x_c at a rate of ionization $\alpha e^{\alpha x_c}dx$ divided by the volume in which they were created, viz., $\pi\bar{r}^2 dx$, or $\alpha e^{\alpha x_c}dx/\pi r^2 dx$. The value of N then becomes

$$(2.53) \qquad N = \frac{\alpha e^{\alpha x_c}dx}{\pi\bar{r}^2 dx}\left(\frac{4}{3}\pi\bar{r}^3\right) = \frac{4}{3}\alpha\bar{r}e^{\alpha x_c}.$$

This concentration of charge leads to a field X′ at a distance of r from the center given by

(2.54)
$$X' = \frac{\frac{4}{3} e\bar{r}\alpha e^{\alpha x_0}}{r^2} = \frac{E.}{r^2}$$

This field parallels and adds to the field X derived from the potential of the anode. Thus along the axis the field is

(2.55)
$$X_1 = X + X' = X + \frac{E}{r^2}.$$

The field E/r^2, in other directions, adds vectorially to the field along the axis. As its value is greatest along the axis, attention will be confined to the influence of this combined field within a cylinder of limited radius surrounding the axis. Accompanying the creation of α ions per centimeter along the field axis, there were created w photons per centimeter of such energy as to be able to photoionize the gas. Then, accompanying N ions in the volume, there are $N(w/\alpha) = f_1 N$ photons. The value of w/α and of f_1, as X varies along the axis, can be considered relatively constant. Of the $f_1 N$ photons created, a fraction f_2 succeed in liberating photoelectrons. Those photons out of the $f_2 f_1 N$ which reach a distance r within a cylinder $\bar{r}$ in radius about the axis will roughly experience the combined space charge and imposed fields $X_1 = X + X'$. Since the rate of absorption of photons is $\mu e^{-\mu r} dr$, the number of photoelectrons created at a distance r from the center of the spherical charge in an element of cylinder $\pi r^2 dr$ will be

(2.56)
$$f_1 f_2 N \frac{n r^2}{4 \pi r^2} e^{-\mu r} \mu dr.$$

Each of these electrons starts an avalanche the tip of which reaches to within $\bar{r}$ of the anode, that is, reaches the surface of the space charge and produces

(2.57)
$$\tfrac{4}{3} \alpha' \bar{r}' e^{\int_{\bar{r}}^{r} \alpha' dr'}$$

new electrons and ions. Here the barred primes are indicative that this ionization occurs in a field $X_1 = X + E/r^2$. Thus $\bar{\alpha}'$ is the first Townsend coefficient in X_1 and $\bar{\alpha}$ is its value in X_1 when $r = \bar{r}$ where it is a maximum. Since the ionization is now produced in a higher field within a distance r_c, which is much less than x_c, the value of $\bar{r}$ will be $\bar{r}'$, which will be smaller. The drift velocity will also increase in the increased field, so that diffusion time t is less. However, as v increases only as $\sqrt{X}$, this increment in v will not be important, while both the enhanced values of X_1 and decreased values of $\bar{r}'$ will be important. To estimate $\bar{r}'$, one can set $\bar{r}' = \bar{r}(r/x_c)^{1/2}$, since the radius varies with the $\sqrt{Dt}$. Combining the expressions 2.53, 2.56, and 2.57 for the number of new electrons, their avalanche values, and the corrected radii, one has, on integration from $\bar{r}$ to x_c, the number of new ions produced by the secondary action of photoionization in the gas:

$$(2.58) \qquad \int_{\bar{r}}^{x_c} f_1 f_2 N \left(\frac{\bar{r}^2}{4r^2}\right) e^{-\mu r} \mu dr \left[\frac{4}{3} \bar{\alpha}' \bar{r} \left(\frac{r}{x_c}\right)^{1/2} e^{\int_{\bar{r}}^{r} \alpha' dr'}\right].$$

This approximation is valid if $1/\mu$ is not appreciably greater than $\bar{r}$. It is at once clear that a streamer will propagate if the integrals above yield a product equal to N, the ionization in the initial avalanche. For then, the first avalanche will perpetuate itself, creating a sequence of new avalanches of N ions that advance the space charge out into the gap by $\bar{r}$ at each step. It is further clear that, since the space charge density augments as long as the photon production is adequate, conditions are favorable for continued advance. If the channel of ions left behind continues to conduct while its high field tip is advancing, the anode potential, less the drop down the streamer channel, could well cause enhanced tip fields that would increase the conductivity of the channel. Despite increased gradients at the tip of the avalanche and increased ionization, its velocity of advance, which depends on $\bar{r}/\tau$ where τ is the avalanche generation and movement time, could well remain fairly constant. This stems from the fact that avalanches produce less ionization as they decrease in duration in enhanced fields. This is one difficulty presented by intense field-ionizing space wave propagation; they advance so rapidly that few repeated ionizing sequences can occur before they have passed on.

Setting the expression 2.58 equal to N, with cancellation of terms, yields the condition for streamer advance in a uniform field as

$$(2.59) \qquad \tfrac{1}{3} f_1 f_2 \left\{\mu \bar{\alpha}' \frac{\bar{r}^3}{x_c^{1/2}}\right\} \int_{\bar{r}}^{x_c} \bar{r}^{3/2} e^{-\mu r} e^{\int_{\bar{r}}^{r} \alpha' dr'} = 1.$$

To solve this equation, α' must be evaluated as a function of X_1 from $\bar{r}$ to r. The value of the initial field at threshold is X_s, and one must assume that $X' = E/r^2$ remains constant in a propagating streamer, so that, once fixed, X_1 and α' are fixed, and that $\bar{r}$ is relatively constant. These conditions are probably not met in practice, as indicated above. However, they serve as a guide in setting up a theory. It is interesting to note that the initial value of N, the space charge density, drops out in the propagation relation. However, the quantities involved, namely $\bar{r}$ and X_1, dependent on N, are included indirectly in fixing α'. Also if N is not sufficient and if n is not sufficient, there will be inadequate field and inadequate photons for propagation.

This relation is interesting in that it fits into the well-known pattern of all threshold equations in simulating Townsend's generic condition

$\gamma_e{}^{ad} = 1$. It also has probabilities akin to γ in the form of $f_1 f_2$. Likewise, it has the perpetuating avalanche multiplying term $e^{\int \alpha dr}$. It is thus subject to considerable fluctuations in single electron triggering events, and these have been observed and studied by Raether's group.

Raether's group has confined itself to the study of breakdown streamers in uniform field geometry.[2.6,2.35] Breakdown streamers appear at around atmospheric pressures in gases investigated by Raether when avalanche sizes at the anode reach on the order of 10^8 to 10^9 electrons. This can be achieved near but above Townsend threshold in air through the increase in avalanche size by space charge distortion in the gap. It can also be achieved by using impulse potentials of sufficient magnitude, $\approx 6\%$ above Townsend threshold for a 1 cm gap at 760 mm dry air, in which case the Townsend discharge does not have time to develop. Raether's group used organic vapors in uniform field gaps, by which γ is reduced to $\approx 10^{-9}$; then single avalanches of 10^8 electrons initiate streamer sparks. Finally, the group has succeeded in projecting enough electron avalanches, starting from an α particle track, to give 10^3 simultaneous electrons that cross the gap to strike the anode within a target area of the diffusion radius r of the single avalanche within 10^{-8} sec.

Under these conditions, at fields which yield an $e^{\alpha d}$ on the order of 10^{-5}, a streamer spark is triggered 2% below the normal Townsend threshold. These have been oscillographically recorded. This last experiment indicates that to launch the breakdown streamers there must be on the order of 10^8 to 10^9 electrons and ions created in a volume of about $2/\alpha$ deep and r in radius. For a uniform field gap this is set by Raether as ≈ 0.5 mm. The radius has been estimated by the author as more nearly 0.2 mm or less. Once the streamer tip is launched, its radius may be considerably less because of the enhanced local fields and the confinement of $d_\mu = 1/\mu$ to distances comparable with r_0 in the ionic space charge field, plus the imposed potential field X. Propagating streamers are certainly less than 1 mm and probably 0.1 mm or less in diameter.

If now one turns to the question of the potentials and avalanche sizes for the pre-onset streamers, more discussion is needed. Recent work in the author's laboratory by Nasser[2.19] and Waidmann,[2.36] using Lichtenberg figure techniques, has thrown much light on this question. Using impulse potentials, Nasser's studies indicate that streamers in a gap in which there is no space charge fouling of the gap appear at about the same potential as do the pre-onset streamers in static corona in air. In a field-free gap the impulse streamers forge further and further out into the gap as potential increases. It appears that they can cross the gap at anode potentials well below spark breakdown, for example, 35 kv, where a spark takes 50 kv. To yield a spark the tip potential of such streamers must be on the order of 25 kv at the cathode. Nasser has shown that, depending on the superposed guiding field, streamers cease their advance when their tip potentials fall between 800 and 1200 volts.

Thus the point-to-plane pre-onset streamer and the impulse streamer can start at much smaller avalanche sizes than do the uniform field breakdown streamers of Raether.[2.35] Various measurements of tip fields by English and others for the pre-onset streamer threshold set $e^{\int \alpha dx}$ as about 3×10^4 to 2×10^5 ions. However, in the converging fields and with values of $r_0 \approx a$ for the point, the diffusion radius of these avalanches is easily 0.1 of that for Raether's avalanches that cross 1 cm at spark breakdown in uniform fields. This would make the positive ion space charge density 100 times that for Raether's avalanches, as a result of a reduced r. In Raether's breakdown streamer avalanches of 10^8, the volume needed is 6×10^{-4} cm^3, which gives an ion space charge density of 1.6×10^{11} ions/cm^3. With 3×10^4 ions for the r_0 at streamer threshold, the volume at $2/\alpha$ for the avalanche will be 7.5×10^{-8} cm and thus the streamer-initiating density of ions will be 3×10^{11}, which appears to be about the same. Raether's breakdown streamers advance in a field of $X/p \approx 40$ and so can cross the gap. The pre-onset streamers can advance only a few millimeters into the gap before the field attenuates to a point where they can no longer advance.

In the case of the burst pulse corona, nothing prevents the convergence of several avalanches into the same volume within 10^{-7} to 10^{-8} sec, since avalanche formation times in these fields are on the order of 10^{-9} sec and the photon radiation time is on the same order of magnitude. Hence pre-onset streamers could start with the order of ten times more electrons through convergence. Raether's proof of convergent avalanches initiating breakdown streamers leaves no question about this possibility.[2.35]

So far a factor which is just as vital to streamer advance as that of ion concentration and number in the space charges, namely that of efficiency of photoionization, has rarely been considered. The studies of Tamura[2.33] and especially of Waidmann[2.36] have shown that these are equally important. Increase in the O_2 concentration in N_2 produces an exponential decrease in range of streamer advance in a given field. The same occurs for H_2O vapor and Freon. The function of these gases is to increase μ and to decrease $d_\mu = 1/\mu$, so that an inadequate number of photons are produced beyond r_0 to permit advance. Such decrease in available photons requires an increase in avalanche size to permit the same advance. Since photon production decreases exponentially with O_2 increase, avalanche multiplication must increase at the same rate. In other words, as with all breakdown thresholds, the magnitude of the secondary mechanism cannot be ignored. When this becomes deficient, enhanced avalanche size is needed to compensate. This is dramatically shown in Townsend discharges, where thresholds are raised by a suppression of the γ. It has always been assumed by Raether and others that with $\approx 10^8$–10^9 ions and electrons there would be more than enough photons.

This can now be verified. Work of Przybylski has shown that in air at 760 mm, the values of μ for the two active photons with largest ff_1 values are 7.7 and 5 cm^{-1}. One comes from O_2 photons, the other from N_2 photons, both acting on O_2. The values of ff_1 have also been estimated and are on the order of 10^{-3} and 2×10^{-3}, respectively. d_μ in these two cases are 1.3 and 2 mm. In uniform fields the equivalent of r_0 after the first avalanche would be that for a 1 mm diameter point at around 2×30 kv if space charge field equals applied field. Here $X/p \approx 75$ and $r_0 \approx 2$ mm. Thus, combined, the ff_1 value is such that $1/e$ or 37% of the 1.4×10^{-3} photoelectrons are produced beyond r_0. The g factor is roughly 0.33. Since the number of ions is $\approx 10^8$, then there are roughly 1.6×10^4 photoelectrons available to advance the streamer. This is certainly adequate, but these are streamers that must have a tip potential of tens of kilovolts at the cathode.

If one regards the pre-onset streamer threshold, there are on the order of 10^5 electrons and ions created in the single avalanche at the anode. Here r_0 is on the order of 0.2 mm. Assume as before that ff_1 is at least 3×10^{-3}. Thus the photoelectrons available, if one chooses a large geometrical factor of 0.5, will be 140. This is just about adequate, especially if converging avalanches are needed. As data of Nasser[2.19] indicate, these streamers will have only two or three branches and will advance only 2 to 3 mm. The tip potentials when they cease to propagate appear to be ≈ 1000 volts.

As point potentials increase, then r_0 increases and the avalanche size increases exponentially. Thus photoelectron yield increases faster than the extension in r_0 diminishes the range of the photons. However, with larger radius points, r_0 becomes too large for the photons in question. Fields at the tip are reduced. Avalanche size increases since potentials become higher, but again in the larger range less efficient photons will be required. There are such photons with $\mu \approx 0.4$ cm^{-1} in air for which $ff_1 \approx 10^{-5}$. These will require larger avalanches and higher potentials. For very small points, though the μ values are adequate to create the Przybylski efficient ionization well beyond r_0, the situation changes, as the *electrons must be created within the ionizing range. Otherwise they arrive too late or do not reach the effective region.* Under these conditions point potentials are so high that electrons have higher average energies. Here many photons of short wave length and much more efficient and smaller μ become important. Thus for needle points Tamura finds $\mu \approx 2500$ for O_2 and 500 for air, but the ff_1 value is near unity. Hence not only does μ play an important role, but, as r_0 and V change radically, the values ff_1, μ, and the nature of the photons needed change.

While this book was being published and especially after it had gone to press, marked advances were made in this area. First came the work of Kritzinger, who, with long sparks, established the presence of repeated return strokes and the reality of the secondary streamer of Hudson, as well as confirming the branching in point-to-plane geometry, to be found in Appendix I. Next came the zero field theory of streamer advance of Dawson and Winn, detailed in Appendix II. Finally, further experimental studies by Dawson and the author's group indicated that the high tip potential observed by Nasser resided in the secondary streamer and that Dawson's primary streamer channel had a high resistance with limited tip potential until illuminated by return strokes. This is reported in Appendix IV.

Important at this juncture is the fact that under asymmetrical anode field conditions a streamer tip consists of a limited volume of plasma with an excess of 10^8 to 10^9 positive ions in a limited volume of the order of 3×10^{-3} cm in radius which advances by its own space charge tip field and limited photoionization in advance, guided only slightly in the low gap field regions by the field. It loses energy gradually by ionization and excitation, as well as by branching. Its initial energy is gained in the high field region of the anode from the field which determines its charge. This is proportional to the point potential and determines both axial range and its total range with all the branches. The photoelectric ionization has been roughly incorporated into the equation for the advance by the author to conform to the findings of Waidmann and the work of Tamura. Branching will be greater for the larger charges. At threshold the space charge starts to build up for avalanches of the order of 10^5 electrons in the tip field with just adequate photoionization. It does not advance as a streamer until successive avalanches have fed into the space charge field of the initiating avalanche head at the anode. This builds the space charge up to the minimum value of 10^8 excess positive ions in a sphere of 3×10^{-3} cm radius which can then advance a few millimeters as a pre-onset streamer. When it begins its advance as such, it has perhaps moved a millimeter from the point tip and has taken a formative time lag of 10^{-7} sec to grow. Above threshold the initial avalanche rapidly increases and the speed of build-up to streamer proportions is much faster. The minimum space charge tip, once formed at some small distance from the anode, advances through the remaining potential drop in the high field region and increases in charge to the point where, for its advance in the low field region, it has energy proportional to the tip potential. In the low field region near the end of its range, it produces just enough photoelectrons to advance and branching ceases.

REFERENCES

2.1   For the basic processes discussed in this chapter, the reader is referred for complete details to L. B. Loeb, *Basic Processes of Gaseous Electronics* (University of California Press, Berkeley and Los Angeles, 1955), and *Encyclopedia of Physics*, ed. by S. Flügge (Springer Verlag, Berlin, 1959), vols. XXI and XXII.

2.2   R. A. Wijsman, Phys. Rev. *75*, 833 (1949).

2.3   W. Legler, Z. Physik *140*, 221 (1955). W. Legler, Ann. Phys. 6–*18*, 374 (1956). W. Legler, Z. Naturforsch. *16a*, 254 (1961). W. Feldt and H. Raether, Ann. Physik *18*, 370, 1956.

2.4   H. W. Bandel, Phys. Rev. *95*, 1117 (1954). J. K. Vogel, Z. Physik *148*, 355 (1957). K. J. Schmidt-Tiedemann, Z. Physik *150*, 299 (1958). H. Schlumbohm, Z. Physik *170*, 233 (1962). J. Pfaue, Z. Angew. Physik *16*, 15 (1963). H. Hoger, *Dielectrics*, 1, 94 (1963).

2.5   A. von Engel and M. Steenbeck, *Elektrische Gasentladungen* (Springer Verlag, Berlin, 1934), vol. 2, sec. 71, pp. 178 ff. W. Bartolomeycyzk, Z. Physik *116*, 235 (1940). P. M. Davidson, Brit. J. Appl. Phys. *4*, 170 (1953). N. Miyoshi, Phys. Rev. *103*, 1609 (1956). P. L. Auer, Phys. Rev. *98*, 320 (1955); *101*, 1243 (1956). P. M. Davidson, Phys. Rev. *106*, 1 (1957). P. L. Auer, Phys. Rev. *111*, 671 (1958). A. L. Ward, Phys. Rev. *112*, 1853 (1958).

2.6   See Legler, 2.3; and L. Frommhold, Z. Physik *158*, 96 (1960); *144*, 396 (1956); *152*, 172 (1958). W. Feldt and A. Raether, Ann. Phys. *18*, 370 (1956). K. Richter, Z. Physik *158*, 312 (1960). R. Kluckow, Z. Naturforsch. *16a*, 539 (1961). H. Schlumbohm, Z. Physik *166*, 93 (1962); *170*, 233 (1962).

2.7   K. Richter, Z. Physik *157*, 130 (1959). H. Raether, Ergeb. exakt. Naturwiss., Flügge and Trendelenburg *33*, 208 (1961). J. K. Vogel, Z. Physik *148*, 355 (1957). J. Pfaue, Z. Angew. Physik *16*, 15 (1963). H. Tholl, Z. Naturforsch. *18a*, 587 (1963). H. Tholl, Z. Physik *172*, 536 (1963).

2.8   A. von Engel and M. Steenbeck, *Elektrische Gasentladungen* (Springer Verlag, Berlin, 1934), vol. 2, sec. 21, pp. 48 ff. R. N. Varney, K. H. Wagner, Fourth International Conference on Ionization Phenomena in Gases (Paris, 1963), paper Vb14. H. J. White, L. B. Loeb, and D. Q. Posin, Phys. Rev. *48*, 818 (1935). R. R. Crowe, J. K. Bragg, and V. G. Thomas, Phys. Rev. *96*, 10 (1954). A. L. Ward, Phys. Rev. *112*, 1852 (1958).

2.9   H. W. Bandel, Phys. Rev. *95*, 1125 (1954). R. Kluckow, Z. Naturforsch. *16a*, 534 (1961). R. Kluckow, Z. Physik *148*, 564 (1957); *161*, 353 (1961).

2.10 E. E. Dodd, Phys. Rev. *78*, 620L (1950).

2.11 P. L. Morton, Phys. Rev. *70*, 358 (1946). G. W. Johnson, Phys. Rev. *73*, 284 (1948).

2.12 G. W. Trichel, Phys. Rev. *55*, 382 (1939).

2.13 S. Werner, Z. Physik *90*, 356 (1934); *92*, 705 (1934).

2.14 H. F. Boulind, Phil. Mag. *18*, 909 (1934). J. S. Townsend, Phil. Mag. *28*, 83, 1914.

2.15 C. G. Miller and L. B. Loeb, J. Appl. Phys. *22*, 499 (1951).

2.16 W. H. Bennett and L. H. Thomas, Phys. Rev. *61*, 42 (1942).

2.17 L. B. Loeb, Phys. Rev. *90*, 144 (1953). L. Colli, U. Facchini, E. Gatti, and A. Persona, J. Appl. Phys. *25*, 429 (1954).

2.18 L. B. Loeb, J. Appl. Phys. *19*, P887 (1948).

2.19 E. Nasser, Z. Elektrotechnik *44*, 157, 168 (1959). E. Nasser, Z. Physik *172*, 405 (1963). E. Nasser, Dielectrics *1*, 110 (1963). E. Nasser and L. B. Loeb, J. Appl. Phys. *34*, 3340 (1963).

2.20 W. Finkelnburg and S. M. Segal, Phys. Rev. *83*, 582 (1951).

2.21 E. Lauer, J. Appl. Phys. *23*, 300 (1951).

2.22 E. L. Huber, Phys. Rev. *97*, 267 (1955).

2.23 J. S. Townsend, Phil. Mag. *3*, 557 (1902); *6*, 389, 598 (1903). Townsend and Edmunds, Phil. Mag. *27*, 789 (1914). J. S. Townsend, Phil. Mag. *28*, 83 (1914). J. S. Townsend, *Electricity in Gases* (Oxford University Press, England, 1915), pp. 376 ff.

2.24 L. B. Loeb, *Fundamental Processes of Electrical Discharge in Gases* (Wiley, New York, 1939), pp. 381 ff.; *Basic Processes of Gaseous Electronics* (University of California Press, Berkeley and Los Angeles, 1955), pp. 758 ff. R. N. Varney, L. B. Loeb, and W. R. Hazeltine, Phil. Mag. *29*, 379 (1940).

2.25 A. M. Cravath, Phys. Rev. *47*, 254 (1934). C. Dechéne, J. phys. radium *7*, 533 (1936). C. D. Maunsell, Phys. Rev. *98*, 1831 (1955). G. R. Bainbridge and W. A. Prowse, Can. J. Phys. *34*, 1038 (1956).

2.26 H. Costa, Z. Physik *113*, 531 (1939). W. Schwiecker, Z. Physik *116*, 562 (1940). A. Przybylski, Z. Physik *151*, 264 (1958); Z. Naturforsch. *16a*, 1233 (1961); Z. Physik *168*, 504 (1962).

2.27 E. Greiner, Z. Physik *81*, 543 (1933). H. A. Stever, Phys. Rev. *61*, 38 (1942). S. H. Liebson, Phys. Rev. *74*, 694 (1948).

2.28 G. W. Trichel, Phys. Rev. *55*, 382 (1939).

2.29 G. L. Weissler, *Encyclopedia of Physics*, ed. by S. Flügge (Springer Verlag, Berlin, 1936), vol. XXI, pp. 304–320.

2.30 L. B. Loeb, Phys. Rev. *97*, 275 (1955).

2.31 L. B. Loeb, Phys. Rev. *73*, 798 (1948).

2.32 L. B. Loeb and R. A. Wijsman, J. Appl. Phys. *19*, 797 (1948).

2.33 T. Tamura, Proc. Phys. Soc. (Japan) *16*, 2503 (1961); *17*, 1434 (1962). T. Tamura, Japan. J. Appl. Phys. *2*, 492 (1963).

2.34 F. Alder, E. Baldinger, P. Huber, and F. Metzger, Helv. Phys. Acta *20*, 73 (1947).

2.35 H. Raether, Ergeb. exakt. Naturwiss., ed. by S. Flügge (Springer Verlag, Berlin, 1960), vol. XXXIII, 176. H. Schlumbohm, Z. Physik *170*, 233 (1962).

2.36 G. Waidmann, Dielectrics *1*, 81 (1963).

2.37 L. Frommhold, Third International Conference on the Physics of Electronic and Atomic Collisions, University College, London, July 22–23, 1963.

2.38 H. Tholl, Z. Physik *172*, 536 (1963).

3

True Coronas: I
The Highly Stressed Positive
Electrode with Negligible
Cathode Influence

A. ELECTRON ATTACHING GASES, NOTABLY AIR OR O_2

1. EARLY OBSERVATIONS OF BURST PULSES AND STREAMER PULSES

The reason for the choice of this more complicated system for beginning discussion seems strange. However, the various complications which electron attachment introduces serve as excellent introductory material for an understanding of the other and seemingly less intricate systems. It also happens that they have been far more extensively studied than the other systems where the *extreme* gas purity needed is inconvenient and has rarely been achieved.

Proper recognition of the sequence of phenomena observed and their tentative explanation in terms of basic processes can be ascribed to the studies of A. F. Kip in the author's laboratory.[3.1,3.2] This study followed the observation of streamer pulses in air by the author and Leigh in 1936[3.3] and the ensuing investigations of Trichel,[2.12] who discovered the burst pulses. Earlier interpretation was hardly possible despite the many engineering investigations and the careful studies of J. Zeleny,[3.4] because, before 1934, the effective photoionization of air by photons of a corona discharge had not been known. Subsequent rapid progress in this study

74

became possible through the availability of convenient cathode ray oscilloscopes with single sweep and time constants progressing from 10^{-5} sec in 1937 to 7×10^{-9} sec by 1955.

2. KIP'S INITIAL STUDIES

Kip's initial study was carried out with a constant high potential source using air at atmospheric pressure.[3.1] Hemispherically capped cylindrical points ranging in diameter from 0.5 to 4.7 mm with gaps ranging from less than 1 cm to 8 cm were used. Currents down to 10^{-10} amp could be measured by high resistance and electrometer. For each gap length there was a definite "apparent" threshold potential, and current in the steady range rose parabolically as potential increased beyond that threshold. These current-potential curves follow the ion space charge controlled current form as expressed by equation 2.24 of Townsend. Onset, when regarded more carefully, was marked by rise of a current from around 10^{-10} amp to around 10^{-7} amp at a sharply defined potential which increased with gap length for a given point. When plotted to an expanded scale, the increase of current immediately above the apparent threshold was linear again in conformity with equation 2.24 and with observations in early Geiger counter studies of Werner[1.7] with coaxial cylinders, as seen in figure 3.1a and b. However, Kip observed, on applications of a potential, that at or even above apparent threshold there was a delay before the current began. This delay was longer the nearer the potential was to the apparent threshold. Thus, it was difficult to determine the threshold or onset of steady current by raising the potential. Once breakdown had taken place, decrease in potential always led to cutoff at the same potential.

This *offset* potential is, in fact, the threshold for onset of steady corona, and figure 3.2 shows the offset potential and corresponding current as a function of gap length. Note first how insensitive offset *voltage* is to gap length and that the offset current increases sharply as the gap length decreases near 1 cm. The current increase is related to the fact that below a critical minimum gap length for each point radius *there is no corona, but instead, the gap breaks down directly* to a spark. Otherwise, there is always a considerable potential range between onset and sparking potential.

The overshoot of potential and delay in breakdown, leading to the necessity of resorting to the *offset* potential as a measure of the breakdown threshold, was at once clarified by use of a radioactive triggering electron source near the confined air gap. With this, the discharge was observed to set in at once when the previous offset potential was reached. Thus, the offset potential is really the starting potential of the steady breakdown once triggering electrons are present in the gap. There is, however,

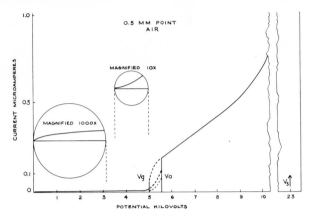

FIG. 3.1a. A schematic diagram of the growth of a positive corona point current as a function of applied potential. The curve below 4 kv, magnified (above) 1000 times, is the saturation ionization current which never collects all the ions in the gap. The asymptotic rise from 3.5 to 5 kv at threshold V_g for Geiger counting, magnified 10 times (above), is the field intensified current caused by avalanche formation and photoionization in the gas near the anode. The intermittent current regime between V_g and V_0 represents the counting regime. Current depends on the photoelectric or other triggering ionization in the gap, as this determines the frequency and duration of breakdown. Above V_0 the current is self-sustaining without outside triggering. At first it rises linearly and later parabolically.

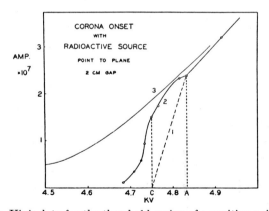

FIG. 3.1b. Kip's data for the threshold region of a positive point-to-plane corona between V_g and V_0 of Fig. 3.1a for a point of 0.025 cm radius. Here the influence of external ionization is seen, curve 1 being any one of a number of curves with a weak ion source. Here the point acts as a sort of a counter. Curve 2 is with an ionizing source of a medium strength. Curve 3 is with a very strong source. Note that such a source can completely obscure the threshold phenomena.

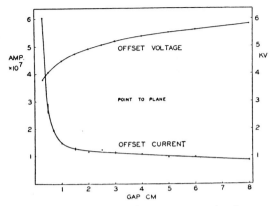

FIG. 3.2. Kip's data for onset, or offset, current and voltage as a function of gap length for hemispherical point of 0.25 mm radius.

an *intermittent current* regime triggered by the radioactive material which sets in at a *clearly defined region below* the *initially assumed* threshold potential. The latter from now on will be designated as the *onset potential of a steady corona*. In that region between rise of current caused by radioactive material and onset of steady corona, the current is proportional to the ionization caused by the radioactivity. The current rise, as will later be seen, represents the *threshold of a type of Geiger counter regime*. This threshold lies below the onset of steady potential. The regime consists of a sequence of small self-sustaining discharges made up of streamers and burst pulses that choke themselves off by local positive ion space charges. They increase in frequency proportionally to the number of triggering electrons until this frequency exceeds space charge clearing time. Thus, if charge in each little current pulse is roughly constant, current increases in proportion to the triggering electron current. Before this can be achieved, a new condition ushers in the steady corona onset. The current in each pulse is not affected by the radioactivity, nor is the current *above* onset affected. If the radioactivity is increased too much, there is a superposed current caused by the very heavy field intensified radioactively initiated ionization current, by multiplication of electrons which create space charges, choking off the pulses. This may set in far below threshold for the intermittent regime. There is, then, no definite onset of corona, and currents will exceed those above onset. This situation is produced by enough external ionization which, amplified by the field, creates a current, the space charge of which so distorts the gap that the normal discharges cannot occur. This teaches an important lesson in experimental use of triggering electron currents: They must be large enough to trigger the breakdown pulses but not great enough that the field intensified current alters the gap fields.

In order to calculate the conditions under which the breakdown occurs, Kip also used a pair of confocal paraboloids with points of radii 0.25 and 0.50 mm, since fields along the axis are amenable to computation. It was established that fields at the point surface and even out into the gap were more than sufficient to ensure detachment of electrons from negative ions, especially in the light of recent knowledge. From the linear portion of the current rise, the effective "resistance" of the space charge in the low field region of the gap could be estimated. It was found that this "resistance" was of the order of 10^9 ohms and varied linearly with gap length for each point. The resistance was less for points of large diameter than for the smaller points. Thus, the larger the point and the shorter the gap, the more effective was the sweeping action of the field on the space charge. If these curves are extrapolated to zero resistance, the intercept for each gap appears at gap lengths which represent the length at which corona ceases to occur before the filamentary spark breakdown. A linear decrease in resistance also exists for confocal paraboloids, and here it is independent of point diameter, which follows from the fact that the field in the weak field region is independent of point size as long as the paraboloids are confocal. This is also indicated by the constancy of onset fields with gap length for a given point in that geometry.

At the time of Kip's study, only the *assumed streamer type of corona breakdown had been recognized*. Thus, theoretical interpretation was not complete. It was, however, indicated that the critical gap length for the replacement of corona by the spark represented some relation to the length of the streamers that initiated the discharge that occurred. This conjecture has been proven to be correct. Unfortunately, both Trichel's earlier work and Kip's first study suffered from inadequate time resolution in their oscilloscopic studies, so that the full nature of the threshold phenomena could not be interpreted. They, however, concluded that below onset of steady corona there were bursts of discharge, presumably streamers, that operated by photoelectric ionization in the gas; they choked themselves off until the space charge of the preceding defunct discharge channel cleared, at which time a new discharge could originate. As fields increased, other spots than the axial spot could start discharges, and clearing speeds increased. Onset of *steady corona* was *assumed* to occur when clearing became so rapid that no more interruptions by space charge occurred. In a large measure this was the accepted explanation, even after the fast oscilloscope was able to indicate that *two discharge forms, burst pulses and streamers*, occurred near threshold. This explanation of the steady corona remained until the work of Hermstein in 1960.[1.16] In Hermstein's extensive investigation of the whole range of breakdown with gaps of great length as well as short ones requiring high potential sources not available to Kip and Trichel, it was indicated that negative ion space charges could not be ignored in air. These charges lead to the steady corona mode and under some conditions to dual breakdown thresholds.

Kip next obtained a Dumont oscilloscope with single sweep, time resolution about 10^{-5} sec, and sensitivity of about 0.02 volt/cm on the screen, to extend the study.[3.2] The scope was best placed across a dropping resistor in series with the ground line from the cathode plate. This arrangement, shown in figure 3.3, disclosed the induced current flow resulting from the liberation of positive ions in the gap and their movement toward the cathode. In later work, a screen grid was placed close to the cathode and between point and cathode. An auxiliary field drawing positive ions from the grid to the plate recorded the arrival of the ion pulse at the plate. The screening of the point was just incomplete enough to allow a pip of current impulse to indicate the start of the strong streamer pulse as the electrons of the discharge entered the anode. The screen grid, so arranged, could not register the weaker current carried by ion movement. With this device the transit time of ions from burst pulse or streamer could be measured.

With the screen Kip discovered that at threshold with adequate but weak triggering there were *two types of pulses*. These were at first sparse at threshold but increased in frequency as potential was raised above threshold. They are nicely illustrated by the oscillogram traces taken from Kip's paper and shown in figure 3.4. Burst pulses are above, streamer and burst pulse following are seen in the center, and a streamer below. Inset is a still photograph of burst pulse and streamers. Figure 3.5 is a succession of streamers. The initial discharge at threshold in a clean gap

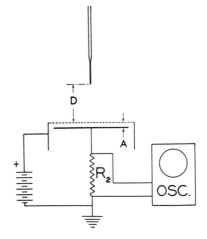

FIG. 3.3. Kip's apparatus for the study of streamers, for a positive point and plane. The oscilloscope and dropping resistor are shown. The screen grid was used to shield the inductive kick at breakdown with the rapid rise of streamer but to register the arrival of the streamer at the grid. Gap length is D, depth of grid field is A. The potential across gap and grid collected the ions from the pulse. The screen covering was not complete, so that a small inductive kick at streamer rise gave a time marker to yield the transit time of the streamer charge across the gap D.

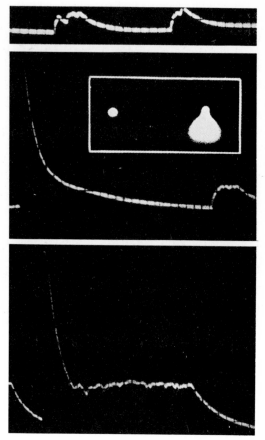

Fig. 3.4. Oscillograms of Kip with positive point. Top line, two burst pulses of some 100 μsec duration. Next lower, a streamer followed by a burst pulse. The time constant of the oscilloscope dragged out the decline of the streamer pulse. Inset is a still photograph of the burst pulse corona just at onset, about 10 times enlarged, left, and a developed mixture of streamers and burst pulses, right. Below, a burst pulse of long duration triggered by a streamer.

gave a single large pulse, the shape of which was determined by the time constant of the oscilloscope. Such a pulse was associated with the streamers which propagate outward along the point axis and the light of which is visible. If the gap is sufficiently short so that these reach the cathode with sufficient vigor, a spark ensues, as Kip indicated in his first paper.[3.1] The length of these streamers increases with the potential and is greater the larger the point radius and the greater the high field region. Once the first streamer has propagated, its space charge inhibits further streamer growth. Thus, following the streamer there are smaller pulses of longer

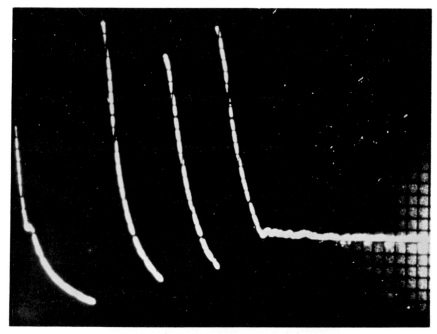

FIG. 3.5. A sequence of four streamer pulses, the last one followed
by a long burst pulse.

duration, which are at once identifiable as the burst pulses of Trichel.[2.12]
These are, in fact, composite and consist of a closely following sequence
of smaller pulses, as will later be seen. These increase in frequency and
duration as potential increases above threshold until they appear to merge
in the continuous burst pulse corona of Trichel. Streamers trigger a
sequence of burst pulses, or a burst pulse may start its own sequence,
but once the gap is fouled by space charge in this *pre-onset regime*, streamers
cannot propagate until the charge clears. The pre-onset streamer corona
is usually characterized by a paintbrush-like form projecting out into the
gap, as in figure 3.4, right. The stem of the brush is caused by the straight

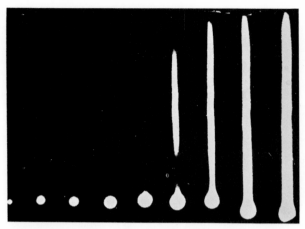

FIG. 3.6. The spread of burst pulse and Hermstein negative sheath glow corona over the point as potential increases above onset, about 10 times enlarged. Low potential at the left.

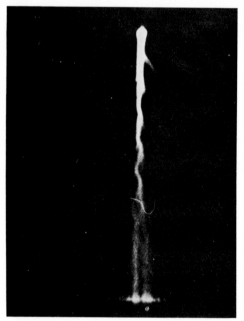

FIG. 3.7. Still photograph of breakdown streamers from a fine needlepoint in a relatively long gap, perhaps 5 times enlarged. Note the presence of some branching and the fact that many hundreds of streamers are needed for this photograph. Note the bright light as the high potential streamer tip approaches the cathode. Contrast this with the highly branching, rapidly attenuating streamers from a larger point in the pre-onset regime.

portion of the streamers near the point, and the flare is caused by the branching of the streamers, as will later appear. The burst pulse corona, once it takes over, shows as a thin layer of velvety glow adhering closely to and spreading over the point surface. As potential and current increase in the steady corona regime, the glow spreads more and more over the surface of the point, as in figure 3.6. Breakdown streamers are seen in figure 3.7. Note the flash of light as they hit the cathode.

By use of the screened cathode plate it is possible to enable triggering electrons, liberated from the grid and cathode by the arrival of the streamer photons and ions in the auxiliary field, to reach the point anode *after the space charge has cleared*. In this fashion it was possible to trigger and observe *only a regular sequence of streamers* very near threshold for different point diameters.

Figure 3.8 gives quantitative data for the ratio between the length of visible streamers and point radius. It is seen that the currents from these diminish as point diameter decreases until in appearance they are indistinguishable from the burst pulses. Figure 3.9 shows oscillograms taken with the screened grid, giving the transit time of the ions. The first pip is the inductive kick produced on the unshielded section of the cathode by the streamer pulse, while the current loop denotes the ion current collected by the plate. Calibration of the scope allows the number of ions per streamer pulse to be measured. This indicated that these were about 5×10^9 ions per cm length of streamer and that burst pulses in the same region of potential gave perhaps one-half or less the number of ions.

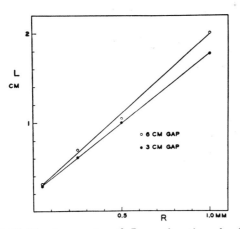

FIG. 3.8. Visible streamer length L as a function of point radius R for two gaps, according to Kip.

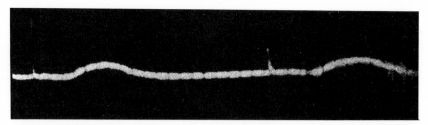

FIG. 3.9. Kip's oscillogram to measure transit time across gap with partially screened plate. Spikes at left are due to streamer rise, while current loops mark arrival of streamers some millimeters long at the grid.

Figure 3.10 gives the positive ion transit time as a function of point radius for two gaps. If the auxiliary field is properly adjusted so that electrons from the cathode reach the point, once the field has removed the space charge, the frequency of the streamers near onset can be predicted by the equation

$$(3.1) \qquad F = \frac{1}{A^2/kV' + t_2},$$

in which A is the length of the auxiliary field, V' the potential across it, k the ion mobility, and t_2 the time of ion transit from point to the grid. Figure 3.11 shows how well this relation agrees with direct observation. If the gap is reduced to 2 cm with a large point, visible streamers cross to the grid but do not penetrate. The photoionization produced by these

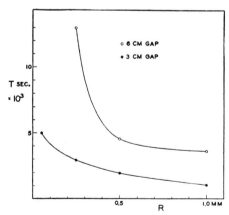

FIG. 3.10. Transit time T as a function of point radius R for two gaps, as measured by Kip.

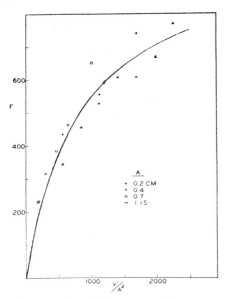

FIG. 3.11. Fit of Kip's equation for frequency of streamers as a function of the auxiliary field potential V' divided by the square of the depth A of that field, for various values of A.

streamers in the gas between grid and plate liberates so many electrons that the value of F is no longer affected by V', which regulates the time of transit of ions in the auxiliary field. If V' be made zero or reversed, so that the photoelectrons do not get out, the streamer repetition ceases. Streamer repetition at regular intervals can also be achieved by placing a large resistance ($\sim10^9$ ohms) in series with the point so that the breakdown by the first streamer lowers the potential, preventing a new streamer from appearing until the space charge clears. This action had been earlier observed by Toepler, who noted the change from a glow to a brush by use of a resistor.[3.5] It must here be pointed out that very early in the studies of Trichel and Kip the author recognized that the burst pulses were in the nature of Geiger counter pulses, of which more will be indicated later. It was also noted that while the streamers disappear in the unscreened gap, as potential increases and steady corona begins, and are not observed when space charge fouling occurs, streamers do reappear just below breakdown to a spark. Reappearance is, however, not always observed if the ratio of point radius to gap length is too large, since the first one to reappear crosses and yields a spark. Actually, as recent work by Nasser[2.19] in the author's laboratory shows, by use of controlled square wave pulses streamers appear at the streamer threshold potential for a given point and increase in length and vigor until they cross the gap to produce a spark. This is possible because they propagate in a clean gap free from space charge as single pulses.

3. THE STABLE GLOW REGIME WITH NEGATIVE IONS

It is proper at this point to consider the findings of Hermstein[1,16] concerning the sequence of pre-onset streamers, burst pulses (formerly designated as steady burst pulse corona), pre-breakdown streamers, and the filamentary spark breakdown. The investigation began using a sphere to plane gap with different radii of the spheres. Gap lengths ranged from 2.5 cm to 35 cm. Sphere radii r ranged from $r = \infty$ to $r = 0.04$ cm. Plotted as shown in figures 3.12 and 3.13 were corona threshold and spark breakdown potentials against gap length for various values of r. For $r = \infty$, $r = 7.5$, and $r = 2.5$ cm there was breakdown to a spark without any pre-existing corona. This was in conformity with Kip's observation and those of Bandel with standard gap geometry. In all cases the spark was filamentary and thus of streamer origin, although Hermstein presented only observations without interpretations at that point. The curves 1, 2, and 3 in figures 3.12 and 3.13 are identical. However, on a continuation of curve 3, when gap length a exceeds 27 cm, there is an inflection and the curve 3' appears but is accompanied by breakdown streamers so close to threshold that they are observed just preceding the discharge. This, as Bandel and Hudson observed, occurs when the ratio of point radius to gap length becomes sufficiently small. For $r = 1$ cm, the corona proceeds along curve 4, figure 3.12, and is the steady burst pulse type. For shorter gaps, on occasion, breakdown to a spark occurs for $r = 1$ cm without the burst pulse type corona. If by chance corona does not set in for gaps from a few millimeters to about 9 cm, there is a spark. If corona sets in, the spark may not appear until higher potentials are reached. This region out to 16 cm is unstable. Between 9 and 16 cm gap length, pre-onset streamers (fig. 3.4) appear, leading to sparks (curve 4, fig. 3.13). For gaps longer than 16 cm the steady corona sets in at a lower potential (curve 4, fig. 3.12) along an extrapolation of curve 4 of figure 3.13. Spark breakdown here requires much higher potentials and follows curve 4''. For $r = 0.04$ cm there is a short range of gap lengths of about 1.5 to 2 cm at which spark breakdown occurs without corona, as seen in curve 5, figure 3.13, which corresponds to the first section of curve 5, figure 3.12. Beyond this, corona threshold is shown by curve 5, figure 3.12, and spark breakdown, preceded by steady corona, occurs at a potential about five times

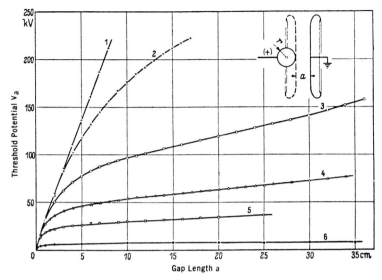

Fig. 3.12. Hermstein's curves for the discharge *threshold* potential of a positive point, V_a, plotted against interelectrode distance a for sphere against plane with the radii of the sphere as indicated below. Here curves 1, 2, and 3 gave a spark and no corona. Curves 4, 5, and 6 apply to a corona.

1. Plate to plate, $r = \infty$ 4. Radius r 1 cm
2. Radius r 7.5 cm   5. Radius r 0.4 cm
3. Radius r 2.5 cm   6. Radius r 0.04 cm

that for corona, as seen in curve 5′ of figure 3.13. For the smallest sphere, corona threshold is very low, as in curve 6, figure 3.12. Spark breakdown lies far above this, but at a gap of 15 or 20 cm there are two forms of breakdown, the first at higher potential proceeding directly from the steady corona (curve 6, fig. 3.13). The other shows a bundle of very long streamers, *stielbuschel*, before breakdown occurs at a slightly lower potential (curve 6′, fig. 3.13). Beyond 20 cm with the discharge replacing the glow, potential drops and breakdown occurs some 250 to 500 volts below the extrapolated glow discharge curve along curve 6′ of figure 3.13. As indicated, these peculiarities had been observed by previous workers including the author and those in his laboratory, but not over a sufficient range of conditions associated with the phenomena to be discussed. The stielbuschel form apparently represents the breakdown streamers which appear in very long gaps with small points before sparkover, as seen in figure 3.7.

The problem was then investigated by Hermstein using concentric spherical geometry. The cathode was a hemisphere of Cu of 10.05 cm

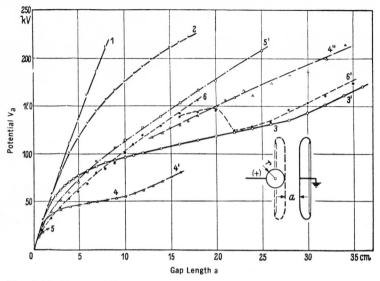

FIG. 3.13. Hermstein's spark breakdown curves for sparking, V_a, and/or corona as a function of interelectrode distance a for various sphere radii r using positive point polarity. Curves 1, 2, and 3 are for r = ∞, r = 7.5 cm, and r = 2.5 cm, and show sparks with no previous detectable breakdown, as in figure 3.12. Curve 3′ at the end of curve 3 shows breakdown streamers before sparking. Curve 4 is for r = 1 cm spark without previous discharge; curve 4′ is for the same radius but with breakdown streamers. For long enough gaps with larger radii, breakdown streamers precede the spark. Curve 4″ shows r = 1 cm, but a glow corona begins first, leading to a higher V_a. In curve 5 r = 0.4 cm, spark without corona. Curve 5′ shows spark after a glow corona started at no great increase in V_d. In curve 6 r = 0.04 cm and corona precedes spark. In Curve 6′ discharge breakdown streamers such as in figure 3.7, *stiehlbüschel*, precede the spark.

radius. The anode was a variety of small spheres and a properly designed stem. The fields before breakdown were calculable. Potentials were measured with an electrostatic voltmeter; currents and current forms were measured by calibrated cathode ray oscilloscope. A charging current went to a 0.1 μF condenser through a 2 megohm resistor, and the point had a series resistance of 50,000 ohms. A protective switch was placed across the dropping resistor of 1000 ohms to protect the oscilloscope, which had a dual sweep for potential across the gap as well as current.

The streamer breakdown or corona was always identifiable by the oscilloscopic current pulses. The visible paintbrush-like branching streamers are identified as pre-onset streamers. The steady corona was identifiable by constant current and by a closely adhering glow over the anode surface.

Figure 3.14 shows the threshold and spark breakdown potential as a function of the radius r of the inner sphere. The breakdown threshold is dashed; the spark breakdown potential is the solid curve. The threshold potential at first increases as the value of r increases. It reaches a maximum at r $\sim$ 5 cm and then declines. This decline occurs when the value of r becomes so large that the field is high across the whole gap length, becoming smaller and in principle causing secondary cathode mechanisms to become efficient, with breakdown approaching that for uniform fields. The spark breakdown curve follows the same trend and nearly fuses with the corona breakdown threshold beyond 1.25 cm. The circles represent corona threshold. The X's represent streamer breakdown as revealed by oscilloscope. The + signs represent a glow discharge breakdown, and the wedge marks represent the visual shaving-brush form of pre-onset corona breakdown. It is noted that between 0.9 and 1.25 cm, breakdown (+ signs above but near curve 1) may occur through streamers without glow discharge at a low value, in keeping with Kip's findings that pre-onset streamers initiate burst pulses but may in short gaps cause breakdown. Otherwise, corona shortly sets in with burst pulses and requires a 25%

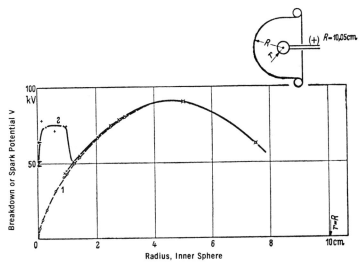

FIG. 3.14. The corona threshold potential V_a or the sparking potential V_d plotted against anode radius r with concentric cathode hemispherical radius R = 10.5 cm. Dashed curve 1 is the threshold, and solid curve 2 is the sparking potential curve. The arrow >, extreme left, represents pre-breakdown streamer discharge characteristic of small radii of curvature. The X's represent pre-onset streamer appearance. The plus signs represent a glow breakdown preceding the spark.

increase in potential. The streamer currents become increasingly more intense as r increases and extend further into the gap, in agreement with Kip's and other evaluations of streamer length and measured charge per unit length. This serves to narrow the gap between threshold and breakdown as noted. With smaller spheres the streamers become shorter and weaker. Thus, in this region the corona breakdown discharge can be studied separately from the spark breakdown. That is, once the burst pulse corona has started, the anode sphere is covered with a glow and increases the resistance to sparkover by increasing gap resistance. Reduction of r below a certain value as seen in curve 2 again reduces the spark breakdown potential, for it is known that a sharp point-to-plane has a constant average value of sparking of 5 kv/cm which, in this gap, is 50 kv as seen in the figure. In this region the shaving-brush corona is apparent, causing sparks at 51 kv.

If it be assumed that the field is little distorted by space charge when the first streamers appear, one may estimate the threshold field strength for appearance of streamers. The field at a distance ρ from the center is

$$(1.9) \qquad E_\rho = V \frac{rR}{\rho^2(R - r)}, \quad \text{with V the anode potential.}$$

The highest field exists at the surface of the small sphere at $\rho = r$. The threshold potential for a spark breakdown, then, is set by

$$(1.10) \qquad E_s = V_s \frac{R}{r(R - r)}.$$

In figure 3.15 this field V_s is plotted as a function of r from r = 0.04 cm to r = 7.5 cm just before breakdown as the solid curve based on the observed V_s. The dashed curves represent the calculated fields E_ρ in mid-gap starting at each r as a function of ρ across the rest of the gap. The curve for $E_\rho = 0.9$ represents the point at which single streamers first cause breakdown, before this breakdown is preceded by a glow.

The theory of these phenomena, as developed by Hermstein, is one that the author overlooked and yet should not have failed to consider. If the streamer is so weak as not to be able to cross the gap and give a spark, but still not too weak for ionization, it will have advanced well into the low field region. Then it consists of a plasma of electrons and positive ions with further electrons and ions in the cloud of photoelectrically created ions which are in fields incapable of continuing the streamer. Furthermore, because of the relatively larger point source, photoelectrons will feed into this laterally, as the author has shown. These electrons are drawn rapidly toward the anode, while the positive space charge drifts away to the cathode. In this separation some, but not a great many, electrons and ions are lost to dissociative recombination. Before reaching

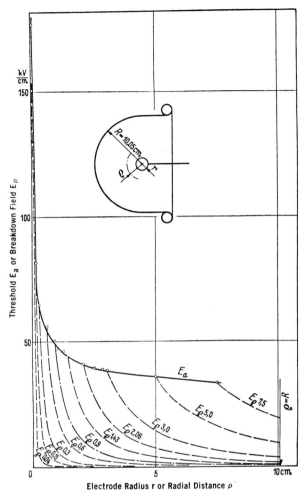

FIG. 3.15. Threshold field strength E_a at the surface of the anode and in the gap E_ρ starting at each r as a function of ρ to the point in question: for outer cathode shell radius R = 10.05 cm.

the high field region, the electrons attach to molecules of O_2 to yield O_2^- ions. Other O^- ions were created by dissociative attachment in the streamer. Thus, between the positive ion cloud and the point there is created a rapidly increasing and nearly equal cloud of slowly moving negative ions, which reduces the field between the positive charge and the point as it moves away. These ions do not shed their electrons until they are much closer to the anode, as fields are too low (see fig. 2.5). Because of recombination, the negative and positive ions are probably considerably smaller in total number than are the ions and electrons initially created by the streamers. The negative ion cloud increases the field near the anode and slows the movement of the positive ions. The interaction between the positive and negative ion clouds will also cause the negative ion cloud to expand laterally as the positive ion cloud expands approaching the cathode. Much more potent in spreading the discharge to cover the surface is the form of the equipotentials created by the sharp drop in field from anode to negative space charge, as was indicated in the discussion of figure 2.5.

The increased field near the anode is, however, of shorter length than the original high field region which started the streamer. Thus, while the initial streamer moving into the clean gap fell short of crossing the gap because of the highly curved anode, the enhanced field between negative ions and the anode is still less capable of propagating a streamer. Thus once a negative space charge of adequate size can form, the streamers diminish in length. As they progressively shorten, they create, however, more photoelectrons beyond the space charge which can feed into the negative space charge as ions. In the steady negative corona region, streamer breakdown does not occur until one reaches a much higher potential. If an impulse potential is applied with a value large enough to create vigorous streamers that cross the gap, but below the glow sparking threshold, a spark will occur as the first streamer crosses and will not give the glow a chance to form. If the anode radius increases, streamers become longer and it takes a higher field to lead to a negative space charge induced corona. The transition to the steady negative ion-created corona discharge will be facilitated by any secondary electron emission from the cathode.

Hermstein points out that the basic factor in the transitions must be the steepness of the decline of the field near the anode. If the equation 1.9 for E_ρ is differentiated with respect to ρ, then

$$(3.2) \qquad \frac{dE_\rho}{d\rho} = -2V \frac{rR}{\rho^3(R - \rho)}.$$

At breakdown, then, dE_ρ/dr at the anode surface at a potential V_s when $\rho = r$ is

$$(3.3) \qquad \frac{dE_\rho}{d\rho} = -2V_s \frac{R}{r^2(R - r)} = -\frac{2}{r} E_s = -kE_s,$$

in which k is the curvature of the anode sphere. The negative sign indicates a decline in field strength. Since E_ρ depends on k, the field is determined strongly by the curvature. Figure 3.16 shows the field gradient as a function of the anode radius r. At a value of $dE_\rho/d\rho$ of 111 kv/cm/cm and above, only the negative ion glow discharge corona is observed. If the

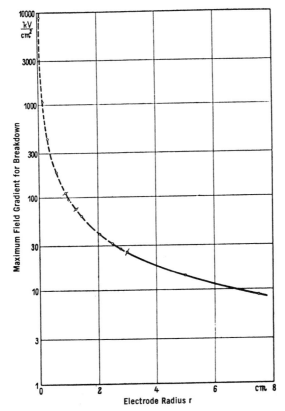

FIG. 3.16. Maximum field gradient for spark breakdown as a function of anode sphere radius for an outer cathode sphere of R = 10.05 cm.

value lies between 111 and 74 kv/cm, both streamers and glow breakdown can occur. Below 74 kv/cm/cm, only streamer breakdown occurs which, below 25 kv/cm/cm, leads only to spark breakdown with no corona.

There are several interesting tests of this hypothesis. Figure 3.17 shows data for a sphere anode-to-plane gap with r = 1 cm and gap length a varied from 0 to 35 cm. Curve 1 for small gaps shows a spark breakdown with no discharge. As 10 cm gap lengths are reached, the streamers do not always cross the gap, and streamers of the pre-onset type preceding breakdown are observed in curve 2. Curve 4 then represents the streamer corona, and curve 5 represents transition from streamers to steady glow corona without sparking. Curve 3 represents sparking potentials with corona glow. There are two breakdown values between 13 and 16 cm. In the one case breakdown is achieved in a streamer (curve 2), and in the other via glow corona (curve 3). If one takes the value of a between 0 and

13 cm, the region of potential above curves 1 and 2 is not directly achieved because sparks always occur. If, however, one reaches the point B by raising the potential at a = 18 cm at A, one may then achieve a glow corona and reduce a, at the same time holding the potential constant. One then arrives at glow discharge operating at values of a below 16 cm and at higher potential. The negative ion glow discharge has stabilized the gap against spark breakdown when it would otherwise have broken down by direct application of a lower potential, or an impulse potential when time is not allowed for the ion sheath to form. If potential is raised, a spark is induced by glow corona, and if lowered, a streamer spark is induced. By decreasing a from B and increasing potential at various points, the glow corona sparking potential curve 3 can be projected backward along curve 3'. The significance of this lies in the proof that once the negative ion conditioned glow corona is achieved, it remains operating until relatively low distances and potentials are reached, when the space charges become so attenuated that they fail by fluctuations, and a streamer spark ensues.

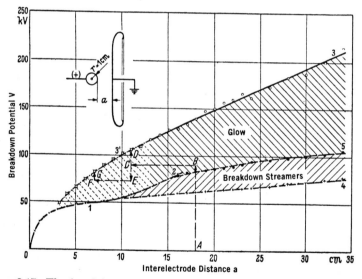

Fig. 3.17. The breakdown regions of a sphere-plane gap with r = 1 cm and spherical anode plotted against gap length a. In the dashed crosshatched region a stable glow corona, reached through the path ABC, is achieved. Region 1, spark without pre-discharge. Region 2, spark from pre-onset streamers. Region 3, spark following a glow. Region 3', a spark after circumventing curves 1 and 2. Curve 4, threshold for breakdown streamers. Curve 5, transition of breakdown streamers into a glow discharge.

In fact for very small points the area available for streamer propagation and negative ion space charge formation is too restricted. Possibly the streamers are not vigorous enough to involve much lateral photoionization. The confined point area gives very high fields just along the axis. Thus, narrow streamers with high radial gradients at their tip develop, leading to the visible shaving-brush corona. Once the streamers cross the gap, sparks follow, as seen in figure 3.7.

The action of negative ions in suppressing streamers and raising the spark breakdown threshold was directly demonstrated by the accidental discovery that a freshly polished copper sphere rinsed off with alcohol, which is known to emit electrons for some time in consequence of surface oxidation, was observed to so weaken streamers in the streamer regime that in a short time a steady glow corona set in, raising the sparking threshold. Then the breakdown was studied between a small sphere with radius 1 cm opposite a large hemispherical electrode of R = 12.5 cm but with *convex* side facing the small sphere, as seen in figure 3.18. The hemisphere had in its center a 5 cm diameter hole with a nichrome thermoelectric emitter which at 1300° C with 4 amp produced a healthy electron emission. The axis of the system was horizontal. The distance between electrode surfaces was a variable a. There is plotted in figure 3.18 the electrode potential for breakdown with a = 20 cm while the heating current in the emitter was varied, controlling the electron emission. The heat of the filament over the 20 cm air gap did not alter gas density appreciably. It is seen that at 68 kv the threshold was set for pre-onset streamers that could not cross the gap and cause a spark. The streamers continued up to 3 amp. From there on streamer threshold increased up to the current 4 amp, when no more streamers appeared but only the steady glow corona was observed. When potentials increased above 83 kv, streamers appeared at 0 to 3 amp. Above 3 amp and up to 4 amp, glow corona appeared *first* at 6 kv, with streamers appearing for a short interval above this, to be replaced by glow corona above 83 kv. Beyond 4 amp, no more streamers were seen. Threshold for the glow corona remained at the streamer threshold, being triggered by weak initial streamers until the proper space charge arrangement for the steady glow corona materialized.

Hermstein considered using the creation of a glow corona to raise the spark breakdown threshold, and suppressing the lower potential breakdown by streamer corona. To achieve this he introduced a very small grounded corona sphere (r = 0.15 cm) with its center 20 cm from the axis of a sphere-to-sphere gap with a large grounded negative sphere of radius 7.5 cm. It could be displaced parallel to the axis, a distance x from the rear of the large cathode sphere. The surface of the smaller positive electrode of the system was placed 1.5 cm from that of the large sphere, as

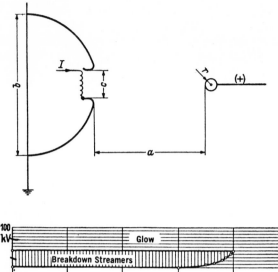

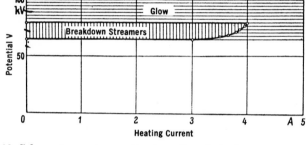

FIG. 3.18. Scheme to create negative ions directly by hot filament at the cathode and thus initiate the spark-preventing Hermstein glow corona at the small anode sphere. This effect is illustrated by showing the influence of a negative ion-creating heated filament on the sparking potential curve with potential plotted against heating current in the ion source. A 1 cm sphere, gap length 20 cm, diameter of the cathode 25 cm, aperture C 5 cm. The vertical crosshatching is the region of pre-onset streamers leading to a glow corona at 68 kv. Above 3 or 4 amps, the streamers are suppressed and a glow discharge takes over.

shown in figure 3.19. Both the large cathode sphere and emitter were grounded. By varying b and x of the small electron emitting sphere, streamer breakdown could be suppressed and a glow discharge threshold could be introduced, increasing breakdown potential for an extended range of potentials by yielding a glow corona with threshold at optimum 10 kv below streamer breakdown, as seen in the figure.

When a = 15 cm and b = 20 cm, so that the small electron emitter did not seriously distort the field of the sphere gap, it was observed that with x = 15 cm flush with the cathode surface or even slightly less, the streamer transition to a glow began to be lowered. At x = 25 cm the influence of the emitter was optimum, lowering glow threshold by 10 kv. Its influence was felt even when x was as great as 25 or 30 cm from the front of the cathode. Such a device could be quite useful in spark suppression under some conditions. This action of an auxiliary electrode under proper conditions might explain some of the sparking potential paradoxes of the past, with chance corona from unexpected points.

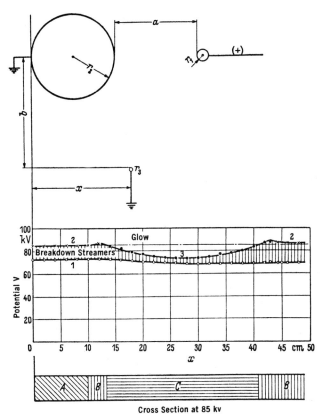

Fɪɢ. 3.19. Hermstein's device for raising the sparking potential in a small anode to large cathode sphere gap by using electrons created by corona discharge from a small sphere attached electrically to the cathode. Anode radius $r_1 = 1$ cm, cathode radius $r_2 = 7.5$ cm, $r_3 = 0.15$ cm, gap length $a = 15$ cm. Distance from center of r_3 to center of r_2 is $b = 20$ cm. The curve shows breakdown or streamer onset at 1, spark breakdown potential at 2, and glow corona at 3 as the position of r along a line parallel to the gap axis goes from a position tangent to the cathode-shielding electrons emitted from gap a to a point where electrons create glow and on past the anode sphere. Below Region A is a region of streamer spark at 85 kv. Region B shows breakdown streamers but no breakdown at 85 kv. Region C shows Hermstein glow corona at 85 kv but no sparks.

4. FURTHER STUDIES OF BURST PULSES AND
   STREAMERS; WORK OF ENGLISH, MOORE,
   AND BANDEL

First, it is best to summarize the sequence of events and various thresholds. The geometry is determinative, especially insofar as the range of the high field region around the positively charged point is concerned. When the positive potential reaches an adequate value in air at atmospheric pressure, a single strong avalanche, or a group of intense smaller converging avalanches, will create sufficient space charge distortion plus adequate photoionization in the gas so as to launch the so-called *pre-onset* streamer. It is so designated since it is one that precedes in potential threshold the onset of the so-called steady corona discussed by Kip and Trichel. These streamers occur down to very small diameter points, as English showed, becoming shorter and more feeble at threshold as point radius decreases. For larger diameter points they can progress quite far into the gap. If they succeed in crossing it with sufficient vigor, as shown by Kip,[3.1] Hudson,[3.7] Hermstein,[1.16] and Nasser,[2.19] they will precipitate a streamer spark breakdown to an incipient arc. These streamers are invisible in shorter gaps but are revealed by the oscilloscope. For small points and longer gaps they appear visible as long filaments before spark breakdown. If, however, the streamers cannot cross and if they are of sufficient vigor, they produce photoelectrons which, given adequate path length, attach to form slow negative O_2^- ions near the point. The secondary cathode emission may also aid in this action in some cases. Some electrons actually form negative ions in the streamer column by dissociative attachment at 3.1 ev to give O^- ions. The negative ions, if numerous and of sufficient radial spread, can form a negative ion space charge sheath opposite and close to the anode. They reduce the field toward the cathode but shorten and enhance the field at the anode so that streamers progressively decrease in size and can no longer grow. The photoelectrons produced in the gas on the cathode side of the sheath also attach to yield more negative ions and maintain this new type of anode glow discharge. This forms at least one portion of the so-called steady burst pulse glow corona region, first observed and named by Trichel[2.12] and previously partly incorrectly explained. Spread over the anode is ensured by the enhanced steepness of the field at the anode by negative ions as well as by photoionization in the gas.

In connection with Trichel's initial study[2.12] of the burst pulses, and before these and the pre-onset streamers had been oscilloscopically resolved by Kip[3.2] in 1938, the author had developed a theory of the burst pulse corona. This stated that if the avalanches at the anode point reached adequate size, then an equivalent of a photoelectric γ_p *in the gas* near the anode could give a self-sustaining discharge. Near threshold it was clear that, if this continued a sufficient length of time, the positive space charge created outward in the gap would reduce the anode point field to such an extent that it would terminate the discharge, giving a pulse on a 100 μsec time scale such as was later observed by Kip in his oscillograms. The velvety glow over the point surface when the burst pulses appeared led to the assumption that these spread along the point surface through the diffusive dispersion of the photons causing the ionization. This action would be enhanced because, as space charge accumulated in the axial high field regions, it would extinguish the discharge in its neighborhood but permit it to continue out over the unfouled surface region. In fact, the author very early described the burst pulses as miniature Geiger counter pulses in air.

The author later developed the more precise mechanism of this action into the theory given in chapter 2, section G.3, which has in part been verified by Tamura.[2.33] If one contrasts the theories of the burst pulse corona noted with the author's and Wijsman's[2.32] contemporary theory of streamer propagation (see in chapter 2, section G.4), the difference between the two phenomena lies essentially in the photoionizing free path $1/\mu$ of the photons and the size of the avalanche active. If $1/\mu$ is much larger than the ionizing zone r_0 at the anode, the discharge will spread over the surface as a burst pulse. If it is commensurate but only slightly larger, or even smaller than r_0 *and if the avalanche is great enough to create enough photons of such $1/\mu$*, then the streamer will propagate outward normal to the anode surface. Thus, depending on gas composition (for example, moisture content in air), on point radius and shape, and on potential, the *initial* breakdown at threshold will either be a streamer or a burst pulse corona.

As geometry, potential, and moisture content varied in air, the initial pulses observed have either been pre-onset streamers or burst pulses, and both precedences of the phenomena have been reported. Both burst pulses and pre-onset streamers near threshold choke themselves off by space charges created. They give intermittence. If burst pulses appear at the first breakdown, streamers may appear at slightly higher potentials and occur in between burst pulses in the clear gap. A streamer may trigger a burst pulse, but a burst pulse once developed will inhibit a streamer until it clears. In some cases streamers appear at threshold, and burst pulses at somewhat higher potentials. High water vapor pressure or high O_2 concentration suppresses burst pulses relative to streamers, since they reduce $1/\mu$ very effectively.

In any event, as potential increases above threshold, either streamers and/or burst pulses become more frequent. The streamers become longer and more vigorous, and the burst pulses increase in duration. As still higher potentials are applied, the increased frequency and duration of the burst pulses point toward an *ultimate threshold* where the applied anode field would clear the space charges as fast as they accumulated and the burst pulse corona would burn steadily with little interruption, as the *steady burst corona* postulated by the author. Since such burst pulses inhibit streamers in their own right, the pre-onset streamer pulses cease with the *onset* of a steady burst pulse corona. In all this, the logic of the author appears irrefutable, and his surmises are in a large measure correct. What he failed to recognize was that it was unreasonable to believe that such a glow corona once established could extend from an onset potential over a range of two to three times that potential before breakdown streamers and a spark materialized. The illogical element arises in that it had been assumed that in a range of some 10% or less above the threshold potential, the anode field could clear burst pulse space charge so as to permit onset of steady corona. In the light of this assumption, it seems strange that a 200% increase in point potential should be needed to clear the same sort of charge for breakdown streamer appearance. This is where the author and his co-workers failed to consider the role of negative ions leading to the Hermstein sheath.

It is now a question as to the point at which the positive ion space charge and burst pulse corona give way to the Hermstein sheath glow corona. There is little question as to the positive ion quenching of burst pulses and streamers near threshold. Burst pulses are observed below streamer threshold. Here negative ion formation cannot be adequate to form a sheath. The question raised as to the role of negative ions therefore could conceivably arise only in connection with their relation to the *onset* of the steady corona. Is this caused by a succession of burst pulses in which the anode field succeeds in clearing the positive space charge to permit continuity, or is it the appearance of the Hermstein sheath of negative ions that accounts for onset? Probably this question is academic. As the burst pulses and pre-onset streamers, or just burst pulses with no pre-onset streamers, become large and frequent enough, negative ions will be created inside and outside r_0 by photoelectric ionization, electron attachment, and dissociative attachment. The most efficient ones are those created outside r_0. Thus negative ion generation increases as burst pulses and streamers increase in magnitude and frequency in proportion to burst pulse and streamer size. These negative ions act to decrease the efficiency of the choking action of the positive space charge on the burst pulse by partially neutralizing it. Together with the enhanced anode clearing field, they bring on the *onset* of steady corona.

Somewhere in the region where the corona has steadied above onset, the Hermstein glow sheath has formed and the stable spark suppressing regime has taken over. It is difficult to place the formation of the sheath just at onset in some cases. Irrespective of this question, Hermstein has contributed a very important mechanism, that of negative ions, (1) in helping to initiate onset of steady corona, and (2) once it is established in preventing reappearance of streamers and spark breakdown for a very extended potential range. This important action was overlooked by the author and his co-workers. Its importance will be realized as one progresses in the study of the positive corona.

The work on streamers, stimulated by the author's conclusions on first observing pre-breakdown streamers from a needle point and following Kip's studies, but before these results were published, led to some theoretical analysis of the problem. This was particularly essential, since as early as the spring of 1936 the author[3.3] had indicated that the streamer propagated along the high field region from effective avalanches and by photoelectric ionization in the air was the elusive mechanism needed to trigger the common filamentary electrical spark in uniform fields as well as in corona gaps. Somewhat earlier, Flegler and Raether[3.8] had interpreted the sequence of events of spark breakdown in uniform fields as revealed in cloud track pictures using square wave form high potential pulses of short duration taken by Raether, in terms of streamers built on about the same model as the author's. Though the author was aware of that work, a confusion in nomenclature in those articles misled the author. It was not until after the publication of Raether's 1937 paper, in which he photographed the process of avalanche crossing in gaps and measured the velocity of the avalanches, that the significance of Raether's cloud track pictures began to be appreciated. Thus when the 1939 papers of Raether,[3.9] clearly showing what appeared to be streamer formation, were received by the author in 1939, the author and Meek recognized that the corona streamers and Raether's streamers, as well as their proposed mechanism, were nearly the same. There was one difference, for Raether used overvoltages in his impulse studies and observed midgap streamers, whereas the corona streamers came from the anode.

In a joint paper summarizing the work done, the author and Kip[3.10] proposed a slightly faulty elementary theory for the avalanche size needed to initiate a streamer, using uniform field geometry for ease of calculation. This theory was corrected in 1939 in the author's book.[3.11] It was that calculation that inspired Meek to develop his semi-empirical equation for streamer breakdown.[3.12] Thus during this period interest focused on some means of obtaining the value of the avalanche size needed for streamer onset. It was thus essential to use a gap form for which the field in a gap undistorted by space charge was known.

A first effort made by Kip[3.1,3.2] in this direction was extended by K. E. Fitzsimmons,[3.13] using confocal paraboloids in dry and normally moist air. The work was begun in the author's laboratory but finished elsewhere without the author's supervision. Two rather important points were established. The first was that water vapor appeared to facilitate streamer formation, lowering the threshold. This conclusion contains a paradox treated at length in section A.9 of the present chapter. The second was that, unless the confined parabolic gap was well ventilated to remove chemical reaction products (such as NO_2) by flushing with fresh air, results, especially current-potential curves, were seriously distorted and thresholds altered so as not to be reproducible. Fitzsimmons used the data to test Meek's semi-empirical equation.[3.13] Data on streamer thresholds taken from Fitzsimmons' paper and later compared with data from point-to-plane gaps will be considered later. Fitzsimmons' data on burst pulse threshold are definitely misleading. As no one familiar with burst pulse forms directed that part of his work, it is suspected that Fitzsimmons was observing individual avalanches with a supersensitive scope at potentials which he ascribed to burst pulse onset. Actually, burst pulse onset is very nearly coincidental with pre-onset streamer appearance, as indicated earlier. It is interesting to note that streamer onset for short gaps of 1 and 2 cm with the 0.009 cm focal length point occurred with 59,900 ions in the avalanche, and with the gap at 3 to 5 cm and with 0.019 cm focal length point, it occurred with 306,000 ions, indicating the increase in length and vigor of the avalanches with point radius and longer gaps. More accurate data on thresholds for burst pulses and pre-onset streamers were obtained for combinations of focal lengths of 0.009, 0.019, and 0.028 cm with F = 1.1, 3, and 5 cm in air at 55% relative humidity at 22° C by English[1.1] and are reported in a later paper on choice of suitable gap forms for point-to-plane corona studies. The values of the true burst pulse thresholds observed by English in that geometry were perhaps 2 or 3% *lower* than those for pre-onset streamers in this careful study. This proves that the *threshold burst pulses are not identifiable* with Hermstein's negative ion corona, which occurs at or above the onset of steady corona. Further proof comes from observations of Tamura on very fine points, privately communicated to the author in December, 1962.

English's careful study of the question of burst pulses and streamers involved the analysis of the breakdown as point size was reduced to 0.002 cm radius of curvature using oscilloscopes of increasingly short resolving times, extending to 2×10^{-7}, which became available after World War II.[3.6] These showed that pre-onset streamer pulses were observed down to the finest points, contrary to Kip's observation with pre-war oscilloscopes and in agreement with Hermstein's observations. Dust-free points finer than 0.001 cm radius in room air gave *no* observable *bursts or streamers*. All that was observed at threshold were slight voltage critical irregularities in the current trace. Fine points dusted with MgO gave bursts or streamers at and above threshold until the onset of the steady regime was reached. These may have been pulses caused by break-down of the insulating solid by field emission. Photographs of the gap with Leica camera with an f2 lens are shown in figure 3.20 for the point diameters indicated. Profiles of the pips, the duration of which was in-strumentally limited at the 6000-fold amplification needed, indicate that streamers initiated a burst pulse with his points. The curve of figure 3.21 indicates a rough linear relation between streamer length as observed photographically and pulse amplitude. Thus the charge created per unit streamer length appears constant. The data yielded an estimate of 10^{10} ions per cm of streamer, in fair agreement with Kip's estimate of 5×10^9 from the screened grid. As point radius increased, the amplitude of the streamers rapidly increased relative to that of the bursts, reaching a factor of 10 at 0.025 cm radius. The duration of the streamer pulses was always below the resolving power of 0.4 μsec of the oscilloscope. The burst pulses had durations of the order of 100 μsec or longer with little reproducibility in successive bursts nor any marked trend relative to point diameter. An analysis of the burst pulse structure had to await the studies of Amin with still faster oscilloscopes.

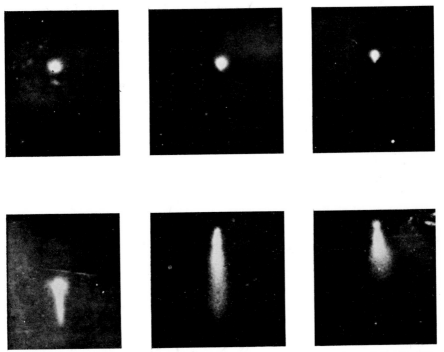

FIG. 3.20. English's still photographs for pre-onset streamers for points of radii 0.002, 0.003, 0.03, 0.04, 0.10, and 0.25 mm, respectively, reading left to right across upper and lower lines.

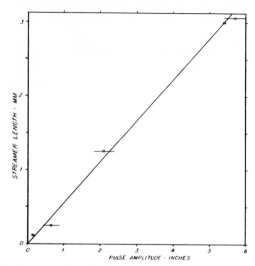

FIG. 3.21. Streamer lengths in millimeters on the photograph plotted against the synchroscope pulse at 6000-fold pulse amplification.

By 1948 the general knowledge of pre-onset streamers and burst pulses at threshold permitted threshold criteria to be formulated by the author[2.31] and the author and Wijsman[2.32] in terms of the photoelectric effect. These criteria are given in chapter 2, sections G.3 and 4. English[3.15] continued his study of positive point corona in air and compared it to the negative Trichel pulse corona. Studies were extended to lower pressures. The necessity of triggering electrons especially for negative corona studies was indicated. Thorianite ore was used in all this work. In general the onset of negative and positive point coronas were fairly close together, from 760 to 200 mm pressure. Accurate evaluation was difficult because of the influence of triggering electrons on the negative threshold, which fluctuated by the order of 100 volts in 5000 to 2000, while the threshold for the positive corona was reproducible within the accuracy of potential setting. At threshold the negative currents far exceeded the positive currents. In general both signs followed the space charge limited current potential law of equation 2.29 with some erratic behavior in the negative current potential curves. English also studied the corona from a water drop, in view of its importance in the mechanism of electrical breakdown in water drop-laden air and the initiation of the lightning stroke.

The studies of English led the author to a reformulation of the nature of the mechanism of breakdown near threshold for positive and negative points.[3.17] In this significant paper are reported visual and photographic studies of the negative and positive corona points at reduced pressure, in which the physical structure of the bright phase of the negative point corona was established to be that of a miniature glow discharge. Further data by English have yielded the magnitudes of the various dark and light spaces at atmospheric pressures by reduction via the inverse pressure law.[3.17] The analysis of the streamer corona from visual and photographic observation showed details of the pre-onset streamer paintbrush out in the gap and the burst pulse glow over the surface. It was from these studies and the quantitative evaluation of the undistorted fields along the point axis in the gap that the explanation of the constrictive and expansive tendencies of the breakdown regions discussed in chapter 2, section F was developed.[2.18,2.20] This and Trichel's calculation on fields also permitted a depiction of the potential ahead of and along a progressing streamer. The picture was later confirmed by calculations from data of Meek and Saxe for breakdown streamers,[3.18] and more directly by the 1959 work of Nasser[2.19] with Lichtenberg figures. The potential distribution in the Trichel pulse corona was confirmed within two or three years by Finkelnburg[2.20] in his study near the cathodes and anodes of arcs. The formulation of these concepts aided materially in interpretation and influenced the direction of research studies as higher time resolution facilitated later study.

English and Moore[3.19] investigated the positive point corona in air at 760 mm, using an impulse generator giving pulses of short duration. The wave form of the H_2 thyratron pulses was essentially square with rise time of the order of 0.1 μsec and lasting 1 and 2 μsec. It was probably a little higher at the end than at the beginning, with a slight dip in the middle. The gap was 2 to 3 cm long, with a 0.005 cm point radius. The pulses were analyzed by synchroscope, and photographic study was made with a Leica camera on Eastman Super XX film as well as a $3\frac{1}{4} \times 4\frac{1}{3}$ plate camera with Eastman plates No. 40 and 11–0. A 5 cm Summar f2 lens was used with an effective aperture of f4.

It was hoped that pulsing would enable viewers to see individual streamers since the repetition rate could be reduced, so that with a short exposure single streamers could be observed. This could not be done with steady potentials. Here again the luminosity was not adequate to observe single streamers. Photographs were obtained of some ten superposed streamers in the pre-onset region and well up into the region where with steady potentials the negative ion-induced corona onset appeared, since the single streamers with impulse move into a gap with no ions. At threshold with the pulsed discharge either a burst pulse or a streamer may appear, but above, streamers predominated. Thus the point surface was covered with a burst pulse glow and one streamer appeared uniformly during each pulse, sometimes early, sometimes later, as indicated by the oscilloscope. Increase in voltage from threshold to 6000 volts (static threshold for this gap lies around 2500 volts) led to streamers 1 cm long occurring within 0.2 μsec of the pulse rise and lasting 0.5 μsec. Only one streamer trace appeared per pulse. Photographs of the streamers from 4200 volts to 9500 volts are shown in figure 3.22.

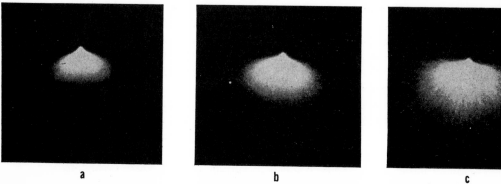

a   b   c

FIG. 3.22. Moore and English's streamer photographs for a positive point with potential impulses of 2 μsec duration for air at 750 mm. (a) 4200 volts with 6000 pulses; (b) 6000 volts with 6000 pulses; and (c) 9500 volts with 500 pulses. Note the excessive branching at overvoltages in the paintbrush corona in a field-free gap. These are confirmed by Nasser's Lichtenberg figures. Note also the great length of pulsed streamers in comparison to static.

This picture is so similar to those taken of pre-onset streamers in steady corona, even though at higher potentials they spanned nearly half the gap, that they caused some concern. This was particularly the case as the pre-breakdown streamers in gaps with small point-to-gap length ratios showed visually single streamers and photographically a few streamers reaching midgap and the cathode, which were narrowly directed along the axis, as shown in figures 3.7 and 3.23, taken from Kip's early studies with

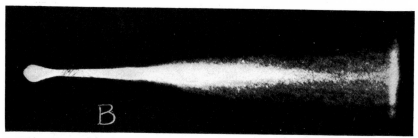

FIG. 3.23. Breakdown streamers photographed by Kip for a needle point in a long gap near but below breakdown. There is little evidence of spreading near the point, and most axial streamers do not cross. The broader haze shows the relative few that do cross. Note the flashes of light at impact points on the cathode.

short exposure. This discrepancy was not clarified until Nasser's recent study in the author's laboratory with pulsed Lichtenberg figure techniques. All streamers into a clean gap from pre-onset to breakdown potentials appear in such gaps and are heavily branched, as is true of steady corona, as shown by Nasser and Loeb.[2.19] The range into the gap increases with gap potential. Branch number increases exponentially starting from near the point surface, reaches a maximum near one-third the gap length, and then declines for longer gaps. However, the vigor and speed of advance depends on the field acting at the streamer tips. As noted in chapter 2, section G, theory indicates that the fields are materially higher along the axis. Thus in gaps long enough to prevent breakdown the more radially directed streamers slow up and fade out. Only the more vigorous axial streamers succeed in crossing as potential increases. In fact, axial streamers cross the gap at potentials well below breakdown but are not able to produce breakdown as their tip fields are too low to invite the required return stroke. When such streamers can develop, only those of sufficient vigor along the axis are readily observed, both visually and photographically. Actually, individual pre-onset and more extensive streamers cannot be individually perceived; only the pre-breakdown streamers along the axis have been seen and photographed. Thus what have been photographed in figures 3.22 to 3.24 are the integrated effects of hundreds to thousands of streamers and their overlying branches. This is shown in figures 3.23 and 3.24, taken of pre-breakdown streamers by Kip for Trichel[2.12] with a longer exposure. Branches as well as the axial components can be guessed at by undulations in profile, as in figure 3.7. The heavy branching near threshold shown in figure 3.22 should have been anticipated from the start, since the ion space charge fields are radial and must be comparable with the applied fields at the anode and since photoelectric ionization in the gas is diffusive.

It has been noted that earlier data differed as to the order of appearance of bursts and streamers at threshold. More data on the question of thresholds of burst pulses and streamers came, as indicated a bit earlier, when

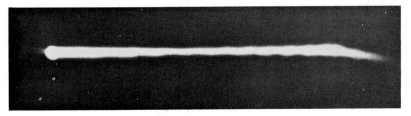

FIG. 3.24. Another photograph by Kip and reported by Trichel on long exposure of breakdown streamers nearer breakdown than those in figure 3.23. This may represent an arc phase (see Goldman[A.I.3]).

English[1.1] undertook a very careful study in connection with the setting up of the standard point-to-plane geometry for which the axial field near the point could be computed. In this, nine different confocal paraboloid combinations were used. These were compared with the burst pulse thresholds for hemispherically capped cylindrical points of 0.0125, 0.025, and 0.050 cm radius and 2, 4, and 8 cm to plane gaps measured under the same conditions by Bandel, giving 3540, 5280, and 7910 volts. The measurements of English showed that, fairly uniformly with the paraboloids, the burst pulses appeared at potentials about 3% below those at which pre-onset streamers appeared. Whether these bursts were triggered by streamers or not appears to be answered in the negative. Streamers were not observed until slightly above burst pulse threshold potentials. This observation is confined to the geometry used, and it involved slight triggering by an external radioactive source. The values of the Townsend integrals $\int_{r_0}^{a} \alpha dx$ were computed for burst pulse thresholds in all cases for the three point diameters and for the three different focal lengths. Correction for the Morton–Johnson effect of field distortion on α was not needed. For confocal paraboloids the values were 11.3, 10.4, and 10.0 for focal lengths of f of 0.009, 0.019, and 0.028 cm respectively, and 10.2, 10.3, and 9.1 for the points of 0.0125, 0.025, and 0.050 cm radii respectively. Considering the extreme sensitivity of the fields to point radius and the difficulties in accurately figuring the point shapes and adjusting the paraboloids, the agreement is good. The average value of the integral is 10.2, corresponding to an avalanche size of 2.7×10^4 ions. This appears to be the size needed to launch a burst pulse. The streamer requires a somewhat greater number of ions, which could run to 3.0×10^5. These ions, in addition, must be concentrated within some critical radius to enhance the field at the surface. In terms of the author's theory for the threshold for a burst pulse, the 2.7×10^4 ions for the burst pulse require that the probability factor for liberation of the self-sustaining photoelectron is $\beta f = 3.7 \times 10^{-5}$. This value, it is noted, is quite small. It is comparable to estimates of γ_p in air needed to initiate a Townsend breakdown at atmospheric pressure in a uniform field gap. This question has been treated in the light of photoelectric yields, combining the data above with more modern concepts, at the end of chapter 2.

After the establishment of the so-called standard geometry for a hemispherically capped cylindrical point-to-plane gap for which fields could be calculated, Bandel undertook a very careful study of both negative and positive point coronas in clean, dry, filtered air over a range of pressures and point diameters.[3.20] This controlled study brought out certain significant factors. As earlier indicated, the standard geometry was of necessity unfortunately chosen, in one respect: since fields at the points are so extensive in the region of *breakdown* streamers, where Hermstein's glow discharge dissipates, once such streamers appear, they cross the gap and lead to a spark. Hence Bandel did *not observe pre-breakdown streamers.* In order to observe them with a 2 mm diameter point, the gap must be twice the standard length in room air and 2.5 times the standard length in dried air. The influence of this geometry may also be responsible for some of the observations in the pre-onset region. Bandel's current-potential curves for pressures from 10 to 750 mm for point diameters of 0.25, 0.5, and 1.0 mm are shown in figures 3.25, 3.26, and 3.27. Here it is noted that burst pulse threshold current jump starts from currents on the order of 10^{-11} amp and that the steady negative ion conditioned glow corona sets in at about 10^{-8} amp to 10^{-7} amp as pressures increase to 750 mm. Visible glow appears about coincident with onset of glow corona. It is only at 760 mm pressure that pre-onset streamers appear in any of these gaps, and then only at potentials above those for burst pulse threshold and for the two larger point diameters. Table 3.1 lists the pressures below which pre-onset streamers do *not* appear for various point diameters.

TABLE 3.1

Point diameters, mm	0.08	0.25	0.5	1	1.47	2
Dry air		(760)	670	630	640	690
Room air	720	610	500	360	290	220

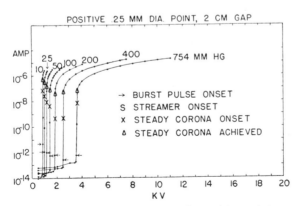

FIG. 3.25. Bandel's current-potential curves for positive point corona in standard geometry, 0.25 mm diameter and a 2 cm gap for pressures from 10 to 754 mm Hg. The various events are indicated by symbols in the legend.

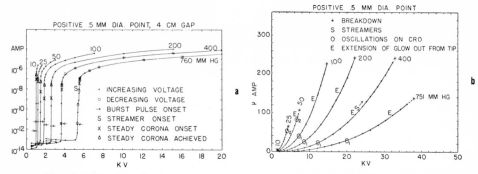

FIG. 3.26a and b. Bandel's current-potential data in standard geometry for a positive point diameter of 0.5 mm at various pressures from below threshold to spark breakdown.

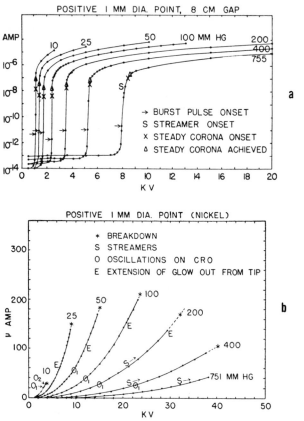

FIG. 3.27a and b. Bandel's current-potential data on standard geometry for a 1 mm diameter point at various pressures from below threshold to spark breakdown.

The relative humidity of the room air ranged from 35 to 45% at about 22° C. This observation is highly significant in that it indicates the need for larger avalanches with sufficiently large ion numbers and concentrations, plus density of photoionizable molecules, to ensure streamer propagation. Water vapor which is readily photoionized by photons from air permits streamers to form at much lower air pressures and thus O_2 densities. This statement is probably an oversimplification of a much more complicated action. Actually water vapor to the extent of 10 mm pressure acts to inhibit streamer advance in air at 760 mm because of its very small d_μ. Thus, according to Waidmann's[2.36] findings, as will be seen in section A.11, water vapor inhibits streams. It, however, facilitates breakdown by causing earlier dissipation of the Hermstein sheath. This action and perhaps the needed increase of photoionization as pressures of O_2 decrease could lead to Bandel's observation.

Extension of Bandel's studies up to breakdown with a positive point revealed the following phenomena. At first, above the onset of steady negative glow corona, the oscilloscope trace smoothed out and the visible, closely adhering electric blue glow became more intense, covering the whole point and even the holder. Before breakdown was reached, some oscillations occurred. They consisted of an up and down displacement of the whole scope trace. They were random in time and amplitude with periods of the order of a fraction of a second. They were initially small but increased in frequency up to a hundred per second as well as in amplitude as potential increased. At all times, however, they remained random. They were more prominent with larger points and higher pressures. The nature of these oscillations is obscure. However, from the Hermstein concept of an anode negative ion sheath conditioned glow discharge with a positive space charge current limiting resistor in low field regions, such fluctuations are not surprising. They could be caused by escape of positive ions between the electrodes to the chamber walls, thus increasing current. The frequencies might indicate such action. A low amplitude oscillation of 10^5 cycles appears at the lower pressures at higher potentials and undoubtedly corresponds to plasma oscillations in this peculiar type of discharge. At certain higher potentials the glow became noticeably brighter on the tip and extended out into the gap. As potential increased, this glow became an intense blue streak extending several point diameters beyond the point. It appeared as the superposition of many streamers. Although occasional streamers were seen, there was no oscilloscopic evidence of streamers in this instance.

It is hard to imagine the change along the discharge axis as the potential increases with a negative ion space charge sheath active. The positive ion space charge is pushed further from the point along the axis. The negative ions are removed more rapidly, and the region where fields are adequate to detach electrons from the negative ions extends outward, increasing the rate of removal of the negative ions and reducing the intensity of the negative ion sheath, as photoelectric ionization in the gas has its d_μ decreased relative to d. Thus avalanche sizes again increase in the expanding high field sheath, permitting streamers to develop. Because of the positive space charge they do not proceed very far. With the high electron current densities localized at the tip, it is not impossible that the streak represents an almost continuous succession of small streamers extending from the point to a region where positive ion space charge still chokes them off. These would not appear on the oscilloscope as separate pulses against the general heavy current background of the glow covering the whole needle. With larger points and higher pressures occasional pulses that resembled pre-onset streamers appeared. Since breakdown streamers have amplitudes many times greater than pre-onset streamers, they can readily be identified. They were, however, not noted. This strengthens the general interpretation of the blue streaks as the beginning of a succession of incipient streamers which fused into the background of continuous current because of their frequency when the positive space charge began to push away from the point. Individual short streamers resembling those of pre-breakdown would only be noted during a brief decrease in current because of the random discharges to the walls. Breakdown streamers, as earlier indicated, were not noted in this geometry and smaller points and longer gaps were required to observe them. When these incipient streamers broke through the space charge to propagate across the gap, the spark followed at once.

5. STUDIES OF BURST PULSES
   WITH OSCILLOSCOPES OF
   HIGH RESOLVING POWER;
   WORK OF AMIN

The next advance in the study of the positive point corona in air came with a further improvement of the resolving power of the oscilloscopes. After leaving the author's laboratory, English[3.21] went to the Chalk River Atomic Energy plant in Canada. There for a day or two he had access to the use of a newly acquired micro-oscilloscope of high resolving power. He at once investigated the pre-onset streamers and the negative Trichel pulse. He was able for the first time to observe what was nearly the true form of these pulses in air at 760 mm and to indicate that the duration of these currents lay in the region of 2×10^{-8} sec with a very high rate of rise for current. Shortly thereafter Amin[3.22] in the author's laboratory had access to a Tektronix 513D synchroscope of rise time 0.02 μsec, and by direct application of the pulses to the deflection plates of the cathode ray tube without preamplifier it was possible to resolve pulses with time intervals of 10^{-8} sec. Analysis of both the electrical pulse as detected across a dropping resistor in series with the plate to ground connection and of the luminous pulse as revealed by a photomultiplier cell became possible.[3.22] The photomultiplier cell was placed in a housing unit mounted so that it could be moved vertically and horizontally by rack and pinion. A lens system focused an image of the area to be viewed onto a rectangular slit placed very close to the photomultiplier. By varying the width and height of the slit, which was normal to the axis of the point-to-plane gap, any section of as well as the whole gap could be viewed and luminosity as a function of time recorded. The slit could be adjusted such that a 0.5 $\times$ 0.5 mm section of the gap could be observed. The signal was sent through a terminated 200 ohm coaxial cable either directly or through a preamplifier to a synchroscope. All cables were so matched that a trace of a 2×10^{-8} sec long square wave impulse was clearly displayed. The point-to-plane gap was of standard geometry and was housed in a bell jar screened on the inside with fine wire mesh. A protective screen was used outside the bell jar as well; both screens were grounded. The plane had a radius of 7 cm, and gaps were correspondingly short. Hemispherically capped cylindrical points were used, the one of 2.5 mm diameter being brass, the smaller ones being Pt. Kip's device of an inductively shielded screen over the plate with auxiliary field was used to measure the number of ions in the several pulses. The dropping resistor across which the synchroscope was placed could be varied from 100 to 10^6 ohms. The electrically induced and photomultiplier signals could be studied either separately or with one of the signals triggering the sweep of the synchroscope while the other appeared on the screen. It was also possible to superpose the two signals, triggering on the rise of the electrical pulse, for a study of their temporal relations. The photomultiplier gave negative pulses, and the electrical signal gave positive pulses. A microammeter could be inserted in series with the plane to study the currents if needed. A block diagram of the arrangements is indicated in figure 3.28.

The first study undertaken was that of the burst pulse corona near onset. As initially observed at threshold and just beyond by Trichel and later by Kip, English, and Bandel, they appeared as low current irregular pulses of duration on the order of 10^{-5} to 10^{-4} sec in contrast to the sharp high current pre-onset streamer pulses of duration less than some 10^{-6} sec as fixed by the resolving power of the earlier oscilloscopes. Characteristic burst pulses as observed by Kip are seen in figure 3.4. The studies were made in room air at 760 mm pressure. The author, as already indicated, had considered the burst pulses as a sequence of Geiger counter-like dis-

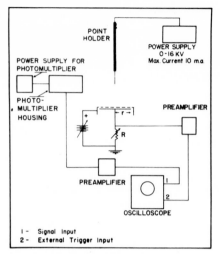

Fig. 3.28. Block diagram of Amin's method of studying burst pulses and streamers with photomultipliers and electrical sweep in a fast oscilloscope. Note the use of screen grid to measure streamer currents, when desired.

charges that spread laterally over the point and extinguished themselves by positive ion space charge accumulation near the point. These were thus assumed to consist of a sequence of pulses arising and extinguishing by space charge, to begin again as soon as some of the nearby space charge at one area of the point had sufficiently cleared to permit the discharge to resume locally. Very near threshold such bursts were expected to terminate very shortly after only a few successive breakdown avalanches because of the longer time needed to clear the accumulated ion clouds further out in the gap. As potentials increased above threshold the series of pulses could burn for longer intervals before the massive accumulated gap charge, in contrast to local space charge, caused extinction. Thus, in analogy to Geiger counter observations, a sequence of primary and secondary pulses were each noted by Amin. Each 1 millisecond sequence of these primary or secondary pulses was extinguished locally on a clearing time for ion movement some 2 mm from the point (of the order of microseconds). Such a sequence continued until the space charge in the gap beyond terminated the discharge. It was such a sequence that created the earlier unresolved burst pulses of Trichel. This required intervals of perhaps milliseconds instead of microseconds to clear, as fields beyond 2 mm were weak. Thus the time between burst pulses was long even compared to the duration of bursts. At higher fields bursts continued for longer times and the time between bursts declined. This, as with Geiger counters, was supposed to result in a merging of pulses leading to the steady but fluctuating corona onset. However, as has been shown, the steady corona above onset in air is in part at least caused by Hermstein's negative ion conditioned glow corona.

The separate bursts in the burst pulse sequence near threshold were observed to follow single triggering events due to cosmic ray or other externally produced ionization. The separate sequence of pulses within the single burst are self-triggered by photoelectrons in the gap near the point. Separate bursts occur in a cleared gap and require outside triggering. Thus bursts were supposed to represent the counts in the region of the well-known Geiger counting plateau in air, with the onset of steady corona representing the equivalent of the self-counting and the end of the counting plateau. The burst pulse corona plateau is seen to be extremely narrow in terms of potential, as noted in Bandel's curves (figs. 3.25 to 3.27). Recent observations of Herreman[5.15] with much improved oscilloscopic techniques, using a coaxial cylindrical counter in air at 200 mm pressure, were even able to distinguish separate small groups of avalanches, the local burst pulse corona, and, for large triggering α particle ionization bursts, the spreading Geiger counter pulses along the wire.

The general interpretation given above, which preceded Amin's study, appears so clear and logical that it is not surprising that Amin's studies sustained it in detail. It is not certain at what point the fluctuating steady corona above onset develops the negative ion sheath and goes over to a glow corona. Where pre-onset streamers trigger the bursts it is possible that the Hermstein steady glow corona with negative ion sheath sets in at once. In studies such as Amin's and Bandel's, where burst pulses precede pre-onset streamers, it is possible that the pulses have no negative ion sheath and that the steady corona which follows extinction of pre-onset streamer pulses develops directly to a glow corona. In the curves of Bandel where pre-onset streamers are absent, as with a 0.25 cm diameter point at 745 mm and at 400 mm and below with larger points in dry air, it is probable that the fluctuating steady burst pulse corona at onset is still only positive ion space charge controlled, but that at the point where the corona smooths out to a steady glow current at higher potentials, the glow corona with negative ion sheath is present. With this orientation it is possible to proceed with Amin's observations.

With 10^6 ohms, imposing greater sensitivity but a long time constant, the electrically induced pulse was observed in order to compare it with previous investigations. No discharges were observed with a 1 mm diameter point and a 4 cm gap with a γ ray triggering thorianite ore source until 7.3 kv was reached. If the time constant was reduced by reducing the resistor, it was observed that this was a single pulse of about 10^{-6} sec duration. At this potential random pulses of similar nature occurred at about millisecond intervals. As the potential increased to

7.32 kv, the initial pulse was followed by a smaller secondary pulse of shorter duration and some 10 μsec later. With further increase in potential the number of these secondary pulses increased and the time interval between them became shorter. In all cases the sequence began with a large pulse followed by secondary pulses at random intervals. The amplitudes of primary and secondary pulses also increased with potential. At about 7.7 kv the streamer pulses began. At this point the secondary pulses followed each other as closely as 2 μsec apart, but on the average they were 4 to 5 μsec apart. The number of secondary pulses per burst therefore varied from a single one near threshold up to a hundred or more as potential increased. With a long RC time constant the conventional early picture of a long, single burst pulse with ripples on top was obtained. In this study the primary and secondary pulses with short time constant could be observed visually on the scope. The traces were too faint to photograph, since they were random, and no intensification with repeated sweeps could be achieved as with other types of more regular pulses. By using the lowest time constants for which traces could be photographed, figures 3.29 and 3.30 were obtained. Figure 3.31 a, b, c, and d are sketches

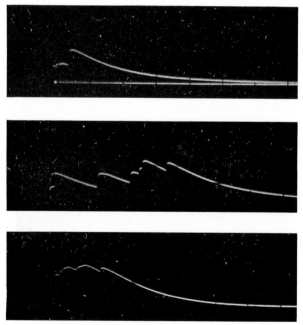

Fig. 3.29. Three single sweep traces with a time resolution which could not separate single burst pulses. However, the first rise is the primary, with secondary pulses superposed on the long time decay.

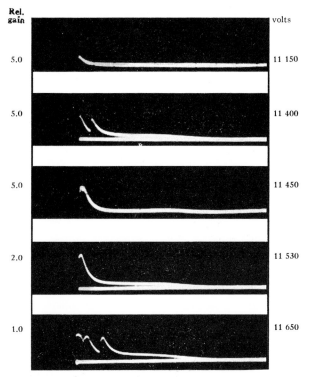

FIG. 3.30. Electrically induced burst pulse signals with 2 mm diameter point, 4 cm gap, $R = 10^6$ ohms, $RC = 10^{-4}$ sec, sweep rate 500 μsec per scale division.

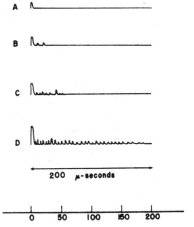

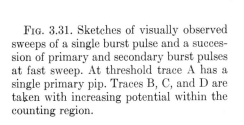

FIG. 3.31. Sketches of visually observed sweeps of a single burst pulse and a succession of primary and secondary burst pulses at fast sweep. At threshold trace A has a single primary pip. Traces B, C, and D are taken with increasing potential within the counting region.

of the visual traces as they appeared with potential increase. These traces are very similar to the observations of Stever[2.27] on a conventional fast Geiger counter. The intervals between primary and secondary and between two successive secondary pulses are readily identified with Stever's Geiger counter dead times. Here on a confined point, pulse duration is shorter than for the long wire in the Geiger counters. However, their dead times are comparable. As will be noted, since actual pulse duration is of the order of 10^{-7} sec, the dead time which is greater than 2 μsec is the shortest time between pulses. This is the time for the ion space charge to clear sufficiently from some part of the point to start a new pulse. The effect of the radioactive source was prominent throughout the burst pulse region, especially near threshold. Without the source no clearly fixed threshold potential could be ascertained. With it threshold was at 7300 $\pm$ 20 volts. It always consisted of single primary pulses milliseconds apart. The appearance of pre-onset streamers at 7.7 kv did not terminate the bursts. They occurred independently or were triggered by a streamer. Where the streamer initiated a burst sequence, the primary pulse was replaced by the streamer and only a sequence of secondaries followed. With increasing potential the streamers increased in frequency and the duration of the bursts increased. At 7.9 kv the pulse duration became longer than the interval between pulses and steady corona onset occurred. Whether this was a Hermstein glow corona or a continuation of the burst pulse corona is hard to state. Since in any event it choked off streamers, the presumptive evidence of negative ion space charge is valid. On the other hand, streamers in the intermittent burst pulse regime never superpose on burst pulses. They occur only in the interval before burst pulses and may trigger them. It is doubtful whether enough negative ion formation to lead to a local glow corona exists in this regime.

The screened grid technique of Kip was used to measure the number of ions in the pulses. Fairly accurate data on the ion currents and time of crossing of ion pulses could be obtained by knowing the sensitivity of the instruments, correcting for the interception of ions by the screen, and using a short screen to plate distance and high fields. Figure 3.32 shows the inductive kick produced by the burst pulse followed by the pip, showing the ion current to the plate. The lower trace shows the electrical pulse without screen at the same sweep rate as above. Here it is seen that the long tail following the pip represents the current induced by the mass of ions in transit. It falls to zero just as the pulse arrives at the plate system in the upper trace. The lower trace is at a slower sweep rate and shows several pulses. The two pulses at the extreme right show that the clearing time is somewhat shorter than the ion crossing time for the whole gap, as two pulses ride up on each other. The new burst started when the last

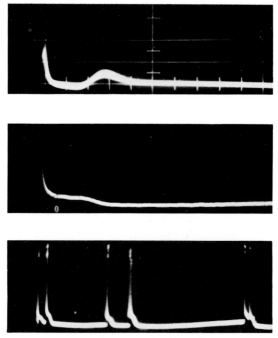

FIG. 3.32. Top trace, inductive kick of rise of the burst pulse followed by current loop when the ions reached the screened grid. The second trace shows the inductive kick followed by ion current for the single burst pulse at the same sweep rate as for the trace above. The end of the plateau falls at the end of the pulse in the trace above. The lower trace at slow sweep rate shows a sequence of pulses. The two pulses at the extreme right show that the clearing time for a new pulse is somewhat less than the ion crossing time for the whole gap.

had crossed some 80% of the gap. The number of ions measured was the number in the whole burst pulse. The burst pulse duration and the statistics of the secondary pulses in a burst were too uncertain to permit an estimate of the ions in a single pulse, primary or secondary. The number of ions in a primary burst pulse did not give a measurable pip with the screen. Streamer pulses in these regions permitted the number of ions to be readily counted per pulse. An estimate of the ion content of the single pulses could be made by comparing the amplitudes of streamer and primary and secondary burst pulses as visually observed and noting that they had about the same duration. The primary burst pulse on that basis had between 0.01 and 0.02 the number of ions in a streamer pulse. The secondary pulses had from 0.001 to 0.005 the number of ions in a streamer pulse. Thus the number of ions in the initial primary pulse of a burst sequence ranged from 10^7 to 2×10^7 ions, and the secondary pulses ranged from 10^6 to 5×10^6. With one hundred secondary pulses in the burst pulse with average time intervals of 5×10^{-6} sec, a burst of 500 μsec duration gave 5×10^8 ions. With only ten secondaries the number of ions was 2×10^7. This range of values agreed with observations using the screen on single bursts.

By using maximum resolution it was possible to observe the primary burst pulse near threshold both with photomultiplier and electrically. The two pulses are shown in figure 3.33 with a 2.5 mm diameter point, 4 cm gap, R = 100 ohms, and RC = 10^{-8} sec, sweep rate 0.1 μsec/cm, the upper trace being the photon pulse, and the lower the electrical pulse.

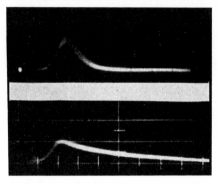

FIG. 3.33. A single primary burst pulse near threshold as revealed by photomultiplier (top trace) and electrically (lower trace). Point diameter was 2.5 mm, and the gap was 4 cm. R = 100 ohms, RC = 10^{-8} sec, and sweep rate 0.1 μsec per scale division.

The initial luminous burst pulse thus has a duration of 1.2×10^{-7} sec, which represents the limit of the resolving power used. The luminosity thus lasts 10^{-7} sec or less. Here it was possible to trigger on the rise of the electrical pulse and to use repeated sweeps. The reproducibility of this pulse, even as influenced by the equipment, is remarkable. Since the secondary burst pulses come at random, the repeated sweep technique cannot be used. A large dropping resistor increases RC so that, although the rise is steep, the decline is long. Secondary light pulses at various potentials are shown in figure 3.34. The photomultiplier slit was decreased to a minimum and the point surface and various distances from the surface were scanned to determine how far the luminosity penetrated into the gap. This was found to be at most 0.5 mm from the surface of a 1 mm diameter point. In theory the rate of ionization should decrease materially beyond 0.2 mm of the surface, which is in agreement with observation as far as it can go. The luminous pulse shape differs from the electrical pulse shape as one would expect it to. The light records the period of intense ionization and radiation of light. This rises at the same rate as the electrical current and falls more rapidly, exactly how much one cannot tell, as the decline is associated with an RC time constant. It is interesting to note that the electrical current pulse continues as the ions move out into the gap. Actually, current decline represents perhaps an extension of the ionizing phase which is too weak in excitation to cause light emission, plus the movement of ions sufficiently from the point to disclose its full space charge, that is, collect the electrons generated.

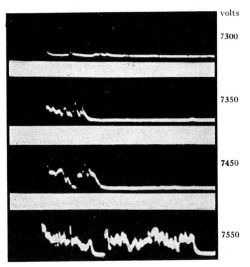

FIG. 3.34. The traces of photon observations by photomultiplier from burst pulses at various voltages. These were for a 1 mm diameter point and 4 cm gap, $RC = 10^{-4}$ sec, and sweep rate 1000 μsec per scale division.

It is clear that the pulse rises to a peak of ionizing activity in 5×10^{-8} sec and declines to zero luminosity in the succeeding 5×10^{-8} sec. In this time near threshold probably half the 10^7 ions and electrons are created. It takes about 9×10^{-7} sec for the current due to ionization to choke off completely. The space charge has been doubled over the intense phase of ionization. The fields at the point surface, X_0, and X_1 at 1 mm with no space charge distortion, would be roughly 9×10^4 and 1.2×10^4 volts/cm with ion drift velocities about 1.8×10^5 cm/sec and 2.4×10^4 cm/sec, respectively. In 1 μsec the maximum displacement of the ions would be 1.8 mm and 0.24 mm, respectively. Thus the field of the 10^7

ions lying at most some 0.5 mm from the point tip is sufficient to choke off the pulse. Since 10^7 ions amounts to 5×10^{-3} esu of charge, the reversed field due to space charge produced at 0.5 cm toward the anode is 600 volts/cm. At 0.1 cm from the center of the space charge it is 1.5×10^4 volts/cm. Thus the field over the first 0.5 mm from the anode is reduced to nearly zero, which is sufficient to terminate the pulse. Incidentally, the pulse shape of the primary pulses is quite analogous to those of Stever's[2.27] in coaxial cylindrical geometry Geiger counters. The minimum time before secondary pulses can occur is 2 μsec at well above threshold. The average time is around 5 μsec in that range. It is probable that dead times as defined by Stever are close to 10 μsec at threshold. Such dead times represent the time for movement of the space charge sufficiently far from some area of the point to permit a new secondary breakdown to start. If the mean speed of the ions at the space charge center 0.5 mm from the tip is 5×10^4 cm/sec, the center would move out to 2.5 mm in 5 μsec. This would reduce fields that choked the discharge by a factor of 25. Even if ion movement reduced the fields only by a factor of 5, it is likely that a new secondary pulse could occur. While such estimates are uncertain as fields are not known, at least they give the order of magnitude of the distance to be covered for clearing during the dead time. When it is realized that at threshold the fields are just enough, according to the equation 2.35, to start a burst pulse, the space charge field to throttle the discharge need not be great, nor need its movement be great in order to allow a secondary discharge to occur. However, the area of high field about the point available for the new breakdown is limited. It will thus not take very much additional space charge to choke it anew. Thus the number of ions in the secondaries are of the order of 0.05 to 0.1 of the primaries. As fields increase, the amplitudes of primaries and secondaries double in the counting regime. Above threshold, where the potentials increase from 7.3 to 7.9 kv for the Geiger counter regime, it is of interest to note that the succession of bursts increases and reaches durations of some 500 μsec with 5×10^8 ions in a burst pulse, with perhaps a hundred secondary bursts, each of 5×10^6 ions. This space charge is sufficiently strong, despite increase of potentials of 600 volts (largely effective in the first millimeter), that the discharge breaks off altogether. The ion cloud must then cross a considerable portion of the 4 cm gap. The direct transit time has been measured in the field at the end of the counting regime, and for the 1 mm diameter point was 4.5 μsec. The average velocity of the top of the cloud thus was about 9×10^2 or roughly 1000 cm/sec, implying fields near the cathode of 500 volts/cm at a maximum. They are probably more nearly 200 volts/cm. The ion transit time is longer than the repeat rate of burst pulses in this region. Thus the minimum time between successive bursts was 2.5 μsec for the point, while the ion transit time was

4.5 μsec. The time between successive bursts corresponds to Stever's[2.27] clearing time. As noted, it comprises about half the ion transit time. Since fields increase as the point is approached, the distance moved by the top of the space charge cloud probably will be around 75 to 80% of the gap length, that is, some 3 cm from the point. The gap then must await a new triggering electronic event before a new burst pulse or, better, a sequence of primary and secondary pulses occurs.

This very significant contribution of Amin on the burst pulse corona may be concluded by giving a table of data taken from Amin's paper. Table 3.2 compares the values for various thresholds for these points in

TABLE 3.2

Point diameter in mm	0.25	0.5	1.0
Bandel's values of onset of burst pulse in kv	3.49	5.06	7.67
Amin's values of onset of burst pulse in kv	3.45	5.00	7.50
Bandel's value of onset of streamers in kv	—	5.26	7.98
Amin's value of onset of streamers in kv	3.55	5.26	7.90

Point diameter, mm	0.50	1.00	2.00	2.50
Range of burst pulse, kv	5.00–5.26	7.30–7.70	11.15–11.73	12.85–13.60
Range as percentage of onset of burst pulse voltage	5.20	5.48	5.21	5.84
Geiger counter region, kv	5.00–5.28	7.30–7.90	11.15–11.87	12.85–13.90
Burst pulse duration at the end of the range, msec	0.4–0.5	0.5–0.7	0.8–1.0	1.0–1.2
Time of crossing at the end of the range, msec	9.5	4.5	2.2	1.5
Minimum clearing time, msec	6.0	2.5	1.5	1.0

standard geometry, taken by Amin using oscilloscopic techniques with those of Bandel. It is seen that agreement is fairly close. The second part of the table gives ranges in potential for burst pulses, potential values for threshold and onset, burst pulse duration at end of the range, ion crossing times at the end of the range, and minimum clearing times, that is, interval between bursts for four points in standard geometry using filtered, dried room air.

6. STREAMER PULSES WITH FAST OSCILLOSCOPE AND PHOTOMULTIPLIER; AMIN'S WORK

Amin next applied his technique to a study of the streamer pulse.[3.23] Here he used room air at about 22° C, with the relative humidity between 40% and 60% in order to induce most effective streamer activity as indicated by Bandel. The 513D Tektronix synchroscope was used. Data

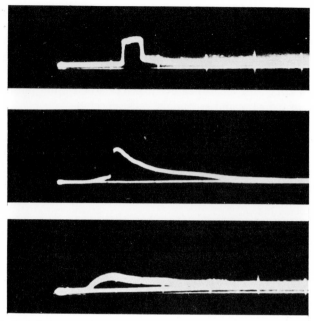

Fig. 3.35. Amin's fast resolution streamer pulses. The top trace is a 2×10^{-8} sec square timing pulse, showing the resolution possible. The second trace shows the luminous pulse as recorded by the photomultiplier viewing the whole gap, that is, the whole duration of the pulse as it moved out into the gap. The third trace shows the electronically recorded pulse during the total time the current could be resolved. Each of these traces represents about ten repeated sweeps.

here presented are for a 2.5 mm point and a 5 cm gap, since different geometries presented no notable differences. Figure 3.35 shows a square wave timing pulse of 2×10^{-8} sec in trace A on top, given by discharging a charged open circuit delay line through its characteristic impedance. As photographed by using repeated sweeps, it indicates the resolving power of the apparatus for such fast pulses. It also served for calibration of the oscillograms. Trace B shows the form of the luminous pulse as recorded by a photomultiplier viewing the whole gap. It gives the light accompanying the streamer as it passes across the tube over its whole lifetime. Trace C represents the electrically induced streamer pulse during the total time that the current induced was resolvable. While individual oscilloscope traces could be seen visually they were too faint, and recourse was had to running a sequence of ten or more light pulses with camera focused on the oscilloscope screen to record the trace photographically. The reproducibility of the light and electrical pulses is remarkable. However, the rather broad traces mark the limits of variability appearing in successive traces. In order to measure the velocity of the streamers, the following method was resorted to. A slit system was set up in front of and very close to the photomultiplier and arranged such that a section transverse to the gap axis 4 cm long and 0.5 mm wide was viewed by the photomultiplier at one time. The photomultiplier (designated as PM hereafter) housing could be raised and lowered by rack and pinion in such fashion that the passage of light across the gap could be observed at various distances measured from the anode point toward the cathode plane. At a later stage in the study the slit could be shortened to 0.5 mm if desired so that only a small region of the gap, either on the axis or off it, could be viewed by displacing the housing horizontally. The movement of the long horizontal slit resulted in the traces displayed in figures 3.36 and 3.37 for a series of positions ranging from 2 mm below the point to approximately 16 mm, at which point traces became too irregular and too weak to record.

The traces of figure 3.36 were obtained by feeding pulses through the amplifiers and delay system of the synchroscope to the cathode ray tube. The sweeping circuit was triggered externally by the electrical pulse with no external delay line. In this figure each frame represents two traces, the upper trace representing the photon pulse and the lower trace the electrical pulse. The amplifiers with a rise time of 2×10^{-8} sec distorted the pulses somewhat. To study these more accurately the photon pulses were directly applied (through an external delay line) to the deflection plates of the cathode ray tube. This is illustrated in figure 3.37, showing a sequence of events similar to the dual trace except that it was not possible to get the electrical pulse on the same film. The sweeping circuit was triggered by the rise of the electrical pulse in all traces. Caution must be used in making interpretations from photon pulse records, since the PM was draining some 20 μA peak current, which was well above its recommended linear response. The overload did not fatigue the PM, since the pulses were short and the traces are qualitatively correct. The power supply and geometrical arrangements limited the study to 4 and 5 cm gaps. The streamers produced in standard geometry were too short for this type of study.

The photon pulse of figure 3.37 is seen to consist of an initial slow rise, followed by a fast rise with a flat top and a lower short plateau which decays exponentially. It is similar to the one high-speed oscillogram obtained by English when he was permitted to use a 5819 PM at Chalk River for one day. In his scope the plateau was more pronounced than with the RCA 1P21 type PM used here. The slow initial rise suggested some sort of pre-discharge on a very short time scale associated with the rise of the streamer. The slit was lowered so that the PM could view all of the gap except the point. The two oscillograms of figure 3.38 show the pulse, viewing gap, and the point in the upper frame, omitting the point

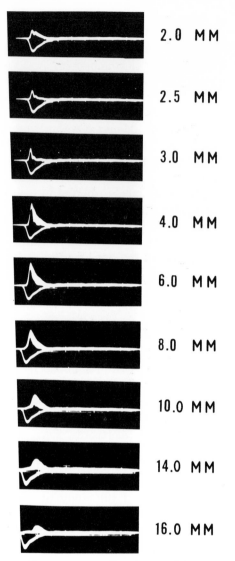

2.0 MM

2.5 MM

3.0 MM

4.0 MM

6.0 MM

8.0 MM

10.0 MM

14.0 MM

16.0 MM

FIG. 3.36. Two traces, the upper one the photomultiplier trace and the lower one the electrical pulse with polarity positive while the photon pulse is negative. These pulses are taken with the photomultiplier slit transverse to the streamer path at the distances from the anode recorded. It is seen that the PM trace seems to show a double peak at the start and that it is delayed relative to the rise of the electrical pulse as distance from the anode increases.

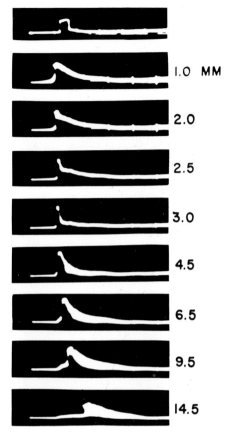

FIG. 3.37. Amin's photomultiplier pulses as recorded at various distances from the anode indicated. The sweeps of the PM were all triggered on the rise of the electrical pulse.

in the lower frame. It is seen that the slow rise is no longer present if the point tip is excluded. This was thought to be a triggering burst pulse. It is probably the growth of the minimal streamer space charge of Dawson and Winn (Appendixes II and IV).

Since it is expected that the brightest part of the advancing streamer is its tip, one is tempted to associate the peak of a curve with the passage of the tip in the photon pulses seen as the tip advances across the gap in figures 3.36 and 3.37. Thus, by noting the time of passage of the peak at various distances out into the gap, Amin could obtain the distance-time curves of the advancing streamer tip. From these the velocity can be estimated by taking the slopes of the distance-time plots. Figure 3.39 shows such curves for three point diameters at 4 cm gap lengths. It is

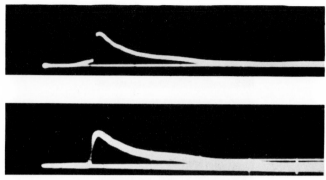

FIG. 3.38. Photomultiplier pulse recorded viewing the anode and gap immediately adjacent to it in the upper trace. Below, the PM trace with the slit in the same position as above but all light from the point surface screened out. It is seen that the small initial rise when anode is viewed is due to probable burst pulse luminosity preceding the streamer. It is not part of the streamer rise, which is sharp and exponential. Recent advances make it seem possible that the slow rise is the development of the Dawson streamer tip before reaching its maximum charge in the high field region near the anode (see Appendixes II and IV).

noted that the initial velocity is high and decreases as the tip advances into the gap. As expected, the initial velocity is greatest for the large diameter point, and streamers advance much further before they fade out with such points. The velocities range from 6×10^7 cm/sec near the point to 2×10^7 cm/sec before they disappear. These velocities are low compared with those earlier observed by Raether[3.27] in uniform fields and also are less than those observed later with improved techniques by Hudson[3.24] for his primary breakdown pulses. Amin observed pre-onset streamers.

It is also possible to plot the relative luminosity of the tip as a function of distance as it advances into the gap. This is shown for the same three points in figure 3.40. There appears to be an initial increase in intensity for the first 1.5 to 2 mm of advance. Thereafter intensity is nearly constant but declines after, respectively 2, 5, and 8 mm. The flat peak could be due to an overload on the photomultiplier, but, since the luminosity of the streamer from the 1 mm point falls far below the limit, it is probable that intensity does remain nearly constant for quite a distance. An attempt at explaining this curve in terms of the streamer theory and the Townsend coefficients was made by the author in an ONR Technical Report in 1954. This explanation applies to the assumption of unbranched streamers and in principle is valid. However, with the recent establishment by Nasser of the branching, another explanation is more likely to apply. That is that the increase in intensity by branches balances decline in intensity for some distance.

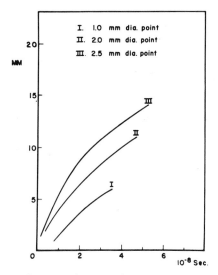

Fig. 3.39. Distance from anode plotted as a function of time for the arrival of the peak of the PM pulses. This is given for the three different points indicated. The slopes of these curves yield the velocity.

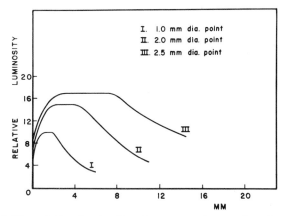

Fig. 3.40. The relative luminosity of streamer tip as a function of the distance of travel from the anode, for three different point diameters. Note the initial rapid rise of luminosity, the constant peak, and the decline as the streamer slows and fades. This will receive its explanation from Nasser's later study.

Relating velocity and distance in these curves leads to the time elements in the luminosity curve. The peak is reached in 5×10^{-9} sec. The duration of the plateau is 2×10^{-8} sec. It then declines. Strangely, the flat top of the peak observed with the slit is much more pronounced than the rise and decline when the gap is viewed as a whole; then there is a rise and a decline but little peak. The rise time represents some sort of luminous development as the streamer moves out from the point. This will be explained later. Viewed as a whole the gap shows the initial rise of luminosity as the tip advances. The channel remains luminous for some time until the excited states have ceased emitting; this could be some 10^{-8} sec or less in any section after the tip has passed. Thus the light seen from the whole gap is the integrated emission from the start to the end of the decline. It is much more intense than the band of light flashing across the slit at each point that represents the flattened profile in the resulting luminosity-intensity curve and the luminosity-time curve of the streamer tip of which it represents the integral. Thus the duration of emission along the channel above the slit alters the profile but is not recorded in the slit observation. This raises the question of the luminosity along the channel after the tip has passed and the luminosity of the tip in a single branch of the streamer, which will be discussed later.

Amin used the photomultiplier to estimate the length of the streamer as a function of point potential and current. From this he was able to compute the number of ions per centimeter advance into the gap by calibrating the oscillograms for current. The number of ions could also be measured by using the screened grid. Both results agreed, but the number of ions per centimeter of streamer length was found to be higher than estimated by Kip and by English. These observers used the length of the streamers estimated by visual observation using telemicroscope and scale, and by length of the photographic image on long exposure. These are more sensitive than the photomultiplier, so that the number of ions per centimeter of path length for different points given by Amin are probably too great by a factor of 1.6.

Amin next viewed the gap with a square aperture of 0.5 mm on a side which could be displaced laterally as well as vertically to obtain the radial expansion of the streamer pattern. Amin found that the radial range of streamers which extended only a maximum of 2 cm axially along the gap was surprisingly large, going out to 7 mm from the axis. At the time the author and Amin interpreted this spread of streamers as being caused by deflection of later streamers by the positive space charge in the gap. They assumed the spread to be primarily a property of the highly asymmetrical gaps with high radial tip fields and space charge fouling. Since the clearing time for burst pulse sequences and streamers is such that streamers must clear to within 0.8 of the gap length before the next streamer can follow, the radial deflection cannot be due to the charge. Thus it will not be surprising to see that the concept must be modified in view of Nasser's[2.19] findings. Amin actually was observing the heavy branching near the point.

As a result of his studies Amin inferred the data shown in table 3.3 for his pre-onset streamers for three point diameters.

TABLE 3.3

Point diameter, cm	0.1	0.2	0.25
Tip velocity, cm/sec, depending on potential	6×10^7	to	1.6×10^7
Streamer length, cm	0.60	1.10	1.45
Number of ions	3.5×10^9	9×10^9	1.2×10^{10}
Number ions/cm streamer	5.8×10^9	8.1×10^9	8.3×10^9

Max and min current amp   0.077 and 0.021
Rate current rise d_i/dt, amp/sec 1.6×10^7
Estimated initial streamer channel diameter, cm   0.02
Max current density at peak, amp/cm^2 2.7×10^3
Ion density in channel if no expansion, ions/cm^3 2.3×10^{13}*

* Branching makes ion densities in streamer branches materially less, since these estimates were confined to a single streamer concept. If Nasser's branch estimates are applicable, this means the densities in a single branch were of the order of 10^{11}.

The discussion and attempted analysis of Amin's curves in his paper are now obsolete. Amin's techniques were taken over by Hudson[3.24] who was able to use improved oscilloscopes and facilities in his study of the breakdown streamers. This was later followed by Nasser's study, using the Lichtenberg figures.[2.19] Detailed interpretation of some of the features of Amin's work must be left until later. Before Amin left the University of California, he and Hudson, using a faster oscilloscope, were able to resolve the light pulses even in the pre-onset region into a fast *primary pulse* followed by a *slower secondary light pulse*. This accounts for the indications of a peak or a plateau before the main rise of Amin's light pulses observed further out in the gap at 9.5 mm (fig. 3.37), in which resolution was still poor. The plateau in the luminosity curves as a function of distance will, as a result of Nasser's study, be related to the branching of streamers and the delay in the arrival of slowed, more radial branch tips at the plane in conjunction with the decay of the excitation.

7. Breakdown Streamers Leading
   to a Transient Arc via the
   Filamentary Spark

a. *Pre-World War II Formulations and Concepts*

It was the author's observation in 1936 of fine electric blue streamers emitted from a needle point in a long gap in air and advancing well across the gap into low field regions that led him to the streamer theory of spark breakdown.[3.3] These streamers were termed the pre-breakdown or breakdown streamers once Kip, by use of the oscilloscope, had discovered the pre-onset streamers at much lower potentials.[3.2] The two regimes are separated by the Hermstein[1.16] negative ion glow corona in the intermediate potential regime. With impulse breakdown, the space charges, negative ion formation, and so on, leading to the glow corona are absent and streamers are observed continuously from pre-onset potentials to breakdown. As potential increases these streamers become more intense and faster and progress further across the gap, crossing the gap long before they produce a spark. A year before, in 1935, Raether,[3.8] using cloud track expansion techniques to make the avalanches visible, had observed phenomena which he and Flegler interpreted as breakdown streamers operating by the same mechanisms as independently invoked by the author[3.3] for the corona anode pre-breakdown streamers. Raether used uniform field geometry in alcohol, water vapor, air, and other gas mixtures from 760 to 100 mm pressure. To get such streamers he was forced to work with square wave impulse potentials at considerable overvoltages. In consequence, $e^{\alpha d}$ for his avalanches was such that they reached streamer-forming proportions in midgap, and the beautiful photographs in his 1939 paper clearly reveal the bipolar nature of the developing anode- and cathode-directed streamers, shown in figure 3.41. Velocities of the cathode-directed positive streamer were higher than for the negative anode-directed streamers and in excess of Amin's values.[3.26] Raether was of the opinion that, *once the anode and cathode streamers had bridged the gap, the arc materialized in a continuous fashion by current increase.*[3.27] His cloud track pictures were very diffuse and not even as well resolved as the still photographs of the positive point corona because of branching and a superposed excessive nucleation by ultraviolet light and chemically active compounds. Accordingly, not much more could be learned from that technique. His square wave pulses were generated in the manner of that time. Their true amplitude was never certain because of possible reflection, since the techniques of matched impedances were not used at that time. One question properly posed by Raether's dipoles was whether the size of the fields between positive and negative charge centers in the developing streamers might not inhibit streamer advance.

Quite recently by the study of avalanches in uniform field gaps using organic gases and vapors such as CH_4 and ether, in which secondary cathode action is reduced to a small value, very large avalanches crossing to the anode in fields just below and at the spark breakdown threshold have been studied with steady potentials. By the use of the clever techniques developed in Raether's laboratory by his remarkably able group of students, avalanches can be electrically measured down to as low as 10^4

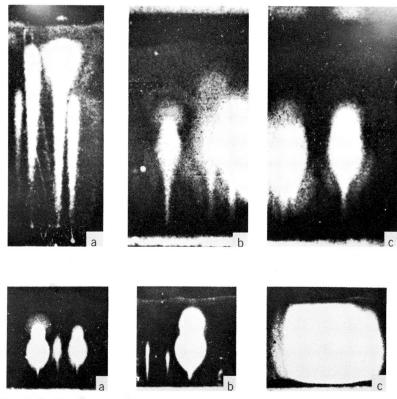

FIG. 3.41. Raether's cloud track photographs of avalanches and streamers observed in 1939. The anode is at the top. (a) Avalanches in transit. The one with the flared head has reached the anode and is starting a streamer. These photographs are about 10 times enlarged. (b) Midgap streamers at higher overvoltage. Note the greater development of the anode streamer in the track to the left as well as the electron avalanches and photoionization on the anode side of the streamer. (c) Later stages of the development. Photographs below, more of the same at reduced magnification. The last one (c) is that of a spark. The great broadening of the streamers is due to photoionization and chemical nucleation by the intense light of the streamers. Note the range of this effect in the more luminous spark channel which usually is less than 0.5 mm in diameter. The water-alcohol mixture under the action of ultraviolet light leads to heavy nucleation and photochemical reaction. These take place in the second between spark and maximum expansion of the cloud chamber.

electrons.[2.7] The study of the larger avalanches revealed that the space charge distortion induced at the higher fields decreased ionization in the avalanche and presumably also slowed advance. Theoretical study by Legler[2.6] and experimental work by Tholl confirmed this and made the appearance of such deviations predictable and quantitative.

As earlier indicated, the author,[3.10] beginning with Kip's second paper, initiated the theory and calculated the approximate avalanche size needed to initiate the anode streamer in a uniform field in air.[3.11] This was made the basis of Meek's semiquantitative theory of the streamer spark.[3.12] In this a streamer was launched from the anode when the space charge field of the positive ions equaled the imposed or acting electrical field. For uniform fields this represents the sparking field. The author and Wijsman[2.32] then derived the self-propagating streamer theory of chapter 2, section G.4, including photoionization. It appears from Nasser's recent study that the magnitude of the tip field may be *one* of the essential elements in streamer propagation in air at 760 mm. However, Waidmann's investigations (chap. 3, sec. A.9) indicate that photoelectric ionization in air is an equally vital factor, and, as this is reduced, the space charge and its density must increase. Meek's theory was promulgated in 1939, and the author and Meek just before the entrance of the United States into World War II published their concepts of the streamer mechanism and streamer spark in an article and later (1941) in a book entitled *The Mechanism of the Electric Spark.*[3.28]

In the meanwhile, Raether had independently arrived at a theory, very closely akin to Meek's, as applied to the negative anode-directed streamer, which was published in 1941.[3.29] Further work done on avalanches studied by means of cloud tracks, leading to some confusing conclusions, made him alter his theory, and he appears to have decided that it was the *number of ions* in the avalanche head and not the space charge density that was of influence. Since all his studies were confined to uniform fields, higher pressures, and breakdown streamers, the figure given by him of around 10^8 to 10^9 ions at the head for such streamers appears to agree with observation. When lower pressures are used and streamers such as pre-onset streamers in corona studies and in Nasser's work are considered, the more rigorous criteria of Meek and of Loeb and Wijsman must apply. Actually, for pre-onset streamers in room air in point-to-plane geometry, streamers start with avalanches of $\sim 10^5$ ions but progress only a few millimeters into the gap. The later studies of Raether and his group have concentrated on the breakdown under uniform field conditions, studies of the avalanche magnitudes, successions of avalanches, photon production of avalanches, the absorption coefficient of the photoionizing radiations, and the build-up of the Townsend discharge preceding the streamer spark, as well as on statistics,[2.6,2.7] and more recently on avalanche field retardation[2.6] and the influence of detachment of electrons from negative ions on spark breakdown.[3.30]

Meek and the author, starting from the concept of the anode streamer either under uniform field conditions *at threshold* or in positive point-to-plane or other small anode breakdown in air, assumed that, *once the streamer crossed the gap, the conductivity of the streamer channel was not adequate to allow growth directly to a transient arc via the spark.* This was contrary to what Raether had assumed.[3.28] The reason for this was twofold. Meek and Allibone,[3.31] in their classical study of long sparks in non-uniform impulse fields using a high-speed moving film camera, had observed that when the anode streamer crossed the gap it was followed by a much more rapid return stroke, after which, given sufficient capacity and low series resistance, a power arc would follow. These beautiful experiments had been undertaken at the suggestion of Schonland, who believed that the spark should exhibit a return stroke, as lightning does. Since the return stroke is so rapid—speeds of the order of 5×10^8 cm/sec and more were expected—a long spark channel was needed in order to record it on a device with a time resolution of 10^{-6} sec.

The second reason was that it was expected that the high tip field of the anode streamer near the cathode, together with the strong photoelectric emission from the cathode consequent on the heavy photon emission from the streamer, must evoke such a return stroke. The light flashes photographed by Kip[3.2] and also by Hudson[3.24] when the pre-breakdown streamers reached the cathode seemed further evidence of this mechanism (figs. 3.23, 3.24, and 3.49). This concept was made more plausible by Westberg's[3.32] study of potential space waves of ionization in very impulsive breakdown of long glow discharge tube columns. The nature of these waves reaching one-third the speed of light was investigated and found to have analogies to the return pulse in streamer spark breakdown. However, for normal gaps such as plane parallel gaps of a few centimeters length, or non-uniform field gaps of some centimeters length, as against meter long gaps, the velocity is too great to permit the detection of a return stroke, so that such strokes have not generally been accepted.

Before proceeding further it is perhaps worthwhile to present the streamer mechanism in pictorial form, using schematic drawings. Figure 3.42 indicates the growth of an avalanche in uniform field geometry, much magnified with disproportionate lateral exaggeration. The anode is on top; the cathode below. Region A represents the exponential multiplication of electrons and ions liberated by a photoelectron proceeding from the cathode. Trace B illustrates the direction of motion of electrons and positive ions, electrons moving more than two hundred times as fast as the ions. Trace C shows an early stage of an avalanche, and trace D shows the stage at which the avalanche head has reached the anode. In a breakdown streamer after 1 cm in air, there would be about 10^8 electrons and ions in the volume and some 10^4 photons escaping, as indicated in sketch F. Figure 3.43 shows the famous photograph, taken by Raether and many times enlarged, of a single avalanche cloud track in an air-alcohol-water vapor mixture. It is seen to resemble closely the schematic of figure 3.42, F.

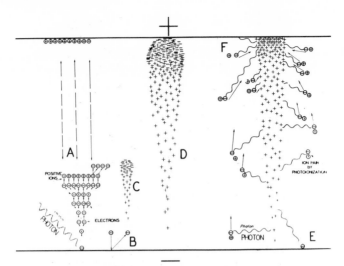

Fɪɢ. 3.42. Schematic drawing greatly exaggerated on a horizontal scale to make it legible. Shows how an external photon A, left, triggers an avalanche. B shows a positive ion striking the cathode and starting an avalanche C. D shows the avalanche as its tip reaches the anode. E illustrates how avalanche-produced photons can liberate electrons from the cathode and in the gas. F shows the streamer ready to start, with positive space charge at the anode adding its field to the externally imposed field. Photoionization in the gas surrounding the space charge is seen.

The further development of the streamers is illustrated in figure 3.44. In this there is depicted the sequence following the arrival of the avalanche head F of figure 3.42, in which as an oversimplification all the electrons have entered the anode, leaving behind the positive ion space charge at the anode surface.

The space charge field added to the uniform field of the gap then draws in photoelectrons so as to extend the positive space charge toward the cathode, sending a stream of electrons up the channel of the streamer moving toward the cathode. Some further few new avalanches are also proceeding from the cathode, because of photons that have crossed the gap from their origin near the anode. However, most photoionization takes place close to the tip of the space charge comprising the streamer.

At I of figure 3.44 the streamer has crossed almost to the cathode. Since it is a conductor with excess positive charge at its tip, it in a measure pushes the anode potential out into the gap and creates a very high field between itself and the cathode. At a distance of a millimeter or less from the cathode, the intense ultraviolet radiation liberates a mass of electrons from the cathode. These multiply rapidly in the high field, leading to an

FIG. 3.43. Cathode below and anode above seen by reflected light from the flash bulb show Raether's famous picture of the cloud track of a single electron avalanche enlarged 10 times or more. The expansion ratio has probably been reduced to get the clear outline here seen.

intensely ionized and accelerating high field region that propagates itself to the anode as a *potential* space wave of ionization at very high speed.[3.32] This completes the spark channel as the return stroke, which is not shown. Figure 3.45 shows photographs of three very long sparks taken on a moving film by Meek and Allibone.[3.31] The anode is at the top; the cathode below. The film is moving from right to left. The streamer and its branches are seen to progress from anode to cathode. The bright return stroke proceeds along the streamer channel from the cathode to the anode, giving the luminous channel, and is so rapid that its lateral displacement cannot be perceived. The film speed was on the order of 1 cm/μsec, and the gap was about 1 m long. What is recorded as a streamer here is probably Hudson and Kritzinger's secondary streamer. The general luminosity in the streamer region here is in part due to the more radial branches.

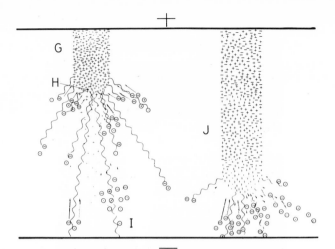

Fig. 3.44. Schematic diagram, a continuation of the sequence of figure 3.42, with G the plasma of ions and electrons comprising the streamer channel which has now advanced to its tip H, one-third of the way across the gap, and with J the stage when the tip approaches the cathode. Here the field between tip and cathode is high and strong photoliberation from the cathode by photons from the tip is just beginning.

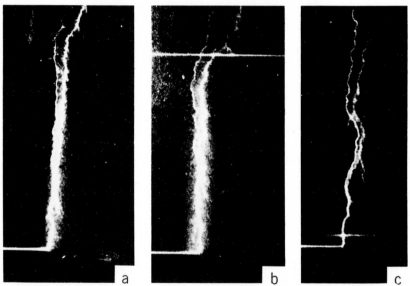

Fig. 3.45. A sequence of three sparks, on the order of a meter in length between anode point and cathode plane, taken on moving film camera by Allibone and Meek in 1938. The film moves from right to left. The streamer proceeds relatively slowly from anode to cathode, as shown by its inclination. The branching of the streamers is clearly seen in traces a and b, and less in c. The return stroke is almost vertical, but follows the crooks and bends of the dominating streamer branch that gives rise to it.

Figure 3.46 illustrates the nature of branching in traces M and N to the left. Traces O and S indicate what happens in Raether's[3.27] experiments when a high impulse overvoltage leads to an avalanche that starts to produce a spindle and a positive and negative streamer process moving toward cathode and anode, respectively, from midgap. Here fields at both ends are high enough to start a new negative streamer avalanche process from the cathode and a cathode-directed positive streamer from the anode at R and T, respectively. In this case breakdown across the gap is very rapid indeed. Actual cloud track photographs of starting anode and mid-gap streamers taken by Raether are shown in figure 3.41.

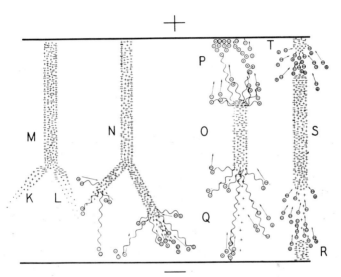

Fig. 3.46. A schematic diagram illustrating how two nearly simultaneous avalanches from the sides lead to two strong branches in traces M and N. Actually, as later data show, branching in breakdown streamers is more frequent. As long as the uniform field in the gap dominates the more radial tip field, the more vigorous streamer channel will follow the field lines with some branching to give it its zigzagged form. Traces O and S show what happens at high impulse overvoltages with the mid-gap streamer leading to an anode-directed negative streamer P and a cathode-directed positive streamer Q. Except at very high overvoltages, the conservative positive streamer is faster and more vigorous. Note in trace S the new anode streamer T induced by photoelectrons from the negative streamer tip that moves down to close the gap.

Figure 3.47 represents the conditions just above the threshold region when pre-onset streamers and burst pulses occur. A streamer has started and projected into the gap some distance and has ceased to advance. Its space charge of positive ions is moving away from the tip of the point. In

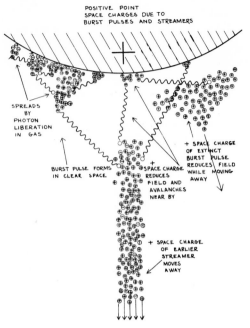

FIG. 3.47. Streamer and burst pulse action depicted schematically for a positive point-to-plane gap, near the anode. Here the defunct streamer, the channel of which extends downward, is moving away from the point in the field, while to the left the burst pulse triggered by radial photoelectrons at some distance from the streamer has caused a discharge to spread along the point surface where the field is lower. An earlier burst pulse which was created to the right of the streamer has quenched itself by its own space charge plus that of the defunct streamer, and moves away from the point. A new burst pulse is seen to be starting at the anode opposite the defunct streamer as it moves away. This accounts for sequences such as those photographed by Kip in figure 3.5.

the meanwhile photoelectrons from the streamer have arrived at the surface with their avalanches both to the left and to the right. The positive ion space charge of the streamer inhibits further streamers, and the burst pulse spreads radially over the point surface in the remaining high field regions. The burst pulse to the right had spread until its positive ion space charge choked it off. The one to the left is in the act of spreading. In terms of Amin's study the initiating streamer replaces the primary or initial burst pulse. Thus the two pulses to the right and left are secondary burst pulses with some 10^6 electrons and ions that have not spread over the entire surface of the anode as in a clean gap. The time for the ion movement here depicted is perhaps 10 μsec for the streamer and 2 μsec and 0.1 μsec for the right and left burst pulses, respectively.

b. *Observations of Meek and Saxe:*
   The Streamer Channel

World War II interrupted studies of streamers and sparks. After the war Meek and his students resumed studies of sparks with improved techniques. Various methods of observation were surveyed, and the moving-film camera of types used by Allibone and Meek and by Schonland proved to have inadequate time resolution for shorter gaps.

For a short but very able study by Meek and Saxe a photomultiplier and oscilloscope were adopted, using a carefully designed point-to-plane gap with lengths from 10 to 50 cm and an impulse potential of rather slow rise time so that high overvoltages were not needed.[3.33] Observations were made using slits viewing the gap normal to the streamer axis but the gap could also be scanned by small apertures to observe the off-axis regions. Currents could be measured as a function of time as well as of the intensity of the light pulses at different points. Thus time-distance and time-current plots were available. The source gave about 20% overvoltage. The impulse was applied in such a fashion that the streamers started at about breakdown streamer threshold, and potential rose as the streamer progressed to its full value before the considerable currents (measured in amperes) caused a decline. The sensitivity of the optics was probably inferior to that of Amin, as was the temporal resolution, coming as their work did five years earlier. However, the intensity of the luminosity was much greater for sparks from larger points at such high potentials and large currents. The currents largely reflected the actual flow of electrons into the streamer tip, and very little went to displacement current consequent on ion motion in such large gaps. The data obtained here and later in Amin's study[3.23] made possible the author's analysis of conditions in the channels, as will be seen.[3.34] This, at the time, was a reasonable interpretation, since the photomultipliers were, if anything, less sensitive than Amin's. The real *primary streamer tips* moving along the axis were probably associated with a certain faint luminosity observed in advance of the bright channel reported by them but not recognized as such at the time.

With all these new data, stimulated by a three-day conference with Schonland on lightning, the author undertook an analysis of the character of the streamer channel behind the rapidly advancing tips. The initial results of this analysis were circulated to an ONR distribution list in the form of a mimeographed report in 1953. The analysis, especially of Meek and Saxe's work, was made part of an ONR Technical Report on "Spark Breakdown in Uniform Fields," dated 1954[3.35] but actually not printed and distributed until 1955. Part of this data was also published in a paper in the *Physical Review*.[3.34] This analysis, in fact, constituted the only analysis and interpretation of Meek and Saxe's studies.

The analysis has many important features, although some of it must be modified because of Nasser's recent study. Essential to it was the assumption of a single axial streamer tip crossing from anode to cathode. This possibly naïve concept derived directly from the long spark observations of Meek and Allibone, as well as Meek and Saxe's study in which a single streamer trace or spark channel, only occasionally forked at the cathode, was observed. It is also in keeping with the pre-breakdown streamers as observed by the author from needle points in long gaps and photographed by Kip.[3.2] How far this applies to breakdown streamers generally is not known. It has been assumed to exist for uniform field gaps and for the long sparks. In Nasser's shorter gaps with larger points, branching is excessive and many tips reach the cathode. As gaps increase in length relative to the point diameter, only the stronger, more axial streamers reach the cathode and these are scarce. As potentials increase toward breakdown, more streamer branches cross, but even at breakdown only two or three strong streamers register on the cathode as of breakdown-producing magnitude. When the spark materializes, it is usually only along one branch. Thus for a breakdown streamer the assumption of a single axial streamer is not basically wrong; branching, however, must now be included.

A second consideration was the apparent *invisibility of the channel behind the streamer*. The inability to photograph individual streamers, while the single pre-breakdown streamers are seen by the eye and are largely unbranched, had led to the conclusion that only the advancing streamer tip had high enough fields and gave enough ionization to be visible. Otherwise, the streamer channel was assumed to be nonluminous. This applied equally to the channel of the stepped lightning stroke. Still photographs of streamers in the pre-breakdown region reveal the superposition of streamer tip traces from repeated streamers or branches. In pre-onset corona streamers they show a maximum effect near the anode, where the stem of the brush is axial, fairly straight, and relatively narrow, as seen in figure 3.48. In the brush itself, intensity along the axis is greatest and fades out toward the edges of the pattern. Hudson's photograph of rare pre-breakdown streamers just below threshold showed, on short exposure, only three or four very broadly expanded, diffuse, luminous patches flaring out from the stem, as seen in figure 3.49. As these moved out to only a third of the gap length, Hudson identified them as probably three secondary streamers. On longer exposure as many as forty or more streamers were observed in an elongated brush emerging from the axial stem near the anode, of which perhaps a third or less reached the cathode, leaving bright flashes on impact with its surface. The work of Kritzinger given in the Appendix confirms this. More complete recent studies in the author's laboratory with impulse streamers and other phenomena make this certain.

Fig. 3.48. English's time exposure photograph of pre-onset streamers from a small point, showing the paintbrush-like form. This has been slightly retouched to bring out the stronger light from the axial stems of streamers.

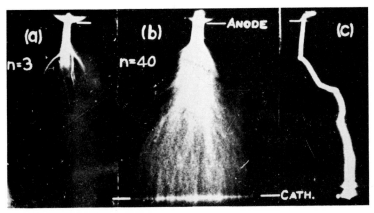

Fig. 3.49. Hudson's still photographs of breakdown streamers very near sparking with a fairly large point. The one marked a shows only three main branches which correspond to Hudson's secondary streamers and which were taken with a short exposure. The photograph marked b shows some 40 or more branches. The primary streamer tip branches reach the cathode and produce bright flashes at the cathode. The main body of the stems of the streamers stop halfway across the gap and may correspond to Hudson's secondary streamers. Trace c to the right shows the spark channel which follows the path of the most vigorous primary branch, reaching it first.

Nasser's Lichtenberg photographs revealed streamers as deflected to horizontal progress along but over the film surface placed normal to the gap axis in the air. In his work, these were usually of greater breadth and luminous intensity nearer the anode, or the point of impact on the film. They were heavily branched and broad near the point but decreased continuously from the point of impact as they advanced. They became very faint and narrow with only occasional branches and finally none near the end. Those taken on color film showed the bright electric blue of the second positive group bands of N_2 characteristic of the high tip fields throughout their length. There was no indication of white or violet near the stem or the anode, which would be expected if they came from an induced current or thermal excitation at low fields. The broadening of the traces is ascribed to halation and diffusion as well as to short heavy lateral branches. The diffusion stems from the fact that the high tip fields of the streamers near the anode possibly cause it to move further above the film in the air than do the weaker streamer tips. This makes the channel appear broader. Increased breadth is also caused by greater luminosity.

There appears at present nothing to indicate that the light from the single streamer comes largely from other than the tip as it flashes by. That is, the channel appears luminous only during the advance of the tip past a point of observation and thereafter remains relatively dark until the return stroke and the growth of the arc channel. While it is possible that the current density in the channel left by the streamer increases along the main branches the nearer one gets to the anode, because of electron influx from the branches, there is no indication that it becomes self-luminous. However, this possibility cannot be ruled out. Photographic and other techniques to date are incapable of giving much more information than stated. Data from Kritzinger (Appendix I) and current studies in the author's laboratory indicate that the bright stems at the anode come from action of return strokes.

While diffuse, Raether's cloud track pictures again indicate single, straight streamer channels with little branching in uniform fields, as seen in a recent study by Tholl.[3.76] However, the diffuse nature of such channels both in Raether's work and the recent work of Tamura mask such branching. Allibone and Meek's photographs show only the single main breakdown streamer with little branching in the anode streamers at 760 mm. Under these circumstances the analysis taken from the author's writings cited above is not seriously modified by Nasser's Lichtenberg figure studies, which apply largely to corona gaps with highly divergent anode fields. Later developments indicate that near threshold only one or two of the branches reach the cathode with enough energy to invoke the return stroke at one time for any spark. Hence the assumptions that follow are justified.

Assuming in consequence that only a single primary streamer is advancing into the gap with a fixed applied potential, having a tip speed v_t, and generating on the average n positive ions per centimeter advance, one may consider the streamer channel in more detail.[3.34,3.35] There must be a flow of electrons up the channel and into the anode to sustain the current. There are certain restrictions, however, imposed on the electron flow and this channel.

As indicated earlier from the field in the gap before breakdown and the distortion produced by the streamer charge, some guess can be made of the tip potential of the streamer as it crosses the gap. If the streamer were a *conducting* filament of length x projecting into the gap from the anode, it would have a tip potential equal to that of the anode; and in a gap of length d, the potential would fall from that at the anode to that at the cathode in d − x centimeters.[2.18] Off the axis at some distance the field would be undistorted; thus the streamer is surrounded by a radial field, as seen in figures 2.4, 2.10, and 2.11. However, the maintenance of a positive charge at the streamer tip requires a flow of the newly generated photoelectron avalanches into the streamer tip and a continued removal of these into the anode through the streamer channel, giving the observed streamer current. There is thus a fall of potential down the streamer channel to maintain this flow, leading to a field X_x at each point x. From the streamer tip at x at a potential $V_x = \int_0^x X_x dv$ there is then a drop in potential in the amount of $V - V_x$ to the cathode d − x distant, where V is the applied anode potential. Now V_x cannot be zero; otherwise there would be no current along the channel. Very near the cathode distant d, $V_x = V_d$, and V_d cannot exceed V the applied potential, but may be less. If the streamer has a very high tip field very near the cathode, as called for by a return stroke, V_d will be less than V by $V_x = V - V_d$. On the basis of currents and conditions in Meek and Saxe's study, the author made some rough guesses as to the nature of X_x, V_x, and V_t which applied to their experiments with a rising potential and which are shown in the Technical Report.[3.35]

Very recently, using the Lichtenberg figure technique, Nasser has made measurements of the tip potentials for the fastest and more vigorous axial primary streamers in his gaps. He has observed that for shorter gaps and higher potentials near breakdown the values of V_x were very high and near V until the streamer crossed to an x of two-thirds of d. Under all circumstances where streamers crossed the gap, there was a value of $V_x = V_d$ close to the cathode less than the applied V unless the tip stopped at x somewhat short of d. The value of V_d near the cathode for streamers that did not cause breakdown was of the order of 2500 to 5000 volts and was higher the shorter the gap and the nearer the potential to breakdown.

Curves are shown in figure 3.73a and 3.73b later in this chapter. These curves represent the values of the tip potentials at various values of x as the streamer crosses the gap. The trace does *not* represent the *fall of potential* along the channel but gives values of V_x at different times t at which the tip reached a point x. What the gradient along the channel behind the tip is one cannot ascertain. But one is in a better position to speculate about the gradient than was the author in 1954. At any point x if at that time X_x along the channel were uniform then $X_x = V_x/x$. This gives a sort of average value. The current along the channel is not constant. It increases with time as the streamer advances, because of the contributions from the branches. For the same reason it also increases upward along the channel at smaller x as the anode is approached. Thus later near the anode the current is high, and it is less nearer the cathode since the electron contribution n_e per centimeter length of streamer increases with influx from the branches. If the channel remained constant in cross section, then current density would increase in proportion to n_e with increasing time t and decreasing x. If the velocity of flow were constant (continuity of current so that space charges did not build up along the channel), X_x would decrease near the anode and with t and increase near the cathode. But the photographs of Nasser indicate that the channel appears to expand near the anode, which might be attributable to more halation from the brighter intense tips there. Hence V_x may be more nearly uniform than one would at first infer. Thus, while it is possible that there may be regions of the channel in which the fields X_x may be temporarily higher than V_x/x at any point in its development, the probability is greater that the fields are fairly well bracketed within the minimum slope of the line joining any V_x at x to the anode where its value is a maximum. From the curves, therefore, X_x ranges between 0.8 to 7 kv/cm. This leads to an X_x/p value ranging from about 1 to 9 volts/cm per millimeter pressure. The high value corresponds to streamers near breakdown. In 1954 the author estimated values of about 5 kv per centimeter. The actual tip potentials from which such curves were derived vary through wide limits. Streamers at the end of their range in the gap have tip potentials of 800 to 1200 v while breakdown streamers near the cathode have tip potentials around 25 kv. In that case X_x may not be great, especially if the gap is long.

At this point it must be indicated that recent data show that the high tip potentials inferred from Nasser's film studies and the discussion above apply to the secondary streamers of Hudson or Kritzinger's leaders with highly conducting channels. Primary streamer tips, as indicated by Dawson and Winn, have tip potentials and fields in proportion to point potential but leave behind channels of low conductivity and high resistance, moving independently of anode potentials in the low field regions. This is discussed in Appendix IV.

We must now return to a discussion of the 1954 paper.[3.34] The question of the limitations imposed on X_x by V_x are directly related to another question that involves the drift velocity of the electrons up the streamer channel. The electron drift velocity in air at 760 mm pressure has been measured directly by Huxley and Zaazou[2.1] as a function of X/p. It ranges from 9.4×10^5 cm/sec for $X/p = 1$ at 0.76 kv/cm through 2.4×10^6 for $X/p = 3$ at $X = 2.28$ kv/cm, 4.4×10^6 cm/sec for $X/p = 8$ at 5.08 kv/cm, and 8.4×10^6 for $X/p = 20$ at 15.2 kv/cm. These values are low indeed, for the streamer tip velocities, v_t, observed by Amin at and around $X/p = 20$ begin to be commensurate with Meek and Saxe's values in long sparks. This led the author to consider the situation in more detail. He indicated that if the electron density in the streamer channel were n_e per centimeter of streamer length while the number of new electrons created per centimeter *advance* of the *streamer tip* were n, as measured by Kip, English, Amin, and Meek and Saxe from the current i_x, then one could, on the basis of current continuity, write $i_x = e n_e v_e$ and $i_x/e = n_e v_e = n v_t$. Thus if $v_t > v_e$ it can be so if $n_e > n$. There were good reasons for thinking that this might be true. In the early days the streamer channel had not been regarded as a dynamic system. It had also been assumed that all the n electrons per centimeter created by the tip advance flowed into the anode to give the current. This was difficult to conceive with the effective space charge of the quasi-stationary positive ions, for it was estimated that with 10^{13} ions and electrons per cubic centimeter, the electrons and ions could not be that easily separated. The excess electrons were assumed at that time to come from a flow of extra electrons into the channel in view of the strong radial fields between the streamer axis with its anode potential relative to the normal gradient in the undistorted gap. This concept has an interesting background.

In 1947 the author became aware of a theory of C. E. R. Bruce,[3.36] who considered the streamer channel as a fine conducting wire and expected it to expand radially by going into extensive corona. He attempted to carry the concepts over to the lightning stroke. When in 1948 or somewhat thereafter, the author explained the shape of the negative Trichel pulse corona glows in terms of the discharge distorted potentials along the axis and the normal field gradient surrounding the channel,[2.18] he inferred a radial influx of electrons into the channel behind the streamer head as it advanced, which added extra electrons to the channel to carry the current at the low fields existing. He, however, did not envision the extreme expansions inferred by Bruce but agreed that there was a radial influx. He viewed this as a more or less uniform influx along the channel while recognizing that individual avalanches would fluctuate widely in magnitude. His reason for not expecting too much expansion was that, so far as observed, the channels remained narrow. Furthermore, owing to the small channel radii, the ionizing fields do not extend out very far. It should be added that in his original concept of the transition from streamer channel connecting the electrodes, Raether[3.27],[3.29] indicated that a strong radial ionization current must occur. This would widen the spark channel, which is usually as narrow as the streamer channel.

If one looks at Nasser's[2.19] streamer channels as revealed by Lichtenberg figures, he will note that, besides the main or relatively vigorous long branches that spread radially because of mutual repulsion, there are obviously many smaller streamer branches. From the broadening and expansion of the photographs, one can well imagine that some of these are possibly large avalanches. Thus the author's and Bruce's concept is in a measure established but in a somewhat altered form resulting from the strong tendency of the streamers to branch. These small branches advance slowly so that the influx of electrons continues well after the streamer tip has passed, as the author postulated.

Thus, making the assumption that n_e was greater than n, the author proceeded to relate the various quantities involved from the data of Amin and Meek and Saxe. In Amin's case v_t ranged from 6×10^7 down and n was about 8×10^9 per centimeter. If then n_e/n were set as 7.5, the value of v_e was 8×10^6, which represented an X_x/p of 20 and an X_x of 15 kv/cm. This was just at the threshold value for ionization by collision of electrons. It would yield *no visible light emission* from the channel. However, the total value of V applied at the anode did not permit an X_x in excess of 5 kv/cm, which meant that the data were not consistent unless the current heated the channel to 900° K or 600° C. This temperature again would not cause it to emit detectable light. This was very well for Amin's study, for an X_x/p of 20 was assumed essential to maintain conductivity of the channel. However, with Meek and Saxe's study, v_e was commensurate with the v_t observed. This meant that if $n_e > n$, v_e had to be materially less than v_t. This led to values of X/p that again were too low for ionization if X_x were to conform to $\int X_x \, dx < V$, thus requiring high temperatures. This was easier to rationalize because Meek and Saxe's values of current ranged between 2 and 10 amp. Nevertheless, it was not satisfactory.

Regarded in the light of present knowledge, however, there were two points on which the author was unwittingly in error in those estimates, assuming that his concept of the single streamer channel was not too naïve. The first was that he was concerned about sufficiently high gradients X/p to make up for losses due to recombination and electron attachment in the channel. Since the time of streamer tip advance across the gap is estimated by Nasser as being about 10^{-8} sec, and in any event the whole breakdown sequence occurs in an interval of a few times 10^{-7} sec, the *loss factors are very small*. Reassessment of the rate of recombination, unless ion densities exceed 10^{15} per cubic centimeter, makes such loss negligible.

Electron attachment to form O_2^- is again too slow to reduce electrons very much. Dissociative attachment is very small at $X/p \sim 1$ but becomes appreciable at $X/p \sim 20$. An $X/p \sim 5$ in air is insufficient to decrease the electron concentration by more than 10% in a microsecond. Thus X_x/p can be anywhere in the range from 1 to 10 without causing undue loss. Again, the *velocities of the streamers observed by Meek and Saxe are not the tip velocities of the primary breakdown streamers observed by Hudson.*[3.24] Meek and Saxe observed what Hudson called secondary, and Amin pre-onset, streamers. These have velocities in the range of $\frac{1}{3}$ to $\frac{1}{10}$ the velocities v_t of breakdown streamer tips. It appears that n_e in Amin's case could well have been $20n$ or even more and X_x/p could have been the more realistic 3 or 5 with no need for an increase in temperature. The same would apply to the Meek and Saxe data, only that tip velocities could have been relatively even higher. There is, therefore, no difficulty about the channels behind the streamer tips, though gradients X_x and n_e may not be known accurately. Kritzinger's long sparks have primary streamer velocities of $\approx 10^8$ cm/sec and more (see Appendix I). He identifies Meek and Saxe's streamers with the slow secondary process.

From Meek and Saxe's study one very important further observation was made. The photomultiplier technique was able *to resolve the return stroke in midgap so that at least down to 20 cm gaps the return stroke* as well as the streamer process was noted. Kritzinger carried this down to 10 cm sparks (Appendix I). This is a significant contribution, as it shows that, within the resolving power of the equipment, streamer breakdown occurs through the sequence of the anode streamer and return stroke.

c. *Significance of the Uniform Field Studies of
   Fisher and Bederson; Space Charge
   Stimulation of Streamer Breakdown*

In the meantime, a new development in spark breakdown studies had taken place. The author and Meek had drawn certain conclusions concerning streamer sparks in their book.[3.28] In his doctoral dissertation, L. H. Fisher had undertaken to test an expected anomaly in breakdown using a large metal chamber with uniform field geometry. The anomaly could not be observed because of the nonuniformity of fields as gap lengths increased and pressures decreased. Fisher's interest had, however, been stimulated in the subject, so that when he resumed research after the war he took the chamber with him. With this he undertook a study of formative time lags in uniform field geometry in air at atmospheric pressure. His technique was to apply a steady high potential near the static spark threshold with adequate triggering photoelectrons released by ultraviolet light from the cathode. His potential source was a particularly highly stabilized high potential source designed by H. J. White for that purpose. With this source Fisher added an increment of potential in the form of a step voltage with sharp rise from a battery source at t = 0. This step potential carried the total voltage to any desired value above the sparking threshold. The time between application and the appearance of the spark could be measured by oscilloscope. Fisher together with Bederson[3.37] discovered that, as potential was lowered from some percentage of overvoltage toward the breakdown threshold, the *formative* portion of the time lag increased from some 10^{-7} sec to a value as great as 10^{-5} sec. The formative time lag is the time between the appearance of the triggering electron in the gap to the materialization of the spark. The statistical time lag is the time needed for a triggering electron to appear in the gap. It can be corrected for in measurements such as Fisher's. Sparking threshold was not clearly defined because of changes in surface and gas composition, but measurements could be carried down to within some 0.2% of the nearest value to a fixed sparking potential attained. This change in time lag was quite significant for the following reason. Since the breakdown streamer tip speeds are $\approx 10^{8}$ cm/sec and return strokes are faster, the *streamer breakdown should occur in an electron transit time some 10^{-7} sec* after application of the potential if adequate triggering electrons are present. If adequate electrons are not present a study of the total time lag statistics permits correction for the so-called statistical lag to be made. The appearance of a *breakdown very near threshold taking 10^{-5} or more sec* signified that the streamer breakdown was preceded by the slower conventional Townsend breakdown of the gap as a whole. This breakdown threshold is set by equation 2.7 $\gamma_p e^{\alpha d} \geq 1$ and proceeds by a succession of generations of electron avalanches followed by photoelectric liberation from the cathode. Here the γ_p was a photoelectric emission from the cathode by photons created in avalanches near the anode. This discovery came as a surprise to everyone, but it need not have done so. From 1934 and 1935 on, it had been shown theoretically by Steenbeck[2.8] and by experiment and theory by Varney, White, Loeb, and Posin[2.8] that, if α/p increases more rapidly than X/p once a Townsend breakdown threshold is reached, positive ion space charges collect. The space charge fields in the gap with the creation of an anode plasma shorten the gap but increase $e^{\alpha d}$, leading to ever-increasing avalanche sizes. If avalanches become large enough they will initiate a streamer spark in the uniform field.[3.38] Thus in air where γ_p is low, $\sim 10^{-4}$ or less, once avalanches reach a value where $(1/\gamma_p) \geq e^{\alpha d}$, the breakdown begins with a Townsend discharge, avalanches and discharge current progressively increase in consequence of the space charge distortion until the streamer-forming avalanches are reached. At this time the filamentary spark breakdown ensues. The long time lag is that needed to build up the space charge to streamer-forming values.[3.37,3.38] The higher γ_p, the lower the value of $e^{\alpha d}$ needed and the lower the Townsend threshold; this means the longer the time needed to build up the space charge to streamer avalanche-creating magnitude. For some gases γ_p is so *low* that $e^{\alpha d}$ reaches streamer-forming magnitudes and sparks occur at breakdown threshold with no preceding Townsend discharge.[2.6] The nearer one approaches the threshold potential from above, the smaller $\gamma e^{\alpha d}$ and the longer it takes to build up the space charge and the spark. Above some 6% overvoltage[3.39] in dry air the values of $e^{\alpha d}$ are adequate for a streamer spark *without space charge* build-up. Thus *at or near threshold for a diffuse Townsend breakdown in air and photoionizable gases with a glow discharge, the spark is filamentary and caused by a streamer.* This point had caused a great deal of confusion in the past. It is, however, only in air that the thresholds for streamer spark by single avalanches and for a Townsend glow threshold lie so close together. In a later study Köhrmann, in Raether's laboratory, showed that indeed avalanches reach streamer magnitudes in uniform field at 6% above Townsend threshold in dry air and 4% in humid air.[3.39] In "spectroscopically pure" grade commercial A, the streamer threshold lies some 100% above the Townsend threshold and the formative lag goes to 10^{-4} or more sec.[3.40] By using gases such as CH_4 and ethyl ether, in which γ_p is very low indeed, there is no Townsend breakdown and, when avalanches reach streamer magnitudes, a streamer spark ensues with no antecedent glow discharge. Time lags are always $\sim 10^{-7}$ at threshold.[2.6]

This general concept of the growth of the streamer spark from the Townsend discharge at the Townsend threshold was later first directly confirmed by Bandel in the author's laboratory.[3.41] Using a stabilized White potential source and step voltage by an ingenious triggering mechanism, he succeeded in observing the growth of the Townsend discharge once the step potential was applied. If this was at too low a potential it gave a nearly constant glow current that eventually extinguished by avalanche size fluctuation. At slightly higher potentials the Townsend discharge began to increase as space charge accumulated and suddenly went over to a streamer spark with a formative lag for the streamer of $\sim 10^{-6}$ sec. Bandel derived a theoretically sound approximate equation for the growth of the Townsend discharge with a photon γ_p. This agreed with observation on the oscillograph except that there was a delay in the observed rate of rise relative to the theory. It amounted to a factor of about three electron crossing times per avalanche generation. The same delay was later observed by Raether's group[2.9] and probably has to do with a delay in the photoelectrically active photon emission. Bandel also showed that the upcurving indicating the space charge growth leading to the streamer set in at the values of current density to be expected from the theory of Varney, White, Loeb, and Posin. Later Kluckow and Mielke[3.42] in Raether's laboratory, using both single and multiple electron avalanches for initiation, also studied the growth of the Townsend discharge at threshold and observed the transition to a streamer spark once space charges had formed. The growth of the Townsend discharge has been the subject of many papers in recent years.[2.5] Hence theory, except for the effects of electron attachment and detachment in air, the photoelectric delay time, and other inaccurately known basic quantitative parameters, permits calculations, although quite tedious, to be carried out. More important, however, are the investigations of Raether's group on the build-up of the Townsend discharge and its development into the streamer, either from single avalanches or a group of simultaneous avalanches.[2.6]

d. *Fast Oscilloscopic and Photomultiplier Studies of Breakdown Streamers by Hudson*

The results of Fisher and Bederson, as well as the later papers of Fisher and his students,[3.40] together with skepticism as to the applicability of the streamer theory to uniform field geometry despite all the work cited, led the author to assign to Hudson[3.24] the extension of Amin's studies to the breakdown region. He was to use the photomultiplier technique and the new fast 517 Tektronix oscilloscope made available for this work. With this sort of study in gaps ranging from extremely fine points-to-planes to as nearly uniform field gaps as possible, it was hoped to achieve the following objectives: (1) prove that the spark breakdown always initiated with an anode streamer; (2) verify the presence of the return stroke; and (3) investigate the peculiar secondary streamer noted by Hudson and Amin. The work consumed much time in overcoming difficulties, perfecting techniques, and in covering the range of gaps envisioned. It was early decided that better records could be obtained if Hudson could eliminate triggering on the rise of the electrical signal in the streamer pulse because this introduced background noise. It proved much better to trigger the rise of the scanning or signal photomultiplier by means of a triggering photomultiplier which operated on the first appearance of the streamer pulse at the anode tip or elsewhere in the gap. This was possible, as it had been shown in the course of preliminary studies that all cases of breakdown appeared to start from the anode, so that triggering always came from the appearance of light at the anode tip.

The work was completed with Hudson's thesis in 1957 but was not published until 1961.[3.24] Two sources of potential were used. One was a carefully stabilized 74 kv high tension rectifier set designed by White but further developed by Hudson. Potentials were stable to better than 0.1% and were read by means of the current in two wire wound 50 megohm resistors in series with a dial resistance box to form a potential divider using a L and N type K potentiometer. A protective resistor of 70 megohms was in series with the anode. For the higher fields required for gaps with the 30 cm diameter spheres against planes up to 9 cm distant, potential came from an air turbine-driven demonstration-type Van de Graaff generator. This could yield as much as 200 kv. Sufficient charge could be stored on the point to give a transient arc-like breakdown (visually a spark of short duration) by connecting a piece of RG-8/U coaxial cable directly to the anode at both ends. This had a capacity of the order of 150 pF, which was generally used.

For use with the Van de Graaff the 35 pF of the generator storage was increased to 135 pF by using a piece of RG-17/U cable which had had insulating oil forced into the space surrounding the central conductor by N_2 at 2000 psi. The anodes consisted of a fine steel needle, Pt wires with hemispherical ends, Ni wires and rods, and brass-, copper- and chromium-plated brass spheres. Anode tip radius varied from the needle point up to 0.62 cm radius. The spheres had radii ranging from 1.9 to 6.35 cm. For the more nearly uniform fields copper spheres of up to 30.5 cm diameter were used. The cathode used with smaller anodes was a brass plate 12.8 cm in diameter. The whole system was housed in a grounded metal box, as shown schematically in figure 3.50. The gap length was varied by moving

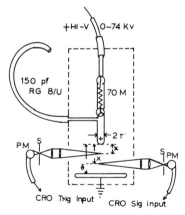

FIG. 3.50. Design of Hudson's apparatus for the two-photomultiplier study. The figure is self-explanatory in the light of the description given in the text.

the anode holder or the cathode vertically. With the Van de Graaff generator, the inside top of the large 64 cubic foot cubical aluminum shielding box was used as cathode, the generator being moved vertically to adjust gap length. Both photomultipliers with their optics were mounted on rigid stands equipped with rack and pinion so that they could be moved vertically. Readings of position were good to 0.2 mm. The trigger photomultiplier was usually focused at the tip of the anode. The signal or scanning photomultiplier was moved vertically to view various points in the gap from anode to cathode. The signal photomultiplier had both vertical and horizontal slits. Generally the slit had a vertical opening of 0.2 to 0.5 mm, and the horizontal slit was at its maximum width of 23 mm. That is, the signal obtained generally came from a horizontal slit ∼0.2 mm wide and 23 mm long normal to the gap axis. A P28 photomultiplier was used for the signal. Considerable study was made of the field of view covered by the optics and its influence on observation. All impedances were carefully matched. Especial care was required in grounding and screening in order to reduce background noise to signal ratio. What was observed was the light intensity of a luminous pulse as it passed across a limited wedge-shaped volume of somewhat restricted depth. As the light pulses, that is, intensity along the gap length at any instant, had a given form, the observations record the integrated intensity of the light in the limited volume of the pulse at each instant t in the section of 23 mm length at a point x exposed to view as a function of time.

If the form of the light pulse along the gap is invariant in time and/or position, then the oscillogram represents the true form of the pulse at all points of the gap. Thus the velocities of toe, peak, or an arbitrary point on the pulse, plotted as a time-distance curve along the gap, gives the velocity of that pulse feature along the gap. If, however, the light pulse changes shape as it crosses the gap, great care must be used in inferring velocities of peak or half-rise from time-distance plots taken from oscillograms. Thus Hudson was soon forced to take a large number of oscilloscope traces at many different points. From the average intensity values at a given t by crossplots for each position of many along the gap, he was able to plot *intensity* as a function of *position* x of the slit in the gap at different times t. In this way the shape of the light pulses at any time could be observed. It was noted within the limits of observation that the primary pulses, within resolving power, had constant shape as the value of t increased. Their relative intensity was low at first and increased to a maximum near the cathode, perhaps remaining constant in the last quarter of the gap. The secondary pulses first noted by Amin and Hudson on pre-onset streamers were observed to be present in all gaps with breakdown streamers. They became more pronounced as breakdown was approached and advanced further into the gap before they faded. They showed a wide variation in shape of individual pulses and were not constant in shape as they crossed the gap. They became rather small in amplitude as they approached the cathode and even disappeared in its vicinity at breakdown potentials in more asymmetrical gaps. As gaps tended to greater field uniformity with larger anode diameters and shorter length, the time intervals became too short to discern the primaries, and in more uniform gaps even the secondaries near the anode were hard to discern. If amplification was increased sufficiently, the primary and secondary light pips could be observed in the neighborhood of the cathode. Under these conditions the secondary was usually observed riding up on the rise of the main pulse of luminosity representing the transient arc which, with the amplification used, was so intense as to overload the photomultiplier. These features are shown in the later figures 3.54 and 3.55.

In order to study pre-breakdown streamers, but just below and at breakdown, gaps with ratio of gap length δ to radius r, δ/r, less than 160 were used to eliminate the many streamers that did not lead to breakdown. Here, often just at breakdown, a sequence of streamers would be observed, the last one of which materialized into a spark. If for the same point the gap length was decreased so that the first streamer could cross the gap with sufficient tip field to cause a spark, then the first or second streamer succeeded in causing the spark. This situation was corroborated by Nasser in his study with Lichtenberg figures.[2.19] That is, streamers cross the gap well below breakdown, but unless the tip potential is high enough, they may not give a spark. Hudson's conditions permitted him to observe this for a limited region only. Nasser showed that the range of potential over which streamers crossed the gap before causing a spark was remarkably great. Hudson's oscillograms at slow sweep showing the succession of primary-secondary pulses needed before a breakdown as the gap was shortened are seen in figure 3.51. Here primary and secondary pulses are not resolved, just one pip being seen for each streamer. This is indicated in the difference between tip potentials for streamers that just propagate, which are of ~ 0.8 kv, and tip potentials of breakdown streamers in the same gap, of 25 kv. Again it takes only of the order of 10^5 electrons in an avalanche to launch a pre-onset streamer with a 0.02 cm diameter point gap, while breakdown streamers in a uniform field gap require of the order of 10^8 electrons.

Hudson's observations then began to reveal unexpected deviations from the preconceived sequence of events discussed earlier. In the first place, the secondary streamers constituted a completely unexpected phenomenon. Allibone and Meek, Saxe and Meek, and Raether had not observed them. Again no obvious return stroke was observed when either primary or secondary pulse reached the cathode. This was disconcerting, as its observation had been expected. If, however, the velocity of the return stroke lies in the range of 10^9 cm/sec, it would not be resolved even in 6 cm gaps.

What was observed, however, was that some 0.2 to 1 μsec after the *primary tip* had reached the cathode there was a rapid rise in luminous intensity across the gap to values which exceeded those of the primary or secondary pulses by at least an order of magnitude. This rise of luminosity

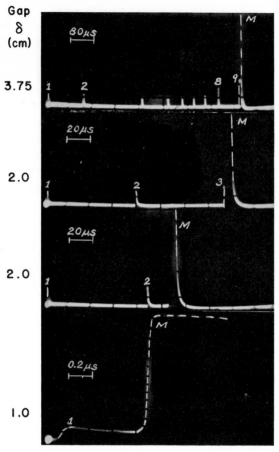

STREAMER MAIN STROKE SEQUENCE

Anode
Tip r = 0.038 cm
Shank r = 0.075 cm

Tip r increased slightly during run

FIG. 3.51. Sequence of pre-onset streamers observed by Hudson with slow time resolution before one led to a spark. At fixed potential with a 3.75 cm gap, the top trace shows nine streamers by photomultiplier observation before the main stroke, M, reinforced here by white dashes. With a shorter 2 cm gap the third and second streamers, respectively, led to the main stroke or spark. For a 1 cm gap the first streamer led to a breakdown. Note that here the time between the initial streamer and spark was on the order of 0.4 μsec.

lasted varying times ranging from 0.1 μsec to microseconds. It obviously represented the transient arc that constituted the filamentary spark channel discharging the 150 pf of the coaxial cable. This was termed by Hudson the *main stroke*, that is, the *spark*.

Owing to the great individual fluctuation in the shape and intensity of secondary streamers in a succession of breakdowns under seemingly identical conditions as well as their attenuation, it became difficult to relate the secondaries in any definite way with the main stroke. It appeared for the more asymmetrical gaps with small anode where time resolution was good, as if *the arrival of the secondary streamer*, either toe or peak at the cathode, was followed by the rise of the main stroke. For more uniform fields at increased PM gain, it was clear that near the cathode the secondary peak was riding up on the toe of the rising main stroke.

One thing was certain: *at no time was the breakdown associated with or initiated by any light pulses except those moving from anode or midgap to the cathode.* That is, for all gaps to very nearly uniform field geometry the filamentary spark, or spark breakdown, was triggered by a luminous pulse moving from anode to cathode. As resolution became poor for shorter gaps with high fields, for example, 30 cm diameter spheres opposite a plane 1 to 2 cm distance, time was so short that the progression of the streamer across the whole gap could not be observed but the rise of the breakdown from a primary and a secondary peak preceding the main steep rise of the spark could always be observed. At 6 cm with these electrodes, progression from anode to cathode could be observed. These phenomena are shown in the traces of figures 3.52, 3.53, and 3.54. One more feature is shown in figure 3.55, in which for a nearly uniform field the region near the cathode is scanned using decreasing gain. This shows the primary and secondary riding up on the main stroke. More information is given by the crossplots of luminosity as a function of distance from anode to cathode at various times. These are shown in figures 3.56 through 3.59, which correspond to data such as shown in the preceding oscillograms. The dashed portions of traces indicate regions in which either resolution or repetition of individual sweeps was not sufficiently close to give the reproducibility of data leading to the solid portions. They represent probable courses. Thus the shape of the primary pulse on the short time scale of figure 3.58 is not accurate enough to draw a solid line. In that figure one significant fact should be noted: when the primary pulse reaches the cathode there is a marked increase in luminosity along the whole gap. This has also been noted by Nasser in another way, and is also indicated in figure 3.59. Another feature of these crossplots is that the average shape of the secondary pulses is not constant across the gap. Thus, measuring the velocity of the peak of luminosity of that wave does not make much sense. Also note that where

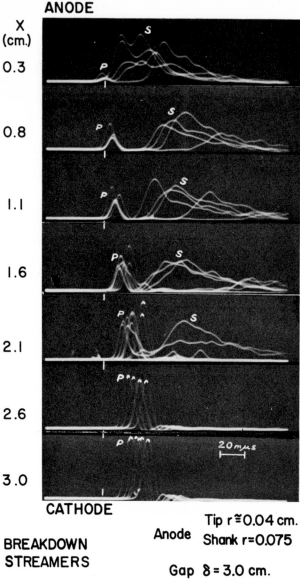

ANODE

X
(cm.)

0.3

0.8

1.1

1.6

2.1

2.6

3.0

CATHODE

BREAKDOWN
STREAMERS

Anode

Tip r ≅ 0.04 cm.
Shank r = 0.075

Gap δ = 3.0 cm.

F$_{\text{IG}}$. 3.52. Hudson's breakdown streamer sequence for a gap with a tip of radius 0.04 cm and 3 cm gap are shown with the signal PM from 0.3 cm to 3.0 cm from the anode. The primary streamer P is seen to rise, grow larger, and progress with time to the right of the initial time signal, given by the white line. Each sequence shows four or five successive sweeps at the same distance. There is some difference in speed of the primary streamers, but by and large they are remarkably uniform in behavior. The secondary pulses differ markedly in form. They are initially quite intense near the anode and are marked s. They lengthen in duration and decline in amplitude as the gap is traversed and are not noted beyond 2.1 cm. The main stroke follows after some 10^{-7} sec and is off the photograph to the right.

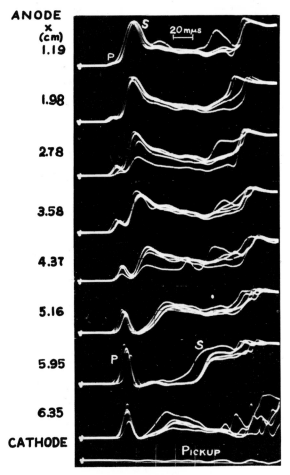

FIG. 3.53. Shows the data taken with point radius r = 6.35 cm and δ the gap length of 6.40 cm. Here the lowest trace is the electrical noise background. The signal PM ranges in position from 1.19 cm from the anode to the cathode. About four sequences are observed at each distance. The primary P is barely perceptible at 1.19 cm, the secondary is quite pronounced, and luminosity appears to continue for the 10^{-7} sec to the rise of the main, which is flattened because at the high gain its luminosity overloads the PM. Very near the anode the primary equals the secondary in intensity, and the secondary is riding up on the rise of the main stroke.

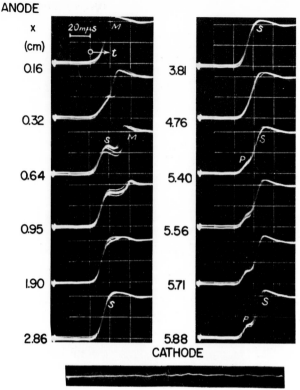

ANODE

x (cm)

20 mμs

M		S
0.16		3.81
0.32		4.76
0.64		5.40
0.95		5.56
1.90		5.71
2.86		5.88

CATHODE

FIG. 3.54. For a gap with r = 15.5 cm and δ = 6 cm with high gain. Shows the primary P at only 5.4 cm from the anode and the secondary, which appears at 0.32 cm and overloads the PM at 0.95 cm. Here we have a condition of heavy, branched streamers in a nearly uniform field crossing the gap so that the primary does not precede them by more than 6 nannoseconds to the cathode.

FIG. 3.55. Shows the sequence of primary, secondary, and main at a point near the cathode, x = 5.9 cm with r = 15.5 cm and δ = 6 cm for a series of PM gains *decreasing* from top trace to bottom. The relative gain at the top is 6, and we see the primary and an overloaded secondary. At 1.5-fold gain the primary is hardly discernible, the secondary is resolved, but the main is overloaded. At gain 1 the secondary and main are resolved. Note the degree of reproducibility in total light output on successive sequences.

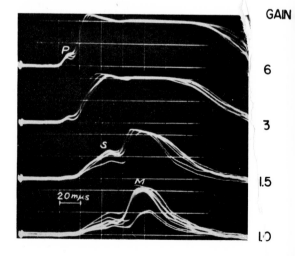

GAIN

6

3

1.5

1.0

20 mμs

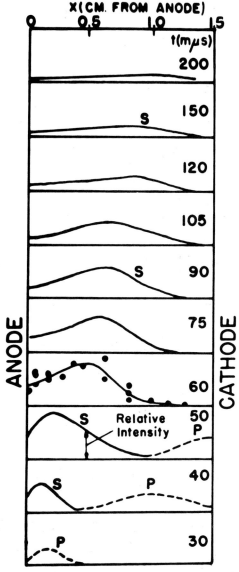

Fig. 3.56. Crossplots of Hudson's time-intensity traces at different distances, showing the luminosity at different times across the gap, that is, luminosity plotted against position at various times. Here r = 0.62 cm and δ = 1.5 cm. The anode is to the left. The lower trace is next, the anode at 30 nannosec. Here the primary is dotted, as separate traces are not clearly duplicated. At 60 nannosec the points plotted show the reproducibility of the data. At that time the primary P has reached the cathode and the secondary toe is halfway across. Note that the peak of the secondary never does reach the cathode even after 0.2 μsec. The secondary declines in amplitude as it crosses the gap and changes its shape as well.

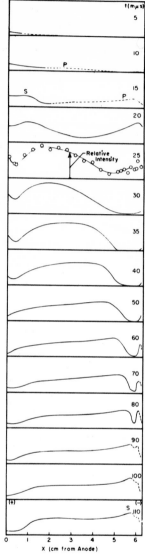

FIG. 3.57. Here r = 6.35 cm and δ = 6.4 cm. The streamers of the primary and secondary pulses are shown for the first 100 nannosec and correspond to the data of figure 5.33. The primary crosses the gap in about 20 nannosec. Note the general increase in light intensity across the whole gap when this occurs. The main stroke is not shown here.

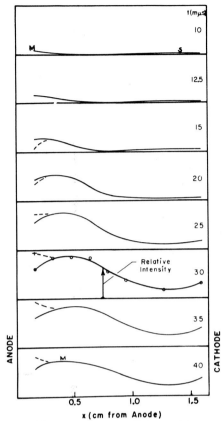

FIG. 3.58. Crossplots with r = 15.5 cm and δ = 1.65 cm approaching uniform field conditions. Here events are so rapid that only the arrival of the secondary at the cathode and the rise of the main can be seen. Here at 40 nannosec it is noted that the luminosity across the gap during the main stroke is not uniform, being high at anode and high again near the cathode.

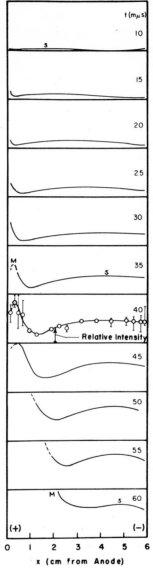

FIG. 3.59. Crossplots with r = 15.5 cm and δ = 6 cm. This shows the last of the primary after 10 nannosec. On its arrival at the cathode, the secondary intensity rises. The secondary peak reaches the cathode in 35 nannosec. The main stroke is at its peak after about 60 nannosec.

the sequence shows the rise and decline of the spark the intensity is not uniform across the whole gap. The main stroke shows maximum luminosity at the anode with a dip about two-thirds across the gap and a small rise in luminosity at the cathode. This confirms the visual observations of these transient arc sparks.

Figures 3.60 and 3.61 show distance-time plots for primary and secondary waves for two gaps. It is seen that the primary peak, which is narrow and has a constant shape, is nearly constant and has two to three times the speed of the secondary pulse. The secondary pulse has a high velocity for the first third of the gap and then slows materially. The change in speed occurs at a distance at which the more axial stem of photographed luminosity begins to flare because of branching; in other words, the velocity of advance is reduced by branching. The effect of varying shape, for example, secondary pulses, is reflected in the different time-distance curves for the primary-secondary minimum, the secondary toe, and the secondary peak in the gap of figure 3.61. The speeds in the asymmetrical gap of figure 3.60 are 6×10^7 for the primary and 1.7×10^7 and 4×10^6 cm/sec, respectively, for the secondary before and after the break. In the gap of figure 3.61 the primary speed is up to 5.6×10^8 cm/sec, and the secondary peak has speeds of 1.1×10^8 and 5.4×10^7 cm/sec before and after the breaks.

One more disconcerting discovery was made by Hudson. In one case where by accident the point was not quite hemispherical in shape, he was able to observe pre-breakdown streamers just below breakdown so that streamers were not followed by an obliterating spark. Figure 3.49 shows still photographs of the breakdown. Trace a shows branching of the streamers after proceeding a short distance from the point; three very diffuse branches are seen. Trace b shows a longer exposure and indicates some forty or more streamers or branches. Hudson called these primary streamer branches "dendrites." They were later identified as streamer branches by Nasser. Hudson unfortunately insisted on his own terminology for the observed phenomena, as they did not fit into his preconceived patterns. Parenthetically it must be noted that such multiple terminology only creates confusion. Trace c shows the ensuing spark in this gap at a very slightly increased potential. Note the finer streamers near the cathode when the exposure was such that superposition revealed the full streamer length. Note the luminosity produced where the streamers of b strike the cathode and compare this with figures 3.23 and 3.24, taken from Kip's study with long and short exposure. Note the relatively straight axial portion near the point in Hudson's photographs. Dawson's recent measurements of currents and luminosity with a 519 Tektronix oscilloscope in the author's laboratory confirm these fine traces as primary streamers and the cathode flashes as signaling their arrival at the cathode.

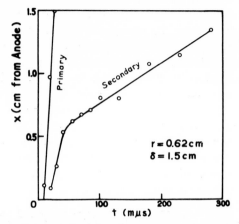

FIG. 3.60. Time-distance plots for the primary and the secondary peaks for r = 0.62 cm and δ = 1.5 cm, for which the crossplots of figure 3.56 gave the data. Note that the primary has a nearly uniform speed within the resolving power of the apparatus. The secondary peak moves rapidly for the first 50 nannosec, but then slows down as later, stronger streamer branch tips reach the photomultiplier. Here the secondary peak has crossed only about one-third of the gap.

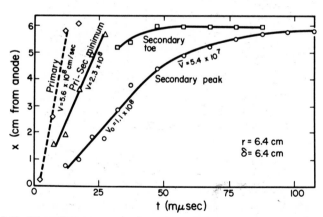

FIG. 3.61. Time-distance curves from crossplots for a gap r = 6.4 cm, δ = 6.4 cm for various features of the intensity pattern. The secondary toe moves more rapidly than the peak. This eloquently gives quantitative proof of the change in *average* shape of the secondary pulse as it crosses the gap.

It had always been known that at higher potentials in the divergent fields there was heavy branching of the streamers. In fact, this was ascribed in part to the dispersing action of the positive space charge in transit in the gap as well as to the radial nature of the high field. This branching is seen in a cloud track photograph by Gorrill shown in figure 3.62. This was taken in 1938 in the author's laboratory with high impulse potential where no space charge was present. However, the extensive branching of Hudson's picture was not anticipated. Hudson's analysis using the oscilloscope on these streamers showed that the short stem and initial flare out to one-third the gap length from the anode might be associated with the secondary streamers. The fine broad branching extending to the cathode of figure

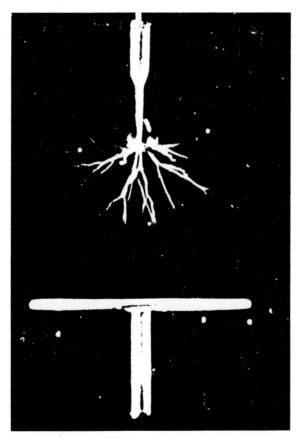

FIG. 3.62. Gorrill's cloud track photograph of 1938 with high impulse potential in a positive point-to-plane gap. Here the extensive branching is seen. The chamber is underexpanded and only heavier branches, corresponding to Hudson's secondaries, are seen.

3.49, trace b, was identified by Hudson as due to primary streamers. Here it is clear that the primary steamers cross the gap below breakdown. Their high tip field plus photon liberation initiates the brilliant flashes of luminosity at the cathode. It is clear that there is less branching close to the point and that the three streamers in trace a appear broadened by branching. They were assumed by Hudson and the author to be secondary streamers which followed only a few of the primary branches.

Use of the vertical and horizontal slits of the signal photomultiplier enabled Hudson to verify the presence of the streamer branches out to a considerable radial distance from the axis, as had been done by Meek and Saxe and by Amin. A point which puzzled Hudson was that, despite branching, *the primary streamers appeared to cross the gap with the same high speed*; that is, they arrived at some horizontal plane from the anode nearly simultaneously. This was surprising in view of the diversity of velocities of axial velocity components of the more radial streamers. It was believed, that despite primary branching or dendrites, the secondaries followed only a few of the primary branches and of them the spark channel developed along only one, as seen in the figure.

It was under these circumstances that the author eagerly accepted the help of the Office of Naval Research in bringing Nasser to this country from Germany for a limited time to put his Lichtenberg figure techniques to the task of clarifying the confusion and contradiction apparently introduced by the very definite and beautiful observations of Hudson.

8. STREAMER STUDIES WITH IMPULSE POTENTIALS USING LICHTENBERG FIGURE TECHNIQUES; THE STUDIES OF NASSER

Nasser[2.19] undertook his doctoral dissertation as a student of Strigel in Berlin. Nearly contemporaneously but slightly later Tetzner started his doctoral dissertation in the same laboratory, using liquids.[3.43] Both studies utilized the Lichtenberg figures in order to investigate the streamer breakdown from a positive point to a plane, the one in air and the other in an insulating oil. Some further work was done under Nasser's supervision by Adipura, a candidate for a baccalaureate, on both positive and negative points.[3.44]

While Lichtenberg figures had been known and studied for many years, the studies had never been properly oriented to exploit the streamer theory which, as will be seen, was a most fruitful approach. Shinohara[3.45] had attempted to approach the problem from the streamer viewpoint. However, perhaps because he was an engineer and not a physicist, or else because of language difficulties, he either did not understand the theory or so misconstrued it that it was impossible to follow his analysis.

Using a point-to-plane gap, with hemispherically capped cylinders of the order of 1.0 mm diameter and various gap lengths up to about 5 cm, Nasser applied square wave pulses of controlled amplitude and duration at potentials both above and below the breakdown threshold for a spark in room air. His pulse duration was that of the crossing times of streamers and shorter. Thus streamers at all times appeared and moved from the anode point into a gap devoid of space charge distortion.

Normal to the axis there was placed a plane photographic emulsion facing the anode. It was usually a film which could be wound from a spool at one end of the plate holder to one at the other end, permitting a series of exposures to be run on the same film. The whole camera and film could be made to traverse the gap from anode to cathode surface by means of a worm gear. Its position as a function of distance from the anode could be read off on a drum. The whole gap was housed in a light-tight box.

The apparatus used in the extension of Nasser's doctoral studies in the author's laboratory was essentially the same as used in his doctoral thesis. In what follows it will be convenient to present past and present results together. The later work differed only in that longer gaps were exploited, with some improvements in techniques. Pulses of longer duration were largely employed in both studies but not in Adipura's unpublished study. Rise time of the pulse was 0.15 μsec or less in most of the work reported. A large variety of emulsions and films were used in the later study, including even color film. The later study was directed in a large measure toward clarifying the discrepancies revealed by Hudson's observations.

Diagrams of the apparatus and the pulse are shown in figure 3.63. When the discharge from a positive point strikes a photographic emulsion placed normal to its axis at some distance, a characteristic pattern is observed. A positive print of one of these is shown in figure 3.64. On the negative of a transparent film or on the emulsion on the surface of a printing-out paper exposed to the discharge, there were dark spots with dark radiating dendrites that appeared to repel each other. In the positive prints from transparent film they reproduced well for contrast and appear white on a dark background, as in figure 3.65. It had been long known that the length l of figure 3.64 of these dendrites from the center of impact on the film was proportional to the potential of the point of impact on the plate. By placing the anode point of known pulsed potential very close to the emulsion, the length of the longest dendrites was found to be proportional to the potential at the point. In this fashion, by calibrating the dendrite length l for a known potential, the potential at the point of impact of the discharge on the film could be evaluated from the length of the principal dendrites. This technique was exploited many years ago in the Kleidonoscope, used for rough measurement of high impulse potentials.

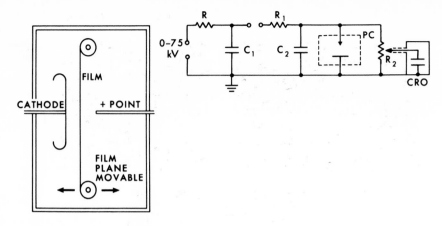

Photographic chamber

FIG. 3.63. Nasser's Lichtenberg figure camera device and the pulser used. The diagram is self-explanatory.

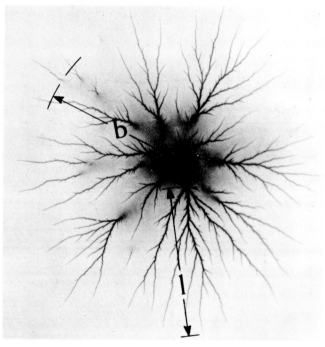

FIG. 3.64. A typical Lichtenberg figure of an emulsion or film placed some 2 cm distant from the point well above threshold. This is like a photographic negative with the streamers black against the unexposed film. Note the impact of ten larger streamer branches. These then parallel the surface, branching still more. The central darkening is from diffuse long wave length ultraviolet from the branches. Tips of some smaller branches can be seen on the left. The dimension b represents the radius of the circle containing the impact points of all streamers. The dimension l represents the length of the longest streamer. When calibrated by known anode potentials with film at the anode, this length l gives the tip potential of that streamer.

FIG. 3.65. This is a positive photographic print of a transparent film, showing a series of figures taken with an anode at 40.5 kv and pulse length of 58 μsec, with 1 mm diameter point and gap of 8 cm. Photograph a has the film at the anode, and the longest streamer branches represent a length l that a tip potential of 40.5 kv would represent. Photograph b is at 3 mm. Here six vigorous branches strike the film. The dark outline about the light traces represents the Clayden photographic effect of an exposure to ultraviolet followed by the more intense light of the streamer some 10^{-7} sec later. Photograph c is at 3 cm, where branching has reached a maximum and started to decline. At 6 cm photograph d shows but few points, and at 7 cm photograph e marks nearly the range of the streamers.

By applying a pulse just long enough for the discharge pattern to propagate across the gap and moving the film from close to the anode down to the cathode at a fixed potential below breakdown, Nasser obtained the typical series of patterns shown in figure 3.65. In this case the point diameter was 1 mm; the gap length 8 cm; the potential was 40.5 kv; and the pulse length was 58 μsec. In his earlier studies Nasser worked with shorter gaps and nearer breakdown so that some of the features of figure 3.65 were not then observed. It is seen that at first (a) there is one central point and branches or dendrites radiate out a long way, there being much branching of each dendrite. At 3 mm or so from the point (b) the center pattern is replaced by a few vigorous branches, each giving a spot with branches radiating from it. As distance increases, the number n of spots with radiating branches increases, branches get shorter, and a great distribution in lengths of the branches becomes apparent. At about 3 cm from the point (c) the number of branches and tips of branches reaching the film is at its maximum. It will be noted that the few spots near the axis have longer branches than do the more radially distant ones. From 3 cm on the number of branches begins to decline, and when the film lies on the cathode, only a few very small spots are noted. In what follows, the radius of a circle enclosing all impact points will be designated as b in figure 3.64.

In view of the branching, one of the first questions to settle was whether the presence of the film across the gap altered the field and occurrences materially. The film is sufficiently conducting that it will very quickly take on its charge appropriate to the region in which it finds itself. As noted in the figures to be shown, no streamers or branches in the plane of the emulsion ever cross each other. This comes from the mutual repulsions of the positive streamer tip and charged channels which arrive nearly simultaneously so that charges can repel.

With emulsion in place two successive pulses following each other in less than 1 sec were placed on the film. Both pictures due to first and second successive pulses with the film at midgap were studied. It was found that with two successive pulses the picture of the traces from the second pulse were recorded just as if there had been no first pulse. That is, the images of the two pulses were superposed on each other, streamer branches crossing each other just as if the first pulse had not occurred. Thus in less than 1 sec the significant charges on the film had dissipated and it had assumed its original potential. When it is considered that after the first pulse the air adjacent to the plate has been quite extensively ionized, the rapid neutralization of all significant charges should rapidly follow.[1]

The action of the film is to intercept the tips of the streamer branches,

---

[1] The film exerts an attraction by electrostatic induction that keeps the tip close to

striking it and preventing their further direct advance toward the cathode, parallel to the gap axis. In virtue of the high tip fields of the incident streamers, the streamer propagation, as will later be proven, proceeds radially from the point of impact on the film through the air close to the film surface. Thus the tip fields of the intercepted streamers act forcefully on the surface for the short time before radial propagation through the air along the surface removes them. While the streamers do not penetrate the film, their intense fields project through the film and may initiate discharges in the air or from not too distant electrodes on the other side. Thus tip potentials of some 48 kv or less intercepted 1 cm from the cathode can produce fields strong enough to launch avalanches or even negative streamers from the cathode surface. This property of the film, of preventing flashover but permitting the streamer process to develop without serious modification, constitutes the chief value of this technique. The influence of the high tip fields in their penetration through the film also leads to interesting consequences.

Most striking in Nasser's earlier study was the extensive branching of the streamers as they crossed the gap. The proof that indeed these branched or dendritic figures constitute streamers in air will come at a later point. It was this branching that disturbed Hudson[3.24] and the author in Hudson's observations. Nasser had made an extensive study of the branching for higher potentials and relatively short gaps in his earlier work. There he had counted the number of streamer tips striking the film as a function of their distance x from the anode.[2.19] He next recorded the radius of a circle b enclosing the point of impact of all tips on the film at each distance x, which measured the extreme radial spread. He also measured the length l of each branch formed and from this inferred the value of the tip potentials. It was of course possible, by changing the pulse length at given potentials, to find the time needed for streamers to reach a given plane, thus obtaining time-distance curves for different streamer tips. All these measurements could be made at different anode potentials. Variations of anode radius and gap length, d, were also possible. However, the amount of data needed for each change of major variable limited the ranges of variation that could be investigated in a reasonable time. Thus initially only shorter gaps with one or two points were studied, and the potentials previously used covered limited ranges in the regions near but below and above breakdown.

In the earlier study the number, n, of streamer branches was found to increase exponentially with distance x from the anode up to 3 cm. The

---

the surface. Again this action causes the launching of just one streamer at the contact point, or when *near* the tip. If no film is present, the higher potential, longer pulses often liberate several streamers from a 1 mm diameter point.

constants for this branching in a limited range were determined. Though branching is a random affair determined by chance, the number of branches λ per centimeter of advance, the average distance of advance per branch, and $1/\lambda$ can be determined and are obviously a function of the potential and field distribution about the point for some distance. As long as branching can continue at a given rate, then since each average branching interval, $1/\lambda$, covered doubled the number of branches, the increase in n with x was exponential.

In the new study gaps were lengthened. Figures 3.66 and 3.67 show n as a function of distance from the anode in centimeters for a 1 mm diameter point for potentials of 40.5 kv and 61.0 kv and values of gap length d of 3, 6, and 8 cm, and 4 and 10 cm for the two potentials respectively. For d = 3 at that potential, the exponential increase in n is clear and extends out to 2 cm and beyond. The quantity n was found to be very large. However, at the lower potential as gap lengths d extend to 6 and 8 cm, it is seen that n reaches a maximum at about 2.5 and 3 cm and then declines. There is still some branching at the cathode at 6 cm. At d = 8 cm branching stops at about 7 cm, the maximum range of these streamers. Increasing the potential for the same point leads to a nearly exponential increase in n out to 3 cm, changing to a linear increase when d = 4 cm. Here n reaches the value of 220. However, a 10 cm gap permits streamers to advance only 8 cm with a peak of n at 3.5 cm.

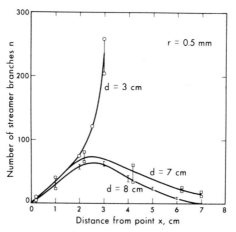

Fig. 3.66. Plot of the number of streamer branches striking a film as a function of distance x for a 1 mm diameter point at 40.5 kv at gap lengths 3, 7, and 8 cm. These are average values for several films. Vertical lines show the range of values observed on individual films.

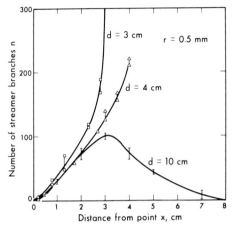

FIG. 3.67. The same as figure 3.66, except for a point potential of 61 kv
at gap lengths 3, 4, and 10 cm.

Figure 3.68 shows the influence of point diameter on number of streamers n as a function of d, the gap length, at point potentials of 30, 36, and
50 kv. Points of radii 0.5 and 1.0 mm were used. As noted before, the value
of n decreases as d is increased, since, because of branching, potential fall
down the channel, and other factors, the tip potentials decrease as x increases and at the same time the gap field rapidly decreases as the tip moves

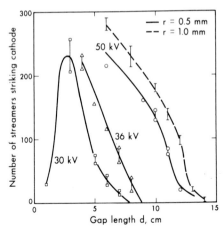

FIG. 3.68. Influence of point diameter on n as a function of gap length d.
Here point radii of 0.5 and 1 mm were used at 30, 36, and 50 kv.

away from the anode. Thus streamers benefit less and less from the external field. With increased tip diameter the surface tip field *at the anode* is directed more along the axis but is decreased near it. On the other hand the high field region extends out much further from the point. Likewise, as the point diameter increases, the field near the cathode increases. That is, the field across the gap tends to greater uniformity but to lower value in the high field regions.

This has a rather important influence on the size of incoming avalanches and on the progress of the streamers. For in high fields such as at the tips of the points used, as X/p increases, α/p ceases to increase rapidly, and in fact for values of X/p in excess of 450 in air ionizing, the efficiency decreases. Thus the excessively high field near the points of very small radius are of relatively little value. Since they attenuate very rapidly, the length of the avalanche-creating zone is very short. In consequence, the $\int \alpha dx$ and the avalanche $e^{\int \alpha dx}$ become rather small. Thus the space charge of positive ions at the anode is small and the guiding field is short. Streamers are weak and do not progress very far into the gap. As tip radius r increases, the value of X/p at its tip becomes smaller, α approaches its peak value, and the field X_0 causing ionization extends farther into the gap. Thus avalanches rapidly increase in size, avalanche space charges at the anode increase, and more vigorous streamers are launched. They travel further because of this and also because the high field region extends out further into the gap to add its quota vectorially to streamer advance. But as this occurs it also results in a longer high field gap region normal to the anode surface at the axis, for the field around a point of 0.01 mm diameter becomes increasingly radial within r_0, the active ionizing zone.

Thus the guiding field is less radial and more axial as point diameter increases. The extreme is reached with uniform field geometry in which the guiding field is parallel to the axis. It has been assumed, with some experimental justification, as shown by Meek's[3.12,3.28] and later by Raether's[3.29] semi-empirical streamer equations, that breakdown streamer advance begins when the positive space charge tip field is about equal to the imposed anode or uniform gap field. If this is indeed true—and, as will be seen, Nasser's study seems to indicate it to be likely—then the direction of the external field vector will play a considerable role in determining the direction of streamer advance. This follows, since the vector sum of ion space charge field and external field determine the avalanche size. Thus the more axial the external field, the more vigorous the axially directed streamers will be relative to the more radial ones.

Branching of streamers was not originally expected to be very extensive except in very high radial fields. However, with the very radial field of the small ionic space charge and the fact that the photoelectrons are liberated at varying distances and random directions about the space charge, branching must be obvious, as seen in figure 3.46. If the space charge is strong— and since the direction from which ideally long photoelectron-initiated avalanche paths comes is a matter of pure chance—the probability of the streamer's continuing along a single straight path in the imposed field direction is small. Thus the more radial the external field and the higher it is, creating a more dense anode ion space charge, the more radial the streamer branches will be. Thus for small points with long gaps the purely radial character of the streamer branching will be as indicated by English's photographs in figure 3.20. As the point diameter increases and the intense field becomes more normal to the point surface and thus along the axis for some distance out, the more the streamers in that region will strive to be axial. If, in addition to this, one realizes that the exponential increase in n gives only a few branches close to the anode, the paintbrush-like pattern of the streamer glow from the larger points in Kip's and English's photographs is obvious (figures 3.4 and 3.48).

There is one more circumstance and this is that given a strong guiding field, the weaker radial branches will tend to branch and progress more in the field direction the stronger the external field. This tends to crowd the positive space charges of the branch tips inward toward the axis, still further increasing axial progress.

However, once the vigorous streamers advance into the regions sufficiently removed from the point so that the field of the point is both more radial and weaker (for example, several point diameters from the anode surface), then vigorous radial expansion will begin. It will be greater the higher the potential. The more vigorous the streamers, the farther they will progress into the gap without much attenuation, and the greater the branching. Increasing point diameter will thus in some measure reduce branching near the anode but enhance it further out by providing more vigorous streamers. Of these, the axial ones will be the strongest and will continue branching farther into the gap. The curves for n as a function of d should probably converge toward the anode, and the relative values of n should diverge toward the cathode as the tip radius increases.

It is important to know what will happen as more nearly uniform field gaps are used. This may prove more difficult to observe, as it involves higher potentials, shorter pulses, and much higher speeds. It would be expected, however, that in a uniform field, branching would occur much farther from the anode, and streamers would follow the field lines more closely while the total number of branches would decrease. Under those conditions anode fields are relatively low, initiating avalanches are long, and ion space charges are high. At all points along the path, especially as streamers weaken with increasing length l, the external field will largely aid the field-directed streamers. Thus chiefly those branches directed along the axis will reach the cathode. In any case there is no reason to believe that the branching will still not be fairly prolific, though not so much as for Nasser's gaps.

As will later be noted, streamer tips may cross the gap to the cathode long before the tip potentials are adequate to produce a return stroke and yield a spark. Consequently, sparks materialize only along that branch of a branched streamer process the tip of which reaches it with an adequate tip potential. Thus sparks as observed are usually along a single crooked or straight filamentary path and are only occasionally forked, that is, branched. However, observation is needed in the uniform field region.

In his earlier study Nasser had shown that all streamer tips do not strike the film surface at the same time. He noted this in using pulses of short variable duration, but it has also been noted in the way in which streamers arriving later are deflected to make their way between other streamers already arrived. The delayed propagation is most important and will be discussed at length later in this section in another connection.

Figure 3.69 shows the maximum range of the streamers propagated along the axis for a gap with a 1 mm diameter point as a function of the applied potential. In this study the gap length should not have been neglected. However, in the low field regions the potential near the cathode does not vary too much with gap length, since most of the energy going into the streamer is given it near the point and is field-dependent. The low field near the cathode may not exert much influence on streamer range. Since the length of the gap seems to be of less and less importance as the gap lengthens beyond some 8 cm, it is not surprising to note that the axial length of the streamers varies in a nearly linear fashion with the applied potential. This is seen to hold from about 9 cm to 20 cm. The earlier studies had shown that at around 12 kv the length of the axial streamers in the clean gap was of the order of 3 cm. Beyond this the streamers require an increase in potential to propagate further. At 22 kv the length is 7 cm, and at 28 kv it is 8 cm.

In his earlier study and to a lesser extent in his later work, Nasser studied the radial propagation of the streamers, measuring b as conditions varied. Figure 3.70 shows the radial range b of the distance x of the film from the anode. It is seen that the radial spread is small near the anode and increases to maximum values of 4, 5, and 6 cm for gap lengths d of 8, 6, and 7 cm, respectively. The region of maximum radial expansion lies around 3 cm. The use of b is not too significant quantitatively, as it depends on statistics and represents a sort of outer envelope which may depend on one or a few fortuitously long radial branches. To get accurate data, a large number of exposures at each x and d are needed. This procedure is more costly than the significance indicates.

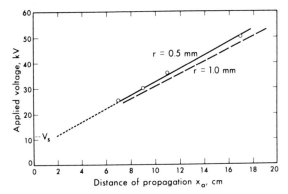

Fig. 3.69. Maximum axial range of streamers as a function of anode potential for a 0.5 and a 1 mm diameter point, to some extent ignoring gap length except that it must have been adequate.

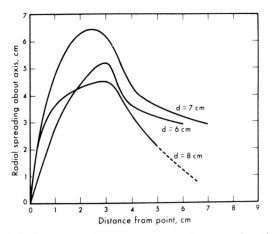

Fig. 3.70. Radial range b of streamers along film as a function of distance x from the anode for a 1 mm diameter point and different gap lengths d at 40.5 kv. The exact shape of the curves is not too significant, as counts were made on too few films and statistical fluctuations are greater in small counts.

In principle, these observations indicate that the envelope of the volume enclosing all streamers is roughly conical, subtending in an angle of some 150° at the anode and perhaps less than 120° at the cathode. Figure 3.71 shows values of b at the cathode for two potentials as gap length d increases. It is again noted that there is a peak value of b which ranges around 3.5 cm to 4 cm from the point as potential increases from 40.5 kv to 51.0 kv. The maximum value of b is about 4.5 cm and 5.7 cm for the same two potentials, respectively. It is seen that b declines markedly as gap length increases.

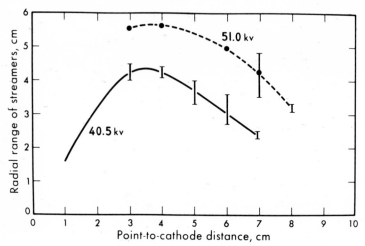

Fig. 3.71. Radial range b of streamers taken at the cathode as a function of gap length d for 40.5 kv and 51 kv.

Thus, one must conclude that at a certain distance from the anode the radial propagation b will reach its maximum value. This distance correlates closely with the value of x or d at which the number of streamer tips reaches its maximum value. It is clear that total length, comprising radial as well as axial extension of streamers, in some fashion causes attrition of streamers as it exceeds certain distances fixed by potential and gap geometry. This is seen and doubtless could be expressed quantitatively from data on the films. The farther a streamer system advances, the weaker and fewer its branches. The shorter the streamer branch, the lower its tip potential. Thus, as a streamer system advances, the branches grow shorter, then tip potentials fall, and finally they cease to propagate. This decrease can be attributed to branching and the consequent increased fall in potential along the channel with decreasing channel current and number of electrons. The attenuation of streamers before they cease to propagate is nicely seen in figure 3.72. These streamers belong to the same sequence as those of figure 3.65. The few spikes seen may even be large avalanches, as these figures are magnified.

Beyond the point where streamers strike the film, a blue-violet sensitive film shows diffuse blackening which presumably comes from the ultra-violet emitted by them. This light goes through quartz but not through glass and does not affect a red-sensitive emulsion. This characteristic will be discussed later.

Next of interest is a study of the tip potentials as the streamers cross a gap. These are given at once by determining l for the longest streamers as these cross the gap. Once the length of streamers on the film when it is

FIG. 3.72. Streamers just at the end of their range as they cease to propagate. The dots and short tracks could be caused by large avalanches. The figure has been considerably magnified.

measured for a series of known tip potentials with x = 0 has been determined, l gives an estimate of the tip potential as it arrives at the plane of the film at any distance x, accurate to about 10 per cent. The axial streamers yield the largest l and represent the important aspect of streamer advance. The earlier of Nasser's studies applied to gaps up to 5 cm. In this study gap lengths up to 8 cm were used.

Figure 3.73a indicates the results. At an anode potential of about 46 kv the tip potentials were determined as f(x) for gaps with a length d of 3, 5, 6, 7, and 8 cm. The value of l in centimeters is there plotted together with the corresponding potential. In assigning tip potentials to this plot, corrections to Nasser's data were applied, discussion of which follows. Nasser had calibrated the potential scale given by his pulser in Berkeley, using the static spark breakdown potential scale for his point-to-plane gap in the absence of the film, with a steady potential source, and with some comparison with impulse breakdown as a spot check. Exceedingly careful repetition of these measurements by Waidmann indicated that the impulse potential for spark breakdown of gaps corresponding to those used in Nasser's work was between 1.4 and 1.6 times higher than by static breakdown.

The potentials used in the measurements under discussion were impulse potentials, and the correct absolute values were essential. The question of calibration of these gaps proved not to be a simple one, since relative sparking potentials with static and impulse are sensitive functions of point radius and water vapor or O_2 content of the gas. The scale on the figure applies to the conditions used in Berkeley by Nasser. This meant that a radical correction had to be applied to Nasser's data as established on his O.N.R. report on the Berkeley work. It must further be recognized that with the film in the gap, the breakdown to a spark occurs at a potential materially higher than that for the gap without a film. This comes from the fact that the breakdown streamers must run over the surface of the film and to the plate, and that thus the breakdown path is considerably longer with the film than in its absence. Hence with the film, potentials well above empty gap spark breakdown can be applied before breakdown. With this in mind, it becomes reasonable that Nasser should have plotted the curves as shown in figure 3.73b.

Here the ratio of the tip potential V_R to the applied anode potential V_a is plotted as a function of the fraction of the gap x/d at each point. This places all data on a relative and comparative basis independent of an absolute scale of tip potentials. Curves 5, 4, 3, 2, and 1 in both figures 3.73a and 3.73b correspond to gap lengths 3, 5, 6, 7, and 8 cm, respectively.

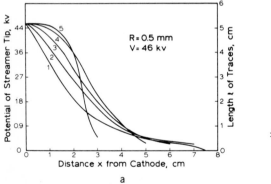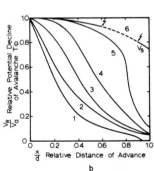

Fig. 3.73. (a) Streamer tip potentials evaluated from calibration of length l of longest streamer with corresponding lengths as a function of x for a 1 mm diameter point at 46 kv for 3, 5, 6, 7, and 8 cm gaps. (b) The same curves as in figure 3.73a, plus one at 2 cm at 1.37 times what the sparking potential would be without film and at breakdown potential with film, plotted in terms of ratios of tip potential V_R to anode potential V_a against the fraction of the gap x/d crossed by the tip. Curve 5 applies to the 3 cm gap, curve 4 to the 5 cm gap, curve 3 to the 6 cm gap, curve 2 to the 7 cm gap, and curve 1 to the 8 cm gap. Curve 6 is for the breakdown potential on a 2 cm gap.

Curve 6 in the latter figure is a curve taken for d = 2 cm and at a potential
1.37 times the static potential for that gap in the absence of the film. Here
one out of every few streamers gave a *spark* with the film present. The
potential acting represents a potential of about 41 kv on the impulse
potential scale. The normal breakdown potential for the gap without film
is 30 kv.

The tip potential of the 8 cm gap falls off almost linearly out to 2 cm,
where its potential is about 18 kv. It still has 2.2 kv at 6 cm, and it falls
to less than 1 kv at 7.5 cm. In the 3 cm gap the tip potential is within 10%
of the anode potential halfway across the gap. Beyond 0.8 of the gap length
it falls sharply. It has only 0.5 kv at the cathode. It will be noted that the
spark breakdown potential of this gap without film is 39 kv. With the film
its sparking potential is close to 53 kv.

Regarding figure 3.73b for the 8 cm gap, when the anode potential is
0.51 of the sparking potential measured without film, the streamer advances
to only 0.95 of the gap length. In contrast, when the applied potential in
the 3 cm gap is 1.5 times the sparking potential of the gap without film,
the tip potential at 0.8 of the gap length is 0.6 of the anode potential,
and 10% of it at the cathode. For the 2 cm gap at the sparkover for the
gap with film, the tip potential at the cathode is above 0.75 of the anode
potential, or on the order of 30 kv. Generally speaking, Nasser found that
streamers with tip potentials in excess of 25 kv yielded sparks on reaching
the cathode.

It would be very nice if an evaluation of the tip potential permitted
one to determine the field X at the streamer tip. All that the curves of
figures 3.73a and 3.73b tell is the *actual tip potential of the streamer* as it
flashes along its path at $\approx 10^8$ cm/sec. One may not infer gradients from
these curves. It is only when the potential just short of the cathode permits
one to evaluate the tip potential for a given V_a/V_s that one can estimate
field gradients within a few millimeters of the cathode. It may, however,
become possible, by measuring the lengths of the δ ray tracks in the
emulsion, to estimate the fields acting on the electrons producing them and
thus to evaluate the tip fields. This point merits study.

Figure 3.74a shows data for the number of streamers n at a given V, r,
d, and x whose length exceeds l for a given value of the length l in centimeters in the region of greatest spreading for a presumably relatively
shorter gap at an intermediate potential. The resulting histogram shows
that the number of streamers exceeding a length l decreases exponentially
as l increases. That is, $n = n_0 e^{l/L}$. Here L is the average streamer length
appropriate to x at which $l/L = 1$. This gives the rate of attenuation of
streamers at a given x, V, d, and r in air at 760 mm. It is to be noted that l
also represents the tip potentials of the branches. Thus attenuation of

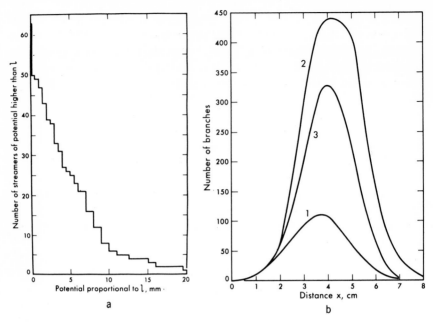

Fig. 3.74. (a) Histogram of the number of streamers n exceeding a length l on a film at some point x plotted against the length l and thus to a corresponding tip potential. This curve is typical of such curves and is of considerable theoretical significance, although the conditions under which this sample was taken are of no immediate concern. A length of 20 mm would correspond to a tip potential of about 18 kv. (b) Curves for n, the number of branches, as a function of distance x from the anode, computed by comparison of the theory with observed curves of n as a f(x) for different gaps. Curves 1 and 2 are computed assuming branching attenuation is inversely proportional to length. Curve 3 assumes an exponential decline of L with branch length, applying to equations 3.4 and 3.5 in the text, respectively.

branch streamer lengths and tip potentials is a statistically determined affair. That is, the greater the branching, the more rapidly the tip potential declines. This signifies that, if a streamer branches into two branches by statistical fluctuations of avalanche size and photoionization by the avalanche,[2.2] each branch will *on the average* have half the tip potential. Such a concept accounts for the rapid decline of branching noted beyond some 2 to 3 cm from the anode. It must be further noted that l projected onto the field axis is related to the variation of n with x, so that the decline of n with x should also be governed by the same equation. However, since all streamers do not have equal tip potentials when they arrive at a given x, the value for L is not really constant for all streamers at x. Since the acting fields are composed of the more radial space charge tip field and the stronger axial field of the anode, the guiding field near the anode is larger for the more axial streamers and less for the more radial ones. For this reason the decrease in radial expansion will be faster than the decrease in axial extension.

Next, it is of value to contrast this declining function with the branching of the streamers, which follows a law $n = n_0 e^{x/L_0}$ near the anode. This increase in branching is statistical for the same reason as that for the decline. At the anode, L_0 is fixed by the anode potential and ample photon production. It has been estimated for uniform field breakdown streamers that there are $\approx 10^8$ ions, 10^{11} to 10^{13} ions/cm³ and 10^4 photoelectrons created beyond r_0. As observed, the guiding field near the anode *appears* to keep streamer ionization, tip potential, and thus L_0 fairly constant for an x of some 2 to 3 cm. Thus the value of n increases exponentially as x increases in that region. Eventually a point x is reached at which the excessive branching and the declining guiding fields cannot maintain the tip potential and thus L_0 at approximately its previous value. Thus L at each x decreases and rate of attenuation by branching exceeds the rate of proliferation by branching.

The situation can best be described by an equation which symbolically represents the picture outlined. Branching is a chance affair and by observation, near the anode it follows a law of the form $n = n_0 e^{x/L_0}$.

As long as tip potential, ion concentration, and photoionization are maintained at nearly the same level, this rate of growth will be maintained. However, at a given value of x, the distribution of streamer lengths and streamer tip potentials appears to follow a law of the form $n = n_0 e^{-x/L}$. Here L is an average length appropriate to that value of x. However, had the film not diverted the streamers, the x components of the streamers would have followed the same law.[2] Thus it is reasonable to write that $n = n_0 e^{-x/L}$ for attenuation. That is, when streamer tips reach any given x, they have an average length L. This means that the tip potential is falling at a certain rate as x increases. This is especially notable after advance in the field and branching have effectively begun to reduce tip potentials and thus ionization and photon production. One can therefore express the rate of branching by the product of two functions, one of the form e^{x/L_0} and the other $e^{-x/L}$, accounting for proliferation and attenuation, both statistical and both due to branching. Thus as a first approximation, one may set $n = n_0 e\left(\frac{x}{L_0} - \frac{x}{L}\right)$. Here L itself is a function of x, which, as the Lichtenberg figures show, decreases with x. An extensive experimental study to

---

[2] It is possible that branching is more prolific because of the film action.

determine L as a function of x under constant V, r, and d has not to date been made and was not considered of value until the present analysis, though enough data do exist to substantiate both of the assumptions and in general to confirm the rough analysis to follow. Accordingly, one must guess at a form that is reasonable. Two forms are suggested. One is that L is inversely proportional to x, which is more or less empirical but not too illogical. The other is that, since branching is statistical, L falls off exponentially with x. Thus one may set either $L = x/L_c$ or else $L = L_c e^{-x/x_0}$. Hence one may write that:

$$(1) \quad \frac{n}{n_0} = \exp\left(\frac{x}{L_0} - \frac{x^2}{L_c}\right), \text{ or that}$$

(3.4)

$$(2) \quad \frac{n}{n_0} = \exp\left(\frac{x}{L_0} - \frac{x e^{-x/x_0}}{L_c}\right).$$

(3.5)

The constants L_0 and L_c or L_c and x_0 need to be defined. L_0 is readily interpreted as the average *branching free path*, or interval, at constant tip potential. It will depend on anode potential, the form of the field, and photoionization. It can be *estimated* from the *observed exponential increase* of n/n_0 near the anode. The quantity L_c is also a distance which depends on the potential at the streamer tip and represents the characteristic distance x at which, owing to *field conditions* and past branching, tip potential falls until branching effectively begins to reduce the average value of L. It can roughly be set as the distance at which the exponential increase of n/n_0 is observed to decrease markedly. The value of x_0 in the exponent for decrease in L is again a sort of average free path for the branching attenuation. It would be near the value of L_0 if tip potentials are high; beyond L_c it is larger as chance of branching is less. The influence of these factors on branching, yielding n as a function of x, assuming $n_0 = 1$, are given for the following values of L_0, L_c, and x_0 in the curves 1, 2, and 3 of figure 3.74b. Curves 1 and 2 use equation (1), and curve 3 uses equation (2).

Curve Number	L_0	L_c	x_0
1	0.4 cm	3 cm	—
2	0.35 cm	3 cm	—
3	0.4 cm	3 cm	3 cm

It is seen that the curves in general simulate observed curves and better agreement could be forced by a more careful choice of the constants, using more complete data available after the theory was formulated. Until a very exhaustive investigation is made, better fitting is futile in view of the comments to follow. The worst deviation of the curves from the observed curves actually lies in the sharp decline of the computed curves beyond the peak, which in the observed curves is less abrupt. It is here that the neglect of the chance statistical fluctuations in the early development of individual branches in streamers becomes important. Wijsman[2.2] and Legler[2.6] have demonstrated an exceptionally wide range of fluctuations in the product of the avalanche size and second Townsend coefficient. It is thus to be expected that individual early branches of an initial streamer may differ widely in tip potential. Thus in a streamer and its first- and second-generation progeny, L_0 and thus also L_c may differ widely. In consequence, the combination of the subsequent multiple branching will deviate from that based on calculations, assuming that average values of L_0 and L_c are sustained throughout. Such action will tend to broaden the peaks and extend the axial branches, particularly since all these are favored by field configuration. The observed patterns will thus usually follow the more exaggerated deviations, resulting in choice of the wrong values for L_0, L_c, and x_0 from observed data, with both L_0 and L_c probably over-estimated. Furthermore, until the influence of the guiding field and its accurate configurations are established, a bias is placed on angular distribution and attenuation factors relative to the axis which makes deductions from statistical observations alone hopelessly difficult.

From these studies Nasser inferred that the tip potentials at which streamers cease to propagate lie between 800 and 1200 volts for the gaps studied, and that the tip potentials for breakdown streamers at the cathode lie above 25 kv, as noted. The potential needed to launch a streamer 3 cm long into the gap lies around 12 kv near the point. For a 1 mm diameter point and an 8 cm gap in dry air, the potential is 8 kv at static streamer onset when the gap is clean. Here the avalanche needed consists of $\approx 10^6$ electrons. The streamer length photographed was probably not in excess of 7 mm. However, these were in a fouled gap. Nasser's 3 cm long streamers in an unfouled gap appeared at considerably above the static streamer threshold. In Nasser's study, the tip potential fell to negligible values of 0.8 to 1.2 kv if streamers stopped short of the cathode. In all cases where they reached the cathode, they had tip potentials very close to that value as the cathode was approached. These were on the order of 3500 to 7000 volts. As the axial streamers observed by the film technique increased in vigor toward those causing the spark on reaching the cathode, tip potentials reached nearly 36 kv, and sparked at 51 kv.

While this book was being written, a very remarkable Lichtenberg figure was obtained by Nasser which is apropos at this point. A gap with r = 0.5 mm and d = 4 cm in air was set up with the film at x = 1 cm. A static potential was applied which rose at the rate of 1 kv/sec. The film was wound past the point at the rate of 20 cm/sec across the gap. As any Lichtenberg figures appeared, the potential could be ascertained by the time scale of movement. The record is displayed on the film of figure 3.75.

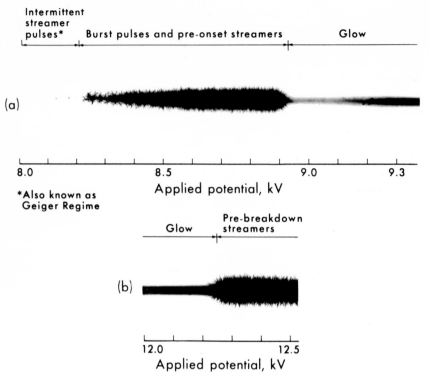

FIG. 3.75. Nasser's moving film trace of pre-onset streamer burst pulses, Hermstein glow corona, and pre-breakdown streamers in static breakdown. 1 mm diameter point, gap length d = 4 cm, film at x = 1 cm. Rise of potential 1 kv/sec. Film moves left to right at 20 cm/sec. The various phenomena observed are clearly labeled on the print.

At just 8 kv a few very faint streamers with two branches about 2 or 3 mm long at most appeared, having crossed the 1 cm and reached the film.[3] These of course were pre-onset streamers. They were intermittent in the Geiger counter regime. There was no marked darkening due to any extensive burst pulse corona at 1 cm from the point. At 8.2 kv the streamers followed one after the other and multiple branches appeared. Streamers were brighter and 4 to 5 mm long on the film, so that they had traversed a total length in air of at least 1.3 cm. Background darkening due to ultraviolet light begins to be noticeable at that point, and one can expect burst pulses to have been interspersed between streamers.

Between 8.2 and 8.3 kv, for example, the elapsed time is 0.1 sec. In this time quite a few streamers and burst pulses separated by a few milliseconds

---

[3] The first pre-onset streamers are not usually 1 cm long. Here undoubtedly the field of induced charge on the film lengthened the streamers.

could occur. That is, from 50 to 100 such events are possible. Actually some 20 streamer tips appear, of which some may be branches, so that there may have been 5 to 10 actual streamers. The diffuse dark background is probably due to the burst pulses. It is not caused by the streamers. Otherwise the white Clayden effect boundaries about the streamers would be noted. This effect will later be discussed. Had the blackening been simultaneous with the streamers, the Clayden effect would have been observed. As it is separated by milliseconds in time from the streamer, the effect is not noted. It is faint as the point is 1 cm distant. Both burst pulses and streamers augment up to 8.9 kv. By this time streamers appear to be about 1 cm long on the film or a total of more than 2 cm long in air, which corresponds to an anode potential of ≈ 9 kv. At this point burst pulses and streamers are more or less following within the dead time of 1 μsec or so. Abruptly the streamers cease as the Hermstein glow begins. This extends from 8.9 to about 12.2 kv. At this point the zone at which electrons detach at $X/p > 90$ has extended so far out into the gap that the sheath can no longer prevent streamers, and, in this short-length large-point-diameter gap, pre-breakdown streamers appear.

It is noted, however, that with impulse potential the streamers should extend 3 cm into the cleared gap at 12 kv. The streamers here noted are not much in excess of 1 cm long on the film. They are also much finer and narrower than those observed with impulse potential. Here, however, it will be noted that the film surface was 1 cm from the point. Thus the streamers are in reality more than 2 cm long at 12 kv. Nevertheless there is a difference, to be ascribed to the space charge between the film and the point and to the surface charge of the film caused by the ion current from the corona, which it intercepts but cannot quite conduct away as fast as it accumulates. With impulse potential such encumbrances are avoided.

However, it is clear that the individual streamers, since they cannot be photographed by their own light, will leave their autograph directly on the film even with static as well as impulse breakdown. This observation does much to establish the identity between the positive Lichtenberg figure and the streamer process, since they appear at the same potentials and are otherwise completely similar in behavior.

It is now essential to consider the nature of the Lichtenberg figures. Nasser in his original papers had correctly interpreted the positive Lichtenberg figure on the film to be a photograph of the track of the streamer tips propagating through the air adjacent to the film. He postulated that the foggy darkening of the film was caused by general ultraviolet radiation from the streamer branches. This light is quite strong but is feeble compared to the ultraviolet emission from the streamers observed by Tetzner in liquid breakdown. Some of these effects can be seen in figure 3.65, exposures a, b, and c, where tip spreading gives the diffuse light in the center background of each main streamer branch system. In the center of traces b and c it is seen that, surrounding the little central branches, the diffuse ultraviolet has been decreased so that each track is surrounded by a black or darker rim. This is an example of the Clayden effect shown by certain photographic emulsions. It is seen much more clearly on the negative, or in direct traces on an opaque film of the light of a more sensitive emulsion. There, all the weaker streamers were black against the unexposed negative. Where the film is clouded by diffuse ultraviolet, the black streamers are surrounded by light patches. This Clayden effect results when a region is exposed to light and then *very shortly* thereafter is again exposed to light. The ultraviolet arrives at the surface some nannoseconds or more before the streamer tip strikes the film and is deflected. This is sufficient to yield the Clayden effect. If exposures are separated by longer intervals, say microseconds, the effect does not appear. The phenomenon is very useful in that it helps accentuate the fainter tips against a cloudy background. It also serves as a time scale for events.

To explore the nature of the figures further, the following experiments were carried out. Figure 3.76 has on its left side a microscope slide of a soda glass 1.2 mm thick in contact with it, held by two thin transparent strips of clear cellophane adhesive tape at its ends. It is noted that those branches that were headed toward the glass passed over it. The light from the streamers still is faintly visible, but diffused and attenuated by the glass and its distance above the film. It is clear that the streamers, having passed over the glass, again come down to the film surface and continue in the air near it. Apparently, the streamers pass through air more readily than through solids. However, they follow the surface of the paper as a result of electrostatic inductive attraction. At one region the glass did not adhere closely to the film, and parts of the strong central pattern penetrated beneath it. The traces clearly show the Clayden effect. The original film shows the streamers through the glass more clearly than the print does. The streamer branches going over the thin tape are reproduced in all detail with little attenuation.

Figure 3.77 shows the streamers seen through a circular quartz disc 1.7 mm thick. Here, again, streamers passing over it are more diffuse than with the glass since the quartz is thicker. However, note the round dark spots which represent the points of impact of the streamer tips on the quartz before deflection. Here the ultraviolet from the approaching tips which is causing the Clayden effect in other traces is seen, and penetrates the quartz but not the glass. The ends of one or two faint streamer tips are seen unattenuated through the cellophane adhesive at the upper and lower edges of the plate. The fact that the streamers do not radiate largely in the ultraviolet as they flash by over the film was shown by the use of a thin microscope cover slide 0.1 mm thick. With this, the streamers were

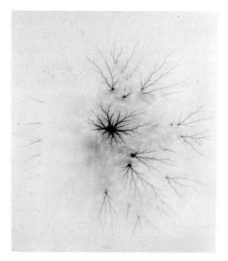

FIG. 3.76. The influence of a 1.2 mm thick glass plate on the Lichtenberg figure. The branches directed toward the glass pass over it and are seen as they drop down to the plate on the far side. The faint, blurred images of the streamers passing over the glass can be seen faintly through the glass. Because of a bulge in the film, the streamer in the center was able to get underneath the glass. At the top end of the glass, the streamers can be seen through the cellophane tape holding the plate.

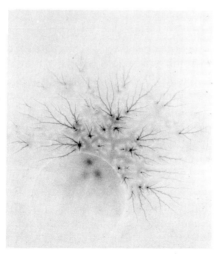

FIG. 3.77. Streamers are seen through a circular quartz disk 1.7 mm thick. Note that more ultraviolet light gets through this than through the glass. Streamers are blurred and diffuse as they are too far away from the film surface to have a clear trace.

reproduced so accurately that the edges of the glass could not even be discerned. Thus the light active here is in the photographic visible. However, the dark ultraviolet patches produced by the streamers seen end-on through thick quartz are absent with the thin glass. The pattern was reproduced nearly as well when the sensitive emulsion side of a transparent photographic film faced the cathode at the cathode surface as when it faced the anode, where it received the direct impact of the streamers. This indicates that the discharge did not occur in the film emulsion but in the gas above it, and that it is the field of the streamers that calls forth any discharges from the cathode, as will later be seen. Figure 3.78 shows the streamers as seen with the emulsion on the cathode side and thin glass cover slides separating the emulsion and the cathode. The edges of the glass can be seen as thin dark lines. These are not drawn in, but represent light from minute discharges from sharp points on the glass under the high electrical fields of the streamer pulse. It is seen that the streamers are recorded even through the glass.

This circumstance is of interest in that little light from the anode streamers penetrated the film, as seen by the faint smooth traces. Instead, the field induced pattern, if looked at with a lens, is *not* the smooth pattern observed with the emulsion on the anode side. It has a structure of grainy dots that outline the high field regions created by streamers. Here, the cathode surface through the image force field of the positive streamers gives a very heavy field across the emulsion. These cause electron liberation or accelerate electrons liberated by visible light in the emulsion. They show small δ ray tracks on the order of 0.05 mm long in all regions of anode streamer impact on the upper side of the film.[4]

Figure 3.79 shows the streamer traces taken with an emulsion insensitive to blue and green light but sensitive to the red. It lies on the cathode with emulsion facing the anode. The usual streamer patterns are barely perceptible, but present. Here one sees clearly the small field-produced avalanches in the emulsion, since they are not observed by the light of the streamer. Sixteen-fold magnification indicates the dots to be small, very tortuous tracks, of lengths on the order of 0.05 mm. They are not seen in the absence of very high fields produced by the metal cathode and the advancing streamer tip acting through the thickness of the film.

It was concluded that the light from the passing streamer tip recorded in the Lichtenberg figures as they branched and spread along the surface of the emulsion was confined largely to the blue and green in the visible. Some quartz ultraviolet was also present but was not prominent. As a final test of these conclusions, the figures were taken on color film, yielding the beautiful picture of the streamers with their striking electric blue color in figure 3.80. It was this color that was seen by the author that evening in

---

[4] The short, tortuous tracks caused by electrons from x rays or other radioactive radiation impact on photographic film were called δ rays.

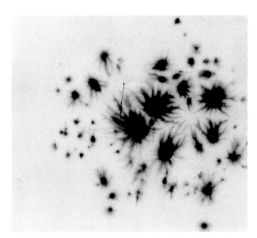

FIG. 3.78. Streamers as noted with the emulsion on the cathode side of the film with thin glass cover slides separating the emulsion side from direct cathode action. The edges of the glass cover slides are marked by sharp black traces. They come from minute point discharges from sharp edges of the glass created by the high field of the streamer tips on the anode side. The fact that streamers are delineated through the thick film, which is not transparent, is important. However, scrutiny of the figures shows that they are not smooth and black and sharp as on the anode side. They are composed of many minute points which on 16-fold magnification turn out to be δ ray tracks of fast electrons liberated in the emulsion and accelerated in fields of ≈ 25 kv or more in the emulsion from streamer tip potentials.

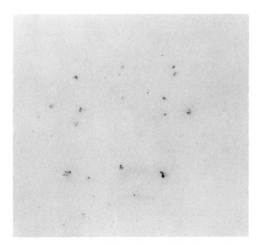

FIG. 3.79. Streamer traces on an infrared-sensitive film exposed with emulsion facing the anode. Here the streamer patterns are hardly discernible, but the δ ray dot tracks appear at the impact points of streamer tips. This indicates that most of the light from the streamer tips is in the blue end of the spectrum.

FIG. 3.80. See frontispiece. This shows that the conjecture stated with figure 3.79 is correct. One sees the brilliant "electric" blue of the streamer tracks such as first observed by Loeb in 1936. The blue color would appear in color film as light from the blue and violet end of the spectrum.

1936 when he first saw the pre-breakdown streamers from a needle point in air. The streamers flashed by from time to time out to a distance of 3 cm from a needle point. Normally the color of a discharge of the burst pulse, negative Trichel pulse, or corona or glow discharge in air is purple or reddish lavender, characteristic of the arc spectrum of N_2 and the bands of NO. The clear electric blue of the electrical spark, so beautifully seen in the color film, is characteristic of the *spark spectrum* of N_2 due to transitions from higher states of the excited N_2 molecules yielding the second positive group bands. These require high electrical fields or else high temperature arc conditions for their appearance.

Thus, in 1936 the author recognized that the streamers seen by him[3.3] represented fine high field regions projected out from the anode at high speeds. The discovery of copious photoionization in and by a corona discharge by Cravath in 1934,[2.25] and consideration of its role in the propagation by lightning strokes, led the author to the interpretation of the anode streamer given. At the 1937 meetings of the American Physical Society at Pasadena he correctly proposed that this was the mechanism leading to the filamentary spark.[3.3]

Nasser had made studies of the velocities of streamer tips by finding the distance to which the tips advance for a given pulse length by moving the film back and forth. Thus, a set of time-distance curves for the more axial streamer tips could be measured. They were remarkably high near the anode, with speeds ranging from 4×10^8 to 10^9 cm/sec, depending on point diameter and potential. As these same streamers advanced across the gap, they rapidly slowed down. In the very long asymmetric gaps, velocities fell down to about 4×10^6 to 10^7 cm/sec before they ceased to advance. In the case of the streamers in the 10 cm gap at 42 kv, the 8.7 cm long axial streamers were estimated to have an average speed of 1.2×10^8 cm/sec and crossed the gap in 5.8×10^{-8} sec. What was more significant was that, as the streamers advanced and branched, the tip potentials fell continuously with a corresponding decline in speed. Thus, the shorter branches advance slowly, and the radial branches advance more slowly than the axial ones. When tip fields are high and remain high enough to cause breakdown as they cross the gap, as shown for d = 3 cm in figure 3.74b, and as was observed as breakdown was approached for longer gaps, the speed of the axial streamers attenuates much less. In a uniform field gap, it is probable that the speed is more nearly constant across the whole gap. In fact, in such fields, although there is some attenuation of tip potential as the cathode is approached, the tip field plus the existing external field may in fact even increase because of the image force field of the positive streamer head in the metal cathode. Observations of velocities in this region require higher impulse fields of short durations than are at present available. However, the inference is borne out by Hudson's photo-multiplier studies.[5]

Hudson's primary streamers are probably to be identified with the most vigorous axial streamers which go straight across the gap with the highest speed of all. As Hudson worked at spark breakdown threshold in all cases, he was dealing with streamers that caused a spark almost every time one reached the cathode. Here the tip potentials and the fields in advance of the tips are such that the velocity could well be constant, as he observed. His velocities were 6×10^7 cm/sec and 7×10^7 cm/sec for a tip diameter of 1.24 cm and gap lengths of 1.5 and 2.5 cm length, respectively. Where the point diameter was 12.4 cm and the gap length 6.4 cm, the primary speed was 5.6×10^8 cm/sec. Actually, the primary streamer tip speeds were not amenable to accurate study over the more asymmetrical gaps, and observational points were not sufficiently numerous to obtain accurate values.

Hudson's observation that all the primary streamers appeared to arrive simultaneously at the same plane, despite branching, can now be explained.[6] Hudson and Amin's observations of secondary streamers are also amenable to interpretation on the basis of the Lichtenberg figure observations, as are the velocity measurements of Allibone and Meek[3.31] and Meek and Saxe.[3.33] Both Meek and Saxe's and Amin's photomultipliers and amplifying systems were entirely inadequate for noting the first faster axial streamer tips which Hudson's later more sensitive system could resolve. Actually, both Amin[3.23] and Meek and Saxe did get indications of the faster primaries, except that they were imperfectly resolved. Thus, once Hudson's 2.3 cm slit was well away from the anode, he got the light pulse of the rapidly advancing stronger axial streamers as the tips flashed by. At first, before they branched, the light was not intense. However, as they branched, largely in the axial direction, the light intensity increased and then remained nearly constant as the branches began to attenuate, since the tip fields increased as they sensed the cathode.

After the tip had passed some plane at x, the light emission continued perhaps some 10^{-8} sec or so and thereafter the channel was relatively, but

---

[5] Recent measurements completed by Waidmann[3.77] in this laboratory show that, as gaps shorten and potentials increase, the velocity across the gap approaches constancy and values are those of Hudson's primary streamers. The increase in speed at breakdown and the vigor of streamers as these approach the anode have been verified by Dawson.

[6] Recent still photographs of pre-breakdown or pulsed breakdown streamers by Kritzinger and by Dawson and Winn in the author's laboratory indicate that the more vigorous primary streamer branches are bent in the field direction so as to arrive nearly but not exactly contemporaneously at the cathode.

never completely, dark, as shown by the traces of figure 3.81 with a high gain. The heavier, more radial branch tips were also advancing. In midgap in a 6 cm long gap, the branching both numerically and radially is nearly at a maximum. However, both because of the angle of the trajectory and because radial external fields are small, these tips advance more slowly than the strong axial ones. Even then the tips of the faster heavily branched streamers began to reach the plane of Hudson's slit. The light intensity then started to rise again and reached a peak as the maximum number of the radially spread branch tips reached that plane. Then it declined. This gave Hudson a slowly rising second light maximum. Owing to the multiplication of tips, this was far brighter than the primary in the proximity of the anode; it also followed the primary pip within a short time in that region. However, as attenuation occurred in longer gaps, the secondary *peak* reaching the cathode region could be weak indeed, and was correspondingly delayed relative to the arrival of the primary peak.

Near the anode the decline of light from the primary was partly smothered by the arrival of the branch tips with little delay, so that it was not resolved. Thus, at only about one-fifth of the way across the gap was the primary really clearly separated. It was most clearly resolved near the anode.

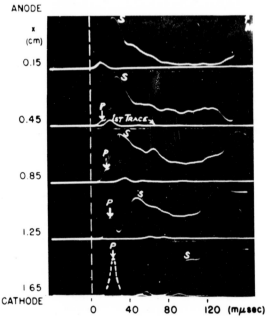

F IG. 3.81. Hudson's primary and secondary streamers at high sweep speed, showing that, when the primary tip arrived near the cathode, where primary and secondary were well separated temporally, the channel appeared dark to the PM.

A succession of primary streamers from the same source will vary relatively little. Since individual streamers show great variation in branching, *the secondary streamer will vary through wide limits,* as Hudson observed. Near the anode the secondary streamers will have high velocities, and these will remain high until the point of active branching is reached. From that point on, the shape of the light pulse as picked up by the slit will change, since the average speed and number of streamers arriving and dying out along the gap is continuously changing. Thus, the time-distance plot of the secondary peak will not show a constant shape but will change with the changing form of the sequence and intensity of the arriving delayed branch tips. Perhaps the beginning of the rise of the advancing secondary is the best reference point for estimating its velocity, since this will change less because it represents the faster of the delayed branch tips. This is seen by the time-distance curve of Hudson's primary-secondary minimum (fig. 3.61), where it is seen that its speed is more nearly constant. The secondary peak in a 6 cm gap advances fairly rapidly up to the point of maximum branching and thereafter progressively slows down.

Hudson's secondary tip speed for the 1.24 cm diameter electrode and 1.5 cm gap was initially 2×10^7 cm/sec in the first third of the gap but fell to 3×10^6 cm/sec in the last two thirds. For the more uniform gap of a 12.8 cm diameter sphere and a 6.4 cm gap, the initial secondary peak speed was high, 1.1×10^8 cm/sec but fell to 5.4×10^7 cm/sec in the last half of the gap. It is to be noted that the values of the speed and the change in speed of the secondaries across the gap are quite analogous to the values in Amin's data. They are also more nearly commensurate with Allibone and Meek's[3.31] and Meek and Saxe's[3.33] speeds. However, the latter used an anode with a rapidly rising potential during the streamer crossing. Thus the tips increased in speed as they crossed the gap. There the rate of rise of potential at the anode was more rapid than the decline in potential because of the increase of potential fall along the lengthening channel.

The presence of secondary streamers, as Hudson called them, may not be entirely a new mechanism.[7] As resolved by instruments of poorer resolution and light sensitivity, the secondary streamer is the only streamer noted, as with Meek, or else it appears as a separate light pulse following the primary streamer crossing as with Hudson. It is, in effect, a phenomenon inherent in limitations on instrumentation. It is partly caused by the sort of action Hudson had looked for when he discovered branching, that is, the delayed arrival of radially branching streamer tips. Hudson,

---

[7] Hudson's secondary streamer has been confirmed by Dawson and Winn (Appendix IV). It is identical with Kritzinger's leaders (Appendix I). The later arrival of branch tips superposes on this and on return strokes, complicating observation and interpretation.

in his observation of the primary streamers, could detect only the faster axial streamer tips. As stated, his curves for these did not permit accurate evaluation of their velocity or any changes other than order of magnitude. He could not clearly discern the tips of weaker and slower secondary streamers until they arrived in mass. This mass arrival gave him the temporally broad secondary light phenomenon which he unavoidably interpreted as a second, or separate, luminous pulse. Thus, his conjecture of delayed radial branch arrival at a plane, particularly of the many small and more radial branches, was correct, but his apparatus was inadequate for their detection and discrimination.

With increasing gap uniformity, as indicated, it was expected that the radial branching near the anode would decrease but that more axially directed branches of fairly high intensity and velocity would develop toward the cathode. Thus, as in figure 3.54 and in the crossplot of figure 3.59, there is little time difference, some 20 to 40 nannosec, between the arrival of the primary and secondary at the cathode. At 15 nannosec on the crossplot, there is continuous luminosity between the primary and secondary peak. The rise in intensity between the arrival of the primary peak at or near the cathode at 15 nannosec and the trace at 20 nannosec is to be noted and is most significant. Contrast this with the less uniform field geometry of figure 3.53. Here the primary and secondary pulses are clearly resolved.

Again, note the rise in intensity of the secondary in 10 nannosec after arrival of the primary peak at the cathode. For the most uniform field at which good resolution was possible, with a 30 cm diameter sphere opposite a plane shown in figure 3.54, and in the crossplot of figure 3.59, the primary, secondary, and rise of the main arc are so closely related in time that the primary can be observed only within 0.5 cm of the cathode, and then only at very high amplification, riding upon the rise of the secondary into which its peak merges. The secondary pulse here is not seen until it is some 0.6 cm from the anode. Even here the gain of the photomultiplier is such that the main arc is overloaded. Figure 3.55, with gain reduced for the same gap, shows the shape of the intensity of the transient arc as a function of time. Here the secondary can barely be discerned at 1.75 cm; it becomes prominent at 4.13 cm and shows traces of the primary at 4.92 cm with little decline in luminosity between primary and secondary. Thus, the interpretation of Hudson's curves, in terms of arrival of delayed branch tips and their increased luminosity as branches that proliferate and arrive at the slit, seems quite reasonable, unless they are related to Kritzinger's leaders (see Appendixes I and IV) in the short gaps used by Hudson.

It is next of importance to consider why, when streamers cross the gap well below spark breakdown threshold, they do not cause a spark at that lower anode potential. It will be recalled that Kip's first study of streamers indicated that in short gaps (≈ 1 cm), when pre-onset streamers crossed the gap, a spark ensued. It will be recalled that the author and Meek had postulated that the arrival of a streamer tip at the cathode led to intensification of ionization in the streamer channel which allowed the spark to develop. This intensification appeared as a light pulse or return stroke which had been verified photographically and by photomultiplier in gaps down to 20 cm. It could not directly be resolved by Hudson. Hudson himself was skeptical of its existence, and Raether had definitely stated that it was not necessary, since the conductivity of the streamer channel once it bridged the gap should lead to the growth of the arc directly. The author invites the attention of the reader to the sudden increase in luminosity along the gap in two or three of the crossplots of Hudson as soon as the peak of the primary streamer reached the anode. This could have been caused by an ionizing potential space wave, that is, a return stroke sweeping up the streamer from cathode to anode at velocities of the order of 10^9 cm/sec.

It is now essential to examine the nature of and circumstances leading to the creation of such space wave return strokes. The physics of these return strokes has not been investigated under normal conditions for the obvious reason that time resolution is not adequate for their study. Observationally they represent a luminous pulse that moves along the pre-ionized channel at a high speed. Since the pulse approaches the speed of light and occurs in fairly dense gases, one cannot assume that it represents ionizing particles such as fast electrons, and certainly not ones associated with ordinary discharges. The role of proposed runaway electrons has repeatedly been invoked but rejected because of the high pressures. The only things these light pulses can represent are rapidly propagating waves or regions of high potential with steep potential gradients that sweep along through the act of exciting luminosity and ionizing as they go, propagating by the field distortion that the ionization carries with it. They have been called potential space waves of ionization by the author.[3.46]

More information on the character of these waves has come from a phenomenon first noted by J. J. Thomson.[3.47] When a high impulse potential is applied to a long discharge tube at low pressures, the luminous discharge propagates down the tube from cathode to anode and back again at high velocities. The phenomenon was studied in some detail by Snoddy, Dieterich, and Beams[3.48] over several years. More recently it has been studied by Westberg[3.49] in the author's laboratory with the benefit of improved techniques and the discovery of a convenient generating device. A 150 cm long glow discharge tube with Al disc electrodes coated with the everpresent layers of oxide was exposed to a glow discharge in the normal mode at some few milliamperes' current. The gas fillings used were air, O_2, N_2, He, A, and other convenient gases. The cathode fall lay in the range of 1.5 to 2.5 kv.

When the potential was raised suddenly to operate the tube in the abnormal mode at some 20 mA of current, a gradual cleanup of the oxide layer at the cathode took place. When the oxide film was sputtered off down to a thickness of some 10^{-5} cm at some small spot on the cathode, a breakdown of the dielectric occurred through a complex field emission process suggested by Jacobs.[1.13] This suddenly liberated such a quantity of electrons that the current, as measured through a thyrite resistor of a few ohms *at the cathode*, rose from 20 mA to about 12 A in 2×10^{-9} sec. The release of such a swarm of electrons, which were accelerated nearly freely through the cathode fall, launched a front of potential distortion, backed by a region of much increased conductivity between it and the cathode, which progressed at the speed of the electrons propagated through the cathode fall *or* faster if photoionization in the gas occurred. Accompanying the burst of current from the cathode, there was a brilliant flash of light which propagated with the front of the ionizing electron cloud down the tube.

By means of a signal photomultiplier placed at various distances from the cathode and triggered either by the rise of the electrical pulse or by a triggering photomultiplier at the cathode, the progress of these luminous ionizing potential space waves was observed. By the use of probes placed at various points along the tube, the arrival of the luminosity at any point was observed to be coincident with the rapid rise of negative potential accompanying the ionization and excitation. No current change was noted at the anode until the luminous pulse reached it. This cathode wave multiplied the pre-existing ionization in the column of the discharge tube by a considerable factor, but not enough to yield a transient arc. The reason for this was that, although the fields X at the front of the wave and the ratio X/p were quite high, the passage of the front across one or two centimeters of column was so rapid as to allow time for only a few exciting and ionizing impacts before the wave passed on.

Thus arrived at the anode, the potential gradient created by the cathode wave was still sufficiently great to lead to a new, equally bright, reflected pulse or wave of potential and luminosity which swept back from anode to cathode. The speed of the return wave was of the same order of magnitude as for the forward one and the luminosity was, if anything, greater. This seemed surprising, since the potential gradients X and X/p were markedly less for the return wave than for the cathode wave. Speeds up to 6×10^8 cm/sec were observed, and in O_2 when photoelectric ionization occurred along the whole length of the tube, the speed went up to 10^9 cm/sec. It appears that the speed of the pulse depends on potential gradient and the degree of ionization in the column. The higher either of these, the greater the speed. Photoelectric ionization in the gap aids the process and increases the speed. On arrival of the second or anode wave at the Crookes dark space near the cathode, the whole column was sufficiently ionized to grow in current intensity to the arc stage by the ordinary diffusive ionization processes.

These observations led the author to the concept of the ionizing potential space waves indicated above. These will occur whenever there is suddenly created in a gas such a high degree of ionization that its disturbance of the potential distribution in the gap cannot dissipate rapidly enough by normal diffusive and conductive processes, but must propagate as a rapidly moving pulse or front of ionization and excitation with high potential gradient. Thus, actually the *cathode- and anode-directed streamers* created at the anode or in midgap by the dense space charges of very extensive avalanches belong to this class of phenomena. In this case photoionization in the gas, especially from the positive streamer, is the agency that gives the higher speed. The anode-directed negative streamer does not initially progress as rapidly. At very high overvoltages in uniform fields, the speed of the negative streamer is very high, as was recently shown by Nasser.[2.19,3.44] The degree of ionization left behind the streamer tip is not high enough to lead to an arc.

It thus stands to reason that, when in a long path discharge a positive streamer tip of high potential approaches the cathode, together with the burst of photoionization from the cathode when the active photons come into range of the cathode, a burst of intense space charge distorting ionization will occur. This burst is very similar to Westberg's cathode waves on emission of electrons from the cathode oxide layer. In fact, both visually and photographically one observes the flashes of light when breakdown streamer tips reach the cathode, as seen in Trichel, Kip, and Hudson's photographs of breakdown streamers. These flashes must launch potential space waves of ionization up the streamer channel. Since ion densities as judged from streamer currents are on the order of 10^{11} ions per cm³, the velocity of the return stroke or ionizing space waves of potential up the streamer channel to the anode must be on the order of 10^9 cm/sec. Even in the long sparks of Allibone and Meek, the return strokes had speeds in *excess* of 3×10^8 cm/sec and were not resolved by their technique. Similar phenomena occur when the positive streamer from the ground, which advances some 10 m from the earth in the usual lightning leader stroke, meets the field of the negative stepped leader. Return stroke velocities of up to 10^{10} cm/sec have been recorded.

Thus, one may consider the streamers, both anode or positive and negative, to be prime examples of potential space waves of ionization triggered by the sudden creation of intense space charges and advancing into unionized air by virtue of ultraviolet photoionization. Here again, owing to their speed and attenuation by branching, the space waves do not have a sufficiently conducting channel to initiate the transient arc called the spark. However, *if the space waves carry sufficient potential at their tips such that tip fields at the electrodes suffice to multiply existing carriers to a density great enough to propagate a new potential space wave, called the return stroke,* the arc may develop. In these streamer channels the ion density is already fairly high. However, the cloud of photoelectrons liberated from the cathode by the strong ultraviolet of the streamer tip at less than 1 mm and the fields of the order of that to be expected from tip potentials can well launch such waves.

It has been suggested by Cravath and the author[3.78] that the speed of the return stroke, v_S, can be determined from D, the distance over which the potential gradients are active, the distance $1/n^{1/3}$ between electrons in the channel, where n is the electron density, and the average speed of the electrons $\bar{v}$ in D. This has been adopted by Schonland in connection with dart leaders and return strokes in lightning.[3.79] The speed is thus $v_S = D\bar{v}n^{1/3}$.

One may estimate the velocity of the return stroke on the following assumptions. At 1 mm from the cathode a breakdown streamer has a tip potential of 25 kv. The field between it and the cathode is of the order of 2.5×10^5 volts/cm, $X/p = 300$, and the static value of α is of the order of 3000. Thus, even if Tholl's correction of a factor of 0.5 is applied to α for avalanche formation, the value of $e^{\alpha d}$ would be e^{166}, an astronomical value. One must then assume that each photoelectron liberated from the cathode could grow to about 10^9 electrons short of 1 mm and before a negative streamer started from the cathode. The 10^4 photoelectrons from the positive streamer which reach the cathode could liberate about 200 photoelectrons from an area of 1 mm². This would produce in roughly 1 mm² of area of the cathode 2×10^{11} electrons and ions. The field at the surface of a sphere of 1 mm radius with 2×10^{11} electrons would be that number times the electronic charge divided by 10^{-2}, the area of liberation. This field is 1000 esu/cm or 3×10^5 volts/cm. While the field of the streamer tip of 0.05 mm relative to the cathode is high, 2.5×10^5, its range is short. The range of the field produced by a volume charge of 2×10^{11} electrons spread over a square millimeter of surface and produced by that electronic charge at the tip of the streamer channel creates a negative space charge field of electrons that can reach over a millimeter up the advancing streamer channel. The field produced by the 2×10^{11} electrons released from the cathode then neutralizes the tip field of the streamer and acts on the 10^{11} electrons per cubic centimeter in the streamer tip and channel to accelerate them toward the anode. Before the burst of electrons these streamer channel electrons were bound in the streamer plasma of positive ions. If the field of the photoelectrically produced electron cloud gives an electron velocity $\bar{v}$ of about 10^7 cm/sec in the channel and extends 0.2 mm up the channel, the velocities of the return stroke will be $v_S = 0.02 \times 10^7 \times 5.8 \times 10^3 = 1.2 \times 10^9$ cm/sec. It is probably much less and could be 5×10^8 cm/sec, as observed by Kritzinger (see Appendix I). The numbers here used give some concept of the cause for the bright flashes at the cathode and the nature of the return stroke mechanism.

Thus it is clear that the return stroke and the materialization of the spark depend critically on the streamer tip potentials reaching the cathode. If there is a lower limit to this for return stroke formation, it is quite clear why, as Nasser observed, streamers cross the gap at potentials well below sparking, and why it is that tip potentials above 25 kv at the cathode are needed.

The extensive information in Appendixes I and IV indicates that Kritzinger's study has demonstrated the existence of a succession of ionizing space waves of potential needed to get the luminous leader stroke of longer duration across the gap. His data extend down to gaps of 10 cm. Some of his photographs of pulsed discharges in the breakdown region show evidence of return strokes up some streamer branches that propagate only partway to the anode. The photographs of the Goldman group in Paris (see Appendix I) show definite evidence that the return stroke starts when streamers first cross the gap but progress only part way up the channel until potential is raised sufficiently high to yield breakdown. Dawson's observation with sparks suppressed by use of low capacities in the author's laboratory has shown that the flash at the cathode marks the arrival of the primary streamer tip. Winn's photographs in the author's laboratory of streamers in N_2 with a small percentage of O_2, with bare cathode and with cathode covered by a film, give evidence of streamer branches in which the return stroke does not reach the anode. These observations account for streamers that cross the gap and yield a cathode flash but do not yield a spark, though return strokes may traverse part of the gap. Whether, with gaps of 3 to 5 cm, more than one return stroke to the anode is needed for spark is not yet determined.

Nasser has investigated the matter, using his Lichtenberg figures. The first evidence in this direction came from the figures observed on films on the cathode as tip potentials increased, as well as on films even 1 cm above the cathode in the gap. Among the films used was an x ray film with emulsion on both sides. Here, as anode potential increased to within 80% of the spark threshold, there appeared evidence of very strong *cathode effects*. This is illustrated by such an emulsion placed on a cathode with anode potential at 51.0 kv. At a gap length d of 7 cm, the streamer tips just reached the film with the pattern shown in figure 3.82a. At 6 cm gap length, well-developed positive streamer patterns appeared, as in figure 3.82b. At d = 4

cm an entirely new pattern of regular stars appeared, as in figure 3.82c, and
at 3 cm many more stars appeared, among them a very large one and three
somewhat smaller ones as in figure 3.82d. It is quite probable that the
large star would have given a return stroke and spark had the film not been
there. These stars appear on the cathode side of the emulsions, as will be

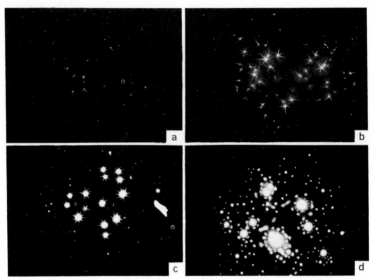

FIG. 3.82. Photographs of figures on a double emulsion film placed on the cathode
for a 1 mm diameter anode, 7 cm gap at 51 kv. The streamer tips as shown on *a*
just reached the cathode. At a gap length of 6 cm the well-developed positive
streamers are seen in b. At d = 4 cm what look like stars appeared. These are
due to the streamers, with the negative or cathode streamer type pattern revealed
on the viewing cathode side of the emulsion. Photograph d, taken with the cathode
at 3 cm, shows on the cathode side many such patterns with one especially large
one. This could have led to a return stroke in the absence of the film.

seen. Next, using the red-sensitive film with a thick paper back (Agfa
silver bromide) with emulsion toward the cathode and adhering closely
to it, Nasser obtained figure 3.83a. Here are noted the tiny dots presumably
produced by electron avalanches liberated and created by high fields *in
the emulsion*. Some sort of faint negative streamer with Clayden effect may
be discerned as well.

This same film was then separated from the cathode by a short air gap.
The spots created by the high fields of the positive streamers are seen in
figure 3.83b, but in addition, characteristic feathery negative streamer
patterns are seen. Using a very fast Kodak Tri-X film with emulsion toward

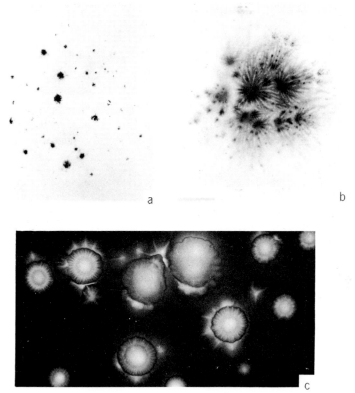

Fɪɢ. 3.83. (a) This photograph was obtained with a red-sensitive film toward the cathode with heavy paper backing. Here the high tip fields of streamers striking the back of the film on the cathode produced the δ ray tracks in the emulsion. A faint negative streamer Clayden effect is seen about them. (b) This was produced on an ordinary paper film sensitive to the visible, with emulsion facing the cathode and placed a short distance above the cathode. Here one sees faintly the positive streamers through the paper with δ rays in the emulsion along the tracks. In addition, one sees the typical feathery negative streamer pattern from the cathode. (c) This is an enlarged photograph with a Kodak Tri-X film double emulsion at some distance from the cathode. The regular positive streamer patterns as seen through the emulsion are blurred and give the branched stems. The circular blobs are a characteristic form of the negative streamer pattern.

the cathode but with an air space, Nasser obtained a typical positive print of a negative Lichtenberg streamer pattern of uniformly round shape with regular fine spokes, surrounded by a dark Clayden effect ring, as seen in figure 3.83c, which is enlarged about three-fold. Since the double emulsion was used, the positive streamer tip pattern on the anode side is seen in the blurred irregular lines projecting beyond the negative rims.

Figure 3.84a and b in positive and negative form were taken using double emulsion film at 1 cm above the cathode. The *a* figures are at potentials leading to return strokes and sparks. The b pictures are below return stroke potentials. Both the b pictures show the normal positive streamer pattern from the impact of anode streamers on the anode side of the emulsion. The *a* pictures show one very heavy negative streamer from the cathode side in the upper print and the typical positive streamers on the anode side in the lower print. Had the film not interrupted the continuity of channel, the heavy negative streamer would have continued as the return stroke and the spark would have materialized.

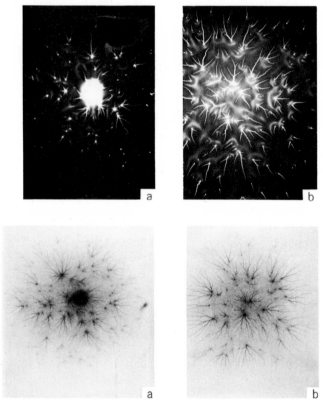

Fɪɢ. 3.84. Two sets of photographs, one set photographic positive, the other photographic negative, on double emulsion film at 1 cm above the cathode. The ones in set *a* show a negative streamer in the center; the others, in set b, under the same conditions but at lower potential where a negative streamer did not occur. The return streamer in *a* in the absence of the film would have caused a spark.

Fig. 3.85. Color photograph (reproduced in black and white) with color *film emulsion facing the cathode* but close to and above it, allowing negative streamers to develop. The positive streamers on the anode side appeared red, and the negative streamers from the cathode side appeared blue with more white than do the positive streamers seen in figure 3.80.

As a further test, a color film was used on the cathode. The positive streamers struck the back of the transparent color film, while the negative streamers impinged on the normal emulsion side exposed in photography. The result is seen in figure 3.85. This color film carries the red-sensitive layer at the center of the emulsion and is reached by the direct light from the positive streamers without penetrating the yellow layer between the red-sensitive emulsion and the blue-sensitive emulsion facing the cathode. Therefore only the red and yellow light from the positive streamer tips records as red anode streamer patterns, and the positive streamers appear faint and reddish yellow. The blue light of the negative pattern from the cathode, however, registers as blue and does not permit the red to penetrate. The light from some of the negative streamers was light blue and some was nearly white, indicating the intense ionization, and in this sense differing from the anode streamers. If the film was moved away from the cathode, the negative streamers declined both in number and intensity. Beyond 1 cm they ceased to reach the film.

Some further study was made of the high field effects on the photographic emulsion when the film was placed with emulsion toward the cathode and heavy streamers struck the anode side. For this the red-sensitive Agfa-silver bromide film was used. Figure 3.78 shows the pattern of small dots that appeared along the regions where positive streamers struck the backing and produced high fields at the cathode, for the pattern is obviously that of the positive streamers but consists of a mass of minute black dots and is not smoothly darkened as in the direct trace of the positive streamers. The pattern is not affected by the microscope cover slides facing the cathode, the edges of which are delineated by the faint dots. These came from discharges from sharp edges of the glass cover slide under the high field.

If now a piece of black paper is placed on the anode side, the light does not penetrate, but the increased thickness reduces the intensity. If next a thin metal foil is laid on top of the film backing on the anode side, the area covered by the metal is quite devoid of any dots or patterns. The reason is obvious; the capacity of the metal condenser is too large to create much field across the film with the cathode surface. The metal foil acts as a Faraday shield. If now a piece of cellophane tape is placed across the back of the metal plate, the patterns of streamers composed of spots on the emulsion side are much enhanced, since the cellophane binds the charge and localizes it. The localized charge acts inductively through the metal and the cathode to enhance the field locally. One may thus conclude that on the cathode surface there is a field effect on the emulsion causing the point pattern which follows the streamer fields. However, an air gap may permit electrons from the cathode, accelerated by the high tip fields, to form a negative avalanche and, if fields are adequate, a negative streamer. This is nicely shown by figure 3.86a (single emulsion, toward the cathode) and b (double emulsion) and by figures 3.83c and 3.85.

The answer to the question of what would happen near sparking threshold, when the negative streamer starting from the cathode to launch a return stroke finds a continuous channel to the anode to complete the spark, was found by accident by Nasser. A streamer system passed by the edge of a film which was not accurately centered in the gap, the film lying in a midgap plane with the emulsion side facing the anode. Those parts of

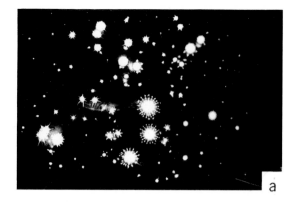

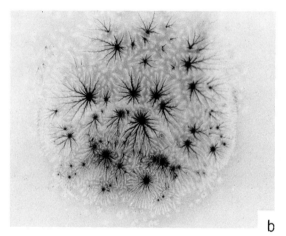

FIG. 3.86. (a) A positive print of negative streamers on the cathode just below spark breakdown, using a single emulsion film facing the cathode. (b) Negative on a double emulsion film at the same position, showing the streamers. The strong photon production gives a marked Clayden effect.

the very intense streamer system just below breakdown which struck the film distant from the edge gave the normal patterns of figure 3.87 at x = 0 and x = 5 mm. In both, the branches that just grazed the film and reached the cathode are enhanced by the return stroke. The two branches at the edge, exhibiting the return stroke, can be seen to be very much intensified. This was observed on several occasions; however, the matter needs further investigation. Waidmann obtained a more conclusive proof, to be related in the next section.

FIG. 3.87. Photographs of two anode streamers that struck the edge of the film near breakdown. Photograph *a* was at the anode. Photograph b was 5 mm from the anode. Note the intense illumination of the branches that reached the cathode. This was presumably created by the return stroke.

9. THE INFLUENCE OF THE CONCENTRATION OF O₂ AND H₂O ON STREAMER ADVANCE; THE WORK OF WAIDMANN

In all considerations of the mechanism of the streamer spark, it has been recognized that both negative, or cathode, and positive, or anode, streamer processes depend on a photoelectric ionization of the gas ahead of the streamer tip for their advance. However, for many reasons most investigators have emphasized the role of the space charge accumulations at the avalanche head as the critical factor. The one attempt at a complete quantitative formulation, both of burst pulse threshold and of streamer advance, has taken account of the photoionization in the gas in furnishing the regenerative counterpart of the γ of the Townsend theory in satisfying the self-sustaining threshold condition.[2.31,2.32]

In this phase of the theory, the reciprocal of the photoionizing absorption coefficient μ_i, that is, the photoionizing free path $\lambda_i = 1/\mu_i$ relative to the range d_0 of the effective ionizing field for avalanche formation about the avalanche head, was recognized as critical in setting the threshold. For nearly fourteen years it was not possible to do much directly to investigate this aspect of streamer advance, because of inadequate techniques. There have been some excellent studies of absorption coefficients in the extreme ultraviolet for the various gases—some of which might be responsible for these effects—by Weissler, Watanabe, and others.[3.50] There have also been numerous studies of the absorption coefficients of the photoionizing radiations in O_2, N_2, H_2, and air, in some cases including cross sections, beginning with the work of Cravath and Dechéne in 1934.[2.25,2.26,2.27] The striking feature about all these studies is the great range of values of λ_i observed extending from 10^{-3} cm^{-1} up to around 1 cm^{-1} at standard density. This discordance of values estimated by the investigators using various sources of photons or inferred from breakdown thresholds, though it unquestionably yields perfectly valid data, has not been useful in the analysis of streamer threshold theory.

Nasser, as indicated, has proved that the Lichtenberg figures from impulse potentials on positive and negative points are photographs of traces of the positive and negative streamers in the gas deflected by the photographic plate and continuing in the gas close to the surface. This observation has led to a new and very powerful technique for studying streamer propagation in gases which are undistorted by ionic space charges. The studies in particular established the presence of streamers for a large range of impulse potentials below that in which the streamers cross the gap with such vigor as to produce a spark.

As anode potential increases from the threshold potential for streamer advance, the streamers move farther and farther into the gap, increase in average speed, and show extensive radial branching. The same applies to the cathode streamer, but its range into the gap is less than for the positive point and its branching is at least an order of magnitude smaller.

During the author's review of the literature on point-to-plane corona, it occurred to him that, since the radial spreading of the anode streamers was a chance affair, depending on the field intensity distribution at the avalanche head and on chance photoionization in the gas, the spreading could be altered by changing the concentration of the photoionizable gas in a mixture. Since O_2 was known to be the photoionized gas in air, he thought it worthwhile to investigate the effect of altering oxygen concentration in N_2–O_2 mixtures on the streamer propagation, using Nasser's

Lichtenberg figure technique at atmospheric pressure. To achieve this the author suggested that the anode hemispherically capped cylindrical point be surrounded by a glass cylinder of large diameter, closed at one end except for the hole that admitted the point, and extending from behind the point to within a few millimeters of the photographic emulsion, as shown in figure 3.88. Through this cylinder there slowly flowed a mixture of air and O_2 from the large reservoir, with the composition of the gas controlled to about 2% by means of controlled flow meters on the tanks of O_2 and N_2 used. Further side studies were made with dry air and air with 100% saturated water vapor at 22° C. The first results obtained by increasing the oxygen content to perhaps double that in air were so dramatic that the full program leading to the results here presented was carried out. While these studies were in progress, Tamura[2,33] sent Loeb a pre-print of a paper in which he had adapted the Loeb equation 2.35 of 1948 for burst pulse corona threshold in air and various gases to the evaluation of the absorption coefficient and the photoionizing efficiency of the photons. In this study it clearly appeared that threshold was governed by both μ_i and the length of the critical avalanche-producing region d_0 near the point, as well as the tip field of this point. In short, the active photon wave length varied with corona point diameter. This led to an extension of Waidmann's study to different point radii.

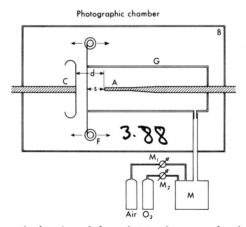

FIG. 3.88. Schematic drawing of the point-to-plane gap showing the Waidmann technique using gases of controlled composition through a flow technique at atmospheric pressure. G is the glass cylinder, M the mixing chamber, and M_1 and M_2 the flow meters.

Figure 3.88 shows experimental arrangements with A, the anode point; C, the cathode plane; G, the controlled gaseous mixture cylinder; F, the photographic film; M, the mixing flask; and M_1 and M_2, the two flow

meters. The whole was mounted in a lighttight box, B. Gases used were
water-pumped air and water-pumped O_2 from the Liquid Carbonic Com-
pany, with relative humidity 28% at 22° C. Experimentally, since the
studies were made with the camera in air at atmospheric pressure, the O_2
concentration was altered by adding O_2 or N_2 to air. Details of the Lichtenberg figure technique can be found in Nasser's papers. Here the impulse
potential had a rise time of $\ll 0.1$ μsec and lasted ≈ 5 msec. Impulse
potentials were determined in two ways. To complete the study, a set of
breakdown threshold values for the streamer-spark transition with static
and impulse potentials for the various concentrations studied are shown in
figure 3.89. Of these more will be said later in this section.

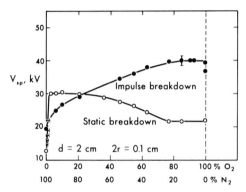

Fig. 3.89. Breakdown potentials in kilovolts for static and impulse spark breakdown of a corona gap at atmospheric pressure as a function of the percentage of
oxygen in an oxygen-nitrogen mixture. Note the rather remarkable behavior of
the values with the two types of potential application. Note that in the dry air
used, the two potentials were nearly equal, static being 10% lower than impulse.
In the room air normally used, static values were 30% lower than here.

The investigations involved the taking of data from the Lichtenberg
figures obtained under the different conditions to be given. Figure 3.90,
traces 1, 2, and 3, show a typical set of figures as observed for air with a
1 mm diameter point, a gap length of 8 cm, 32.5 kv at a distance s from the
point of 1, 3, and 4 cm for air. Traces 4, 5, and 6 show a similar set of
figures at 1, 3, and 4 cm, but with 36% O_2 instead of the 21% of air. It is
noted that the range of the streamers with increased O_2 is much decreased,
as are the number of streamer branch tips reaching the plate at any given s.
By varying gap length d, applied potential V, and the film distance s for the
various mixtures, the following data could be obtained: the number n of
branch tips reaching the film, the radius b of the circle in which the tips
fall, the range a of the longest branches on the film, and s_m the maximum
range of the tips along the axis under the set conditions. The ratio b_m, the

maximum radius about the impact points relative to the value of s_m, gives a sort of spreading ratio index. Evaluations of these quantities from the films have permitted a summary of the results in terms of the curves.

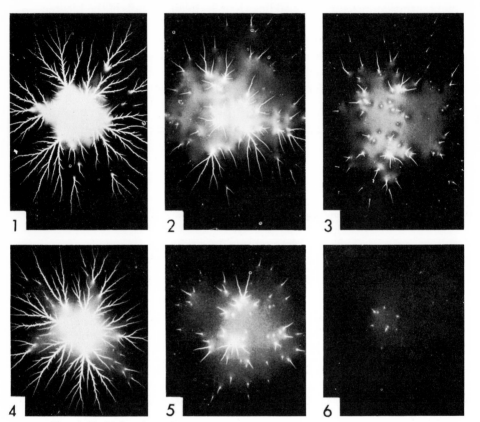

FIG. 3.90. Lichtenberg figures at 1, 3, and 4 cm from the 1 mm diameter point in air with 8 cm gap, and in N_2 with 36% oxygen, showing the strong suppression in branching and axial range.

Figure 3.91 shows the number of streamer tips striking the film at various distances s for different concentrations of O_2 in air, with d = 8 cm and V = 32.5 kv for a 0.1 cm diameter point. The top curve for air agrees with Nasser's findings. It is to be noted that the rise of the n count to the maximum in all curves follows that for air before the curve falls off to a maximum. Striking is the rapid decline in axial range s_m of the streamers as O_2 content increases. Note that the number n also decreases and that the peak shifts more to shorter distances lying between ½ and ⅓ the range. If the O_2 concentration is decreased, streamers become even more vigorous and increase their range. This is indicated in figure 3.92, where it is seen that the range s_m and n are much increased. It was impossible with the equipment to carry studies of n and b down to lower concentrations of O_2, as the value of b so much exceeded the breadth of the standard film used that measurements were unreliable, though s_m could be determined by increasing gap lengths beyond those used at higher concentrations. Figure 3.93 shows the curves for n as functions of s for different percentages of O_2 at d = 8 cm and V = 32.5 kv for a point of 0.01 cm diameter. It is to be noted that the family of curves is about the same for the smaller point. However, n is in general reduced by some tens of percentages and the range at high concentrations is also reduced for the smaller point.

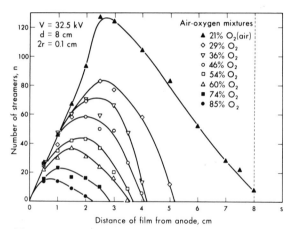

Fɪɢ. 3.91. The number of streamers for a fixed set of gap conditions at various distances from the anode as a function of percentage of O_2 in N_2. Note the rapid decline in radial branching and in axial range as oxygen content increases.

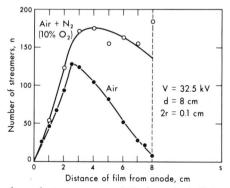

Fɪɢ. 3.92. Number of streamers n plotted against distances of film from the anode for air and for nitrogen with 10% oxygen. Radial spreading below this became so great that the film width did not permit accurate count of n.

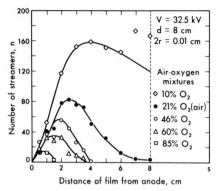

FIG. 3.93. Number of streamers n plotted against distances of the film from the anode for a point diameter one-tenth that of the series shown in figures 3.91 and 3.92 with various concentrations of oxygen.

Figures 3.94 and 3.95 bring out the influence of point radius in more detail. Figure 3.94 is for air at 32.5 kv and d = 8 cm, data being shown for points of diameters 0.01, 0.1, 0.2, and 0.3 cm. It is interesting to note that the 0.1 cm diameter point gives the largest number of streamers. Similar curves are shown in figure 3.95 for d = 8 cm, V = 39 kv with 46% O_2, for diameters of 0.01 cm, 0.1 cm, and 0.2 cm. Here the dominance of the point of 0.1 cm over the 0.2 cm diameter point is not so pronounced. It should here be noted that the configuration of the point as a hemispherically capped cylinder must be quite accurate to obtain reproducible results. Figures 3.96 and 3.97 compare the amount of radial spreading relative to axial spreading. Here are plotted s_m and b_m for the 0.1 and 0.01 cm points, as well as their ratio for the various O_2 concentrations over the range, where these can reliably be observed. It is noted that b_m/s_m *declines* as the percentage of O_2 increases for the 0.1 cm diameter point, and *increases* for the 0.01 cm diameter point. This indicates that the radial field about the smaller point adds its component vectorially to the similar components of the field about the ionic space charge at the avalanche head, so as relatively to increase the probability of large radial avalanches in contrast to the conditions for the 0.1 cm point. In that case the axial component is relatively more powerful, depressing the development of radial branches.

Figure 3.98 shows a semi-logarithmic plot of the values of s_m as a function of the percentage of O_2. Plotted as a function of the pressure of O_2 in mm, this indicates that the maximum axial range is determined by the absorption coefficient of the photoionizing photons. These follow equations of the form $s_m = s_0 e^{-\mu_1 x p}$, where s_0 is the extrapolated range at $\approx 0.2\%$ O_2. The N_2 used is already contaminated with small amounts of H_2O and O_2 and

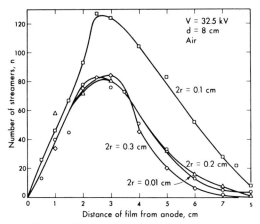

FIG. 3.94. Further illustration of the influence of point diameter under constant potential and gap length for air.

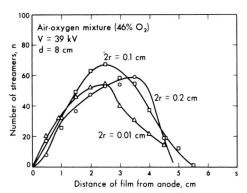

FIG. 3.95. The influence of point diameter with constant potential and gap length in an air-oxygen mixture with 46% oxygen. Contrast this set of curves with those of figure 3.94.

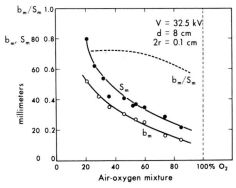

FIG. 3.96. The branching and branching ratios for 0.1 cm diameter point for air-oxygen mixtures.

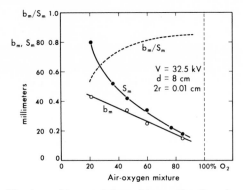

FIG. 3.97. The branching and branching ratios for air-oxygen mixtures under same conditions as in figure 3.96 but for a point diameter one-tenth that of figure 3.96.

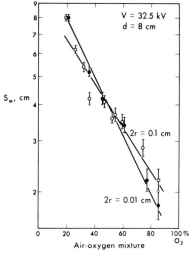

FIG. 3.98. Semi-logarithmic plot of maximum axial range of streamers as a function of oxygen concentration at constant potential and gap length for two point diameters. The slopes of these lines yield approximate absorption coefficients for the photoionizing radiation.

does not reflect the behavior of pure N_2. The quantity s_0 is evaluated from the curves as $s_0 = 12.5$ cm for 0.01 cm diameter point, and $s_0 = 9.6$ cm for a 0.1 cm diameter point. The quantity μ_1 is expressed for 1 mm pressure, so that the pressure p must be placed in millimeters in the equation. Since μ_1 is the reciprocal of a length, that is, the absorbing free path for photo-ionization, one must place the distance x into the relation where x is a path length pertinent to streamer advance. The value of the coefficient of p evaluated from the slope of the curves from the data is not μ_1 but μ_1 multiplied by some critical absorbing distance d_0 for streamer propagation. Since the field at the streamer tip is not known, an approximate critical field has been set as equal to the applied field at the point. One can use d_0 as the length of the active ionizing zone at the point anode for the field at this point. This is a distance from the point surface at which the increase in $\int_{d_0}^{a} \alpha \, dx$ has fallen virtually to zero. For the fields at which streamers start at the point surface, this distance d_0 can roughly be set for the two points as 0.1 and 0.06 cm. Had the field been twice as high as indicated below, these values would be little changed. Using these values and dividing them into the quantities derived from the slopes of the curves, as $\mu_1 d_0$ of 2.33×10^{-3} and 3.07×10^{-3}, respectively, one obtains μ_1 as 2.33×10^{-2} cm^{-1} Torr^{-1} and 5.12×10^{-2} cm^{-1} Torr^{-1}. At 760 mm this makes μ_i equal 18 and 39 cm^{-1}. Tamura indicates that the change in μ_i effective is an average value that must come from the influence of the point fields on electron energy distributions and thus on the quality and quantity of photons excited.

One must therefore conclude that at a constant potential and gap configuration, the range of the streamers, as a function of point potential and field strength, is governed by the photoionizing photons produced in adequate intensity to create ions beyond a critical length representing that of the maximum effective ionizing range of the avalanches feeding into the head of the streamer. This in first line must depend on the field strength and radius of the space charge at the avalanche head superposed on the field produced by the corona point. The semi-empirical equations of Meek[3.12] and Raether[3.29] set the axial component of the tip field of the avalanche head as about equal to that of the anode tip for streamer advance. The Meek equation that applied to observed breakdowns in air has often appeared to be more than qualitatively successful in accounting for many diverse breakdown observations.[3.28] The calculations above, using the field at the point surface instead of double the value at the surface to include the unknown space charge, are very little in error, as indicated by comparison for one case of higher field. As the value of the effective absorption coefficient for a gap $\mu_1 p = \mu_i$ increases with the partial pressure of O_2, the number of photons that produce ionization at a given V, point radius r, and value of d_0 will decrease. Thus the distance to which a streamer can advance into the gap, being dependent on its inflowing electron avalanche current and space charge field, will decrease. In fact, Nasser's studies have shown that the tip potential of the streamer as it advances into the low field regions of the gap maintains a value which is a not inconsiderable fraction of the applied point potential, and ceases to advance when its value falls to some 0.8 or 1.2 kv. In fact, tip potentials determine the range for adequate photoionization. Thus the distance to which a streamer propagates starting at a tip potential V depends on the ability of the advancing streamer tip to sustain itself by photoionization as it advances until losses by branching and attenuating field in the gap reduces the tip potential to $\approx$0.5–0.8 kv. Accordingly, the maximum range along the axis at a tip potential begins to decline because of branching and appears to depend on the fraction of photoelectrons ionizing in advance of the tip. Since, as O_2 content increases, the number of photoelectrons decreases nearly exponentially, the range will fall accordingly, other factors being equal.

The question of radial extension or branching of streamers will depend on the shape of the vector field about the head of the space charge and the tip of the point. For a point of radius 0.1 cm or more, the radius of the point is perhaps fifty times the radius of the space charge. The strongest field will therefore be the combined field along the common axis; that is, the axial component will be near the scalar sum of the two fields. Thus the radial field component at right angles to the axis in the strongly ionizing zone will be largely due to the space charge field, since the point field within d_0 is nearly normal to the point surface. In consequence, the effective radial ionizing field will be something like one-half that of the axial component. The avalanches feeding normal to the axis will be much weaker than along the axis. Accordingly there will be far fewer occasional large avalanches, or avalanche groups, feeding in at 90 deg than along the axis to advance these radial streamer branches. Branching will tend to be in a more axial direction, as observation indicates. As the streamer point advances into weaker field regions of the gap, however, the radial branching should increase for vigorous axial streamer tips, since the radial field begins to dominate over the point field with large s. This will continue until the potential of the streamer tips diminishes so that the advance of weaker branches ceases. Thus branching will at first increase nearly exponentially as the tip advances, that is, with s, and then will rapidly decline, as indicated earlier. This is precisely what Nasser observed.

In the case of the 0.01 cm diameter point, the point diameter begins to be only a few times the space charge tip diameter. Were the two radii approximately equal, the two fields would be more nearly concentrically radial and streamer branches at 90 deg to the axis could be nearly as vigorous as those along the axis. Thus branching radially near the point surface should show a relative increase with decrease in r. The branching will increase exponentially close to the point but will decline rapidly in all directions as s increases. As observed, the profile of the small point streamer should be rounder and shorter as tip radius declines. Because of the poor statistics this influence is not evident in figure 3.94, but it can be seen in photographs of the glow in figure 3.4. With V, r, and d constant, then as

$\mu_1 p = \mu_i$ increases and λ_i decreases with O_2 content, the effective photons penetrate to shorter distances, so that fewer avalanches are created outside d_0 and those that are so created become weaker. This will reduce both the over-all axial and radial advance, but will reduce the axial advance more rapidly than the radial, as the guiding axial field attenuates very rapidly for fixed d with the small point, while the radial field is more sustained in streamer advance. Hence the ratio of b_m/s_m will increase with O_2 content. This is seen to be the case for the smaller point. Where the point radius r is fifty times the tip radius and absorbing free paths are initially great, as in air, increase in O_2 concentration will reduce the number of effective incoming avalanches. Both axial and radial advance will thus be reduced in the same measure, but, radial branching having initially been weak relative to the axial, the radial branches will be reduced slightly more rapidly than the axial with reduction in available tip potential. Thus b_m/s_m will remain nearly constant or decrease slightly with O_2 content, as observed.

In the light of what has been said, increase in point diameter to 0.2 and 0.3 cm diameter will still further enhance axial propagation relative to radial propagation. In the measure that this is effective, the number of branches should decrease and the maximum n should occur at greater distances from the anode. This is rather nicely shown in Amin's pre-onset streamer luminosity–distance curves at poor time resolution (fig. 3.40). The maximum range for the axial streamers should be increased for the larger radius under similar point surface fields.

It would thus be expected that the range of the streamers might be longer with larger diameters. As point fields at the surface in the Lichtenberg figure studies were not the same, potential being constant, the ranges do not appear to differ too much, especially since s_0 was not easily determined with sufficient accuracy. In Amin's curves this is more clearly indicated as point fields are more nearly comparable. It is to be noted that in air in the four curves of figure 3.94 the 0.1 mm point shows the greatest number of branches and that the maximum number at other diameters is less and the peaks are much broader and shifted to relatively larger s values, as expected. At 46% O_2, the potential had to be raised to yield curves comparable to those in air. The reduction in n, as well as the broadening, is marked in all curves, and the relative effectiveness of the 0.1 cm diameter point has been reduced by increase in $\mu_1 p = \mu_i$ or decrease in λ_i, presumably because higher average electron energies have excited more highly absorbed but effective short wave-length photons.

The involved problem of streamer range and branching as related to absorption coefficient is treated in connection with the Dawson–Winn theory of zero field streamers in Appendix II and especially in Appendix IV.

10. Static and Impulse Breakdown Values
       as a Function of Percentage of O_2

It is now of interest to discuss the breakdown potentials for static and impulse breakdown in the various mixtures. The N_2 used was not pure. It had a partial pressure of H_2O of about 4.7 mm or 0.62% H_2O. There was O_2 present, perhaps to the extent of $\approx 0.1\%$. Thus the gas was readily photoionized by its own photons and active states; that is, it constituted a streamer-forming gas. There was not enough O_2 present to form negative ions in any appreciable numbers or to alter the Townsend first coefficient. Again, negative ions are too few to give significant space charges. The impulse corona breakdown occurs in such a short time, in the order of microseconds, that positive ion space charges do not form in the gap. Thus streamers appear early (at pre-onset thresholds) and persist, as well as grow, throughout as potential increases to breakdown. The cathode may emit photoelectrons to help initiate the discharge once ionization begins. Calculations show that, when the O_2 concentrations reach a few percentages, it is possible for negative O^- ions created in the shorter streamers to form a Hermstein sheath. This would require higher spark breakdown potentials with values increasing as O_2 content rises.

N_2 plus its impurities is readily photoionized by its photons and, in consequence, shows the lowest breakdown potential on impulse.[8] As O_2 concentration increases, the potential rises through the joint influence of a Hermstein sheath and the effect of absorption of the active photons in N_2 by the O_2 at increasingly smaller values of λ_i. In a smaller measure the increase in potential with O_2 comes from the fact that the first Townsend coefficient in N_2 is reduced for air and O_2 because of reduction in Townsend's α by electron attachment. In synthetic air, 21% O_2 in N_2 mixtures with about 4.7 mm partial pressure of H_2O, the potential for impulse and static breakdown is about the same. Since Hermstein's negative ion glow corona with *static* potentials is known to determine spark breakdown, it would appear that this equality indicates that the glow is active in impulse breakdown though produced by O^- ions. The breakdown potential appears to become independent of the further concentration of O_2 at about 70% O_2. Above this concentration the Hermstein sheath could become less effective at the potential needed, in consequence of photoionization within the sheath. It is also possible that this represents the stage at which the ionization of O_2 by its own photons from states above the ionization potential, which begin to be present at the high potentials, predominates over that by N_2 photons. In impulse breakdown, in contrast to static, little O_3 and O

---

[8] Recent results by Winn in the author's laboratory indicate that streamer propagation reaches its peak in N_2 with between 1 and 5 per cent O_2.

are formed and accumulate to give a two-component photoionizable gas. It is also conceivable that, with the 2 cm gap used, the electrons needed for the avalanches at higher O_2 concentrations come from the cathode, for the aluminum cathode could well have liberated photoelectrons. At any rate, above 70% O_2 the impulse breakdown requires the highest potentials in order to give streamers, by whatever mechanisms are active.

In the case of static breakdown, the threshold in the tank N_2 might be expected initially to be higher than for impulse breakdown in view of the positive ion space charge resistance in low field regions built up by the corona discharge which sets in well below spark threshold. Notwithstanding this inference, it appears that the static spark threshold for the impure N_2 used lay markedly *below the impulse value.* Partial clarification must be sought in work by Weissler[1.8] on the corona thresholds and the pre-breakdown streamer thresholds. (The latter lie very close to the sparking threshold and are accurately determinable.) Weissler worked in pure N_2, $N_2 + 0.2\%$ O_2, and $N_2 + 1\%$ O_2 with a 0.25 mm diameter point and 3.1 cm gap. In pure N_2 a pulseless corona set in at 4800 volts, whereas 0.2% O_2 gave a corona with "incipient pre-onset streamer pulses" at 4500 volts, and at 1% O_2 gave pre-onset pulses at 4800 volts. Because of point damage, sparking potentials which parallel pre-breakdown streamer pulses could not be observed, but the appearance of pre-breakdown streamers was observed. These occurred at 11 kv for pure N_2, 9.3 kv for 0.2% O_2, and 12.5 kv for 1% O_2.

Pure N_2 with clean cathode gives a γ_i cathode-determined corona threshold of low current density. This does not suffer from local positive ion space charge choking, but requires a fairly high potential. Electrons remain free. Photoionization in the gas can occur only at high current densities. As noted, the ultimate spark breakdown occurs via a streamer spark which requires photoionization in the gas at high current densities and high potentials, conditions under which much atomic N and other active photoionizable constituents are present. With 0.2% O_2, the cathode γ_i is suppressed but a γ_p occurs as photons get through, and very little negative ion formation takes place. However, the O_2 is readily photoionized in the volume of the gap by N_2 photons and furnishes a second gas-conditioned γ_p type of source of electrons. Threshold for the appearance of a corona is lowered. Currents are higher and the "incipient pre-onset streamer pulses" of Weissler could well have been positive ion space charge interruption. This would account for the lower threshold. The negative ion concentration is too slight to lead to the formation of a Hermstein glow. Thus, despite the positive ion space charge with γ_p and more effective photoionization, especially with the H_2O vapor, a low static streamer spark threshold should occur. This gas appears to parallel the behavior of the N_2 gas used in the present study.

With 1% O_2 the γ_i action at the cathode is probably eliminated or much reduced. Photoelectric action at the anode and in the gas is much enhanced. It is probable that negative ions form a weak Hermstein sheath. Positive ion space charge is present as before. The corona threshold is raised to that of pure N_2. The pre-breakdown streamers now require much higher potentials than before because of the Hermstein sheath, despite more effective photoionization in the gas. Thus it is to be expected that even with 1% O_2 the static threshold should rise sharply and with more O_2 rise above that for the impulse potential. With 5% O_2, therefore, the static spark threshold reaches its highest value, despite efficient streamers.

As noted, the breakdown potential when the O_2 reaches 20% is then about as high as the impulse potential. Increase in water vapor pressure from 1 mm to perhaps 6 mm appeared to *lower* the spark breakdown potential perhaps 20% in static but not in impulse corona breakdown.

What is very interesting is the decline of the sparking potential with static corona somewhere around 10 to 20% of oxygen, which continues to around 80%, at which point a constant breakdown potential appears. The two factors which raised the sparking potential were the inhibiting effects of the positive ion space charges and the negative ion Hermstein sheath. The influence of both of these on decreasing or preventing streamer production and, therefore, spark breakdown is the cause of the raising of the potential. As oxygen concentration increases from 10 to 20%, several factors can be suggested which contribute serially to a decline in potential. With increase in oxygen concentration above 10%, the photoionizing free path is so reduced that the negative ions which created the dense sheath at lower concentrations are being formed so close to and within the sheath area that they are no longer as effective. Then the Hermstein sheath with its suppression of streamers becomes less effective as O_2 concentration increases. Hence the sheath is more readily dissipated to allow breakdown streamers to appear at lower potentials, which seems to have occurred in impulse at 70% O_2. Furthermore, with the higher concentrations of oxygen, the Hermstein corona discharge will cumulatively produce large quantities of atomic oxygen as well as ozone. This action cannot occur with the impulse breakdown. As indicated by the increase of breakdown potential with impulse potentials, at lower O_2 concentrations nitrogen photons are failing to produce the streamers. However, with creation of much atomic oxygen, especially in the sheath zone, the chance of creating streamers in oxygen through photoionization of the molecular gas by highly excited atomic oxygen photons may lead to a much more effective photoionization of the oxygen by its own radiation and, therefore, further the streamer advance under static breakdown conditions. Thus one will have to explain the decrease in sparking potential with static breakdown in terms of a weakening of the Hermstein sheath through various changes produced by increased oxygen concentration and the production of new streamers by photoionization in a dissociated oxygen mixture.

11. The Influence of H_2O Vapor on
    Streamers and Sparks in Air

It had been the belief of the author and others that water vapor present in the air stimulated streamer formation by adding another photoionizable impurity. This belief came from observations that streamers appear in moist air more readily than in dry air. Köhrmann,[3.39] in uniform field geometry, showed that streamer-forming avalanches leading directly to a spark appear at 6% above the Townsend threshold in dry air but at only 4% above the Townsend threshold in humid air. In consequence, streamer progress across the gap was investigated with the present technique for thoroughly dry air and air which is at 100% relative humidity at 22° C or with about 10 mm partial pressure of H_2O vapor. Figure 3.99 shows the Lichtenberg figures for dry air at 2, 3, and 4 cm in traces 1, 2, and 3, and for the same distances in the humid air in traces 4, 5, and 6 at 32.5 kv with a 0.1 cm diameter point with d = 8 cm. It is at once seen that water vapor, contrary to expectations, acts very much as does O_2 in that with a relatively small quantity (around 1.2%) it produces a marked decrease in streamer propagation. This is compatible with Köhrmann's and Schröder's[3.39] observed 5% increase in sparking potential in uniform field static breakdown with saturated H_2O vapor in air. It is also in agreement with Przybylski's[2.26] recent studies of the absorption coefficients. This apparently contradictory behavior is still further indicated by Waidmann's observation, in trying to get a pre-onset streamer corona in air for a study of its spectrum, that with a point in room air the burst pulse glow corona predominated. If, however, one breathed, or blew, humid air onto the point, the pre-onset streamers appeared.

The data on streamers in this work clearly indicate an inhibition of streamer propagation both in axial advance and in branching by the addition of a few millimeters of H_2O vapor, owing to photon absorption. It is essential to clarify the contradictions leading to this finding.

The threshold for the static spark breakdown in uniform field geometry coincides with and is set by the theoretically lower threshold for the initial Townsend breakdown of the gap. The initial, or antecedent, Townsend discharge builds up a positive ion space charge that distorts the gap in such a fashion to facilitate the transition to an arc via a streamer spark. The potential for the appearance of a streamer spark without intervention of the antecedent Townsend breakdown is established by means of formative time lag studies. By such studies Köhrmann[3.39] observed that direct streamer

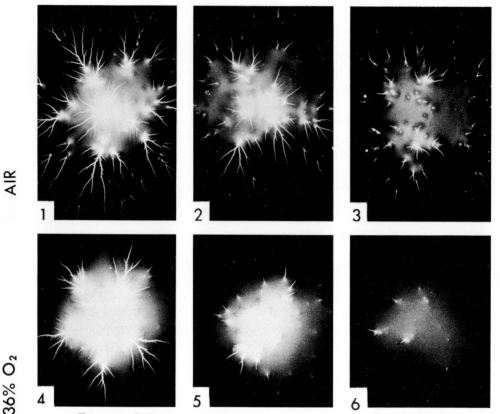

FIG. 3.99. Influence of water vapor content on the range and branching of Lichtenberg figures. Upper series for dry air with film at 2, 3, and 4 cm. Lower series for air saturated with water vapor at 22° C at the same distances.

breakdown occurred at 6% above static breakdown threshold for dry air and 4% above static breakdown for moist air. However, what Köhrmann and Schröder also observed in conformity with earlier work was that the threshold for Townsend spark breakdown in moist air lies about 5% above that for dry air. This means that in their work, not only did water vapor raise the Townsend threshold and hence streamer spark threshold by 5%, but it *actually raised the potential needed for direct streamer spark breakdown by 5% − (6 − 4)% = 3%*. Thus the uniform field data are consistent with the influence of H_2O vapor on streamers here reported. Note that in such fields there is no Hermstein glow, and H_2O vapor does not materially alter the Townsend threshold but acts on streamer development.

There is, however, the paradox of the *decrease* in static spark breakdown with water vapor which involves breakdown streamers and the appearance of pre-onset streamers (absent in dry air) when water vapor is added to the air in *positive point-to-plane corona*. This paradox is clarified when it is realized that *space charges positive and negative* in the highly asymmetrical fields of positive point-to-plane geometry *act to suppress or inhibit streamer formation*. The action of the positive space charge in corona is thus opposite to that in uniform field geometry. The negative ion space charge also inhibits streamers through the formation of the negative ion sheath glow corona, the high field of which is too narrow to permit streamers, as shown by Hermstein. Thus despite the fact that anode streamers in dry air are strongly influenced adversely by absorption of photons by H_2O vapor, the appearance of such impaired streamers may still be stimulated, leading both to appearance of pre-onset streamers or presence of breakdown stream-ers at lower anode potentials through some action of H_2O vapor on the positive and negative ion space charge sheaths.

It has long been known that the appearance of pre-onset streamers is adversely affected by the lateral spread of the Geiger counter-like burst pulse over the point surface leading to a streamer-inhibiting positive ion space charge sheath in the high field region of the point. The value of λ_i has been estimated as ≈ 1 mm. Reduction of λ_i by water vapor to around 0.1 mm will inhibit burst pulse spread. How the water vapor affects the Hermstein sheath development is not known. That the sparking potential is lowered about 2% or more in the breakdown streamer region, while streamers are weakened and perhaps require an increase in potential, follows from the fact that the point is then *already overvolted* for static streamer breakdown if no Hermstein sheath is present. The Hermstein sheath thus requires higher potentials to dissipate it in dry air. It is not known if moist air weakens the Hermstein sheath by absorption of photons inside the sheath instead of permitting photoionization outside the sheath, or by creation of negative ions of a less stable form than those of O_2. One must assume that the sheath is weakened materially by H_2O vapor and thus permits an earlier appearance of vigorous breakdown streamers. That absorption may be the factor is indicated by the action of higher concentra-tions of O_2 in lowering the static threshold by a considerable percentage. In uniform fields the Hermstein sheath does not form, so that an increase of potential occurs in that case.

12. Przybylski's Studies of Photoionizing Radiations

A more recent paper of Przybylski[3,52] has extended his earlier studies. For creating his photons he used a spark gap consisting of two conical points of 0.04 cm radius at the tip with a gap length of 0.1 cm. The spark was used to give greater luminous intensity. The spectroscopic data in air, recently obtained in the author's laboratory, as will be seen in section 14, indicate that in sparks there are many lines and bands of good intensity which are *not* in evidence in streamers, for example, the lines from excited N_2 and from N as well as lines and bands from NO. Certain O and O_2 lines and bands are also more prominent.

The currents used by Przybylski ranged from 0.01 to 1 mA. Plotted were the absorption coefficients k, observed as pressure increased from 1 mm to 200 mm. Coefficients k for oxygen increased linearly with pressure to a peak at 50 mm, the slope giving $\mu_i = 38$ cm^{-1}. From there on the coefficients declined rapidly to a minimum, to rise again to a new slope with increasing pressure to $\mu_i = 2.5$ cm^{-1}. This value happens to be the one reported first by Dechéne[2.25] and Raether[2.26] as being active in spark and corona discharge, respectively, in air. Cravath's[2.25] value observed in corona discharge in air at 760 mm gave μ_i as 10 cm^{-1}, and μ_i reduced to O_2 as about 50 cm^{-1}. Undoubtedly this was a necessarily crude evaluation of Przybylski's 38 cm^{-1}. Przybylski also measured the ratio of photoionization current in the measuring chamber to the discharge current for his two sources, a corona discharge from a cathode wire and the spark, as a function of pressure in O_2. Between 10 and 200 mm the ionizing efficiency ratio of photo current to discharge current fell from 10^{-6} to about 10^{-10}, the spark efficiency being 5 to 100 times stronger. The two effective radiations clearly appear in the spark curve, and the 2.5 cm^{-1} appears to be on the order of 10^2 to 10^3 times weaker than the 38 cm^{-1} radiation in relative ionizing efficiency. For the 38 cm^{-1} radiation from the corona, Przybylski found a relative ff_1 value for photoionization, which he designates δ/α, the number of photoionizing events per ion in the gap of $\approx 10^{-3}$. He gives for the pressure dependence of δ/α the value $\delta/\alpha = 1.5 \times 10^{-3}(1 + p/p_0)^{-1}$. Here p_0 is the quenching pressure at which half the excited atoms radiating the 38 cm^{-1} wave length lose their excitation to radiationless collisions. From the data he obtains a p_0 of 36 mm of O_2. This gives the product of $w\tau_0 = 3.1 \times 10^{-9}$ sec. If the lifetime of the state τ_0 is taken as 10^{-8} sec, then $w = \frac{1}{3}$, or quenching occurs in about one out of three impacts with O_2 molecules.

Przybylski determined the variation of δ/α, the photoionizing efficiency, for the 38 cm^{-1} radiation as a function of X/p at the surface of the corona-generating electrode between the values $200 < X/p < 500$. While the results showed a change by a factor of 2 between 11.7 and 19 mm pressure, the variation with X/p was no more than 20% at any one pressure.

His work was next extended to artificial air, $N_2 - O_2$ mixtures, both moist and dry with corona discharge and spark. As expected, k with water vapor was greater by 25% than for dry air. From 5 mm to 250 mm pressure there was not much difference between spark and corona. Efficiencies fell off more rapidly with pressure for moist than for dry air.

It appears that k as a function of pressure is linear, but does not pass through the origin. Thus k is a function of p. At higher pressures μ_i seems to approach the value 5 cm^{-1}. This, in conformity with Przybylski's earlier studies of N_2 and O_2, leads to a relation for the absorption in air of

$$\mu = \mu_{N_2} + \frac{\mu_{O_2} p_{O_2}}{760},$$ when μ_{N_2} is the coefficient of absorption of the radiation

by N_2 and is $\mu_{N_2} = 0.3$ cm^{-1}. Under the observational conditions this was independent of pressure. The value of $\mu_{O_2} = 25$ cm^{-1} for the N_2 ionizing radiation in air and p_{O_2} is the partial pressure of O_2. Since for air $p_{O_2} = p/5$, the value of $\mu_i = 5$. This applies to the photoionizing radiation from N_2 in N_2-O_2 mixtures. Thus the N_2 absorbs very little of the radiation, while O_2 absorbs most. The value of δ/α falls parabolically from 6×10^{-3} at 25 mm to 1×10^{-3} at 235 mm. Previous work had given μ_i in air as between $1.8 < \mu < 2.2$ for pressures 136 to 273 mm, and $0.5 < \mu < 0.8$ between 460 and 760 mm.

In moist air, the water concentration was 1.4%. The value of μ_i for H_2O vapor was estimated as $\mu_i > 200$ cm^{-1} at 760 mm pressure. Summarizing his work on a variety of gases, Przybylski gives the values of the quantities determined in table 3.4. It is to be noted that at lower pressures where

TABLE 3.4

Gas	Pressure Range, mm Hg	μ cm^{-1} p 760	δ/α	Source of Radiation
O_2	10–30	38	$\approx 10^{-3}$	Corona
	3–10	250		Corona
	1–3	550		Corona
	40–220	2.5	$\approx 10^{-5}$	Spark
	10–30	38	$\approx 10^{-2}$	Spark
N_2	1–5	≈ 750	$\approx 10^{-3}$	Corona
Air	10–200	≈ 5	$\approx 2 \times 10^{-3}$	Corona
CO_2	1–3	200–800	$\approx 10^{-4}$	Corona
CH_4		970		Corona
C_2H_5OH		400–700		Corona
H_2O		> 200		Corona

electron energies are much higher and measurements can be carried out to greater μ values, there are present very effective radiations of much larger μ. These appear to predominate. Thus when d_0 becomes very small, as for a 10^{-3} cm diameter point, μ can be very different from that for a 0.1 cm diameter point, as Tamura[2.33] has indicated.

13. The Observed Data on Avalanche Size
    for Breakdown Streamers; Work of
    Raether's School

When Raether assumed his duties as Director of the Physical Technical Institute at Hamburg University some years after World War II, he instituted a series of investigations on the production of breakdown streamers. These studies largely centered about the streamer-forming avalanches and were carried out with exceptional skill and ingenuity. Observations of avalanche growth were made using electronic detection systems capable of resolving as few as 10^4 electrons in avalanches.[3.51,3.53] The avalanches were studied with photomultipliers[2.6,2.7] and even observed using an image converter. Photon production was observed in avalanches.[3.53]

The statistics of avalanche development and those of avalanche sequences were theoretically and experimentally investigated in detail.[2.3] This study involved as well the growth of Townsend discharges from single and multiple avalanches. Studies were made of the radial diffusion of the avalanche head.[3.51,3.54] Photoionization in a gas by discharge photons was investigated, as indicated in the previous section of this chapter. Currently work is being carried out on the influence of electron attachment and detachment in avalanche growth.[3.30] Studies are also being made of the retarding influence of the dipole field of positive ion and electron space charge clouds for large avalanches as the anode is approached.[2.7] Avalanche tip velocities and energies of the electron in the cloud as well as diffusion coefficients have been roughly evaluated.[3.51,3.54]

Unfortunately, as far as coronas are concerned, Raether has confined his study largely to uniform field gaps. However, valuable lessons can be drawn from such studies. As earlier indicated, uniform field breakdown in gases like air, N_2, O_2, H_2, and even A is confused by the near coincidence of the threshold for the Townsend glow type of discharge, V_T, and the threshold V_s for the transition of such a glow, or of a nonconductance of the gap, to a filamentary spark via a streamer. These thresholds in air lie remarkably close together, $V_s/V_T \approx 1.04$ to 1.06, depending on whether the air is moist or dry. The thresholds are still fairly close in relatively pure H_2 and N_2, but in A $V_s/V_T \approx 2$. However, if one exceeds the Townsend threshold V_T, given by $\gamma e^{\alpha d} = \gamma e^{\eta V_T} = 1$, by very little, a space charge plasma of positive ions accumulates. This effectively extends the anode into the gap, increases the field, X, and decreases the gap length, d. This has the consequence that if $d(\alpha/p)/d(X/p) > 1$, the value of the avalanche size rapidly increases from one yielding a V_T to one yielding a V_s. At the V_T threshold the time of growth of the glow discharge lies in the 10^{-5} to 10^{-4} sec range, depending on gap length and whether a γ_p or γ_i is active. The streamer spark materializes from a single avalanche or a simultaneous group of avalanches in about two avalanche crossing times, that is, the time for an electron avalanche to reach the anode and the time for the resulting streamer to go from anode to cathode. This lies in the range of 10^{-7} sec or less. It is now essential to estimate the size of the avalanches yielding a V_T and those yielding a V_s directly.

Raether[3.26] in his earlier studies had arrived at the conclusion that when an electron avalanche reached a magnitude of the order of 10^8 to 10^9 electrons and positive ions, a breakdown streamer would ensue. This prediction appeared to agree with the earlier estimates of the author and Meek[3.28] for air at atmospheric pressure in a 1 cm gap. Raether's conclusions were drawn from phenomena derived again for air and other gases, largely in uniform field geometry at pressures from 100 to 760 mm.

It will be noted that Raether in his original theory calculated the *space charge density*[3.29] needed to give the streamer-initiating space charge field, a criterion also used by the author and Meek. Later he seemed to abandon the *density* concept for that of the *pure number*. Phenomenologically, however, he has recently had to come back tacitly to a density concept, as will be seen. The theory of the author and Wijsman,[2.32] as noted, considers both density and photoionization yields, and these will ultimately prove to be the criteria. Actually, because of photoionization, the threshold depends on both density and total number of ions.

At this point one may accept the number concept, as much can be learned by setting a numerical limit of 10^8 electrons for a *spark-producing streamer avalanche*. One must now regard a gas like air, N_2, or H_2, and realize that, while a streamer spark sets in at V_s, which permits an avalanche to grow to 10^8 electrons at the anode, such sparks with a time lag of 10^{-7} sec will not be observed directly in a static breakdown study near threshold. The reason for this is simple.

In most uniform field gaps with metal cathode, the value of γ_p or γ_i is on the order of 10^{-3}, and for poor cathode surfaces at lowest γ_p will not be less than 10^{-5}. Thus, since at V_T, $\gamma_p e^{\eta V_T} = 1$, if γ_p is 10^{-5}, $e^{\eta V_T}$ need only be slightly above 10^5 for a single avalanche to lead to a self-sustaining Townsend glow discharge. This discharge may or may not, after a long build-up time, yield a streamer spark through space charge action, as shown by Bandel.[2.4] Thus streamers are not observed to yield sparks directly in 10^{-7} sec in most uniform field gaps. With great ingenuity Raether has shown that one can overcome the difficulty introduced by γ_p or γ_i in various ways and observe a streamer spark directly when V reaches V_s, with time lags of 10^{-7} sec. These will be listed and discussed below.

1. Using organic gases such as CH_4, C_2H_6, C_2H_5OH, ethyl ether, and so on,[3.51] Raether found that the values of γ_p and γ_i are very low, and of the order of 10^{-8} or 10^{-9}. This comes from the many inelastic impacts that electrons experience. Thus energies in electron swarms are relatively low.

Highly excited states that can photoionize are not readily created. Any photons yielding a γ_p are readily absorbed and degraded in energy by such molecular gases and their fragments. The ions produced are cumbersome, slow, and have very low ionizing potentials leading to little γ_i. Finally, carbon deposits on the cathode, producing a very inactive surface of high work function, are quickly formed through breakdown of the gas in Townsend discharges.

In such gases avalanches of 10^8 and 10^9 electrons can occur after the cathode is inactivated, and a spark materializes in 10^{-7} sec without a Townsend discharge.[3.51,3.55] This has been observed directly on an oscillograph with a rise of a single avalanche leading to a streamer in 10^{-7} sec with no succession of avalanches. The data on avalanche size have been verified by electronic recording in such gases.[3.53]

2. Since in air it takes many generations to build up a Townsend discharge through a γ_p or γ_i, even though γ is as high as 10^{-3}, if a potential pulse of duration of the order of 10^{-7} sec or certainly of less than 10^{-6} sec is applied to a gap with a cathode weakly illuminated by ultraviolet light, V can be raised above V_T to V_s before the Townsend discharge develops. This then yields a streamer spark with build-up time of 10^{-7} sec, before the Townsend breakdown can materialize. Thus streamer sparks can be produced directly by use of impulse potentials at overvoltages not far above V_s/V_T in a given gas, with no intervention of a Townsend discharge.[3.39] If time lags can be avoided, this could establish the value needed for V_s in the gas in the absence of space charge. This assumes that the streamer spark can occur in a gas that is not modified chemically by the Townsend pre-discharge.

3. Obviously if the gap is of such form that the anode is highly stressed and isolated, so that a γ_p or γ_i cannot come from the distant cathode, the spark will occur by streamers, though a corona discharge may and usually does precede the streamer spark and thus alters V_s.

4. The most convincing proof of all, however, comes from a recent consideration.[3.56] In the 1 cm uniform field gap, avalanches of streamer-forming proportions have a diameter of the order of 1 mm when they reach the anode, as estimated by Raether from avalanche crossing time and diffusion coefficients, by observations of cloud tracks, and by image converter.[3.54] The author has estimated this as nearer 10^{-1} mm in past analyses.[3.28] The diameter lies within these limits at 760 mm pressure. Now if γ_p is 10^{-5}, $e^{\eta V_T} = 10^5$, and a Townsend discharge should follow when V reaches V_T with 10^5 electrons per avalanche. If, however, the space charge for streamer growth comes from 10^8 electrons within a circle of 1 mm diameter or 7.5×10^{-3} cm^2 area at the anode within an avalanche crossing time, it is immaterial whether the electrons come from a single electron avalanche or from 1000 avalanches of 10^5 electrons each. Thus a Townsend gap, in which 1000 avalanches of 10^5 electrons each land on an area of 7.5×10^{-3}

cm^2 within 2×10^{-8} sec of each other at the anode, should at V_T or even possibly at a lower V yield a streamer spark which normally requires a potential V_s giving 10^8 electrons in one avalanche.

With a very small hole in the center of the cathode, it was possible to trigger the discharge of a gap with single α particles projected parallel to the field axis. Now an α particle creates about 2000 electrons per mm path length in a gas. It creates them along a track of some 10^{-3} cm diameter and in 10^{-9} sec. Thus it is possible to inject the 2×10^3 electrons from 1 mm α particle-path into the cathode region in a minute area in 10^{-9} sec. Under these conditions streamer sparks materialized as if a single avalanche had created them at values of V equal to V_T and sometimes as low as 2% below V_T. This was a remarkable achievement. It now leaves no doubt that streamers can be created in Townsend gaps in air by a suitable number of ions produced in a limited region of the anode. What is more, such sparks can occur *below* V_T if enough electrons hit the target area. (One problem still remains concerning single avalanches with approximately 10^8 electrons in air, CH_4, and so on. There is a delay of some nannoseconds between arrival and streamer, caused by dissociative electron attachment near the anode.)

The consequences of this crucial experiment are many:

1. It reconfirmed what the author has always insisted upon, that it is not the *number size* of the avalanche alone that determines streamer advance, but the number of electrons, and hence of *positive ions within a given volume*. This volume in essence is given by the diffusion radius of the avalanche, or $\approx 10^{-2}$ cm^2 in air at 760 mm, and at least $1/\alpha$ cm deep, since one-half the ions in the avalanche are created in $1/\alpha$ cm. At $e^{\alpha d} = 10^8$, $\alpha d = 17$, and, as d = 1 cm, $1/\alpha = \frac{1}{17} \approx 0.05$ cm. Thus the volume using Raether's radius is 4×10^{-4} cm^3. This makes the ion density of the order 1.3×10^{12} ions/cm^3, since only half the ions lie in $1/\alpha$. It is clear that Raether must again revert to the concept of density for streamer advance if he ever really abandoned it. However, the number as well as the density must be adequate to ensure photoionization.

2. The experiment established the fact that a streamer can be formed as the result not of a single avalanche of $e^{\alpha d}$ electrons and ions, but of n_0 avalanches, which can be of smaller size, such that $n_0 e^{\alpha d}$ if they land within a given target area at the anode. Thus where it was estimated that 10^5 electrons and ions initiated the pre-onset streamers at threshold for a positive point, the streamer may not actually have been caused by 10^5 electrons of one avalanche, but could well have been produced by 10^6 electrons in a limited area by simultaneous converging avalanches.

One must therefore be careful about conclusions on this score. However, again in the case of the corona point, the avalanches travel at most 2 mm for a 0.1 cm diameter point. The diffusion radius for the 10^5 electrons estimated by English is thus at least one-half, and doubtless much less than the 0.1 cm estimated by Raether for the uniform field. Thus streamer densities may be achieved by a smaller number of ions. At any rate, convergence of *more* than ten avalanches into the reduced target area within some 10^{-8} sec, even in this confined area, is not likely. Hence the true value of the threshold number of electrons for the points studied by English cannot deviate by more than a factor of ten from the value given, but this leaves open the question of the density needed, which at present is not open to calculation. That pre-onset and less vigorous streamers start with smaller avalanches and hence smaller point potentials than the 10^8 posited by Raether for spark breakdown streamers in uniform geometry is indicated by the wide range of point potentials below sparking observed for impulse streamers in the studies of Nasser. Nasser even estimates the tip potential for streamer advance as ≈ 1 kv once it has started (p. 108).

3. The critical factor in many cases, especially with corona, may be the number of photoelectrons produced. If one adopts the figures of Przybylski, there may be three photoionizing radiations active in air, one coming from O_2 at $\mu_i = 38$ cm^{-1} in O_2, one coming from N_2 acting on O_2 at $\mu_i = 25$ cm^{-1} in O_2, and one from O_2 at $\mu_i = 2.5$ cm^{-1} in O_2. These have values of μ_i of 7.7, 5, and 0.5 cm^{-1} in air with ionizing free paths in air of 1.2 mm, 2 mm, and 4 mm. The values of their photoionizing efficiency were estimated at 5×10^{-5}, 10^{-4}, and 10^{-5}, respectively. Thus for a critical *spark*-producing Raether avalanche in uniform fields in air, there should be $\approx 10^4$ photoelectrons, largely produced outside two ionizing free paths, or 1 mm, from the anode, with more electrons on the order of 1000 to 100 produced still further out. If all of the 10^4 photoelectrons produced avalanches of 10^4 electrons, then the streamer would advance. In uniform field geometry this can only occur if the positive ion space charge field at the anode is of adequate magnitude, and added to the static field it extends out less than 1.2 mm from the space charge. In the more convergent point-to-plane geometry, the range of 1.2 mm and the 100 photoelectrons from avalanches of 10^5 to 10^6 electrons and ions will readily suffice for propagation. This sort of speculation at the present state of knowledge is stimulating, but cannot be accurate until much more data on photoionizing coefficients and efficiencies is at hand. In any event the results of the last few years have gone far to clarify the whole mechanism, and it is clear that adequate ion densities are needed to give the fields an adequate total number of ions to ensure enough photons (see Appendixes II and IV).

14. SPECTRA OF STREAMERS

As can readily be imagined, the spectra of streamers are exceedingly difficult to obtain because of the feeble intensity of the light emitted. They have been photographed for air in color by Nasser,[2.19] and show the same electric blue color that the eye notes in the pre-breakdown streamers from a fine point. During the studies of Kip and English,[3.2,3.15] the pre-onset streamers in air were resolved with a small quartz spectrograph. The analysis was probably done under the supervision of F. A. Jenkins. He stated that the light came largely from the second positive band system of N_2, that is, from highly excited states of the N_2 molecule.

During the spring term of 1962, Robert L. Byer, working under the direction of Professor S. C. Davis, collaborated with Waidmann in an attempt to obtain more complete information on these spectra. The impulse-produced streamers gave a chance to observe the luminosity of the streamer tip as it moved out into the gap without the burst pulse corona glow. The slit of the spectroscope with suitable light-gathering optics was set parallel to the anode axis and focused along the axis. It thus got the maximum light from the most vigorous tips. Even here it took a succession of thousands of streamer pulses to give sufficient light to resolve the spectra clearly enough to ensure identification of the bands. In addition to this, the pre-onset streamers near the anode were analyzed, as were pre-breakdown streamers. In contrast, spectra of one or more single sparks gave more extensive and informative spectra. The pulsed discharge streamers ranged in potential from 15 to 75 kv. The pulses lasted on the order of 30 μsec. Discharges occurred at the rate of one every 5 sec. The apparatus was that of Waidmann, and was housed in a nearly light-tight wooden box. The point used was about 2 mm diameter with a gap length ranging from 2 to 15 cm.

Two spectrographs were used. One was a 90 degree constant deviation quartz spectrograph by Hilger. The slit was placed at the focus of a 17.5 cm focal length quartz lens. The axis of the horizontal gap was the source of light. The light thus had to be taken from a hole in the box and reflected by a 45 degree mirror to bring the beam leaving the box from the side into alignment with the axis of the spectrograph. An Hg arc and later an iron arc, placed back of the point and axis, could be focused on the slit to give comparison spectra. The iron source proved essential for accurate analysis. The spectrograph was adjusted to cover the range from 3500 Å to 4500 Å. A test with a glass plate indicated that the use of quartz was desirable, and later work extended from 2000 Å to 5500 Å. Kodak 103a–F plates were used. The Hilger spectrograph proved to yield inadequate resolution, so that a faster quartz spectroscope later replaced the Hilger. This still did not have adequate resolution, with a dispersion of 30 Å/mm at 3000 Å, but it was the best that could be done.

The first clearly discernible bands were obtained with the quartz instrument after 2000 pulses at 30 kv and gap length 2.1 cm. On the same plate exposures of three spark breakdowns were made, as the gap was too short, so that spark breakdown interrupted some of the long pulse sequences. Lengthening the gap reduced this difficulty. The pre-onset streamers studied with a static potential at threshold required four-hour exposures, while the pre-breakdown streamers required only one-half to one hour. These gave the best spectra for analysis, though they did not differ materially from the pulsed ones.

The results of the analysis may be summarized as follows: In all three, the band spectrum was that of the second positive group. With the longest exposure a N_2^+ band head was observed. This was the (0,0) band that lies between the (2,5) and (3,6) band heads of the second positive system. No other band heads of the negative system were observed, because of their low intensity and their proximity to those of the second positive band heads.

The bands were more intense near the anode, as the streamers are most intense at this region because of later light from branching, which does not reach the slit further out in the gap. The accurate wave-length determination was made on the more intense spectra with the Fe arc lines. The identification of the bands could then easily be carried over to other spectra where Hg had been used as a standard.

In contrast, the spark spectra were much richer in bands and showed many lines. N and O lines were observed. The oxygen was ionized in the region of the anode tip and extended to one-third the gap length. The N ionization spread out over the whole gap length in the spark. As might be expected with the limited condenser stored charge to be dissipated, the arc that resulted was too short-lived to bring lines of the anode or cathode metals into the arc channel. However, the spark did melt the anode surface and pit the cathode plate. A few A lines were also present in the spectrum. Some lines that appeared were, however, not identified. The pre-breakdown streamer discharge was, as expected, the brightest of the streamer spectra. All second positive band heads, from (0,5) at 4667 Å to (2,0) at 2976 Å, were observed.

No bands of the NO β system were definitely identified in streamers. They were either not present or too faint to observe. Here again they were near the second positive bands. The NO γ bands were also not observed, but as these are so faint in any case, they could have escaped attention. Actually, unless there was a strong, low, and sustained corona glow discharge about the streamer channel, NO bands would not be expected. The streamer process is too transient and fast to create sufficient NO from O_2 and N_2, and to excite its spectrum. NO could have been observed at the anode surface in the pre-onset streamer region with the burst pulses present. The burst pulse spectrum was, however, very faint, because of attempts to suppress it in favor of streamers. One band head of doubtful identity was noted near the anode in the breakdown streamer process. It was measured to be 2855 ± 5 Å, and the wave length given for the (0,5) band head of NO is 2859.5 Å.

Visual observation during a pulsed streamer discharge revealed more detail. Light was seen in the yellow and green, but was too weak to affect the plate. It was possible that some bands of the first positive band system of N_2 were present.

The pre-onset streamers were too weak to show more than the second positive group. Neither NO nor N_2^+ bands can be expected to be seen if present under such short exposures. The burst pulse or Hermstein sheath glow was brighter, but it was scrupulously avoided in this study, so that it did not produce enough light.

It is clear that much more work must be done with greater intensity, longer exposure, more resolution, and a study of the visible, in order to answer all the questions to be raised. However, as so far revealed, the propagating streamer tip has an energy distribution such that in its 10^{-8} sec or less in passing the region in front of the slit, it excites primarily the higher states of N_2 and may produce a few bands of the N_2^+ system, but in a relatively smaller measure.

There is no time for secondary excitation and ionization in any intensity. Doubtless many O and N atoms are created as the tip passes, but most of these are not in the excited form and time does not permit their excitation before the tip has passed. The creation of NO by chemical reaction of the dissociation products of O_2 and N_2 is too slow, and bands of NO should not be strong. In contrast, the spark channel which has been pre-ionized and dissociated by the streamer tip is then further *highly* ionized by the return stroke. This will ionize and excite the atoms of O and N left by the streamer. It finally becomes highly luminous, because of an increasing arc current lasting into the microseconds, which with thermal ionization is capable of creating many emitting species not seen in the streamer. Here all chemical and secondary ionization processes as well as thermal ionization may appear. The spark presents therefore a very different luminous picture from that of the transient streamer tip.

B. SPARK-SUPPRESSING GASES—FREON
AND FREON–AIR MIXTURES

No one has made extensive studies of the positive and negative point corona in the strongly electronegative gases, or gases suppressing sparks by raising the sparking potential. Only one study was made by Mohr and Weissler using Freon, CCl_2F_2, and Freon-air mixtures.[3.57] At the time (1947) too little was known about coronas to permit interpretation. They used a 0.5 mm diameter, polished, hemispherically capped Pt cylindrical anode and a 4 cm diameter circular Pt plane with contoured edges 3.1 cm distant. The latter was enclosed in a Pyrex glass envelope. A rectified 70 kv source was available, and a 50 megohm corona shielded wire wound resistor, with accurate microammeter, recorded the potential. Hg vapor was excluded by cold traps. Air dried over CO_2 snow at 745 mm at 22° C was used, and various mixtures from 10^{-4} to $10^2\%$ Freon, by volume, were used as well as air. A 30X telemicroscope was used to view the discharge. Current potential measurements were made. Oscilloscopic observations across 3×10^4 ohms between plate and ground were made to observe pulse shapes. The RC time constant did not allow better resolution than some 10^{-5} sec. A weak Ra source was used to trigger the pulses.

With air, as indicated, the intermittent corona regime began at 5 kv with pre-onset streamers and extended over a range of 200 volts. Some burst pulses were observed. At 5.2 kv or above, onset of the continuous glow corona current of Hermstein, due to negative ions, was observed. This regime continued until, between 20 and 25 kv, occasional breakdown streamers occurred. Weissler did not go to sparking but stopped just short of this. The only visual manifestation, except for the occasional breakdown streamers, was the thin layer of glow over the point. He had previously observed, as had Fitzsimmons,[3.13] that prolonged use of the same air in a confined space led to changes in the breakdown properties as a consequence of impurities created. The effect of the created N_2O, NO, and O_3, and so on, was the appearance of streamers in the region of onset of the steady glow corona. As soon as clean air was substituted, the glow corona reappeared without streamers. The same result could be produced if 10^{-4} to $10^{-1}\%$ of Freon was added to *clean* air. Streamers appeared in the region between 5.2 kv and 10 kv superposed on the glow corona.

In air, the negative ions are of two sorts: O_2^- formed by a three-body attachment process from *slow* free electrons to O_2 molecules, and O^- formed by dissociative attachment of electrons with 3.6 eV of energy, creating O^- and O atoms on impact with O_2 molecules.[3.58] This has a large cross section, 3×10^{-18} cm^2 for values of X/p between 10 and 15, and the O^- ion concentration declines above an X/p beyond 20. The formation of O_2^- by attachment takes some time in air and is highly pressure- and field-dependent. The formation of O^- is concomitant with the avalanche creation. If X/p remains high and above a value of 20, the O^- ions are eventually converted to O_2^- by charge exchange collisions with O_2 molecules at relatively low energies.[3.59] The O_2^- ions begin to shed their electrons appreciably at an X/p of 90.[3.60] Thus, the ions are destroyed at fields with values of X/p above 90. If X/p values of 20 or more persist, all negative ions are possibly O_2^- ions.

Now the influence of the negative ion formation on Townsend's first coefficient α is different in the formation of O_2^- and O^-. In an avalanche head at high X/p, the negative ions created immediately together with the electrons are O$^-$ ions, and these reduce the effect of α in the avalanche by an effective value η. Thus for the same X/p, the avalanche is decreased in size by dissociative attachment, as shown by Geballe and Harrison.[3.58] Fields diminish after the avalanche head has passed, but as long as X/p > 20, O$^-$ goes over to O$_2^-$ with a lower mobility. In any event, electrons created in the avalanche, especially in its negative head, will slow down and form O$_2^-$ ions after many impacts. This delayed attachment is relatively slow on avalanche and streamer tip time scales and is probably trivial compared to the initial dissociative attachment. But the electrons created out in the gap are not trivial. In the streamer channel after a microsecond or so, electron loss by attachment to O$_2$ can become of importance.

There are several consequences of the negative ion formation. First, it reduces avalanche size and reduces the first Townsend coefficient. Second, it produces delayed reactions. That is, in non-uniform fields with high anode fields, the negative ions *detach* in the anode region and liberate electrons which can then, even after milliseconds of delay, perform the destined ionization denied them in the avalanche proper. Certainly the recent studies of Phelps and his associates on negative ions as a function of temperature indicate that at appropriately high values of X/p the electrons that attach do not remain permanently attached, but attach and detach so that ionization by collision in a delayed form is still active.[3.61] Thus, two effects of negative ion formation are reduction of avalanche size for a given X/p and delayed ionization.

A third effect has not been considered. If avalanches become very large, that is, in fields where avalanches reach streamer spark propagating proportions on crossing to the anode (for example, magnitudes at 760 mm of $\approx 10^8$ ions), the backfield between positive space charge and negative space charge centers reaches such proportions as to retard streamer formation and avalanche advance.[2.7] The influence of the negative ions formed by dissociative attachment and left *in situ* along with the positive space charge will reduce the retarding fields in some measure and will thus tend to facilitate streamer advance in the pre-spark region.

It is thus clear that in higher field regions where X/p is high and one deals with streamers capable of yielding sparks, (a) negative ions will raise the potentials needed to yield these by decreasing α, (b) this will be offset if photoelectric yields are correspondingly effectively increased, (c) they will reduce retarding effects of space charges, and (d) they may produce delayed ionization.

One is now in a position to attempt to analyze the paradox which appears when small amounts of N$_2$O, NO, H$_2$O, Freon, and other strongly photo-ionized and thus absorbing gases are added to the air. These appear to stimulate the appearance of streamers, and yet are known actually to reduce axial and radial expansion of streamers and even to raise the sparking potential. Little was understood about this until very recently.

It had been originally assumed by the author that these gases increased photoionization in the region just outside the critical distance r_0 from the anode surface needed for avalanche growth. This was assumed to stimulate streamer development, as it yielded free electrons at the surface of r_0 and not inefficient negative ions through attachment in air. In consequence of this reasoning, Waidmann, at the author's suggestion, studied the effects of Freon vapor on streamer growth.[2.36] This was done using the Lichtenberg figure technique of Nasser. Increasing the concentration of any of these gases reduces both the radial and axial growth of streamers. H_2O vapor and Freon are far more effective than O_2 in this respect. In O_2, the axial range decreased exponentially with the partial pressure of O_2, yielding two rough average absorption coefficients for the effective photoionizing radiation under constant field conditions for each of the two points, but different for the two points. The values were 18 cm^{-1} and 39 cm^{-1} reduced to 760 mm O_2 pressure for point diameters of 1 mm and 0.1 mm, respectively.

This indicates that the increase in absorption, or decrease in absorbing mean free path, reduces the range of the streamers. These coefficients applied to the O_2 in air make μ_1 about 4 cm^{-1} and 7.6 cm^{-1}, leading to photoionizing mean free paths of 0.25 cm and 0.13 cm, respectively. The larger coefficient may be identified with Przybylski's coefficient of 38 cm^{-1}, and the 18 cm^{-1} could be a sort of an average value or else an improperly estimated value near the one Przybylski identified as that of an N_2 photon ionizing O_2.[3.52] For this radiation the coefficient in O_2 is 25 cm^{-1}, and thus 5 cm^{-1} instead of 4 cm^{-1}, inferred by Waidmann for air. Both of these are very effective photoionizers, having an efficiency of around 10^{-4}, the latter perhaps being the more efficient. The value of μ_i for H_2O vapor given by Przybylski is >200 cm^{-1}, and the data used to estimate this agree with Waidmann's. Waidmann observed that Freon was perhaps even more effective than H_2O vapor, although the two are comparable. Freon, however, added another feature in that, besides suppressing the streamer by reducing its range, it showed a very wide spectrum of ultraviolet capable of penetrating quartz, but filtered out by glass, which is not present in air. These photons have energies ranging from around 3.7 to 5.9 eV. These and other short wave-length photons may play an important role when the concentration of Freon and its dissociation products reaches higher values. They are probably not of influence in what follows.

It is now of interest to regard the depth of the ionizing zone about the points in question. These are roughly calculated by Waidmann as 1 mm and 0.6 mm for 1 mm and 0.1 mm point diameters. If the 25 cm^{-1} radiation is the effective one, instead of the 18 cm^{-1} computed with the zone $r_0 =$

0.6 mm, the effective value of r_0 would equal 0.45 mm. It is best to confine discussion to the 1 mm point which Weissler used. It is seen that in air the photoionizing free path, that is, the range from the ionizing region at which the photoelectrons are reduced to $1/e$ of their number, and within which 67% of the photoionization occurs, lies between 2 mm and 2.5 mm depending on whether 25 cm^{-1} or 18 cm^{-1} is used for μ_i. As the ionization and photons are created close to the anode surface, it is seen that the photoelectrons are being effectively created beyond the ionizing zone in air. Thus streamer growth is fairly effective.

Now Weissler's studies used static potentials. This meant that in addition to streamers in the pre-onset region, the burst pulses also appeared. Burst pulses develop and spread laterally over the surface under conditions in which streamers may not be able to propagate, since streamers require avalanche sizes sufficient to give positive ion densities leading to streamer-propagating space charge fields. Conditions for burst pulse spread are facilitated if the photoionizing free path is too large to yield effective streamers, but not too small to propagate the Geiger counter-like burst pulses over the surface. The author has pointed out that, on the basis of experimental evidence of Huber,[2.22] the reduction of the ionizing free path from 2 mm to 0.4 mm will suppress burst pulse spread. Instead, in a small region a sequence of avalanches are created which extinguish themselves by local positive ion space charge build-up. If burst pulses cannot propagate along the surface and foul the whole high field region with positive ion space charge, then the localized pulses will grow in avalanche size as potential increases until they propagate streamers. Under such conditions only streamers are observed, as is the case in pure O_2.[2.22] Where burst pulses can form, they may be created so as to preclude any streamers. Otherwise, they will be triggered by streamers and temporarily interrupt the streamers. Under conditions where both appear, the streamers occur after burst pulse space charges clear.[3.2] This happens in air depending on its water vapor content, point diameter, and pressure. The *burst pulses are thus streamer-suppressing phenomena*, which occur at corona threshold in air under appropriate conditions. Suppression of burst pulses facilitates the appearance of streamers.

Streamers are also effectively suppressed in the region of onset of steady corona and by a new phenomenon discovered by Hermstein.[1.16] If a streamer corona exists and if one can inject enough negative ions into the gap so that a sheath of slow incoming negative ions can form, the ionizing zone is much altered. Since the O_2 ions lose electrons to give avalanches only under conditions where $X/p > 90$, and more effectively when $X/p > 150$, then the active ionizing zone next to the anode is confined to a region in which $X/p > 90$ or more. Owing to the negative ion space charge sheath, this region has very high fields, but they are confined to distances much less than r_0. Furthermore, electron energies are very high and diffusion is great. Thus the decreased depth of the region of ionization and increased electron diffusion do not permit avalanches to grow to streamer-generating proportions. Again the Hermstein glow corona, once established, prevents all streamer formation, and Hermstein has shown that this steady glow corona is responsible for the extensive potential hiatus between preonset streamers and breakdown streamers.[1.16] (See chapter 3, section A.3.)

The static corona breakdown in air leads naturally from the burst pulse interference with streamers to the Hermstein glow interference at the onset of steady corona. In fact, it may still be debatable as to how much of the burst pulse interference may not be attributed to the development of incipient Hermstein sheaths in the burst pulses themselves. The Hermstein negative ion sheath requires that enough negative ions be created in the region beyond r_0 and below $X/p = 90$ to accumulate near the anode faster than they are removed from the ionizing zone by reionization. Within r_0 the O^- ions created in the avalanches are relatively few, and convert to O_2^- and are removed in the high field regions. The O_2^- ions formed beyond r_0 are in fact the ones that accumulate if created in large quantities. These come from two sources, according to Hermstein. The first are photoelectrons created by the photoionizing photons sufficiently far beyond r_0 to attach to form O_2^-. The second source, he asserts, comes from electrons created by the streamer tips which cease advance, leaving their electrons and positive ions behind. Many of the electrons in the dead streamer channels, whether they form negative ions or not, recombine dissociatively or in ion-ion recombination with the positive ions within tens of microseconds of time. Nevertheless, a certain number escape.

Those streamer-produced electrons liberated within a 2 or 3 μsec transit time in the field to reach r_0 may escape recombination, but can form negative ions. Also, O^- ions will convert to O_2^- in these regions and yield a few unrecombined negative ions. It is thus electrons and negative ions created beyond r_0, but within perhaps a few millimeters beyond r_0, by the streamers, that contribute with photoelectrons to the creation of the Hermstein sheath. Electron drift velocities are on the order of 10^7 cm/sec in this region. The average distance for electron attachment to form O_2 under these conditions is on the order of 1 mm or even less. Attachment increases with partial pressure of O_2 or Freon.

It is thus seen that the formation of the Hermstein sheath is critically dependent on the photon absorption and conditions of electron attachment and recombination within the column left by the streamer. Any factor increasing electron attachment, so as to make the streamer channel in the critical region of an electron transit time from r_0 an all-ion channel, will increase recombination and reduce the number of ions reaching the sheath, since ion mobilities are relatively low. Too little is known about the relative rates of these processes.

Unquestionably the *photoelectrons* created far enough beyond r_0 by photons at the anode and from streamers have a much greater chance to escape recombination than the electrons in the streamer plasma, as they are more sparsely distributed. These will therefore be an important factor in establishing the glow. The glow growth will then depend on the density of photoionization beyond r_0 and on the density of streamer channels that have penetrated beyond r_0 for its formation. It must thus not only depend on photon and streamer production, recombination, and attachment, but on the *density of the current resulting from the burst pulses and streamers together*, which furnish photons and electrons. The onset of the steady corona or the Hermstein sheath glow corona is thus determined by the factors mentioned, but *cannot* be achieved *short of a certain current density* which varies under different conditions. It is therefore improbable that Hermstein's suggestion, that the pre-onset streamers reach a sufficient length to form negative ions by electron attachment, constitutes the real criterion for the growth of the glow. Pre-onset streamers exceed 5 or 6 mm in length almost from their threshold, and long before the Hermstein glow is reached. In air at 750 mm, the attachment of all electrons is assured within a very few millimeters distance. Hermstein failed to take into account the heavy recombination loss of the 10^{12} electrons and ions/cm^3 of streamer channel. He also failed to recognize the importance of the rate of ionization per cm^2 of anode surface on the negative ion production, by whatever process. Finally, Waidmann's observations of the reduction in streamer length by $\approx 1\%$ Freon and by H_2O vapor were not sufficient to alter the process which Hermstein envisioned, while the effect on photoionization was.

The foregoing has established that the range, rate, and density of photon production in the zone inside r_0 relative to r_0, as well as the production of ions from streamers, is vital to the creation of the Hermstein sheath; that the reduction of the amount of photoionization beyond r_0 inhibits the growth of that sheath; that the burst pulse corona is likewise vitally dependent on the photoionizing free path; and that when absorption renders this too small, burst pulses cannot propagate; and finally that both Hermstein's glow and burst pulses inhibit streamers. With this it is possible to clarify the paradox that, although N_2O, NO, H_2O, O_2, and Freon in appropriate concentrations *depress the advance of streamers*, they *appear to facilitate the appearance of streamers* by more rapidly destroying the streamer-inhibiting agencies, burst pulses and the Hermstein sheath. Where *Hermstein's sheath raises the spark threshold, the reduction of photoionization which suppresses the sheath may actually lower spark threshold slightly*.

Noting the magnitudes of r_0 and $1/\mu_i$ for the gap used by Weissler, one sees that conditions are ideal for the action shown by H_2O and Freon, for which at the concentrations used $1/\mu_i$ is of the order 0.01 mm. To produce the same effect, the partial pressure of O_2 would need to be increased by a considerably larger amount. Actually, in going from 20% to 70% O_2, the static breakdown potential shows a relative decrease. Part of this may be ascribed to the suppression of the Hermstein sheath. However, under these conditions the N_2 photons may no longer be active in the O_2, and some other photoionizing radiations may be involved. Thus the apparent paradox is clarified.

Thus, it is not surprising that streamers persist much further into the Hermstein glow potential regime if the right impurities are present in trace amounts. As potential increases above threshold, these streamers that are no longer prevented from forming by the Hermstein negative ion sheath become more vigorous. Still, when the current density becomes high enough, the photoionization by less completely absorbed but less prolific photons, such as those of Costa, Schwiecker, and Przybylski,[2.26] with $\mu_i = 2.5$ cm^{-1} in O_2 and $\mu_i = 0.05$ cm^{-1} in air, with $1/\mu_i = 20$ cm^{-1}, and efficiency of 10^{-6} may become the agents.

With 1% Freon in air, the corona was markedly altered. The onset potential was about 5 kv, as before, so that α was not much affected,[9] but the range over which streamers appeared was extended to 9 kv. Currents fluctuated between 0 and 3 μA. This resulted from suppressed bursts, reduced streamers, and no Hermstein glow, the streamers coming in bursts as space charge cleared. Continuous corona set in at 9.15 kv, but streamers were still active. The continuous corona which with less Freon had been a closely adhering sheath on the anode changed to a very narrow, intensely localized, luminous channel or spindle. The contact point at the anode was very bright and 0.05 mm in diameter. The maximum diameter of the spindle was never greater than two or three times the diameter at contact. That is, unlike air, radial branching was much reduced, no doubt by reduced μ_i and electron attachment. As the potential was raised, the intensity of the spindle increased; it appeared to decrease in diameter and attained a length of 6 mm at 21 kv. The tip swayed laterally and gave the impression of streamers feeding into it. The current-potential curve lay only very little below that for pure air. A very significant observation was made that after the discharge had run for a very short time, say a series of ascending potentials, and the run was repeated after only a few minutes from 10^{-2} to 100% Freon, threshold potentials of the intermittent corona were shifted several thousand volts higher than those of the first run.

---

[9] Such action could come from the compensation for a decreased α in heavy avalanches by the action of negative ions in assisting avalanche growth.

The effect seemed to depend only slightly on concentration. In the lower part of the intermittent regime, the currents were far below those of the first run in the intermittent regime but converged toward the point of the intermittent corona.

A tentative explanation of some of these features may be given in terms of the suggested influence of the photoionizing free path in this gas mixture. The inhibiting effect of negative ion formation at the beginning of a run is not sufficient to raise the starting potentials. After discharge has run, Cl_2 and F_2 alter α radically. Attachment of electrons to Freon appears not to be important until fields produce dissociative attaching impacts. Thus, the initial influence of Freon must be ascribed to photoelectric ionization changes. This is indicated by two symptoms. First, the failure of the glow to spread over the surface of the anode as at lower percentages. This is quite analogous to what happens when pure O_2 is substituted for air at 760 mm, as will be noted in Huber's[2.22] coaxial cylindrical studies. That is, when the value of the photoionizing free path falls to 0.25 mm or less, the Geiger counter pulse does not spread over the surface, as shown in chapter 5, sections A.4, B.3, and C.2. Instead, very local burst pulses form that choke off the discharge, or else push outward as streamers. Thus streamer pulses are observed but require higher potentials than in air. With Freon, the burst pulse spread over the surface is *prevented* and is replaced either by localized small bursts that choke themselves off by their own space charge or else by streamers. These are intermittent because of the choking effect of the localized space charge. As potentials increase, the streamers push out further into the gap. However, the same circumstance that restricts burst pulse spread also precludes the formation of the uniform Hermstein glow discharge with negative ion sheath. The streamer pulse creates a localized space charge that stops the further streamers until it is removed. Any negative ions created in the streamer probably recombine with the positive space charge as they move in low fields and create no sheath. The photoelectrons are produced in the high field regions and no Hermstein negative ion sheath forms. There is one more peculiarity of these coronas, as depicted by Weissler and Mohr in figure 3.100. That is their failure to take on the paintbrush-like shape characteristic of air. That shape is caused by the heavy branching of vigorous streamers once they get away from the point, as noted by Nasser. The same short-range photoionizing absorption will act to inhibit excessive branching. Decrease in branching favors longer axial streamers, which will be aided by electron attachment in longer avalanches. Thus, streamers will be much more axial. The swaying and apparent branching of the streamers as they project well out into the gap might be attributed to the fact that the very nearly axial fine individual branched streamers in the sequence which the eye does not temporally resolve are deflected this way and that by the residual

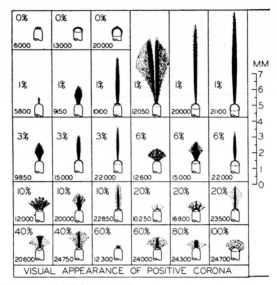

FIG. 3.100. Weissler and Mohr's sketches, drawn to scale, of burst pulse corona at streamer threshold and higher potentials in air-Freon mixtures at the percentage of Freon by volume in clean dry air and at the tip potentials indicated. The point diameter was 0.5 mm.

space charges of the antecedent streamer channels that have not adequately cleared.

The effect of a run of some duration is quite obvious. The discharge naturally breaks down the Freon to its constituent *radicals* such as CCl_2F, $CClF$, CF_2, CCl_2, and liberates quite a lot of Cl, Cl_2, F, and F_2. These may acquire electrons or appear as negative ions, but are soon recombined with positive ions. These products, especially Cl_2 and F_2, spontaneously attach electrons, since the electron affinity of Cl^- and F^- is greater than the dissociation energy of Cl_2 and F_2. These, therefore, act to raise the breakdown threshold materially by reducing α and thus the avalanche size. What they do to absorption coefficients is not known. In any event, until they are removed by secondary chemical actions in the gas or by absorption by glass and electrodes, the whole discharge character will alter. The detection of a very strong radiation in the region of 3.7 to 5.9 eV photons with 1% Freon in air by Waidmann indicates that some other radical changes are taking place.

When the potential reaches a high value, say around 9.1 kv, with a sufficiently high current density, it is probable that the intermittence ceases and a steadier current corona burns. This is apparently not entirely a Hermstein space charge sheath corona. As drawn by Weissler, it appears that, at 1% Freon, there is above 9.15 kv a glow that spreads over the surface corresponding to the Hermstein sheath. Thus, the space charge glow may be present over the surface maintained by negative ions created out in the gap by photoelectric ionization, either because of heavier currents or through the agency of the new photons acting in the dissociated gas. Still, single streamers were present and followed each other sufficiently rapidly as to add to the Hermstein glow to give the continuous corona observed. The single streamers, perhaps with less branching, produced such small current increments as not to spoil continuity. It is probable that, with the increasing intensity and heavier currents, enough photons of longer ionizing free paths became active to lead to the glow. This could occur, for the photons observed by Waidmann are copious and have energies close to 4 and perhaps 6 or 7 volts. These can perhaps act on some of the newly produced radicals to give photoelectrons.

Increasing amounts of Freon above 1% began to raise the threshold for the intermittent streamer breakdown. The value was increased from 5 kv for air to 13 kv for 100% Freon. The intermittent regime increased in range, extending to 25 kv for 100% Freon. The currents decreased for the same potential as the Freon content increased. This, as indicated, militates against formation of the glow, which requires more photons and heavy currents. Thus, at 20 kv the current in Freon was $\frac{1}{15}$ that in air. This reduction must in part be ascribed to decreased ion mobilities in such mixtures. The intermittency in all corona phenomena also leads to small currents, and this is again governed by ion mobilities. While in a burst of streamers currents may go up to 3 μA, there are long intervals during which the currents are zero. Decrease in the size of avalanches for a given field situation, that is, point potential, reduced the vigor of the resulting streamer currents. This current decrease is to be anticipated.

Increase in Freon content beyond 1% appeared to inhibit full development of streamers. The spindle shape was retained especially at higher potentials, but they decreased in length and increased in diameter. The discharge lost its well-defined shape, and became diffuse. The color of the discharge changed from the pinkish purple characteristic of air to a bluish white color. Aside from the increase in negative ion-forming constituents and the inhibition of avalanche growth, it is not clear what other changes greater quantities of Freon produced. It is probable that, as the concentration of chemical products and molecules of these foreign species increased, the photons changed and more photoionization was created further out into the gap. This would tend to cause more spread toward the Hermstein glow, suppress streamers, and increase branching in the streamers. At any rate, this is what appears to have occurred. A great deal more study is needed in this area to interpret fully these strange forms.

C. CORONA FROM WATER
AND ICE POINTS

1. CORONA FROM WATER POINTS

The experiments of Macky[3.62] in 1931 on the influence of water drops on the sparking potential between metal plates had been undertaken at the suggestion of C. T. R. Wilson in order to clarify certain important questions concerning the lightning stroke. Here it was observed that water drops falling in a uniform field in excess of 8000 volts/cm (depending slightly on the size of the drops) became distorted and lowered the spark breakdown potential of the normal air from 30,000 volts/cm to the value of 8000 volts/cm. The induced positive end of these drops was drawn out to a fine point which emitted spray and a characteristic streamer-like corona process. The induced negative end showed some pointing but most of it was clearly defined and had a faint glow superficially resembling a negative point corona. That a negatively charged water surface should yield a negative point corona appeared strange indeed. Such Trichel pulse coronas in air require that the surface emit electrons by positive ion bombardment or else by photoelectric effect. Since water does not have free electrons and binds its electrons firmly, such a discharge seemed strange. Despite earlier studies by Zeleny[3.63] done under conditions in which lack of modern techniques failed to reveal the facts known about coronas in 1947, it was thought imperative to study *corona* from both positive and negative water points separately under controlled conditions such as those used for metal points. This study proved to be exceedingly fruitful and later led to very important conclusions by the author[3.64] about the nature of the cloud to ground lightning stroke.

The study presented formidable experimental difficulties, since the control of curvature of the water droplet electrode surface, that is, water point, which is critical, was exceedingly troublesome. It was difficult to obtain uniformity and stability of the droplet surface without electric field at atmospheric pressure and below. Application of the field produced immediate distortion of the drop, ending in its eventual rupture at the potential at which surface tension was overcome. Thus, the available range of drop sizes was small, and constant change of drop shape with potential made it impossible to determine more than an approximate average value of the curvature of the drop and thus of the field. The problem was very skillfully carried out by W. N. English[3.65] in the author's laboratory.

Figure 3.101 shows the disposition of capillary tube and reservoir to give a liquid "point." The remainder of the corona study technique was standard and is described in English's paper on coronas from metal points.[3.15] For a study at reduced pressures which became necessary, both the bell jar and water reservoir were pumped. In arrangement b, the capillary was tried with and without a coating of paraffin wax. At low pressures the arrangement b gave a less stable drop than a. Zeleny used the a method. The results were, however, the same for both a and b arrangements. The resistance of the water column in a was 3 megohms for tap water and 30 megohms for distilled water. Adjustment of the reservoir height to give satisfactory results over the whole range was essential. Generally, the hydrostatic pressure was varied until a nearly hemispherical drop was obtained just below corona threshold. The reservoir level would then be about 1 cm above the capillary tip. Although boiled distilled water was used, air bubbles would often block the capillary at pressures below 300 mm Hg. At lower pressures the fields for corona onset were lower, so that the drop was not pulled out beyond the capillary. In consequence, negative corona took place from the corners of the *glass*. Parenthetically, that a negative corona should develop from glass is not surprising. Pyrex glass such as used in laboratory envelopes is a photoelectric emitter of electrons for photons in excess of about 4 ev.[3.66] Thus, excitation in air by electron impact is capable of yielding a photoelectric γ_p and second Townsend coefficient of value 10^{-3} to 10^{-4}.[3.66,3.67] This probably comes from liberation of electrons from nonbridging O^- ions at the surface. When these lose electrons, the Na^+ ions bound to them are released to enter the conduction band and increase the conductivity of the glass.

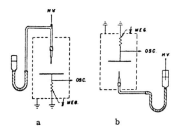

Fig. 3.101. English's two forms of apparatus creating water drop points of controlled diameters at the end of capillary tubes.

Adjustment of the reservoir under such conditions caused the drop to become unstable and fall off. These effects limited measurements to pressures above 200 mm. The observations made were as follows:

a. Corona gap visually observed with 70X telemicroscope.
  (1) In the dark.
  (2) With steady illumination at 135 degrees to the light beam.

    b. Photographically with Leica camera at $f2$ using Kodak Super XX film.

      (1) In the dark.

        (i) Time exposure.

        (ii) Moving film record by winding the film slowly by hand.

      (2) Illumination at 90 degrees from an Edgerton flash lamp (50 μA at 1800 volts) with condensing lens. The flash time was about 300 μsec and rise time of the flash 60 μsec.

    c. With Dumont 241 oscilloscope.

      (1) Visual.

      (2) Photographic.

    d. Current-potential characteristics.

In what follows, the drop end of the capillary will be referred to as the "main drop" or as the "point"; the water spray emitted in discharge will be called "droplets."

Failure of control of main drop curvature resulted in some variation of the values of the observed potential thresholds for corona and thresholds such as onset of steady glow corona. Values given are for a typical run. Slight variations of conditions caused variations of these thresholds of several hundred volts in some thousands for otherwise seemingly similar conditions. Important were the *relative* values of potentials for the various phenomena in a given run, not the absolute values. With great care to maintain conditions precisely constant, these relative values could be observed with an accuracy not far short of that reached with a metal point.

One effect noted was that described by Zeleny in which the spray threshold potential increased steadily at first and reached a steady value some hundred volts above its initial value. This was ascribed to cleansing of a contaminated surface of lower surface tension by the discharge. Hydrostatic pressure had a great effect on the spray threshold. Lowering the reservoir 0.5 cm changed it from 6900 to 7500 volts in one run. Vibrations were troublesome. The drop was not always *seated* in the same way on the capillary, presumably because of trace contamination spots on the glass. This caused differences in repeated runs and led to variations of the order of 200 to 300 volts between successive runs under otherwise constant conditions. If the point was not unduly disturbed, a consistency of ±50 volts could be obtained over a run of several hours.

A capillary of 0.13 cm outer diameter giving a point of about the same diameter of curvature was used in arrangement b (fig. 3.101) with a 3.2 cm gap and atmospheric pressure. Corona set in at 6800 volts. Groups of positive pulses were noted on the oscilloscope. They were of about 1000 μsec duration. Their forms were complex and of the three following types:

(1) smoothly increasing amplitudes with almost no streamer-like pulses superposed; (2) constant amplitude with streamer pulses; (3) decreasing amplitude with a succession of streamer pulses superposed. Under illumination, a very fine spray of drops could be seen shooting from the tip of the main drop, causing an almost imperceptible wetting of the cathode plate. In the dark, streamers were seen out to a few millimeters from the point. These were associated with the spray, for the luminosity was coarser and more intermittent than for the metal point. It was confined to the region of fine spray.

Increase in potential gave an increase in spray, in luminosity, and in pulse frequency up to 7100 volts. The spray pulses changed in shape with decrease in the smooth portions and increase in number and amplitude of the streamer pulses. Above 7100 volts, only separate streamer pulses occurred. At this point the usual paintbrush luminosity of the metallic point streamer corona was manifest. The tip was now covered with the uniform burst pulse glow. With illumination, the drop was now round and stable, no more spray being seen. Zeleny also had noted the stable regime. Further increase in potential did not alter the picture except to decrease the streamer pulse amplitude and increase the glow. At 9000 volts only the glow corona remained and became so bright that it could be seen in the lightened room. At 10,500 volts the drop was suddenly violently disrupted. Large chunks of water were torn off and large drops crossed to the plate, forming a puddle in a short time. In the dark, bright ragged luminosity followed the broken water surface and at times shot out into the gap. On the oscilloscope, large changes in potential were observed with intermittent streamer pulses. Increase in potential up to 16,000 volts merely increased the rate of dissipation of water from the capillary. Figures 3.102 and 3.103 show the oscillograms through this succession of events. Figure 3.102d represents on a much slower time scale, the disturbances caused by the large chunks thrown off. Figure 3.104 shows the glow from the spraying positive point near the threshold before stabilization; unfortunately, details showing the roughness due to fine spray are largely lost. The large paintbrush (b) represents streamers in the steady regime. Traces c and d are of steady and moving film exposures in the violently disrupted drop region. In contrast, trace e shows the steady trace of the metal point glow corona with a still photograph of the tip at the right. Figure 3.105 shows flash photographs of the capillary and points in the three regions. Exposures a show the distortion of the water surface in the threshold region. Exposures b show the droplets emitted in that region. Exposures c and d show the steady glow corona point at 8600 and 9300 volts. Exposures e show the droplets being formed and in transit at 12,500 volts.

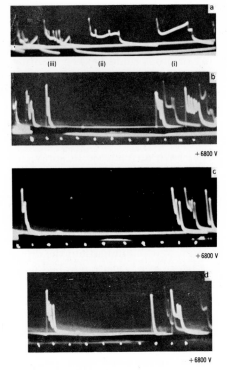

Fig. 3.102. Positive point spray pulses at 6800 volts. Timing interval in b, c, and d is 10^{-3} sec. The three types of pulses are displayed in the top trace and are designated i, ii, and iii. Trace d shows the pulses at a slower sweep rate when large chunks of water are torn off the surface.

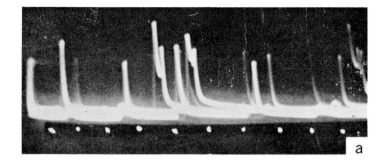

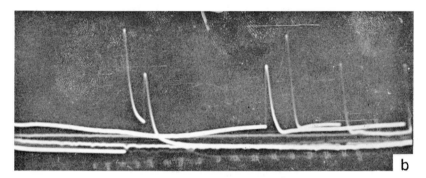

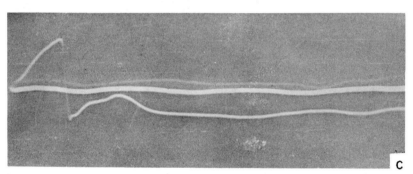

FIG. 3.103. (a) Spray pulses interspersed with streamer pulses, timing dots at 10^{-3} sec, 6900 volts. (b) Spray pulse streamer pulses and burst pulses, timing interval 2×10^{-4} sec, 8000 volts. (c) Electrostatic pulse due to a large droplet leaving the point in the drop regime at low amplification and 16,000 volts.

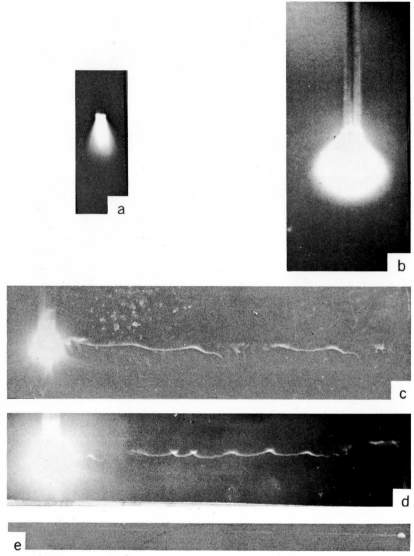

FIG. 3.104. (a) Spraying positive point, 3 min exposure, 7300 volts. (b) Stable positive point with streamers, 5 min, 8400 volts. (c) Second trace photograph of positive point in drop regime, left end. Right, moving film record winding the film in 1 min. (d) The same winding the film in 4 min, 13,900 volts. (e) Moving film record of a positive Pt point showing steady corona, 8750 volts.

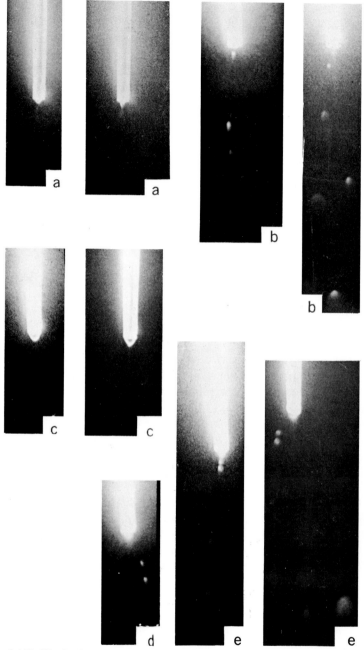

Fig. 3.105. Flash photographs of a positive drop point. (a) Distortion of spraying point, 6700 volts. (b) Spray droplets from spraying points, 68,000 and 7600 volts. (c and d) Stable drop at steady corona, 9300 volts. (e) Droplets ejected in drop regime, 12,500 volts. The formation of droplets is seen.

What is observed can readily be interpreted in the light of what we know. At and above 6800 volts, electrostatic forces overcome surface tension and electromechanical disruption of the surface begins. It is accompanied by creation of smaller points of water and fine spray, as noted by Macky, and these yield positive corona streamers in this region. As with a metal point, the pre-onset streamers appear first. It is difficult to interpret the rather varied types (fig. 3.102, i, ii, and iii) of pulses seen. In part they represent slow changes in potential owing to the drawing out and departure of droplets. These account for the larger pulses. In addition, droplets created are not uniform and length and size of these pulses vary. Different-sized droplets are also accompanied by different degrees of streamer activity. It is also likely that the type i pulses represent a burst pulse regime, with glow current increasing in time as the droplet approaches the cathode. The streamer pulses also appear superposed on the burst pulse corona. At the sweep rate used, the irregularities noted by Kip on metal point burst pulses would not be resolved.[3.2] Streamers superposed on burst pulses are generally weaker. Actually, some traces may represent simultaneous events at different regions of the disrupted surface which superpose. The streamer pulses probably appear as the fine droplets break off and are distorted in the high field close to the point. This causes the streamers to be projected straight outward away from the point.

The presence of water vapor, as indicated earlier, suppresses burst pulse and Hermstein glow, while it inhibits branching. The large drops with more extensive high field regions favor streamers relative to the Hermstein glow if photon absorption and photoionization are appropriate. When the range of high fields shrinks as for small points, the photons again form outside the high field and Hermstein's glow sets in readily. In any event, the patterns of luminosity, evidence of disruption, and occurrence of streamers and burst pulses account for observations in this troubled and turbulent regime. As potential increases, burst pulses and streamers become more vigorous and currents increase. As soon as current density is adequate, the space charge from the streamers and corona reduce the electrostatic forces at the water surface. Thus, disruption is prevented and the point soon becomes a smooth hemispherical point. With the *stabilized surface the discharge resembles that from a metal point*, and the *usual paintbrush-like pattern of the streamer corona appears*. Note, however, that *streamers persist to much higher potentials*, and, although there is a glow corona, the negative ion formation of a Hermstein sheath has not yet become important enough to suppress streamers. Here the very extensive potential range of pre-onset streamer occurrence due to H_2O vapor is again to be noted, as with Freon and with impurities. Branching in the vigorous higher potential streamers is still notable, since H_2O does not attach electrons as readily as Freon, but only reduces μ_i. Particularly interesting as potential increases and the Hermstein glow corona develops is the degeneration in magnitude of the streamers before they are suppressed. Eventually at around 10,500 volts, even with the space charge reduced water surface fields, electrostatic forces again lead to disruption of the surface. Here the disruption is so violent as literally to tear it to pieces. It produces large slow current fluctuations on the oscilloscope record. As the drops tear off, shatter, and cross the gap as larger charged masses of water, potential varies slowly. This action is complicated by an irregular streamer corona from the tips of the droplets.

After the study with the large water point, a very fine water point was used. This came from a capillary of outer diameter of 0.041 cm with a water point of that diameter. The gap length was 3.2 cm and pressure atmospheric. In this instance, no *streamers* were observed at any potential. This was a consequence of the suppression of streamers by water vapor *and the low potentials causing disruption*. At 4000 volts disruption in the form of pure spray, consisting of single almost invisible drops, was noted. These drops snapped off the main drop at regular intervals. The frequency increased uniformly with increase in potential. It is probable that here the droplet size was such that distortion incident to the formation of drops struck some oscillatory uniformity of the main drop. Thus distortion, disruption, and recovery occurred in a regular fashion. Increase in potential allowed regular recharging of the drop to disruption potential. It is clear that 4000 volts was below the electrical breakdown threshold, so that all that was noted was disruption. No luminosity was noted. At 4450 volts electrical breakdown began with a steady glow corona at the tip. Streamers did not propagate. With establishment of the glow at these low potentials, streamers could not form and the disruption of the drop ceased. At 9000 volts, the drop was drawn out to an elongated shape twice the diameter of the capillary. At potentials beyond this, it again broke off into drops. Here the breaking off was much more regular and increased in frequency as potential increased. The breaking drops gave oscilloscope pulses akin to those observed with disruption of the larger point, only more regular. Pulses can be created below corona threshold when small drops are forced from the large point by raising the reservoir. These pulses are essentially the same as those created by disruption separation above onset.

Here again interpretation is not too difficult. With a small point, fields producing disruption occur at lower potentials, since drop curvature is greater. The high field region about a finer point is much more restricted. The drop point at the smaller radius appeared unusually stable; ripples and irregularities were absent. Thus, disruption began at well below any breakdown threshold and the shape of the drop led to regularity. The

pulses were so like those caused by pressure-induced small droplet formation that it is clear that the pulses were not breakdown pulses but pulses
due to transport of charged droplets across the gap.

Again when surface fields reached breakdown values, the restricted
length of the fields with the water vapor present were not adequate to
create streamer avalanches. Instead they led to burst pulses. This trend
together with the small radius of curvature favored the creation of negative
ions beyond the high field portion. Thus the Hermstein glow corona set
in very early and no streamers materialized. Disruption at 9000 volts is
not surprising on the smaller drops. It is surprising that the disruption
did not lead to a streamer corona at this potential. This presumably must
be ascribed to the very *orderly* disruption even at 9000 volts. Apparently
drops were drawn off the diameter of the point in a regular fashion. There
was an observed elongation to twice the diameter, followed by a necking
in at the glass end as higher potentials further elongated the drops. Such
large regular drops were stable and spherical in the field and at best could
give rise only to a glow corona at their cathode side. Thus, no water points
or spray was formed, and streamers did not occur. This was probably a
fortuitously chosen geometry but could well apply to the smaller-sized
droplet regime.

Current-potential curves are shown in figure 3.106. Here the transition
from the shattered surface streamer regime, with its larger charge transfer
by electrostatic drop transport to the steady Hermstein glow plus
streamers, is clearly seen by a current decrease. Thereafter, the current
increases monotonically and smoothly. Currents are small compared to
those with metal points, owing to the high series resistance of the water.
The fine positive point shows a current-potential curve that is smooth.
Currents are higher at the same potential for the small point than for the
large, which is to be expected, since at 8000 volts it is at twice the starting

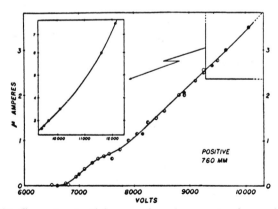

Fɪɢ. 3.106. Current-potential curves for a large water drop point corona.
The data are averaged for three separate runs.

potential while the large point at 8000 volts is at less than 1.14 times the starting potential.

Here some few words must be said about the negative point corona. The reason for this is that it relates to, but differs from, its positive counterpart in a significant fashion and it is needed for the discussion of the ice point studies.

The threshold for the breakdown from the negative point was, as might be expected, the same as that of the positive point for both large and small water points. Electrostatic surface forces are independent of the sign of the charge. Two types of pulses were observed. Both lasted some 1000 μsec, as did the large positive point pulses. These were flat pulses with multiple peaks, and large pulses of shorter duration. In fact, the latter strongly resembled streamer pulses but moved in a negative direction and were much more numerous. Traces of them are seen in figure 3.107.

FIG. 3.107. Traces of the negative water point corona. (a) Near threshold showing the two types of pulses. (b) At a higher potential. (c and d) At the highest potential.

They did not represent the regularity of the normal Trichel pulses from a metal point in air. With steady illumination, fine drops of spray could be seen projected from the point as for the positive point. In the dark, no luminosity could be *seen*, although *it could be photographed* in a 10 sec exposure. This characteristic differs materially from the negative Trichel pulse corona from a metal point, which is localized, very steady, and very bright. Increase in potential produced more frequent pulses and more spray. At 7100 volts a ragged fluctuating luminosity became visible in the dark. It occurred intermittently all over the surface of the main drop and was not concentrated at its outer tip. It extended only a very short distance into the gap and did not show the structure of the negative Trichel pulse corona from a metal point.

At 8000 volts the spray emission and ragged glow had spread all over the end of the capillary. Above this, spraying gave way to the separation of larger drops. Visually the transition was fairly abrupt, and at times the discharge would fluctuate between spray and larger drops. The current potential curves rose monotonically and smoothly for both large and small points. This is not surprising as currents fluctuated considerably all through, and at this point the oscilloscope screen was filled with pulses. Beyond 9200 volts, water was ejected from the capillary as with the positive point at slightly higher potentials. Figure 3.108 shows photographs of the corona made in the dark with still and moving point, and with a metal point in the two lower short traces. Note the familiar shaving brush fan with the metal point and the fixed location of the discharge points on the line. The irregularities in spacing of the *metal* point Trichel pulses come from the non-uniform speed of film advance. Note the random spatial location of the fine points of light both on the moving film and on the circular light patches for the still photographs at the ends. The differences in corona pattern are clearly seen in the flash photographs from a metal point *a* and *c* of figure 3.109 at 8100 and 9100 volts and in the pattern from the water point showing droplets at 8100 volts in *b*. The dark space characteristic of the luminous negative point corona from the metal can clearly be seen, while the light from the water point is not seen at all, until drop formation appears. The currents for the negative drop corona lie above the corresponding currents for the positive corona.

It appears clear that the negative point currents come primarily from drop formation. The Hermstein glow corona does not appear, and no stabilization of the surface by field changes from electrical discharge and space charge formation occurs. The current can largely be accounted for by spray and droplet charge transfer. For this reason currents for negative corona are larger than for the positive corona, since spraying is inhibited for both large and small positive points by the Hermstein glow corona. The currents from the glow corona are smaller than those created by droplet formation, as seen by the low voltage positive point behavior.

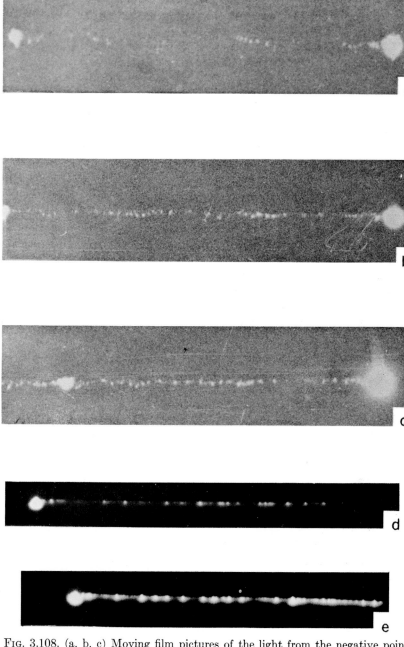

FIG. 3.108. (a, b, c) Moving film pictures of the light from the negative point water drop corona. (d, e) Those for a Trichel pulse corona from a metal point. Non-uniform speed of the film makes the traces seem irregular. Note shape of the Trichel pulses from metal and compare with the irregular vertical disposition of water drop corona luminosity, caused by droplets torn from all over the water point surface.

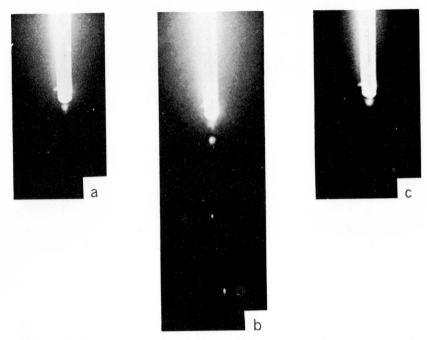

FIG. 3.109. Flash photographs of the negative corona (a) from a metal point at 8100 volts, (b) from a water point showing droplets, and (c) from the metal point at 9100 volts. No light is seen until the water drops separate.

Still to be explained is the feeble luminosity of the negative point noted by both Macky and English, since it appears that, according to prediction based on the nature of the water surface, a negative point discharge was unlikely. In interpreting English's results the author accounted for the feeble luminosity as being due to a streamer or a burst pulse corona produced by the high field of the ruptured water point on the cathode side of the droplet as it separated from the point. That is, the negative main point induced such a high field on the cathode side of the receding ejected droplet that a short-lived positive streamer or burst pulse projected positive charge from it to the negative point. This accounts for the streamer pulses and burst pulses and the ragged luminosity observed close to the negative water point. As the high field region is only near the main point, the streamers will be confined to this region. Since the regions disrupted are from all parts of the point, they cause quite a jitter in distance of appearance of the small flashes relative to the tip of the capillary, as seen in the moving film traces.

To verify this, a crucial experiment was performed. It is known that the starting potential of the true corona for a given gap will decline almost in proportion to the decrease in ambient gas pressure. This led to a study of the starting potentials of positive and negative point discharges as a function of the pressure. It was a difficult experiment and one limited in range, since below 200 mm Hg the edges of the negative *glass* capillary forming the water droplet undergo a corona discharge of their own. English observed that, as pressures decreased from 760 mm, the starting potentials of negative and positive water point breakdown were the same and were always accompanied by spray formation until pressure of 505 mm were reached. Below 629 mm the positive corona started *without* spray formation. That is, a stable droplet and steady corona set in with burst pulse corona. As pressures decreased to 200 mm, the starting potential of the positive corona fell in a nicely linear fashion from 7200 volts to 3400 volts, as shown in figure 3.110, taken for a fortuitously larger drop than that mentioned below. The scatter of points is remarkably small, considering the difficulties in adjustment of the point. This behavior is just

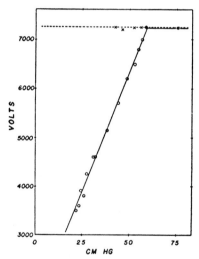

FIG. 3.110. Onset potential V_0 against pressure for corona from positive and negative water points. The full line is for the positive drop. Its potential falls linearly as p decreases, beginning at 505 mm. The negative corona potential remains at 7300 volts irrespective of pressure, which is the potential for droplet disruption by electrostatic forces.

the same as for a metal point of the same diameter. The spraying does not occur for the positive water point at pressures below 500 mm because below this pressure the discharge begins before disruption at 5800 volts and stabilizes the drop. Thus, the spray disruption potential for the positive point lies between 5800 and 6400 volts. For negative it lies around 6100 volts. Data are difficult to get, as points are not exactly reproducible. In contrast, the negative point had a threshold potential which remained constant at 6100 to 7200 volts, depending on point radius, with spray formation down to 200 mm pressure.

Thus it is clear that the negative water point does not give a true corona. It does disrupt, and, if fields are adequate on disruption, a discharge takes place, close to the point surface from the induced positive end of the receding droplets, which may be a streamer or a burst pulse.

Calculations made by Macky, Zeleny, and English indicate that electrostatic forces become more important than surface tension forces for droplets of the diameter used at about the potential observed for threshold, that is, 6000 to 7100 volts in the droplets studied. This leads to the fine spray at 760 mm on both positive and negative points, which continues down to 500 mm with the positive point observed. As the droplets break off they are in fields of the order of 5×10^6 volts/cm at the droplet surface at a radius of 0.06 cm. At a distance of 3.1 drop radii (that is, at 1.8 mm), fields are 5×10^4 volts/cm. These are of streamer-forming magnitudes, provided the fine droplets are stable. Stability depends on their size and the induced surface charges. In any event, a positive corona from the advancing end of the ejected positively charged droplets is possible. A similar discharge from the induced positive ends of the receding droplets from the negative points must occur within a distance of 0.5 to 1.5 mm of the negative droplet. Again the streamer discharge from a positively charged droplet emits ions which can positively charge, or neutralize, the charge of negatively charged droplets ahead of them. The ions have a high mobility compared to the velocity of droplet ejection. Thus, negatively charged droplets from the disrupted negative point will be neutralized and positively charged by preceding drops, and so more readily enter into corona on their trailing ends. This would account for the luminosity seen near the negative point, especially as discharges become more violent with increase in point potential.

If potentials on positive points reach values such that the ejected droplets and the points torn out from the disrupted surface create a positive space charge, the field at the droplet surface is reduced by its space charge. Such a reduction and alteration of field configuration is known to suppress streamers by reducing avalanche size. Whether the negative ion sheath of Hermstein forms that early or whether it is positive space charge alone, the surface field at the main drop is reduced while that further in the gap is enhanced. This reduction is such as to prevent surface rupture. Therefore, higher pressures and larger radius positive water points shortly after rupture and streamer formation yield space charges that reduce the field, stabilize the drop, and give a normal glow corona. Thus, such points act essentially like metal points. At still higher potentials, when the Hermstein sheath breaks down, the disruption of the surface again appears and at the higher potentials is more violent than at rupture threshold. At lower pressures, or for some smaller radius positive drops, the electrical break-down may occur at potentials and surface fields below the rupture point.

Thus, the initial spray phenomena are absent. One must use care in interpretation, since, as was apparent from English's studies, point cur-vature and surface fields do not *alone* enter into the question of rupture. It is clear that possible modes of oscillation and deformation of the surface by the fields may cause such distortions and changes of surface configura-tion that calculations are not possible. Again certain radii of points may be quite unstable at a given potential, whereas an intermediate radius may be stable. The modes of rupture will certainly vary with point radius and field strength, especially if deforming oscillations occur. With a very small water point at which no streamers occur, the pulses from positive and negative points are identical. Since for the positive point the discharge is unquestionably a burst pulse corona, it may be assumed that the corona from the fine negative point is a burst pulse corona from the induced positive side of the receding droplet.

The argument against the Trichel pulse type of corona from the negative water drop is strengthened by the form of the glow seen. The glow is sep-arated from the negative surface for some distance by a dark space. That is, the negative droplets must recede a distance of ≈ 0.5 mm from the positive point before they are neutralized and before the air gap is long enough to yield avalanches giving a burst pulse or streamer corona on the induced positive end. However, this corona never assumes the shaving brush-like appearance of the typical Trichel pulse corona. Furthermore, the width of the dark space and the dimensions of the glow are unaffected by decrease in pressure, whereas a characteristic feature of the Trichel pulse corona is the increase in these dimensions as pressure decreases. In contrast to the failure of the water points to give a negative corona, the corners and sharp points on the surface of the ground end of the glass capillary give a very bright corona consisting of myriads of sharp Trichel-like pulses on the oscilloscope. Visually there was a bright brushlike corona that was rather coarse and streaked. This type of corona from a glass surface is now believed to be most likely, since Townsend discharges occur at glass electrodes in uniform field geometry.[3.67] Glass has a photo-electric γ_p on the order of 10^{-3}. As it is a poor conductor, it will, when properly overvolted, lead to a negative point breakdown of considerable magnitude. The breakdown will have a very short duration even in non-electron attaching gases such as clean A. Its surface rapidly acquires a neutralizing positive charge from the ions which are not conducted to ground as quickly as they diffuse to the point from the plasma. Potential thus drops sharply. Note that even glass shows a more characteristic Trichel pulse-like form and behavior than does the water point.

2. Corona from Ice Points

The absence of a corona from a negative water point led naturally to the conclusion that coronas would not propagate from negative ice points, since ice, like water, would be expected to be a poor secondary emitter. It appeared in a 1950 conference on thunderstorms that the question was not trivial. The important role of water drops in the electrical cloud drainage and funneling of the charge from a cubic kilometer of cloud through the branching and spraying of the positive point corona from water drops raised the question of the mechanism of lightning strokes from snow clouds.[3.64] The question was raised at the conference, and the author assigned to H. W. Bandel[3.68] a study of the corona from ice points. This was not an easy study, since the ice point had to be created and maintained at low temperatures. Points of 1 mm and 2 mm in diameter were cast in ice molds, melted out, chilled, rounded by dipping in water, and "fire polished." They were tested in air at atmospheric pressure. The temperature was maintained at $-78°$ C by dry ice in alcohol. A Pt point 1 mm in diameter was used for comparison. The water used was tap water and distilled water. Ice, in contrast to water and Pt, has a high resistance. For tap water ice the resistance of the ice point ranges from 10^{11} to 10^{13} ohms. For distilled water it is 10^{14} ohms. This will influence corona manifestations because discharges can become intermittent. The breakdown will discharge the capacity of the point, and the RC time constant for charging is important. The gap length was 8 cm. The current-potential curves for positive and negative points of 1 mm diameter, composed of distilled water, tap water, and Pt in the 8 cm gap are shown in figures 3.111 and 3.112.

Both sets of curves show that there is a sharp rise from low order currents of 10^{-14} to 10^{-13} amp to currents from two to more orders of magnitude higher in both cases. The steady state currents are materially higher for Pt than for tap water ice; the lowest in the group are the steady currents for distilled water ice, as one might expect from the relative resistance.

The starting potentials of the coronas represented by the sharp rise appear to differ for the three points. This is accidental and has no basic significance. They merely indicate the difficulty of getting accurately contoured hemispherical points of the same diameter with the ice. The starting potential is extremely sensitive to point shape and radius. There is also the fact that the air at 76 mm pressure at $-78°$ C or $195°$ K is at a density equivalent to that at a pressure of 1150 mm at 22° C. If the pressure is 500 mm, the density at 195° K is the same as at 760 mm pressure and 22° C. This means that the starting potential for the 1 mm point is correspondingly higher, that is, some 10,000 volts instead of 6500 volts.

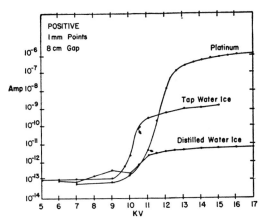

FIG. 3.111. Current-potential curves from a positive ice point corona for distilled water ice, tap water ice, and a Pt point. Except for decreased currents due to the high resistance of the ice and small inequalities in point radii, the curves are similar.

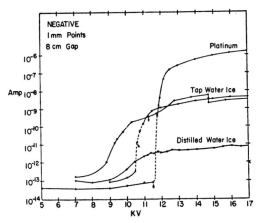

FIG. 3.112. Current-potential curves from a negative ice point corona from distilled water ice, tap water ice, and a Pt point. Note the difficulty of duplicating results with tap water ice, owing to varied conductivity and difficulty of diameter and shape control.

Of importance is the fact that at the rise, an intermittent corona occurs which is nearly the same as that for the metal point with both *polarities*. The positive Pt point, because of increased gas density, yielded only a streamer pulse corona at threshold. Below 500 mm pressure, specifically at 400 mm, both burst pulses and streamers were observed. Pulses were influenced by the low conductivity of the ice point. In fact, for the pure water ice no pulses could be seen up to 19 kv. The energy stored was too small. Pulses were random in time for both ice points. They increased in amplitude and frequency as potential increased. There were seldom more than four in a single 0.03 sec sweep. Decreasing resistance by shortening the point increased pulse amplitude. With the 2 cm diameter point, the pulses definitely appeared to be streamers which could be seen visually extending out to 6 mm into the gap. The pulses from the ice point were smaller than with the Pt point and were rounded off. With the smaller points the discharge could not be seen visually below 10^{-9} amp but could be photographed on long exposure.

With a negative point it is noted that there is a corona setting in at about the same region of potential as with the Pt point. It was particularly hard to fix accurately the threshold potential for the rise, because of the small triggering volume for a negative point and the erratic shapes. The 1 mm point did not reveal much to help identify the discharge as a Trichel pulse corona. However, the oscilloscopic traces were of that character. With the 2 mm point, the discharge was visible as a diffuse brush with a *fainter* glow visible over the point. The faintness and shifting made it impossible to identify negative glow and Faraday dark space as had been done with the Pt point.

With an ice point there was not any question of the shattering of a water surface. The breakdown of the negative point was unquestionably an electrical breakdown of the gas in the gap and not a shattering of the surface leading to an induced backward positive point corona. There seemed no escaping the conclusion that an ice point yielded a negative point corona in air, while an analogous water point was unable to do so. In fact, this paradox was so striking that Bandel hesitated to publish his observations. Fortunately the author happened to discover that in 1912 in Lenard's laboratory, W. Obolensky had studied photoelectric emission from various surfaces.[3.68] Obolensky reported that measurable photoelectric emission did not occur for water surfaces above 1300 Å wave length, while such emission occurred for ice in the region between 1800 and 2000 Å. These currents had a value on the order of 10^4 times that for water. The emission for this wave-length region with ice was 0.7 that for an oxidized Cu surface. Thus ice has a γ_p of sufficient magnitude to initiate a glow discharge with highly stressed cathode, while water fails to do this.

The conclusion is that we can expect positive and negative coronas from ice points. The currents from these coronas are low, because of the high resistance of the ice. Thus currents are normally very low. They will therefore not contribute materially to electrical breakdown as do water droplets.

D. TIME LAGS IN POSITIVE
POINT CORONA

Only one study of this phenomenon has been made. It was an excellent one carried out by Menes in a doctoral dissertation and published under his name and that of L. H. Fisher.[1.11] The study was very completely and carefully carried out. In this study, as in Fisher and Bederson's work,[3.37] a potential impulse causing breakdown is applied across the positive point-to-plane corona gap at zero time, and the time is observed from the application of the potential until the corona appears. The gap consisted of an 11 cm diameter brass plate with a polished W needle making the anode point. Radii were 0.007, 0.02, and 0.03 cm with gaps of 0.5, 1, and 1.5 cm. The gap was in a 14 cm diameter brass tube of 12 cm length with windows for illuminating the cathode with quartz ultraviolet light, and a window by which the discharge could be viewed visually and with a photomultiplier to scan the gap. Pressures of air dried chemically and passed through a liquid air trap were 700, 500, 300, 200, 100, 50, and 30 mm. The gap was fired by a stabilized voltage supply, steady to one part in 5000, which was set above threshold *bucked* by a potential of 1000 volts. When this bucking potential was removed through a 5D21 switching tube, the breakdown potential was suddenly applied. There was a dynamic drop in the switching tube of 55 volts at 1000 volts bucking potential. This was reproducible to within 3 volts. The time to switch from the approach voltage, that is, the applied potential less the *bucking* voltage to within 5 volts of the final voltage, was found to be of the order of 0.03 μsec. The current pulses from the corona wire were amplified, delayed 0.3 μsec, and displayed on the oscilloscope tube. The sweep of the latter was triggered by the switching tube. The switching tube also sent a small voltage pulse through the capacity of the gap to the amplifier. This yielded a time *reference mark*. Pulses as small as 10^{-6} amp could be observed across an input resistor of 1000 ohms.

One of the first things that Menes did was to study onset, or threshold voltage, with steady and with the pulsed supply. This study was essential. As has been noted before, current rises quite abruptly in the gap from values of the field intensified photocurrent of the order of 5×10^{-11} amp to the order of 5×10^{-7} amp or less as potential increases. This region of steep rise is marked by the appearance of occasional streamer and/or burst pulses, and it is very hard to determine the *exact* threshold. There is a regime of some 20 to 200 volts, below steady or Hermstein glow onset, depending on the threshold voltage of the intermittent pre-onset discharges, and the exact relation of the steep rise to this regime is difficult to fix. With weak ultraviolet illumination the onset potential of the steady glow corona is more accurately fixed.

Menes apparently has miscalled the *onset* potential of steady corona the threshold *value* for steady corona. He chose the impulse threshold as the lowest impulse voltage at which he could observe discharge pulses on the oscilloscope. It is clear that his oscilloscopic instrumentation was too insensitive to observe burst pulses. The absolute accuracy of the impulse voltage was ± 15 volts. Thus with an applied impulse voltage of *nominally* 4000 volts, the potential could be 3985 or 4015 volts. The steady potential threshold was much more accurately established. At times steady and impulse voltages differed by as much as 60 volts. This he assumed to be far beyond any known experimental error. At higher pressures for the larger point radius, he found that the observed impulse threshold was some 60 volts *above* the onset voltage with steady potential. At lower pressures they were nearly equal. This difference resulted from the width of the Geiger counter regime (between pre-onset pulses that were not detected with impulse and onset of steady corona), which is very much smaller at smaller voltages. The difference was greatest for the 0.03 cm radius point, and with the 0.007 cm radius point the impulse threshold was, if anything, 20 volts lower than the steady corona onset potential.

Actually, Menes should have made a comparison between thresholds for the intermittent corona in both cases. Both thresholds should be below onset of steady corona, and, barring errors in estimating the true value of the impulse potentials active, the impulse threshold should coincide with the steady potential threshold. It is easier to locate the threshold with steady than with pulsed application, and this might in general lead to observing higher pulse thresholds in this manner. The systematic variation of the difference with point diameter the higher the pressures indicates a possible difference of space charge conditions in the gap with pulsed and steady potential application, which could raise the streamer threshold under pulsed conditions.

Menes also attempted to calculate the current rise as a function of potential below onset. This calculation assumes an initial rise of the field amplified current across the gap, but ignores space charge effects. Thus observed currents lie below the computed ones. At higher potentials where the current rises sharply, Menes again appeared to forget that the sharp rise is the true beginning of a self-sustaining breakdown that interrupts itself by local space charge near the point. This region is not one of a field intensified current. It is one not open to exact computation, as it consists of isolated pulses (streamer or burst) of self-sustaining discharge which increase in frequency as the potential increases. They are longer than the field intensified photocurrents. Thus the calculation cannot be carried out in this region, initially because of space charge inhibition and because an intermittent space charge limited breakdown pulse sequence is not a field intensified photocurrent. The *onset* of steady corona (probably the Hermstein glow corona) is *not* the breakdown threshold. Menes' attempts to ascribe the failure of the observed curves to measure up to the computed curves in terms of electron attachment may not be correct, because in the zone of electron multiplication by the field, the electron attachment is primarily included by the lowered value of α. The photoelectrons which start from the cathode are negative ions long before they reach the ionizing zone near the point where they again detach in the high field ionizing zone.

Menes' time lag measurements showed much scatter. This decreased with increasing photoelectric emission by ultraviolet light and with increasing overvoltage. Too high a photoelectric emission wiped out the threshold and the pulses in consequence of space charge distortion of the gap, as Kip[3.1] long ago observed. Thus it required twenty-five measurements of time lag for each setting. Figure 3.113 shows a Laue diagram of

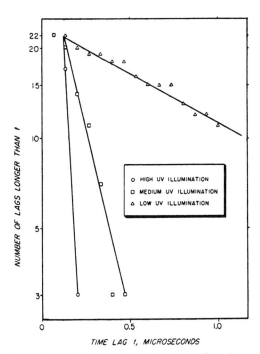

FIG. 3.113. Laue plots of observed time lags for three intensities of ultraviolet triggering illumination of the cathode. The overvoltage above onset for steady corona was 0.18%, point radius 0.3 mm, gap length 1.0 cm, pressure 200 mm Hg. Each curve represents 22 observations. The common point of intersection of the lines represents the formative time lag of 0.15 μsec.

the number of lags exceeding t, plotted to a logarithmic scale relative to the time lag in microseconds, for three different ultraviolet illuminations. Note that all plots give straight lines that intersect at one point. Here at 0.18% overvoltage, for a 0.03 cm radius point at 200 mm pressure with a 1 cm gap, the time lag is 0.15 μsec.

If the mean scatter time is computed from the mean statistical lag, as shown by the linear plots, the initial photocurrent from the cathode yields the initiating probability of one photoelectron. The initiating probability for electrons from the cathode was calculated from observations and starts about 0.7% at threshold, increasing to 7% at 3% overvoltage. These values generally represent those observed for various gaps. Figure 3.114 shows the formative time lags for the 0.03 cm radius point at 1.5 cm gap setting over the whole pressure range, from 30 to 700 mm, at from 0 to 4% overvoltage. Note that the curves are relatively flat. From 700 to 100

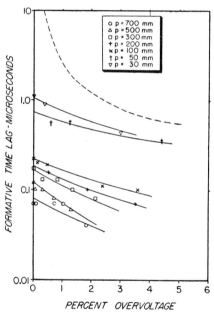

FIG. 3.114. Formative time lags for Menes' positive point corona with point radius 0.3 mm and gap length 1.5 cm as a function of percent overvoltage and a range of pressures. The dashed line represents Fisher and Bederson's curve for a uniform field of 1 cm gap length. The time lags remain short and decrease very little as overvoltage increases. For a uniform field gap at 0.5% overvoltage, there is an increase of 10- to 100-fold as 1.2% overvoltage is approached. The longest lags are for low pressures, as they should be with large $1/\mu_i$.

mm pressure, the lags bunch around 0.1 μsec to 0.3 μsec. At the two lower pressures the lags appear to be near a microsecond, which may indicate an increased diffusion and raises the question whether the streamers are the mechanisms active. Note the different curve of Fisher and Bederson[3.37] for a uniform field gap in which a glow discharge precedes the streamer mechanism and initiates it 6% below the normal streamer spark threshold.

These data are plotted in figure 3.115 for the 0.03 cm radius point as time lags at threshold for a series of different pressures. Note that the formative time lag above 100 mm pressure decreases linearly as pressure decreases, and, except for the two points at 30 and 50 mm, they are relatively independent of gap length. For comparison, the electron transit times of the whole gap for the gaps of different lengths are shown, as well as the transit times of the high field ionizing regions. Note that formative time lags are on the order of or less than electron crossing times, except below 100 mm where the corona discharge may be a glow. Data for the 0.02 and 0.007 cm radius points are shown in figures 3.116 and 3.117. Here there is more variation of the lags as pressure goes up at lower pressures, indicating possible successive generations because of the large $1/\mu_i$ as points become smaller. It is concluded that the formative lag *is, in general, shorter than the time of electron crossing from point to plane.* This shows conclusively that cathode secondary mechanisms play little role in the breakdown, except at lower pressures and for the smaller points as photoionizing free path increases. That is, the corona comes through ionization and regeneration very near the point, except for smaller points and very large pressure. However, the *high field* transit time is one order of magnitude or more less than the formative time lag, indicating repeated cumulative photoelectric corona build-up. This might indicate a succession of electron avalanches contributing to the breakdown in the high field region. This is as it should be. All that the time lag determination measures is the time between the application of a potential and the time which one electron, presumably liberated from the cathode or existing in midgap, takes to cause a streamer current of such magnitude that it shows on the oscilloscope trace as a sharp pip or blob of light. It does not really say anything about how the streamer arises.

We do not know which electron starting from what point triggered the streamer. The field intensified photocurrent at the approach voltage $V - 1000$ volts could have had ions or electrons in transit that could have triggered the start of the breakdown at any time after the voltage was applied. Thus, the formative lag could have been accounted for by electrons in the high field region only in repeated ionizing acts, or by a single electron or group of electrons leaving the cathode at impulse time.

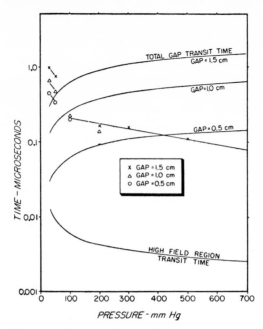

FIG. 3.115. Formative lags for the positive point corona for r = 0.3 mm and gap lengths of 0.5, 1.0, and 1.5 cm as a function of pressures. Down to 100 mm all lags fell on a single curve. Below 100 mm lags are greatest for the 1.5 cm gap and least for the 0.5 cm gap. Computed electron transit times for the whole gap and for the high field region are also shown. Ion transit times lie in the region of 200 times the electron transit times.

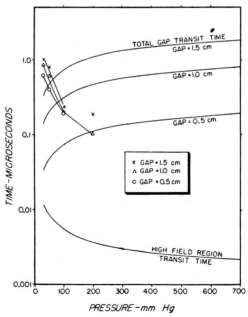

FIG. 3.116. The same as figure 3.115 except that the point radius is 0.2 mm.

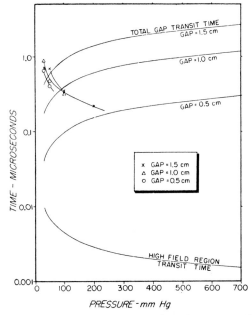

Fig. 3.117. Same as figure 3.115 except that the point radius is 0.07 mm.

But the breakdown can involve no more than the one electron crossing the gap, and probably only a small portion of the gap, except at the lowest pressures and for smaller points. The insensitivity of the lags to the gap length indicates that the electrons involved presumably did not cross the whole gap. In any event, these observations without a doubt establish the streamer or burst pulse theory at higher pressures.

E. FREE ELECTRON GASES

1. INTRODUCTION

Relatively little has been done on coronas in pure non-electron attaching gases. Townsend's[2.23] group did some work with coaxial cylinders at various times. Certain anomalies in relative threshold potentials of positive and negative coronas observed by Huxley[2.23] and Boulind[2.14] prompted Penning,[3.69] using the very clean gas techniques of the Philips research laboratory, to investigate the phenomena. They indicated in that region the very powerful influence of the Penning effect caused by traces (down to one part in 10^5) of impurities ionized by metastable states of one gas.

This was particularly true for Ne with A as an impurity and for Ne and A with Hg and Cl_2 as impurities in trace amounts. These studies on both positive and negative coronas, done at pressures near the minimum potential, will be reported in the chapter on coaxial cylindrical coronas.

It is logical that, after the studies of coronas in air, the author should have assigned to Weissler a study of coronas in such gases.[1.8] At that time, the need for extreme purity was recognized. However, there were no adequate induction furnaces for outgassing metal electrodes in vacuum, and the technique of using electron bombardment later used by Miller[1.8] was not possible. The metal valves had not as yet been developed, so that stopcock greases were always a source of trouble. Gases were reasonably pure, but not of Penning purity. Later work by Miller[1.8] and Lauer[2.21] with coaxial cylinders improved the situation by using electron bombardment to outgas metal parts by heating for ten hours at above 800° C. The study of Weissler was later repeated and extended using Penning effect gases as well as additions of CO_2, O_2, and N_2, with point-to-plane geometry by Das in Schulz's laboratory in Karlsruhe.[3.70] Here Das worked with reasonably clean gases. He did *not outgas* his electrodes by electron bombardment and again used glass stopcocks. His gases were commercial grade purity gases of the same quality as Weissler used. Weissler did not study the Penning effect mixtures, but Das did. However, Das never used impurities in the range below one part in 10^4 and usually worked with one part in 10^3 and higher. He thus missed the most characteristic sensitive Penning effect ranges. In fact the purity of his Ne was such that his data really apply to a neon with 10^{-5} and more parts of A. Thus he was recording *as characteristic of pure Ne the most efficient Ne-A Penning mixture.* Consequently he obtained thresholds much too low for what he called pure Ne. These factors must be considered in reporting and evaluating the work of Weissler and Das. Only He is easily obtained pure to better than one part in 10^8, and that is in an Alpert type system using He from a super fluid leak at liquid He temperatures. Today high purity in Ne, He, A, and probably N_2 can be obtained by Alpert techniques and gas cataphoresis cleansing. Commercial grade A can only be purified of Kr and N_2 by Ca arc treatment or, more effectively, by gas cataphoresis at pressures below 50 mm. Ne requires gas cataphoresis to rid it of A, and He to rid it of Ne.

The work of Das suffered from other deficiencies. He used a confocal paraboloid formula for calculating the field along the axis of a hemispherically capped cylinder. This is a poor approximation in the very critical region near the point. He was also apparently unaware of the important Townsend[1.6,1.7] relation (equation 2.29) for the space charge limited current for a continuous corona in which the current is exactly given by an equation of the form:

$$(2.29) \qquad i = \frac{2K}{p} CV(V - V_s).$$

Here K is the ion mobility at 1 mm pressure, p is the pressure in milli-meters, C is a constant dependent on the dimensions of cathode and anode, V the applied potential, and V_s the threshold potential. The only change created by the change in geometry lies in a change of the way in which dimensions enter into the constant C and the absolute value of the con-stant. The dependence on K, p, and V should be unchanged. Where $V - V_s$ is small and changing rapidly relative to the change in V, the cur-rent i is approximately linear, as observed by Kip,[3.1] Das, and by the early Geiger counter workers.[1.7] As V becomes very large compared to V_s, then i is proportional to V^2. In between it varies along a curve that can be approximated by a convenient power of $V - V_s$. As observations of i over a range of V are bounded by V_s and the value of V for sparkover, the range is limited. Thus Das empirically set i as:

$$(3.6) \qquad i = C_1\delta^{-C_2}(V - V_s)^{1.4}.$$

This purely empirical relation has no theoretical significance and conceals the real relations of the variables. The quantity δ is the relative gas density to that at 760 mm at 20° C. The constants C_1 and C_2 depend on the gas used. As he used fixed geometry, Das was not aware that C_1 and C_2 would change with geometry. Neither Weissler nor, apparently, Das was aware of the Hermstein glow corona, as it was discovered about contemporaneously with the work of Das. Finally, in interpretation of his results, Das tended to oversimplify his explanations, being unaware of the considerable com-plexity and body of antecedent knowledge. With these reservations one may proceed to discuss the data of Das and Weissler.

2. WEISSLER'S STUDIES: PURE H_2, N_2, AND A

a. *Techniques of Weissler*

Weissler used Pyrex tubes outgassed by heating to 400° C in vacuum with metal electrodes, which were heated by an induction furnace so that the metal parts glowed dull red for three hours. Electrodes were Pt hemispher-ically capped cylindrical points of 0.5 mm diameter polished with SnO_2, and gap lengths ranged from 3.1 to 4.6 cm. The walls of the tube were rendered conducting with a thin evaporated gold film. The tubes were filled with the gas through standard purifying trains followed by liquid air traps, Hg being carefully excluded. With N_2 a W filament was burned to remove the last traces of O_2. The gases were tank gases. The tubes were sealed off after fill-ing. A stabilized power supply was used and the discharge was viewed by a 70X telemicroscope. An oscilloscope was used to detect pulses. This had a resolving time of some 10^{-5} sec.

b. *Results in Pure H_2*

With a 3.1 cm gap, threshold for the discharge occurred at 3500 volts with 4 μA. A feeble localized light spot of 5×10^{-3} cm diameter appeared at the tip. In contrast to air, this is a very confined discharge at the tip of the point instead of spread over the point as for a burst pulse corona. At threshold potentials streamers were not observed, and the discharge seemed to be due to a succession of field intensified avalanches. The field intensified avalanches gave small irregular kicks with four stage amplification because of irregular sequences of avalanches, as noted by Lauer. At 9 to 10 kv, weak streamers appeared which increased in intensity and length with further increase in field, and at 18,000 volts streamer sparks occurred. The current-potential curves were of the conventional Townsend form.

Little comment is needed here. As pure H_2 does not attach electrons except dissociatively at high electron energies, negative ions do not form. H_2 on outgassed Pt subjected to a glow discharge in H_2 gives off copious photoelectric emission, leading to a high γ_p at lower pressures and low X/p. At short gap lengths these were the source of electron avalanches. Thus the corona is a Townsend cathode conditioned breakdown, as observed by Lauer, in coaxial cylindrical geometry. Photoionization in the gas is very small, and the photons responsible are very highly absorbed. Thus burst pulse spread over the point surface is precluded. Avalanches increase in size until they can produce enough photoionization to launch streamers. These appear at nearly three times the threshold for detectable avalanches. The fluctuations observed in the initial stages of breakdown could have been caused by bursts of avalanches separated by crossing times of electrons from the cathode. Positive ion space charge accumulation from a group of avalanches will also interrupt the succession of avalanches. The H_3^+ ions have a relatively higher mobility than in most gases (13.4 cm^2/volt sec relative to <4 cm^2/volt sec), so that fluctuations could well be more rapid. Such fluctuations were observed by Lauer in coaxial geometry in H_2.

Needless to say, addition of as little as 0.1% O_2 led to the appearance of pre-onset streamers. Threshold was raised from 3500 to 5000 volts. The increase in threshold is to be attributed to a reduction in γ_p, possibly by photon absorption or action on the cathode. But the presence of streamers and marked intermittence points to the suppression of the Townsend discharge with the introduction of O_2 on the clean cathode. Then the increase in threshold can be ascribed to a transition from a Townsend breakdown to a streamer breakdown, which even with O_2 has a higher threshold. These streamers increased in intensity and led to breakdown streamers at 8000 volts, followed shortly by a spark. The amount of photoionization resulting from the long photoionizing free path in the O_2 mixtures presumably propagated a discharge over the surface and led to a burst pulse corona. This first appeared with 1% O_2. As indicated in the theory of spread of Geiger counter pulses, there is a very limited range of conditions in which this can occur. The absorption coefficient for the streamer producing radiation must have been relatively too large in the 0.1% O_2 − H_2 mixture.

c. *Results in Pure A*

In this gas there is a marked difference between results reported by Weissler and by Das.

Weissler used a gap of 4.6 cm. His electrodes were relatively well out-gassed and his A was relatively pure. Das used a recirculating system over heated calcium metal flakes and a cold trap. It is thus very difficult to compare the purity of the gases. Heated Ca getters can be a source of contamination as well as of purification. It all depends on the Ca. As commercially purveyed, Ca is immersed in a liquid hydrocarbon and has a crust of oxide and reaction products on its surface. The hydrocarbon must be removed and the Ca flaked. *Pure* Ca can be had only by *repeated vacuum distillation.* Unless this is done under vacuum conditions much CaO is present and when heated is a continuous source of O_2. How pure Weissler's Linde A was is uncertain. His purification train consisted of passing it over reduced hot Cu in which the absence of O_2 could be ascertained during H_2 reduction when H_2O ceased to be evolved. Then further heating and baking removed the H_2, and the gas passed over a P_2O_5 tube through a liquid air trap into the chamber. His A was probably as pure as that used by Lauer in coaxial cylindrical studies, and both used really outgassed systems. Das had electrodes that were not outgassed, and his Ne and A may have been of inferior purity as the results will show. Both Lauer and Das used electrode arrangements in which the secondary emission from the cathode played a significant role, Das because of his short gap length, Lauer because of his geometry.

Weissler, with his long gap and his somewhat larger point diameter, observed that *there was no corona in A*. A streamer spark occurred at 3500 volts. Das, with his finer point and shorter gap of 2.5 cm at 760 mm of A, got a corona at 3400 volts. This followed a conventional current-potential curve leading to a spark at near 10,000 volts. Lauer used a Pt wire 0.13 mm in diameter in a cylinder of 1.45 cm radius. His positive corona from extrapolated data at 400 mm set in as a photoelectrically conditioned γ_p corona at some 2700 volts. It had a current for a 5 cm long wire of about a microampere at threshold. It went over to a streamer spark at around 4300 volts. With a point diameter of 0.1 mm and 2.5 cm gap at 400 mm pressure of A, Das's corona threshold was 2500 volts with a spark at 6300 volts.

Thus Weissler did not get a corona before spark breakdown in his pure A, while Lauer got one of limited range that broke down to a streamer spark at about 1.6 times threshold, and Das had his corona start at about 3400 volts and break down to a spark at nearly three times threshold. Das does not dwell on the appearance of streamers or burst pulses in his positive corona. He probably could not detect these with his small currents. However, Das's currents were of the order of 10^{-8} amp at threshold, while Weissler generally observed currents starting at 5×10^{-7} amp in all but pure A. It is also to be noted that Das's A differed from Weissler's in that he got a spark at 10,000 volts whereas Weissler's spark occurred at the threshold of 3500 volts, at which Das observed a corona. There was thus a real difference in the gas.

It is clear that Lauer got a corona because of a Townsend type discharge in which the cathode played a considerable role. This was indicated by his studies with α particle triggering in the same tube. This was *not* a positive point corona with photoionization *in* the gas. Das used a short gap with a possibly very sensitive cathode to a γ_p emission because of his failure to outgas the cathode. Thus it is likely that Das also got a Townsend discharge such as Lauer had. Weissler had a clean cathode and one insensitive to a γ_p with a gap length of 4.6 cm. Thus, until he achieved streamer forming potentials at his point because of photoionization in the gas, he did *not* get a corona, as his conditions were adverse to an effective Townsend threshold. A similar situation has been used by Raether's group,[3.55] who deliberately desensitized the cathode in their uniform field studies to suppress Townsend breakdown and enhance streamers. Once streamer thresholds are reached in A, a spark occurs. With a Townsend discharge in the asymmetrical point-to-plane geometry, the positive ion space charge in transit in the lower field regions reduces effective point fields and suppresses streamers. This raises the sparking threshold relative to the corona threshold. Thus Das got his spark above the Townsend threshold with its low currents beginning at 10^{-8} amp. Weissler observed no burst pulses in A. One must assume a Townsend discharge in Lauer's case, as shown by his time analysis study, and quite probably in the work of Das. This is the more likely, as Das's gas purifying procedures were closely analogous to those used by Colli and Facchini[3.71] in coaxial cylindrical geometry where such a discharge occurred at 760 mm.

d. *Results of the Influence of H_2, N_2, and O_2*
   on the Corona in A

In order to test the influence of metastable atoms on the A corona, Weissler added small quantities of H_2 and N_2, known to destroy these states. This action would prevent or at least raise the threshold for Townsend discharges in A. It had been believed that photoionization of metastables in A by photons from excited states might be the mechanism leading to streamer formation. This was a mistaken notion, as was finally shown in 1959 by Westberg and Huang.[3.72] Thus N_2 should not influence streamer formation in A for that cause. *Really pure* A does not yield a streamer spark in a point-to-plane gap such as was used here.[3.72] The photoionization comes from traces of Kr and N_2 in the spectroscopic grade A which are ionized by A photons. With 0.3% N_2 Weissler observed a very few strong streamers at 3500 volts followed by a spark at the same potential. An increase of N_2 to 0.5% gave a streamer corona threshold at 3600 volts which persisted to 4100 volts. At this point positive space charge was such as to suppress streamers, but a localized corona with occasional burst pulses was observed. At 4600 volts streamers reappeared again and led to a streamer spark at the potential. Increase in the amount of N_2 did not alter pre-onset streamer threshold, but led to a breakdown to a spark at higher potentials. This would indicate that the normal impurity N_2 and Kr in the A sufficed to give very vigorous axial streamers that led to spark breakdown on appearance. Addition of 0.3% N_2 did not alter this very much.

Addition of more N_2 appeared to lead to a change in the range of photoionization which reduced the streamers so that they were not able to cause a spark at the threshold. The first streamers yielded space charges that suppressed the streamers for a potential range which increased as the amount of N_2 increased. The increased spread over the surface and the burst pulses might indicate appearance of photons that permitted burst pulse spread. Thus, as recently observed by Waidmann[2.36] on addition of O_2 to air, which reduces the ionizing free path, streamers are reduced in range and vigor. This reduction and the influence of space charges from accumulated streamers would explain the burst pulse region. N_2 does not form negative ions and thus does not yield a Hermstein glow corona. This clearly proves that initial burst pulse choking in *air* is also a positive ion space charge phenomenon and that only later does the negative ion sheath intervene in that gas. The contamination of A with H_2 showed essentially the same phenomena as with N_2. Destruction of metastables in A by addition of these gases serves to suppress the Townsend glow discharge at higher pressures, since the γ_p in A comes from an excited molecular state created in triple impact between metastable A atoms and ground state A atoms. This radiates a wave length of 1250 Å which is not absorbed by A atoms and gives a γ_p.[3.71] The A resonance photons are so absorbed that they give no γ_p in reasonable times. The other transitions in A are too long in wave lengths to yield a γ_p. The action of H_2, however, has nothing to do with such action where streamers occur, and the explanation must be the same as for the action of N_2.

Addition of 0.1% O_2 to A at once raised the corona threshold by 300 volts and led to a streamer corona with a spark at 4100 volts. No burst pulses were observed, as would be expected with too little O_2. Streamers could form. Addition of 0.4% O_2 still further raised threshold and gave a small burst pulse corona with a spark at 7000 volts; 1% O_2 reduced the threshold for corona to 4000 volts with only small burst pulses and a uniform glow corona. This raised the sparking threshold for streamer sparks to 15,000 volts. Above 1% O_2 the Hermstein glow corona apparently sets in very early.

Weissler also observed the coronas in the three gases H_2, N_2, and A at reduced pressures. The sequence of phenomena at 300 and 100 mm A were essentially the same as at 760 mm, except that threshold potentials were correspondingly lower. The rate of current increase at lower pressures and the currents in general were greater at corresponding potentials because of the higher ion mobilities as pressure decreased. The starting potentials for the three pressures and current-potential curves for N_2 gas are shown in figure 3.118.

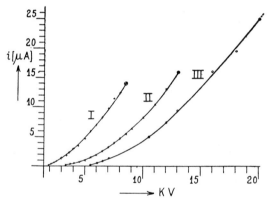

Fɪɢ. 3.118. Current-potential curves for positive point corona in N_2 at p = 100 mm (I), at p = 300 mm (II), and at 760 mm Hg (III). The end points of the curves represent sparking threshold. Curve III ends at about 23 kv with a 4 cm gap.

e. *Results in Pure N_2*

The sequence of events and data in pure N_2 are quite parallel to those in H_2. Corona threshold of the Townsend type occurred at 4800 volts in pure N_2. No burst pulses were observed. The currents for H_2 and N_2 at threshold were 4 and 3 μA, respectively, which is much larger than with the streamer or burst pulses present, where threshold currents were about 0.5 μA. Breakdown streamers in pure N_2 occurred at 11,000 volts, and a spark appeared at 20,000 volts.

Addition of 0.2% O_2 to N_2 gave either a threshold potential for incipient streamers or fluctuating avalanche currents at a potential lower by some 300 volts than that for the Townsend glow threshold in pure N_2. Breakdown streamers appeared at 9300 volts, and a spark at 10,000 volts. No burst pulses appeared at this concentration. Studies with coaxial cylinders show that γ_p in pure N_2 with outgassed Pt electrodes does not appear to be strong, and a γ_i is active.[2.22] This accounts for a high Townsend threshold. Once O_2 is added, γ_i is reduced by a factor of five and the streamer corona can occur at a lower potential than that for the Townsend discharge.[2.22] With O_2, a γ_p from N_2 could also replace the γ_i. Thus a Townsend threshold with γ_p plus photoionization in the gas could give a lower threshold. Addition of 1% O_2 again raised threshold for streamers to 4800 volts, possibly because of the elimination of γ_p action and the beginning of a streamer corona. Burst pulses appeared and a Hermstein glow corona set in. This raised the threshold for breakdown streamers to 12,500 volts and led to a streamer spark at 20,000 volts.

The observations today are consistent and explicable in terms of the findings of recent years, especially as regards the Townsend and the Hermstein phases.

Weissler's measurements are summarized in table 3.5.

f. *Results of Das on Ne and Ne–A mixtures*

As in argon, Das observed a corona in "pure" neon that set in at some 280 volts at 760 mm and rose with potential along a typical Townsend positive space charge controlled corona curve until a streamer spark occurred at 1250 volts. The currents were very small at threshold, lying in the 0.01 μA range. This was unquestionably a feeble Townsend discharge, and there was no question of burst pulses or streamers. In contrast, the A threshold at 760 mm was around 3500 volts, as in Weissler's case, and yielded a streamer spark at near 10,000 volts. The very low threshold for Ne, in contrast to A, is quite surprising. The current near spark threshold was far greater in Ne than in A again because of the much higher mobility of the Ne ions than those of A. It is certain that the Ne of Das was far from pure, its principal contaminant being more ionizable gases like A. Thus, the very low threshold for Ne corresponds to the more efficient Penning gas Ne + 10^{-3}% A. Actually, the threshold for *really pure Ne would probably have been above 800 or 1000 volts*. The influence of small traces of A in Ne in lowering the sparking potential is remarkable. Addition of 0.1% A to the Ne of Das resulted in only a slight lowering of the threshold potential, if any, and lowering of the sparking potential to about 800 volts. This means that the Ne was already at the lowest threshold potential because of A. The sparkover was lowered because at high X/p the 0.1% A in Ne gives

TABLE 3.5

Essential Data of the Positive Point-to-Plane Corona

Gas	Pre-onset Streamers	Onset Potential	Corona		Current,* micro-amperes	Onset of Breakdown Streamers	Approximate Breakdown Potential
			Visual Character	Burst Pulses			
H₂ pure†	None	3500 v	Localized‡	None	4	10,000 v	18,000 v
H₂ + 0.1% O₂	Yes	5000 v	Intermittent streamer-corona	None	—	8,000 v	—
H₂ + 0.5% O₂	Yes	4500 v	Streamer-corona	None	3	9,000 v	—
H₂ + 1% O₂	Yes	4000 v	Uniform§	Small	3	15,000 v	18,000 v
N₂ pure†	None	4800 v	Localized	None	1	11,000 v	20,000 v
N₂ + 0.2% O₂	Incipient streamers	4500 v	Streamer-corona	None	1	9,300 v	—
N₂ + 1% O₂	Incipient streamers	4800 v	Localized, for higher v uniform glow	Yes	1	12,500 v	20,000 v
Room air†	Yes	5500 v	Uniform	Yes	1	15,000 v	20,000 v
A pure‖	None	None	None	None	—	—	—
A + 0.1% H₂	Yes	3300 v	Localized	None	0.5	—	—
A + 0.5% H₂	Yes	3500 v	Localized	Small	0.5	—	—
A + 1% H₂	Yes	3450 v	Localized	Small	0.5	3,500 v	3,500 v
A + 0.3% N₂	None	None	None	None	—	—	—
A + 0.5% N₂	Yes	3600 v	Streamer-corona	Small	—	4,600 v	4,600 v
A + 1% N₂	Incipient streamers	3500 v	Localized	Small	0.5	—	—
A + 0.1% O₂	Yes	3800 v	Streamer-corona	None	0.5	6,500 v	6,500 v
A + 0.4% O₂	Yes	4500 v	Streamer-corona	Small	0.6	4,100 v	7,000 v
A + 1% O₂	Yes	4000 v	Uniform	Small	0.6	—	15,000 v

* Current at a potential which is about 20% above the corona onset potential.
† Gap length 3.1 cm.
‡ Localized glow, due to electron avalanches approaching the point in the highest field region.
§ Uniform glow, caused by the burst-pulse corona.
‖ Gap length 4.6 cm.

the lowest value of V_s. With 0.5% A the starting potential *increased* to about 750 volts with spark breakdown at 1100 volts. This is sufficient evidence that the most efficient Penning Ne–A mixture was the gas which Das believed to be pure Ne and that addition of more A did not alter the first Townsend coefficient but did increase photoionization in the gas, leading to a lower streamer breakdown threshold.

As stated earlier, Das ignored Townsend's equation 2.29 and developed a physically meaningless empirical expression (equation 3.6) for the current as a function of applied potential. He plotted the point tip field at threshold for the threshold of the corona as well as the threshold potentials for breakdown as a function of pressure of A, Ne, or the mixed gases. He did the same for the sparking threshold values. Other than this and his speculations based on inadequate background knowledge of the phenomena observed, there is little more of value to his paper in the positive corona study.

F. STREAMER BREAKDOWN IN PURE SINGLE COMPONENT GASES; THE WORK OF WESTBERG AND HUANG ON A

Vital to the streamer theory of the filamentary spark transition is the pure single component gas, in which photoionization of the gas by its own photons is meager. If a gas has a single atomic component which can only be photoionized by excited states above the ionization potential, the materialization of a filamentary streamer spark is most unlikely. This assumption stems from recent data of Raether,[2.35] Przybylski,[2.26] and Waidmann,[2.36] brought together by the author. Small avalanches on the order of 10^5 ions at the pre-onset streamer threshold require ≈ 100 or more photoelectrons, and the breakdown streamers of 10^8 ions have around 10^5 or fewer photoelectrons available. However, to date all gases, no matter how assumedly pure, have been observed to yield filamentary streamer sparks in transition from glow to arc, or in direct breakdown at pressures of sufficient magnitude to yield arcs.

It was the author's assumption that all gases used to date were sufficiently contaminated to yield a two-component gas. This appears to have been justified very recently. However, since the late 1930's at the Philips laboratory in Eindhoven and in post-World War II years at the Bell Laboratories and at Westinghouse, and more recently in many other laboratories, the inert gases have been assumed to have been studied in reasonably pure form. Assuming adequate gas purity, the only means the author could see to permit streamer sparks in such gases was through the accumulation of high concentrations of metastable states across the gap, which could be photoelectrically ionized by the resonance lines. This seemed the more rational since such states are built up in quantity by the corona, or Townsend pre-discharges, and accumulate. Concentrations as high as 10^{10} per cm^3 metastable atoms are not uncommon in glow discharge.

Of the gases studied, the one of most interest seemed to be A, since its metastables are apparently not readily photoionized by the resonance radiation. Only one study seemed to show that, when the A gas was sufficiently pure, it did *not* yield a streamer spark. This was a study by Menes[3.73] in an outgassed metal chamber under rigorous Alpert vacuum techniques. Menes studied the Townsend build-up at various pressures in uniform field geometry. He found, in conformity with Colli, Facchini, and Lauer, that the build-up was by a γ_p above 200 mm A, but a mixed build-up by γ_p and γ_i below 100 to 50 mm. How high his power source with its limiting resistors permitted him to go in current is not known. However, he obtained only a Townsend discharge build-up of the type developed by Bandel for air, with *no transition to a streamer spark.*

It was thus thought urgent that A be studied using Hudson's[3.24] two-photomultiplier technique with truly pure A. Until that time no one had gone much further in purifying A than using the purest "spectroscopic" commercial grades of A, and subjecting them to a possible "clean-up" in a Ca or Mg metal arc. These arcs will remove O_2 and N_2, but not Kr or Xe. The author, an ex-chemist, regarded with misgiving the use of such arcs, as the metals used are never pure. Running an arc with them while removing O_2 and N_2 is equally likely to add other contaminants. At the time in question, the possibility of further purifying inert gases by means of gas cataphoresis seemed to promise a better means of cleansing. Thus it was decided to use this technique with A. As evidenced by several papers at the Fifteenth Annual Gaseous Electronics Conference at Boulder in October, 1962,[3.74] subsequent events have proven that the former so-called pure gases, including even He, are not pure but contain sufficient traces of gases such as Ne, Kr, and N_2, as to yield spurious ion species and ionization probabilities. Appropriate gas cataphoresis treatment has sufficiently cleansed them so that past complications and discrepancies have been resolved.

It was with this in mind that the author assigned to R. G. Westberg[3.72] the problem of using the two-photomultiplier technique on supposedly pure commercial A and on A purified by gas cataphoresis to see whether the streamer spark materialized in those gases.

The problem was not an easy one, as successful gas cataphoresis in A operates best at 10 to 20 mm pressure with low currents. It was essential to purify A at a minimum of 300 to 400 mm to ensure adequate streamer sparks. Furthermore, in contrast to N_2, the light from the A discharge streamers and sparks is largely from the atomic lines, and these, though intense because of their limited width, do not give the luminosity that the second positive group band spectrum of N_2 yields.

The apparatus used is shown in figure 3.119. The streamers were observed in a point-to-plane gap. It had a 0.236 mm diameter hemispherically capped cylindrical point, and a thin Ni cathode of 7 cm diameter, 3 cm distant. It was mounted in a glass cylinder of 3 liter capacity. The vacuum system consisted of three parts. A vacuum pumping system consisting of an oil pump, an Hg vapor diffusion pump, and cold trap were used to evacuate the apparatus. It was cut off by a bakeable Alpert metal valve (1).

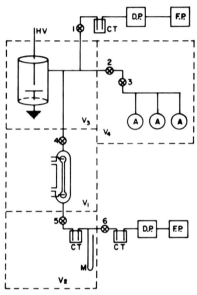

FIG. 3.119. Westberg's arrangement for cataphoresis purification and the observation of streamer sparks in commercial pure grade and specially purified argon.

Three 1 liter flasks of "spectroscopic" grade Linde A with means of breaking the seals by steel shot were connected through bakeable valves (2 and 3) to the system. The cataphoresis tube was isolated by bakeable valves (4 and 5), and could be evacuated through a pair of bakeable valves (5 and 6) provided with an Hg manometer and a second mercury vapor diffusion trap pumping system. The whole system was mounted on a Marinite base and was outgassed by baking to 360° C. The manometer was kept isolated by cold traps, and valve 5 was opened only during evacuation and for the reading of the A pressures as needed.

Electrodes and metal parts could be outgassed by induction heating to 950° C. The cataphoresis tube was an oil-cooled discharge tube. At 300 mm pressure it was expected that at least a 300 mA current would be required for cleansing. The gas volume of the cataphoresis tube of 6 mm inner diameter and 20 cm length was 5% of the volume of gas in the chamber V_3 to be cleansed. The seals to the A flasks could not be baked. Therefore valves 2 and 3 were used as a bleeder space to flush impurities out of the system before admitting the A.

After bake-out at 360° C for 24 hours and induction heating of metal parts for 6 hours with pressures down to 10^{-9} mm and less, the A was admitted and the pressure read. Then with valve 5 closed but 4 open, all other valves being closed, the discharge was run for 3 to 4 hours. The presence of impurities and the cleansing action were at once evident. The discharge was at first bluish over the whole tube. It quickly changed to pink at the anode, with the gas near the cathode assuming a green color. The spectrum of the bluish glow showed only A. The pink glow showed only A. The green glow showed both Kr and N_2. Valve 4 was then closed and valves 5 and 6 were opened, removing the gas in the cataphoresis tube with its contaminants. Then valves 5 and 6 were closed and valve 4 opened, and the current was again run for 3 to 4 hours. This process was repeated five times. After the first run the gas from the corona gap chamber no longer showed the bluish cast but was pink. However, as time went on, the green again collected at the cathode.

An attempt was made to monitor the decline in intensity of the green light at the cathode using a photomultiplier focused on a Kr line, but the light was too feeble. At the end of the fifth run, the pressure of the gas in the discharge chamber had dropped to 240 mm. Using extrapolated data from Riesz and Dieke's[3.75] work at 20 mm, Westberg estimated that the impurity in the gases could have been reduced by 0.33 at each run, leading to a reduction by a factor of 0.041. Probably this was optimistic, and the Kr and N_2 were reduced by a factor of about 0.1. It seems surprising that diffusion of the gas from the chamber should have been sufficient for this cleansing. However, reports of the cleansing without repeated removal of contamination, as was here done when He and other gases were cleansed by cataphoresis at 20 to 30 mm with currents of tens of milliamperes, indicate that streaming processes in the cataphoresis column and diffusion generally are more active than are suspected.

One more word needs to be said. Corona processes and sparks are electrically very noisy. Therefore excessive shielding must be resorted to in order to reduce background noise in the photomultiplier traces. This was not easy for either Hudson or Huang.

It is now essential to describe the observations. Actually Westberg had an exceptional number of delays, owing to malfunction of one or the other of the many valves and to undue breakage of his elaborate apparatus. Part of the trouble was ascribed to the use of an extensive Marinite base without adequate support for the large bake-out area required. Marinite warps on heating unless strongly supported, and the warping caused much breaking of seals. He therefore was not able to check the results as much as desired.

Westberg's initial observations on the "spectroscopic" grade A of Linde indicated that the positive point corona led to a streamer spark breakdown on a time scale of 10^{-7} sec. This was in conformity with Weissler's conclusions. To further investigate this, Dr. Huang undertook a detailed study of the gas in an outgassed and clean system with A that had, or had not, undergone partial purification by what later proved to be an inadequate cataphoresis cleansing. The investigation began with a sparking potential curve for both types of A, indicated by crosses and circles, shown in figure 3.120. The one triangular point represents the breakdown potential of the A with 1% O_2 added. Figure 3.121 shows the luminous pulse traces as a function of the distance x of 0, 5.3, 10.8, 16.1, 21.4, and 26.7 mm from the anode, with 0 at the bottom and 30 mm at the cathode at the top trace.

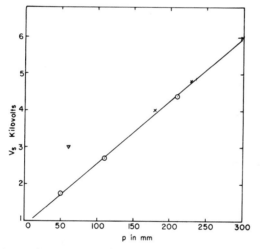

FIG. 3.120. Sparking potential curve as a function of pressure for Huang's spectroscopic grade A.

It is seen that the whole spark luminosity curve is shifted to the left as x increases. This series of traces permitted crossplots of the luminosity at various points x to be made for various times ranging over 75, 100, 150, 200, 250, and 300 mμsec. These crossplots reveal curves similar to Hudson's, with a secondary peak followed by the main stroke moving from anode to cathode. The speed of the peak is 3.7×10^7 cm/sec, while that of a point of 0.17 peak value on the rising toe is 1×10^8 cm/sec. Here one encounters

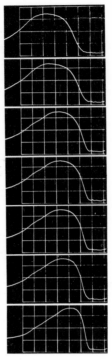

FIG. 3.121. Luminous pulse traces for spectroscopic A as a function of distance x of the signal photomultiplier from the anode. The distances are, beginning with zero at the bottom, 5.3, 10.8, 16.1, 21.4, 26.7, and 30 mm. Time goes from right to left at 300 mm pressure and a sweep speed of 50 mμsec/cm. Each graticule marks 1 cm.

the difficulties arising when the luminous intensity is low. The primary peak is not clearly delineated, and only the secondary peak, which seems to usher in the main luminosity—using Hudson's terminology—is seen. It would appear here that, in terms of Nasser's interpretation, the velocity of the streamer and its branches is more nearly uniform across the gap, and the first and rapid more axial streamers are too weak to show, in contrast to the streamer tree with all its branches. In all the traces that follow, x increases from bottom to top, and the motion in time is from right to left.

More resolution in Hudson's sense is seen at maximum photomultiplier gain at 180 mm pressure, with 50 mμsec/cm sweep speed, seen in figure 3.122. Here what may be a primary or a secondary is seen preceding the main stroke. It is hard to separate near the anode, as its speed is not very different from that of the branches. Figure 3.123 shows a series of traces at increasing photomultiplier gain from the lower traces to the top traces on each film, with the highest gain badly overloaded and flat on the main

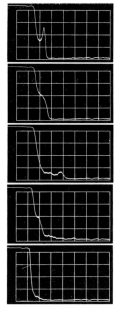

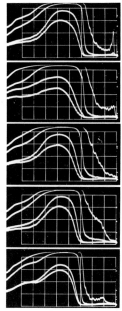

FIG. 3.122. Same as figure 3.121 at 181 mm pressure with 50 mμsec/cm.

FIG. 3.123. Same as figure 3.121 except that at each x there are four traces with increasing amplifier gain. The lowest gain shows the main stroke, and the top gain the rise of the main stroke and secondary or primary and secondary with overloaded peak. This is at 110 mm with 100 mμsec/cm sweep speed.

stroke. These are taken at 110 mm of Linde A with 100 mμsec/cm sweep rate. In this case the intensity was such that the primary peak could be seen near the anode. Its structure with increasing gain is more and more what one would expect for the sort of streamer branching, with a slit and poor light resolution. The speed of the primary at this low pressure is $\approx 5 \times 10^7$ cm/sec. At 50 mm pressure with 100 mμsec/cm sweep, the oscillograms are as shown in figure 3.124. Here the gap breaks down without any sign of streamer motion or motion of luminosity. The luminosity across the gap rises in a similar fashion with time. It seems that with the few impurities present at 50 mm pressure, the density of photoionization is inadequate to cause a streamer breakdown. The time scale of breakdown is still $\approx 10^{-7}$ sec. Addition of 1% air at 60 mm pressure with 20 mμsec/cm sweep rate yielded the traces of figure 3.125. At 5.3 mm from the anode, an anode streamer is discerned, the luminosity of which increases so that it fuses with the luminosity of the slower branches as the cathode is ap-

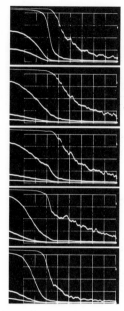

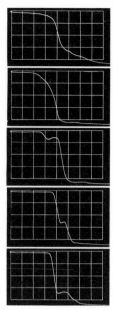

FIG. 3.124. Same as figure 3.121 but at 50 mm pressure at four gain values. Note that there is no pronounced motion and that anode streamers, if present, are not indicated.

FIG. 3.125. Same as figure 3.121 but at 60 mm pressure A with 1% air, with 20 mμsec/cm sweep speed. Here the streamers reappear and have a high speed as well as increased luminosity.

proached. The speed here is 1×10^8 cm/sec. It is noted that the air restores streamers, increases luminosity, and yields a higher speed.

When Westberg on his final attempt succeeded in getting 240 mm of relatively purer A, the oscillograms were markedly altered. Apparently the sparking potential of 4800 volts for the pure A was virtually the same as for the less pure A. Figure 3.126 shows a series of traces, starting with the anode at the top trace and having values of x from 0 through 0.35, 0.70, 1.21, 1.89, 2.11, 2.93, and 3 cm at the cathode. The sweep rate here was 1 μsec/cm. Time here progresses from left to right. The slow rise time of the breakdown rendered triggering of the breakdown difficult. In the first trace at the anode, there are two sweeps. The declining traces noted below are unavoidable retriggering of the scope during later phases of the sparks. The difference in starting times, coming from the poor triggering, makes any attempt at discerning movement impossible. This breakdown is obviously very different from the streamer breakdown on a time scale of 10^{-7} sec or less. Here there is a gradual rise to an overloaded peak, in all cases

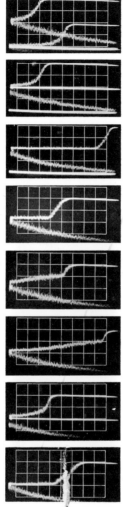

Fig. 3.126. Westberg's traces for cataphoretically purified A. Here time moves from left to right. The anode at x = 0 is at the top, and traces correspond to 0.35, 0.70, 1.21, 1.89, 2.11, 2.93, and 3.0 cm from the anode. The sweep speed here is low, 1 μsec/cm. Two traces are shown at the top. Triggering is bad because of slow pulse rise, and no motion can be seen. The rise of current is gradual and on a time scale ten times that of the spectroscopically pure argon.

consuming more than 10^{-6} sec for the steep part, with a gradual rise from zero extending over several microseconds.

The rise of luminosity was faster at the cathode and anode than in mid-gap. In midgap there was a slow rise to half intensity in 4 to 6 μsecs, and a final sharp rise to a peak in 0.5 μsec. The long duration of luminosity in the channel, extending beyond 6 μsec in some cases, indicates that the discharge current was of an entirely different order, taking a long time to discharge the capacity. Figure 3.127 shows a sequence at 200 μsec/cm sweep speed at maximum gain, with conditions otherwise the same as in figure 3.126. The traces were much wider than those of Hudson and Huang, with much background noise. There is, however, no structure seen anywhere at high gain. Rise to overloading even at high gain occurs on a scale of 1 μsec, relative to a twentieth of the time on Huang's traces. Figure 3.128 shows two sweeps taken at 5 μsec/cm, the light being observed just at the anode tip. The first sweep at low gain shows the full pulse shape, indicating that most of the light output is in 2 to 3 μsec. If the photomultiplier gain is increased to be comparable with the high gain used by Huang, the high trace shows that there is some light output for 35 μsec at the level of Huang's secondary pulses.

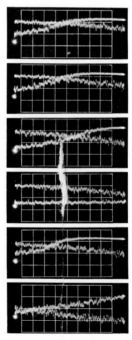

Fig. 3.127. Westberg's traces at 200 μsec/cm sweep speed at maximum gain conditions, as in figure 3.126. There is no structure seen, and traces are noisier. The rise is very slow.

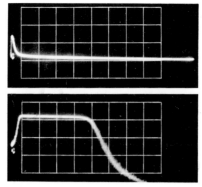

FIG. 3.128. Westberg's traces at 5 μsec/cm, the light being observed at the anode tip at x = 0. At low gain the full pulse shape is seen, and most of the light output is over in 3 μsec. If the gain is increased to that used by Huang on the spectroscopically pure A, the high trace is overloaded but shows light comparable with primary and secondary streamer intensity out to 35 μsec.

It is clear that even relatively pure A in the absence of an adequate photoionizing streamer mechanism does not break down as a filamentary spark. It breaks down in a fairly diffuse channel of some 2 mm diameter. This may be a constriction of an axially confined Townsend discharge mechanism to a localized glow in some 3 μsec of time. The current is probably on the order of one-tenth as heavy as that at the peak of the streamer spark, and lasts on the order of ten times as long. The relative reduction of impurity removed by the cataphoresis can be estimated to have been on the order of one-tenth the amount present in Linde A, as judged by the pressure of 50 mm at which streamer sparks failed to produce streamers, though they gave a short breakdown time with only a slight blurring of the filamentary channel.

REFERENCES

3.1   A. F. Kip, Phys. Rev. *54*, 139 (1938).
3.2   A. F. Kip, Phys. Rev. *55*, 549 (1939).
3.3   L. B. Loeb and W. Leigh, Phys. Rev. *51*, 149A (1937).
3.4   J. Zeleny, Phys. Rev. *25*, 305 (1907); *26*, 129 (1908).
3.5   M. Toepler, Ann. Phys. *2*, 560 (1900).
3.6   W. N. English, Phys. Rev. *71*, 638 (1947).
3.7   G. G. Hudson and L. B. Loeb, Phys. Rev. *123*, 32 (1961).
3.8   E. Flegler and H. Raether, Z. f. Techn. Physik *16*, 435 (1935). E. Flegler and
      H. Raether, Z. Physik *99*, 635 (1936); *103*, 315 (1936).
3.9   H. Raether, Z. Physik *107*, 91 (1937); *112*, 464 (1939).

3.10 L. B. Loeb and A. F. Kip, J. Appl. Phys. *10*, 142 (1939).

3.11 L. B. Loeb, *Fundamental Processes of Electrical Discharge in Gases* (John Wiley, New York, 1939), p. 426.

3.12 J. M. Meek, Phys. Rev. *57*, 722 (1940). L. B. Loeb and J. M. Meek, J. Appl. Phys. *11*, 438, 459 (1940). L. B. Loeb and J. M. Meek, *Mechanism of the Electric Spark* (Stanford Univ. Press, Stanford, Calif., 1941), chap. III, pp. 107–171.

3.13 K. E. Fitzsimmons, Phys. Rev. *61*, 175 (1942).

3.14 H. M. Herreman and L. B. Loeb, J. Appl. Phys. *34*, 3160 (1963).

3.15 W. N. English, Phys. Rev. *74*, 170 (1948).

3.16 W. N. English, Phys. Rev. *74*, 179 (1948).

3.17 L. B. Loeb, J. Appl. Phys. *19*, 882 (1948).

3.18 L. B. Loeb, Phys. Rev. *94*, 277 (1954).

3.19 D. B. Moore and W. N. English, J. Appl. Phys. *20*, 370 (1949).

3.20 H. W. Bandel, Phys. Rev. *84*, 92 (1951).

3.21 W. N. English, Phys. Rev. *77*, 850 (1950).

3.22 M. R. Amin, J. Appl. Phys. *25*, 210 (1954).

3.23 M. R. Amin, J. Appl. Phys. *25*, 358 (1954).

3.24 G. G. Hudson and L. B. Loeb, Phys. Rev. *123*, 29 (1961).

3.25 L. B. Loeb and W. Leigh, Phys. Rev. *51*, 149A (1937).

3.26 H. Raether, Ergeb. exakt. Naturwiss., ed. by S. Flügge and F. Trendelenburg (Springer Verlag, Berlin, 1949), vol. *22*, 100.

3.27 H. Raether. See reference 3.26, p. 97. E. Flegler and H. Raether, Z. Physik *99*, 641 (1936). H. Raether, Arch. Elektrotech. *34*, 56 (1940).

3.28 L. B. Loeb and J. M. Meek, *Mechanism of the Electric Spark* (Stanford University Press, Stanford, Calif., 1941), chap. II and especially chap. III.

3.29 H. Raether, Z. Physik, *117*, 375, 524 (1941).

3.30 H. Schlumbohm, Z. Naturforsch. *16a*, 510 (1961). H. Schlumbohm, Z. Physik *166*, 93 (1962); *170*, 233 (1962); G. A. Schröder, Z. Angew. Phys. *13*, 296, 367 (1961). J. Pfaue, Z. Angew. Phys. *16*, 15 (1963).

3.31 T. E. Allibone and J. M. Meek, Proc. Roy. Soc. (London) *A166*, 97 (1938); *A169*, 246 (1938).

3.32 R. J. Westberg, Phys. Rev. *114*, 1 (1959).

3.33 J. M. Meek and R. F. Saxe, Nature, *162*, 263 (1948). Allied Brit. Ind. and Research Assoc. Report, sec. L, Dielectrics, L/T 183, unpublished. See also 3.35, pp. 111 ff.

3.34 L. B. Loeb, Phys. Rev. *94*, 227 (1954).

3.35 L. B. Loeb, Spark Breakdown in Uniform Fields, Office of Naval Research Tech. Report (Washington, D.C., July, 1954), pp. 114 ff.

3.36 C. E. R. Bruce, Nature *147*, 805 (1941); *161*, 521 (1948); Proc. Roy. Soc. (London) *A183*, 228 (1944).

3.37 L. H. Fisher and B. Bederson, Phys. Rev. *81*, 109 (1951).

3.38 L. B. Loeb, Phys. Rev. *81*, 287 (1951).

3.39 W. Köhrmann, Z. Angew. Physik *7*, 183 (1955); Ann. Phys. *18*, 379 (1956). G. A. Schröder, Z. Angew. Physik *13*, 296 (1961); Ann. Phys. *18*, 385 (1956).

3.40 G. A. Kachickas and L. H. Fisher, Phys. Rev. *88*, 878 (1952); *91*, 775 (1953).

3.41 H. W. Bandel, Phys. Rev. *95*, 1117 (1954).

3.42 R. Kluckow, Z. Physik *148*, 564 (1957). H. Mielke, Z. angew. Physik *11*, 111, 410 (1959).

3.43 V. Tetzner, Arch. Elektrotech. *44*, 56 (1958); *44*, 69 (1959).

3.44 E. Nasser, Dielectrics, *1*, 110 (1963), gives data from Adipura's unpublished work.

3.45 U. Shinohara, Memoirs Faculty Eng., Nagoya Univ., *9*, 201 (1957).

3.46 L. B. Loeb, Report of Third International Conference on Ionization Phenomena in Gases, Venice, June 11–15, 1957, published by Italian Phys. Soc. Report, pp. 646–674, Oct., 1957.

3.47 J. J. Thomson, *Recent Researches in Electricity and Magnetism* (Clarendon Press, Oxford, 1893), p. 115.

3.48 L. B. Snoddy, J. B. Dietrich, and J. M. Beams, Phys. Rev. *50*, 469 (1936); *51*, 1008 (1937); *52*, 139 (1937). F. H. Mitchell and L. B. Snoddy, Phys. Rev. *72*, 1202 (1947).

3.49 R. J. Westberg, Phys. Rev. *114*, 1 (1959).

3.50 G. L. Weissler, Encyclopedia of Physics, ed. by S. Flügge (Springer Verlag, Berlin, 1956), Vol. XXI, pp. 304 ff.

3.51 H. Raether, Ergeb. exakt. Naturwiss., ed. S. Flügge and F. Trendelenburg (Springer Verlag, Berlin, 1961), vol. *33*, 176.

3.52 A. Przybylski, Z. Physik *151*, 264 (1958); Z. Naturforsch. *16a*, 1233 (1961); Z. Physik *168*, 504 (1962). W. Legler, Z. Physik *173*, 169 (1963).

3.53 K. J. Schmidt, Z. Physik *139*, 251 (1954). See also Raether, 3.51, pp. 177 ff.

3.54 See Raether, 3.51, pp. 195 ff. See also Loeb, 3.35, pp. 74 ff. See also L. B. Loeb, *Basic Processes of Gaseous Electronics* (University of California Press, Berkeley and Los Angeles, 1955), pp. 251 ff. It is probable that the later measurements of Raether applying to gases with organic molecules are correct for such gases.

3.55 J. Pfaue and H. Raether, Z. Physik *153*, 523 (1959). W. Franke, Z. Physik *158*, 96 (1960). See also Raether, 3.51, section 7, pp. 237 ff.

3.56 H. Raether, Lecture Before Naval Postgraduate School, Del Monte, Calif., Oct. 5, 1962; Fifteenth Annual Gaseous Electronics Conference, Boulder, Colo. Invited paper Oct. 10, 1962. H. Schlumbohm, Z. Physik *170*, 233 (1962).

3.57 E. I. Mohr and G. L. Weissler, Phys. Rev. *72*, 294 (1947).

3.58 R. Geballe and M. A. Harrison, Phys. Rev. *85*, 372 (1952); *91*, 1 (1953).

3.59 H. Eiber, Z. Angew. Physik *15*, 103 (1963).

3.60 L. B. Loeb, Phys. Rev. *48*, 684 (1935).

3.61 L. M. Channin, A. V. Phelps, and M. A. Biondi, Phys. Rev. *128*, 220 (1962).

3.62 W. A. Macky, Proc. Roy. Soc. (London) *A133*, 565 (1931).

3.63 J. Zeleny, Proc. Cambridge Phil. Soc. *71*, 18 (1914); Phys. Rev. *16*, 102 (1920).

3.64 L. B. Loeb, pp. 66 ff. in *Atmospheric Explorations*, ed. by H. G. Houghton (New York: John Wiley, 1958).

3.65 W. N. English, Phys. Rev. *74*, 179 (1948).

3.66 V. K. Rohatgi, J. Appl. Phys. *28*, 951 (1957).

3.67 J. M. El-Bakkal and L. B. Loeb, J. Appl. Phys. *33*, 1567 (1962).

3.68 H. W. Bandel, J. Appl. Phys. *22*, 984 (1951).

3.69 F. M. Penning, Phil. Mag. *11*, 961 (1931).

3.70 M. K. Das, Z. Angew. Physik *13*, 410 (1961). Doctoral Dissertation, Tech. Univ. Karlsruhe, Jan., 1960.

3.71 L. Colli and U. Facchini, Phys. Rev. *96*, 1 (1954); *80*, 92 (1950); *84*, 606 (1951); *88*, 987 (1952). L. Colli, Phys. Rev. *95*, 892 (1954).

3.72 L. B. Loeb, R. J. Westberg, and H. C. Huang, Phys. Rev. *123*, 43 (1961).

3.73 M. Menes, Phys. Rev. *116*, 481 (1959).

3.74 Emphasized in Papers Nos. A1 and F1 by H. J. Osborn and No. E3 by P.

Patterson and E. C. Beaty, Fifteenth Annual Gaseous Electronics Conference, Boulder, Colo., Oct. 10–12, 1962.

3.75 R. Riesz and G. Dieke, J. Appl. Phys. *25*, 196 (1954). G. Mierdel, *Handbuch der Experimental Physik*, Akademie die Verlage gesellschaft, Leipzig, Germany, 1929, vol. 13, pp. 519 ff. M. J. Druyvesteyn, Physica, *2*, 255 (1935). C. Kenty, Bulletin of Am. Phys. Soc., *3*, 82 (1958). L. B. Loeb, J. Appl. Phys. *29*, 1369 (1958). Also Gen. Elec. Co. Scientific Paper, 1958.

3.76 H. Tholl, Z. Naturforsch. *18a*, 587 (1963).

3.77 G. Waidmann and L. B. Loeb, Sixteenth Annual Gaseous Electronics Conference, Pittsburgh, Pa., 1963, paper A3.

3.78 A. M. Cravath and L. B. Loeb, Phys. *6*, 125 (1935) [now J. Appl. Phys.].

3.79 B. F. Schönland, Encyclopedia of Physics, ed. by S. Flügge (Springer Verlag, Berlin, 1926), vol. *22*, sect. 55, p. 622.

4

True Coronas: II
The Highly Stressed Cathode
with Anode Playing
a Minor Role

A. ELECTRON ATTACHING GASES, NOTABLY AIR OR THOSE CONTAINING O_2; THE TRICHEL PULSE CORONA

1. EARLY STUDIES OF TRICHEL, KIP, AND HUDSON

The study of the physics of these coronas properly starts with the advent of the oscilloscope having a time resolution better than 10^{-5} sec. In the early spring of 1937 the author assigned G. W. Trichel[2.12] the task of analyzing the signals produced by streamers from a positive point-to-plane corona gap which the author had visually observed in the fall of 1936 in connection with Leigh's study. In the course of this study in the electrical engineering department at Berkeley,[2.12,2.28] Trichel analyzed the positive and negative point-to-plane discharge, both with shock excited circuitry and with a mechanical oscilloscope of 10^{-4} sec resolving time. Toward the end of the investigation a brief study of the negative pulses was made with the department's Dumont oscilloscope of 10^{-5} sec resolving time. This latter analysis came in the early spring of 1938. In the late summer of 1937 the author had participated in the University of Michigan Conference on Electronics and Electrical Breakdown in Gases. There Dr. Marcus O'Day

told the author that he had observed some very regular relaxation oscilla-
tion-like pulses by means of an oscilloscope placed across a dropping resistor
in the plane to ground circuit for the negative point-to-plane discharge in
air at atmospheric pressure. He asked the author to attempt to explain
these.

The author and his group had, up to that time, had little interest in, or
knowledge of, corona phenomena. It was not until Trichel had spent about
a year and a half in observations on the study of corona pulses that he
rediscovered the negative pulses with the oscilloscope. By this time, Trichel
had gathered enough data on coronas in general to permit the author to
make some guesses concerning mechanisms. The very complete physical
study made by Trichel[2.12] on the negative pulses, using varied techniques,
in contrast to the superficial prior observations made by O'Day, induced
the author to name these pulses Trichel pulses to differentiate them from
other threshold pulse phenomena studied by Trichel. However, to O'Day
goes the priority of having first observed them and having recognized them
as an interesting phenomenon. Trichel finished his work and wrote a thesis
before he was called back to active duty in the Army. This left to the author
the task of preparing Trichel's paper for publication, and the paper appeared in December, 1938.

Notable in this study were several factors. Trichel, with the help of M.
Sitney, had in his positive corona studies calculated the fields, Townsend
coefficients, and ionization along the axis of a confocal paraboloid gap
configuration. This is not the same in numerical values as that for the point-
to-plane gap but is quite similar. Computations of this sort at once brought
out the difference in the ion distribution and the fields between positive and
negative point under ionizing conditions such as shown in figures 2.3 and
2.6. These were of immense value in clarifying the interpretation of the
Trichel pulse corona. Trichel also studied the frequency of the regular
relaxation oscillator-like pulses as affected by various parameters. First he
observed that the current which showed regular pulses at around 0.5 or 1.0
μA had a pulse frequency which increased linearly with the current increase.
The frequency for the same current increased the more rapidly and was
the higher the finer the point. This is shown in figure 4.1. A study of the
change in gap length for a given point radius on the frequency-current
curve revealed that the length exerted little influence between a 0.75 cm
and a 6 cm gap length. This very significant observation indicated that
whatever caused the regular interruption of the pulses was confined to a
region very close to the point. The variation of frequency with current as a
function of gas pressure is noted in figure 4.2 for a given point diameter.

Decrease in pressure decreased the frequency such that, when the pressure was decreased from 760 mm to 20 mm, the frequency at about 20 μA
changed from 100 to 20 kc. The structure of the visible glow representing

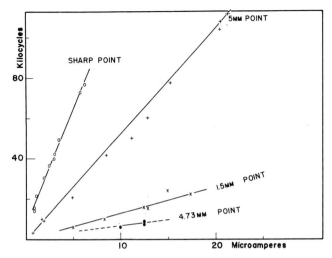

FIG. 4.1. Trichel's curves for frequency of Trichel pulses
vs current for four different points.

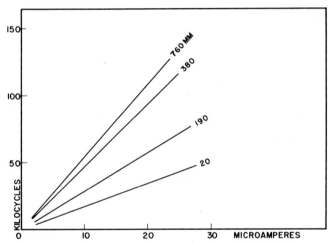

FIG. 4.2. Trichel pulse frequency vs current for various pressures
for a 0.5 mm point diameter.

the luminous phase of the changing breakdown sequence could be observed
with the telemicroscope. The discharge was confined to a very small area
at the tip of the point near threshold at low currents. It consisted of a
narrow short disc of luminosity, either whitish blue or blue which, unlike
the positive point burst pulse corona, did not appear to adhere closely to
the cathode surface. At lower pressures it was seen to be clearly separated

from the cathode by a thin dark space. Beyond this glow, there was a more pronounced and larger dark space beyond which lay a purple shaving-brush–like fan of luminosity. The fan became more pronounced as the current increased. The discharge had all the earmarks of the conditions at the cathode of a glow discharge tube. The first dark space could be identified with the Crookes dark space, the bright blue glow with the negative glow, the second dark space with the Faraday dark space, and the fan with the positive column. Later study has confirmed this designation, which will hereafter be applied. The bright transient phase of each pulse thus represents a miniature glow discharge operating at atmospheric pressures and hence reduced in dimensions. The width of the negative glow depended somewhat on point radius, on current, and on pressure but was fairly constant and lay between 0.1 and 0.2 mm for the points usually used. A typical corona as depicted by Trichel is shown in figure 4.3.

As current increased with increased potential, a region was reached where

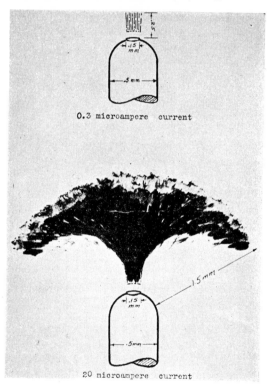

0.3 microampere   current

20 microampere   current

Fig. 4.3. Trichel's sketches of his pulse coronas near threshold for a 0.5 mm diameter point with 0.3 μA and at a higher potential with 20 μA of current. The dimensions are given in the enlarged sketches.

the regular pulses gave way to irregular pulses and current fluctuations. This, it appears, came from two types of action. The first was some alteration of the surface as current density increased, causing the spot to wander on the surface; the second was that the frequency of pulses from a given spot, and thus the current, appeared to be limited to the neighborhood of 10^6 cycles or so. When this limit was reached on larger surfaces, a new second corona spot appeared at a fresh spot where potentials are adequately high. If the point was too small, it appeared along the *shank*. If the point diameter was large enough to accommodate two or more spots which repelled each other, they formed symmetrical patterns over the surface. The overloaded spots on the surface tended to extinguish in favor of a new spot, but several spots could appear simultaneously. However, spots occasionally went off and on, making pulses incoherent and yielding current fluctuations of a larger magnitude, as shown in figure 4.4.

FIG. 4.4. Kip's photographs of the Trichel pulse corona as potential increases, much enlarged below and going from left to right. Above, multiple spots about the shank and at the tip of a fine wire cathode.

With the aid of the potential and derived ionization diagrams of Sitney and Trichel, the author attempted to explain the pulses as follows. When an electron is liberated from the point surface at threshold or above, it creates an avalanche with most of the ionization localized in a zone of some width and with the center distant from the point surface, as shown in figure 2.6. The photons from this ionizing zone reach the cathode and liberate a further group of electron avalanches. These augment the ion space charge and the rate of ionization. This process continues for a number of generations of electrons, causing an exponential or even more rapid rise of current. The positive ions created at some distance from the surface remain practically stationary on the time scale of 10^{-8} sec, since their mobility is low compared to that of photons and electrons. The electrons, generated with their residual velocities and drawn by the field of the anode, proceed out into the gap beyond the positive space charge. Here they were, at the time, assumed to form negative ions by attachment to O_2 molecules and thus to build up a slowly moving cloud of negative ion space charge. This negative space charge with the positive ion cloud creates a dipole field between cathode and anode. The field between *positive ion cloud* and the *cathode* becomes very strong as the positive ion space charge accumulates and exceeds the original cathode field. This rapidly enhances the initial ionization and speeds build-up of the avalanches in number and size.

In time, however, the positive ions are drawn to the cathode. These also liberate further electrons to form new avalanches by impact, or γ_i, on the cathode. The positive ion movement shortens the effective high field, although it increases its strength. It *was then believed* that, as this shortening becomes sufficiently severe, the electrons could not be accelerated and collide sufficiently to be effective. That is, it was *then* believed that the positive ion space charge eventually acted to reduce ionization, as does the negative ion space charge in the Hermstein glow corona. This action, together with the influence of the negative ion space charge accumulating beyond but near the positive space charge, so reduced the effective field near the cathode that further multiplication ceased and the space charges choked off the discharge. The field between the negative ion space charge and the anode was assumed to be enhanced and to cause some excitation and ionization, producing the fan. However, this was *unable* to *prolong* the discharge and it ceased shortly after the cathode discharge quenched. The discharge then remained extinguished until the space charges dissipated. The positive ions went into the cathode and the negative ions moved out into the gap to distances from which they no longer influenced the fields near the point. The last of the positive ions were extracted from the residual plasma and reached the cathode as the field was restored, to trigger a new sequence of avalanches so that the process repeated itself.

This general description appeared borne out by the fact that all properties listed seemed to be functions of the field close to the point and independent of gap length. The localization of the discharge in one spot and its constancy of area as potential increased accounted for linear increase of current with frequency. The repeat rate depended on the clearing time. This should decrease as point radius decreases, and distances of ion movement are reduced as the high field regions about the point shrink. By a similar token, the shorter the ionizing zones, the smaller the amount of charge was needed to build up a quenching field and the shorter the time required to build up and quench the discharges. Thus the charge per pulse was less for small points. For a 0.5 mm diameter point the number of microamperes of current that flowed per pulse was 2×10^{-4} μA, while for a 4.7 mm diameter point they were 1.4×10^{-2} μA. That is, the current decreased roughly as the point radius squared, indicating primarily the decrease in length of the ionizing zone from the point surface for similar point potentials.

As delineated in this first analysis, the charge was assumed to build up mainly by positive ion impact on the cathode, though it was recognized that the photoelectric effect on the cathode was needed to initiate the build-up. At that time the duration and rise time of the pulses were not resolved. The quenching was then assumed to be largely caused by the *reduction in field by the positive ion space charge movement*, though the negative ion space charge was recognized as a factor. In outline, this picture was correct, but later study forced modifications in detail.

The next advance came in a joint paper by Loeb, Kip, Hudson, and Bennett.[1,2] Bennett called the attention of Loeb to the fact that if dried clean air were used in place of room air, the regular sequence of Trichel pulses did not occur near threshold. The pulses would start, run for short bursts of a few pulses (3 to 10), and then go out. Bennett, who had studied field emission from negative points in Millikan's laboratory, showed that the discharge was different when fine points of field emitting diameter were used and that it went over to Trichel pulses in air if the diameter of the point were large enough to yield ionization by avalanches in the air. The intermittence of the Trichel pulse corona in dry air was a puzzle. At somewhat higher potentials persistence was *achieved*. This observation was confirmed through extensive studies by Hudson.[1,2]

It was believed that in the creation of the positive space charge and its extinguishing motion to the point after the clearing time, there was no positive ion residue left to start new electron avalanches in order to continue the sequence by initiating a new pulse. As was later learned, since the actual duration of the ionization and its extinction last an exceedingly short time (between 10^{-7} and 10^{-8} sec) while the clearing time for negative space charge between pulses at threshold is on the order of the reciprocal of the repeat rate (10^{-3} sec near threshold), this surmise was perhaps correct in part. When the repeat rate reaches 10^5, continuity of pulses can be expected. The electrons to start a pulse must be created in the exceedingly small volume of 10^{-6} cm^3 at the tip of the point. The chance of the appearance of an electron in this region in 10^{-3} or so seconds is remote. Thus it was found that the difference between the action of room air and the filtered clean dry air of Hudson and Bennett lay in the absence of dust and perhaps a binding moisture film which served to trigger pulses in room air.

Logically, it should have been possible to trigger the corona in clean dry air and on a dust-free point by providing electrons. However, both Hudson and, before him, Kip, had tried to do this by illuminating the point with ultraviolet light, Kip using room air.[1.10] As used by these workers, the ultraviolet light focused by a lens from a quartz Hg arc led to difficulties, such as raising the threshold voltage for the pulses and giving steady corona. It did not help. English later encountered the same difficulty.[1.10] It was thus thought that the light caused enough ionization to lead to negative ion space charges which inhibited the corona and destroyed the pulses. It was only much later that Bandel[1.14] discovered that the illumination used by Kip, Hudson, and English, especially in room air, was too intense, leading to altered conditions in the gap near the point. By reducing the intensity, Bandel was able to produce regular triggering and pulses in filtered, clean dry air with ultraviolet light on the point at threshold.

The action of dust was investigated by Hudson,[1.2] who discovered that very fine specks of insulating dust such as MgO, Al_2O_3, or even silica on the point acted to restore the regular pulsing. Moist air permitted adherence of the dust particles on the point for longer periods. If the air was dry, the dust particles were gradually removed by the discharge. The function of the dust was to become charged by the positive ions of the first pulse. The fields across this thin ($\approx 10^{-4}$ to 10^{-5} mm thick) insulator reached field emitting proportions and, via what is known as the Malter or Paetow[1.13] effect (later investigated in detail by Jacobs[1.13]), gave rise to electrons for seconds or more after being charged. It was noted that, in the filtered dry air, the discharge glow usually appeared very close to or adjacent to such dust specks and could be seen with the telemicroscope. In room air, dust specks were at all times present, hence the regular corona.

Kip studied the negative point corona in more detail. It always occupied a single spot never more than 0.2 mm in diameter. On larger points, if larger spots appeared, they consisted of one or several single spots close together in the confined high field region. When one spot was active, the others were extinct for a period, and the discharge fluctuated from one to the other. The corona current under these conditions fluctuated and was less than for a steady corona from one point. On large points at high currents, several spots were active simultaneously and were well separated in regular patterns over the high field surface regions.

Kip next made a study of the influence of the metal point material on the starting potential of the Trichel pulse corona. It was clear that this depended on secondary emission, either by photon or ion bombardment, of the metal and this mechanism was generally regarded as dependent on the work function or condition of the metal surface. To his surprise, Kip, using room air, found no measurable difference between Cu, brass, Al, Fe, and Pt;[1,2] and Bennett found the same for W, Rh, alloys of Cu, Ni, Zn, Sn, and Sb, provided the points were properly "conditioned."[1,2] What conditioning implied was never properly understood, except that it required running the points for some time in room air. English observed the same thing after World War II.[3,15]

The question of conditioning and the differences of material are associated with another phenomenon studied by Kip. Trichel had observed that, below about 1000 cycles per second, the Trichel pulses ceased to be regular and occurred randomly and separated in time. That is, a pulse would start but not be able to sustain itself at a frequency less than 1000 cycles per second. This again was unquestionably a matter of triggering electrons and of the emission time or rate of the average activated dust spot that furnished the electrons. It was thus clear that some sort of discharge other than the pulses was occurring in this potential range. The matter was investigated by Kip.[1,2] He used a 3.18 mm diameter brass point and a 4.2 cm gap in room air. A preamplifier was used to measure the feeble currents. At 11.7 kv current fluctuations first became noticeable. There was no sharp onset when they appeared; with the system of measurement used they became detectable at about this potential. This pre-pulse threshold had a faint glow associated with it at a distance from the point. There was no negative glow spot near the point. Increase in potential to 13.89 kv increased the size of the irregular kicks, and the current increased from less than 10^{-9} amp at 11.7 kv to 10^{-8} amp at 13.8 kv. Above this the first regular Trichel pulses were observed, and at slightly higher voltage they became regular. Hudson also noted these faint discharges below pulse threshold, and in all cases the glows were associated with fairly large dust specks seen in the microscope.

The existence of pre-threshold currents with positive corona in very pure dry gases in coaxial cylindrical geometry has been established by Miller in later work, and these pre-discharges, whether caused by dust specks or otherwise, must influence the threshold for appearance of the regular pulses, or even govern them.[1,8] They do this by altering the metal surface by sputtering, cleaning, or oxidizing it, or by creating triggering electron currents via the Malter effect. Thus, the Trichel pulse threshold for sporadic as well as for regular bursts is governed by conditions other than the simple Townsend threshold expression $\gamma e^{\int \alpha dx} = 1$, for which the coefficient γ is sensitive to cathode material. Its appearance is determined by the increase in potential to a point where actions lead to the positive ion space charge build-up, thus increasing fields leading to a self-choking discharge. There thus appear to be low order pre-pulse breakdowns leading to weak discharge currents that alter the point. While these may have a threshold set by a γ_i, such thresholds in air will not be fixed or too characteristic of the metal. The reason lies in the presence of oxide films which may differ from surface to surface and alter as the pre-discharge continues. This matter is further discussed under discharge threshold hysteresis phenomena.

It must be added, however, that quite generally even the Townsend glow discharge threshold under uniform field conditions shows relative insensitivity to cathode material at higher pressures, as we know today. This stems from the fact that many discharges are initiated by a photoelectric γ_p and not by γ_i. The photo effect is not as sensitive to surface composition as γ_i. Further, the really sensitive surfaces giving a large *change* in γ_p or γ_i are effective *only* in the pure outgassed state. These states occur primarily with surfaces in vacuum or in pure inert or inactive gases. Some very high values of γ_p and γ_i occur with absorbed gas films. Only carbon surfaces have a truly low γ_i and γ_p. The second, and dominant factor, in reducing the sensitivity of threshold values of potential for breakdown is the effect of high atmospheric pressure. Since at most γ does not vary by much more than a factor of 10 for normal surfaces or those with initial oxide films, and since $\gamma e^{\int \alpha dx} = 1$ depends exponentially on $\int \alpha dx$, it is clear that at higher pressures where $\int \alpha dx$ is relatively large, small changes in applied potential can alter $\int \alpha dx$ by amounts sufficient to compensate for changes in γ. Thus such potentials are not sensitive to changes in γ, since small changes in applied potential can readily compensate for 10-fold changes in γ.

It is therefore not surprising that the influence of γ and cathode material on the threshold for Trichel pulse corona was not noted by the various observers. It is thus clear that the threshold for Trichel pulse corona really represents a potential at which rate of ion generation reaches such a magnitude as to build up a positive space charge that leads to an unstable condition of an autocatalytically increasing discharge current. Thus, a runaway discharge inherent in the geometry of any negative point (and which also appears in uniform field geometry) sets in. In this discharge in air, the growing arc is eventually interrupted by space charges near the point. Any low order pre-Trichel pulse currents can facilitate the appearance of pulses by cleaning the point.

Hudson observed that, after his negative point corona was run in air for a while, the place where the glow centered appeared as a bright, clean spot of metal surrounded by a dark crater-like ring. The nature of the deposit appeared to be an unstable oxide or compound of Pt. Photomicrographs of the spot and the deposit are shown in figure 4.5. It was clear that the high energy of impact of the positive ions must have been such as to sputter off a number of atoms of Pt. These reacted with nitric oxides in the air to form Pt oxides. Sputtering is well known in low pressure glow discharges where ions strike the cathode with hundreds of volts of energy. Since at

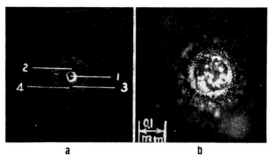

a b

FIG. 4.5. Hudson's photomicrograph of the tip of a 0.45 mm diameter Pt point after a 2.25 hour run in partially saturated room air at 6.5 kv and 3 μA. (a) A dark reflection is seen in the central smooth spot (1). This is surrounded by a heavy oxide ring (2), a narrow clear zone (3), and the thin outer dark ring (4). (b) The negative point after a 3 hour run in dry air at 10 kv and 17.8 μA. Several small bright pits appear instead of a single bright spot as in (a).

760 mm the ionic free path in air is quite short, it was surprising to observe sputtering as evidenced in this study.

The positive ions were thought to gain only about one volt of energy before striking the surface at 760 mm. This point was later investigated in Weissler's laboratory.[4.1] Here the negative point consisted of a thermocouple such that the energy of impact of the positive ions could be determined by the heating of the point. The average energy of the positive ions striking the surface was found to be about 10 ev, and this accounted for the sputtering. How the ions got such energies was a mystery until the work of Morton.[2.11] The high pressure resulted in the deposition of the sputtered and oxidized Pt very near the sputtering surface, causing a crater. With 1.5×10^{18} ions received and electrons emitted from the surface, there were, according to Hudson, 7×10^{14} atoms sputtered. That is, roughly one atom of Pt was sputtered for every 2000 ions striking the surface.[1.2]

Hudson also studied the influence of ultraviolet light on the corona. It appeared to increase the current at a given voltage by ensuring triggering electrons. It further appeared to decrease the current supposedly by low mobility ions in air. The effect that predominated depended on size and surface conditions of the point. The maximum increase appeared when several spots were active.[1.2]

Hudson heated the point to liberate electrons in dry air and thus regularize the pulses. This was done by heating a hairpin-shaped wire; the bend at the end served as a point, and the constriction there caused maximum heating. At first, heating gave regular pulses. As the point warmed up, it apparently shed its dust and corona became very irregular. When it reached temperatures at which electron emission could be expected, the pulses became nicely regular. This again showed the importance of triggering electrons. With increase in temperature, the frequency increased. Why this occurred is not clear, since increase in temperature decreases gas density, and decreased gas density decreases the frequency of the oscillations at fixed current. It is possible that Hudson did not hold current constant. However, the density change produced by the heated point was not uniform, as was the density change resulting from Trichel's change in pressure, such that very different conditions prevailed. Cooling the wire coated it with oxides, and the corona, on cooling, became very erratic.

2. Theoretical and Post-World War II Studies; Contributions of Loeb and English; Relative Starting Potentials

As a consequence of the Hudson-Bennett paper,[1,2] the theoretical interpretation was modified in one important respect. This was based on Bennett's observation[1,2] that there are no pulses in pure H_2, a gas in which *negative ions do not form*. This had been forgotten, for Weissler's[1,8] corona study with pure A, H_2, and N_2 had shown that there were no pulses, only a steady corona, in N_2 and H_2 and a spark in A, where negative ions do not form. Miller found in fact that with clean electrodes N_2 gave no glow but went directly to a spark.[1,9] In all cases there was a very low order conditioning current that cleaned up the oxide layers on the cathode. At threshold, the frequency of the Trichel pulses indicates a clearing interval on the order of the crossing time for negative ions in the gap, as observed by Kip[3,2] for positive ions from streamers. This was a definite step forward.

It is true that the positive ions do increase the strength of the field at the point and that they also shorten the gap at very high X/p where α/p no longer increases more rapidly than X/p, so that to some extent it throttles the discharge, as in the case of the Hermstein[1,16] negative ion space charge sheath with positive point in the glow corona. However, as long as the positive ions continue to move in, it appears that, while the ions may reduce the current, they do not choke off discharge completely, as with Hermstein's glow. The reason for this failure was not suspected until the time of Morton's study after World War II.[2,11] Thus, the *termination of the discharge must be ascribed to the rapid accumulation of negative ions just outside the positive space charge, and not to the positive ion space charge movement.* This negative space charge will grow more rapidly than the positive space charge, as positive ions are rapidly withdrawn in a high field while the negative ions pile up in a region of rapidly decreasing field. While Bennett's suggestion was at once adopted by the author, it presented one serious difficulty. This came from the circumstance that the rate of attachment of electrons to O_2 molecules to give O_2^- ions was so slow that the negative ion space charge could accumulate only at a point about five times or more too distant from the positive space charge to be most effective. This difficulty was soon recognized by the author, but it was not clarified until 1952.[4.2]

One more observation made in this study was that, when the potential was raised to near spark breakdown value with a single spot, bright blue-white streaks appeared at the cathode side of the edge of the shaving-brush–like positive column. As potential further increased, it extended toward the cathode. It was assumed that this was an incipient positive streamer and that when it was able to penetrate to the cathode, the spark materialized. A sketch of this is shown in figure 4.6. The phenomenon was also reported by Bandel[1.14] and by Greenwood.[4.3]

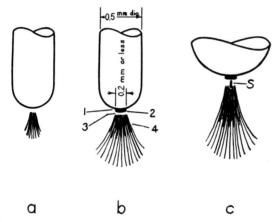

FIG. 4.6. Kip's sketches of the Trichel pulse corona on a larger point at higher currents. (a) Pre-onset corona. (b) Typical corona with regular pulses. (1) Crookes dark space; (2) negative glow; (3) Faraday dark space; (4) positive column. (c) Glow with incipient retrograde or possibly negative streamers before spark-over.

Little further advance in the study of negative point Trichel pulse corona was made until after World War II, though Weissler[1.8] had studied negative coronas in pure free electron gases, H_2, N_2, and A. In this work, no Trichel pulses were observed until O_2 was added to the pure gases, forming negative ions. As little as 0.1% O_2 added to pure H_2 at 760 mm pressure led to Trichel pulses. The same occurred with the addition of 0.1% O_2 to pure N_2. No Trichel pulses appeared in A when N_2 or H_2 was added, although a glow discharge was present. Again, addition of 0.1% O_2 to A yielded Trichel pulses.

With the resumption of research activities in his laboratory following his absence in the Navy during World War II, the author had, as a result of some wartime and immediate postwar work, foreseen possible difficulties in the theory of the Trichel pulse corona as interpreted in the foregoing. These led to a short paper on the mechanism of the negative point corona in relation to the first Townsend coefficient.[4.4] Morton, and later Johnson,[2.11] had shown that in highly divergent fields at adequately low pressures, the Townsend coefficients may not be applied to calculate the Townsend integral, as they lead to values of the integral giving only a sixth of the ionization actually occurring. A study of conditions indicated, however, that for air at 760 mm, the wire or point diameter would have to reach values as small as 10^{-3} cm in the gap undistorted by space charge before the Townsend integrals failed. On the other hand, in Morton's step by step calculation of the ionization from a high field point or wire, a remarkable similarity appeared between the ionization in a Townsend glow discharge at low pressure and the ionization pattern calculated by him. This was even more exaggerated than that indicated by figure 2.7. It correlated at once with the structure of the bright phase of the Trichel pulse. It occurred to the author that through the positive ion space charge *initially* created by a sequence of avalanches calculated under the Townsend integral with photoelectric ionization at the cathode, the movement of positive ions to the cathode would *shortly yield a cathode field condition that did fall into the Morton regime.* This would produce the Crookes dark space and negative glow region beyond it during the bright phase of the Trichel pulse. It would endow the *electrons* with the several tens of electron volts of energy that would permit them to penetrate the positive space charge and ionize and attach beyond it. Such a high cathode space charge field would also give the *positive* ions the ten volts of energy needed, *and observed*, to produce sputtering, since the normal cathode fields before positive ion space charge build-up could only endow positive ions through their short free paths with approximately one volt of energy.

A second question which arose was that of the relative starting potentials of positive and negative coronas in air, H_2, and N_2. These are shown in Table 4.1. Weissler's data for H_2 and N_2 are given as well as English's determination for the same points in air. It will be noted that in H_2 and N_2 the *threshold* of the corona was on the order of 700 to 1000 volts *lower* for the *negative* point than for the positive point. In air, the two potentials differ at most by 100 volts in 5000, and in fact, as will be later shown, the two curves of starting potential for positive and negative points in air as a function of point diameter cross at 1 mm, the negative being higher for smaller points and lower for larger ones by a very small amount. This has a simple explanation. In pure gases such as N_2 and H_2, photoionization of the gas by its own photons is not very prolific. Thus, a positive point corona requiring this type of ionization will need large avalanches with many ions to start a surface glow. The transparent gases N_2 and H_2 can yield photo-

TABLE 4.1

DATA ON POTENTIAL IN VOLTS FOR BEGINNING OF BURSTS FOR POSITIVE AND
NEGATIVE POINTS (Vg±) AND FOR ONSET PHENOMENA OF WEISSLER (V$_{onset}$±)

Gas	Vg+	Vg−	Self-Sustaining Corona V$_{onset}$+	V$_{onset}$−
H$_2$	3500	2800	—	—
N$_2$	4800	3700	—	—
Air (Weissler)	—	—	5500	8000
Air (English)	5100 ± 50	5000 ± 50	5550 ± 50	8000–10,000 Indefinite oscillographically, individual pulses still resolved

ionization at the cathode rather readily, especially in H$_2$ and a γ_i with the high point fields. Thus the negative point corona with $\gamma e^{\int \alpha dx} = 1$ should set in at lower values of V than would the positive. In air the effect of O$_2$ and its oxides on the metal point certainly reduces γ_i, as studies of Parker showed; the reduction is easily of the order of 10- to 100-fold. The value of γ_p is not very much lowered unless really thick oxide layers are present. On the other hand, O$_2$ and air are readily ionized by their own photons, especially at a partial pressure of 150 mm, as in air. Accordingly, the burst pulse and streamer threshold is lowered considerably. Thus, in air the two potentials almost coincide.

There is another circumstance that acts in this connection and probably applies to all negative ion forming gases. Electrons readily attach or form O$^-$ and O$_2^-$ ions in air. Before any ionizing events and avalanches can occur for either the negative or positive point, the X/p ratio at a sufficient distance outward from the point must exceed 90 to cause detachment of electrons from O$_2^-$ ions. In many cases O$^-$ ions change to O$_2^-$ ions at X/p $\approx$ 20, so that detachment applies only to the O$_2^-$ ion. Thus, not only is the threshold for the negative point corona raised while that for streamer and burst pulse corona is relatively lowered, but the condition X/p > 90 at some distance from the point surface makes the positive point threshold higher. Therefore, in air both thresholds are higher than is either one for N$_2$ and H$_2$, as seen in the table.

With the improved synchroscope-type oscilloscopes, photomultiplier cells, and pulsing circuits of World War II available, English[3.15] began anew the studies of coronas with the hope of determining the true duration and shape of Trichel, burst, and streamer pulses. In this he was to be disappointed, since the actual duration of the elementary pulses in air at 760 mm was on the order of 10^{-8} sec, and his instruments had a resolving power of 10^{-7} sec. He was, however, able to calibrate his synchroscope readings so that he could determine quantitative data as to currents and those few current variations that could be resolved. Thus, he observed that the Trichel pulses maintained their shape but changed in amplitude with increasing voltage. Their rise appeared to be slower than that of pre-onset streamers in some cases.

He next investigated the pre-onset and onset regime for all the pulses, obtaining current-potential characteristics as well as comparing starting potentials of positive and negative point coronas, as he reported in a later paper.[4.10] In this work, he was once again faced with the triggering problem. Here, owing to the small ionizing volume about the negative point, radioactive triggering was useless. In the end he did as Hudson did and used fine specks of MgO stuck to the point with Duco cement (nitrocellulose) diluted in acetone. Even with such points, current variations were annoyingly great. Occasionally, a smooth curve would be obtained for several successive readings, but suddenly with constant potential the current would jump to higher or lower readings, returning to the former value after some seconds. These changes were accompanied by sudden changes in amplitude and frequency of the pulses. The glow would also shift its position and flicker. A current increase implied an increase in frequency and a decrease in amplitude of the pulses. Often the corona would favor just two such modes in a given voltage range which could be identified with pulse amplitude. Sometimes the mode would stay constant for the recording of a whole curve; at other times it changed. This is nicely illustrated in the two curves of figure 4.7, taken under the same conditions. Transition between modes was not always abrupt, as can be seen in figure 4.8. In the first run of that figure with the hump above 6800 volts, two modes were present simultaneously, of 0.8 ± 0.05 and 1.0 ± 0.05 inches in height, respectively, as indicated by pulses. The transition progressed gradually until at 7600 volts the second mode had completely replaced the first. On the two succeeding runs, only one mode occurred. Double moding occurred most frequently on ascending voltage runs. As will later be realized, this proved to be due to alteration of the surface and change of work function.

As voltage increased over a larger range, even with a single mode, pulse amplitude slowly decreased with increasing voltage and current; at the same time pulse frequency increased. Before this trend had gone very far toward a steady negative corona current, a mode jump would occur giving reduced frequency, and the process would start anew. The freshness of the MgO had no influence; only when it was worn off did irregularities become troublesome. The pre-Trichel pulse onset currents of Kip were investigated by English and found only when MgO was used. They began a hundred volts below Trichel pulses and merged with the pulses above onset. They were not associated with the negative corona proper and were associated only with the dust specks. The light out in the gap seen here always emanated from a dust speck.

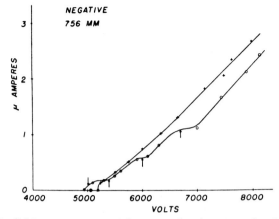

FIG. 4.7. English's current-potential curves, showing steps due to mode changes. The one without steps corresponds to a single mode.

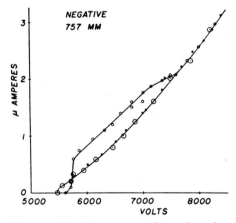

FIG. 4.8. Here the transition from one mode to the other is gradual. The first run from low to high potential leads to a gradual mode change, and the second run has only one mode.

Threshold observations were not good largely because of the triggering vagaries. However, the onset of Trichel pulses began more or less abruptly with currents on the order of a thousand times the field intensified avalanches below this point. Later work with satisfactory ultraviolet triggering by Bandel[1,14] indicates the appearance of the pulses at around 10^{-9} amp. The exact onset was not easy to ascertain. Occasionally it would consist of an isolated burst, the current contribution of which appeared negligible when divided by the time before the next burst. In some cases, one or two isolated pulses would be followed by regular pulses with a microampere of current. Where only isolated bursts occurred, they became more and more frequent as potential increased to about 100 volts above the 5000 volt threshold. Trichel looked for such pulses but never found them. These pulses had an amplitude two to four times greater than regular pulses because they moved in a gap devoid of space charge. The larger currents with the regular pulses result merely from continuity and number in a given time interval. The potential range of random pulses lay between 50 and 250 volts. Offset of regular pulses was usually a little below the onset value.

The detailed difficulties and irregularities can be ascribed to two influences. The first is the triggering problem from using dust specks. These initiate small discharges of their own. The dust specks are not uniform in size or efficiency, and they come off with use. Once Bandel eliminated their use and substituted ultraviolet light of weak intensity, thresholds were clearly defined. He used two different intensities of ultraviolet. The only difference was in an increase of the field intensified current below breakdown to the pulsed corona, so that the current transition from field intensified corona to pulses was less prominent.

The second difficulty leading to the different modes and the allied fluctuations in frequency and amplitude of pulses stems from the one thing which all workers, including English, looked for but could not observe at the threshold. This is changes in secondary emission coefficients with material and cathode condition. Initially, all metal cathodes in air are coated with layers of oxide. They may also have salt deposits from fingerprints, organic materials, and so on. As the field intensified currents increase, ions bombard the surface. Such bombardment tends to clean up the surface by sputtering off surface layers by ion impact. Oxidation of organic materials ensues through action of O, O_3, and N_2O_4 in air. On the one hand, the surface is cleaned to the bright metal; on the other hand, O, O_3, and N_2O_4 will react to re-form oxides, and N will form nitrides with some metals. Thus, the surface is subjected to simultaneous opposing action by cleansing and oxidation. Which action predominates depends on gas pressure, current density, and electrical fields at the surface, that is, ion transport and impact energy, sputtering, and diffusion of oxygen to the surface. A fairly rapid bombardment at not too high energies cleans the surface. High potentials and heavy currents deliver ions to and *into* the surface faster than they can be accommodated; gas diffusion at high pressure brings in O_2, O, O_3, and N_2O_4. Thus, actual bursts of gas evolution may disrupt the surface. Some surfaces show evidences of heating and melting under ion impact. In some low boiling point materials, actual vaporization may occur.

One characteristic of many surfaces under such bombardment is the flickering and wandering of the discharge over the surface. This can be ascribed to one of the several actions described above. If a proper gas layer increases γ—and there is ample evidence of this experimentally—then denudation by bombardment and sputtering causes the discharge point of impact to shift to regions where a layer is still present. There is evidence that γ_p on Pt is highest when the surface is covered with what probably is a monolayer of oxygen gas as O^- ions. Clean Pt or heavily oxidized Pt have lower yields. The photoelectric energy threshold is not altered. γ_i is increased by adsorbed gas films in some cases, decreased greatly in others. H_2 reacts strongly with some metals. It is heavily absorbed. Weissler[4,5] studied the change in work function of a clean W surface subjected to bombardment by gas ions. O_2 bombardment increases this quantity by 1.2 ev. N_2^+ bombardment *decreased* it about 0.3 ev below a surface cleaned by heating by electron bombardment. Bombardment by A^+ ions did not alter it. H_2 reduced the work function by about 0.5 ev. The value of the work function will alter γ_i but will not alter γ_p in the same fashion and to the same extent. Al_2O_3 on the surface of an Al cathode is gradually reduced by sputtering at adequate current densities. As the layer thins to about 10^{-5} cm thickness, composed of loose aggregates, a field emitted breakdown of the Al_2O_3 grains leads to bursts of electrons from the surface which can trigger powerful shock waves of ionization down the tube, leading to arcs. If a local discharge current density establishes an arc at some small point, the arc can lead to melting, and vapor jets of magneto-hydrodynamic nature occur. These jets can suffice locally to extinguish the arc, causing wandering and fluctuating currents.

It is clear that not too much is known and less is predictable about the behavior of surfaces under ion bombardment. This is largely because all such action is quite specific to the chemistry of the metal and gas used. Lower current densities will alter or "condition" a surface by removing what is on it and cleansing it, or substituting a new surface in combination with the ambient gas. In general this low current density bombardment increases both γ_p and γ_i, reduces the threshold, and increases the currents. Heavier bombardment will alter the surface by sputtering and possible chemical action. In many cases, such action disrupts the discharge and causes wandering. Arc-like discharges melt, vaporize, blast, and completely disrupt surfaces.

The principal Trichel pulse threshold, or onset phenomenon, probably consists of a cleansing in the pre-threshold and threshold region. This accounts for the lack of reproducibility, and depends on the metal and its past history. Double moding is readily explained by the coexistence of two surfaces, one of lower, the other of higher, γ. The higher γ spots will lead to larger pulses and will start first if present. Their discharge may be interrupted by chance or may be desensitized by prolonged bombardment. Then the lower γ spot may take over, giving smaller pulses at the same potential. As this spot cleans up, the higher mode takes over again. This situation is aggravated by the use of dust speck triggering, which anchors the region of continued discharge to the immediate proximity of the specks. It thus limits breakdown under altered conditions to certain localities, and when one speck fails, some other speck takes over. Conditions were much more uniform with moderate ultraviolet triggering. Here one mode prevails. However, when the current density at a spot becomes too high, the spot begins to wander. Depending on the pressure, this occurs at around 20 μamp at 760 mm for 1 mm diameter point, at 30 μamp at 400 mm, at 80 μamp at 200 mm, and at 100 μamp at 100 mm. At lower pressures, wandering also reduces the current for a given potential slightly below that which it would have had if no wandering had taken place. A smaller point diameter increases the current density needed for wandering. For the 1 and 0.5 mm diameter points, multiple glows set in at somewhat above the current density for wandering and at an increase in potential. It is clear that overloading of current at the limited area of the Trichel pulse spot leads to wandering and eventually to the activation of two or more spots, with one spot at first taking over the current carrying function for some milliseconds, while the other spot presumably recovers. All this points to deactivation by excessive bombardment. Such actions lead to an apparent modification of the pulsed corona in the form of the ring discharge observed for larger points, discovered somewhat later by Greenwood[4.6] and to be discussed in section A.3 of this chapter.

Following the paper of English, the author[3.17] summarized the progress made in corona studies up to that time (1948). The article was especially significant in its day, since it analyzed the observations of English, as has been done here in the last several pages. At the time, the Trichel pulses had not been adequately temporally resolved, although English believed that they had. Thus, some of the data in the article are no longer valid. On the other hand, use of a telemicroscope and reduction in pressure allowed English to analyze the structure of the Trichel pulse corona at its maximum brightness.[3.17] It enabled the author to calculate the dimensions of the various regions and to correlate these with the potential distribution along the axis as a function of distance. This, when analyzed by means of the Townsend coefficients, led to a great deal of information about the growth and form of the discharge. These data are presented as follows. Figure 4.9 shows an actual photograph of the Trichel pulse corona at 760 mm with the point traced in. Figure 4.10 shows the author's sketch with dimensions for the Trichel pulse corona at its height at 760 mm for a 0.38 mm diameter point. The Crookes dark space is 2×10^{-3} mm thick, the negative glow is 5×10^{-2} mm deep and about 0.1 mm wide. As pressures decrease, these dimensions increase proportional to $1/p$, but the width of the spot merely increases from 0.1 to 0.2 mm. The

Faraday dark space is 3×10^{-2} mm deep, and the brush extends out to some 1.5 mm or more. It should now be possible to use a Kerr cell, or other shutter, triggered so as to observe the luminosity after the bright phase is over, to see if the luminosity of the fan persists during the early part of

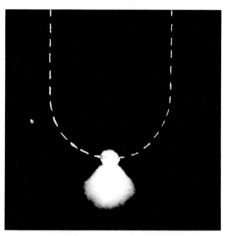

Fig. 4.9. English's photograph of a Trichel pulse corona in air at 760 mm with the point sketched in and greatly enlarged.

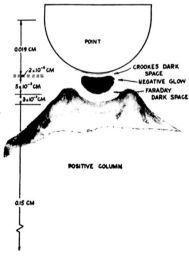

Fig. 4.10. Author's sketch of the Trichel pulse corona at its brightest phase at 760 mm for a 0.38 mm diameter point. The dimensions, inferred from English's measurements at lower pressures and scaled down to 760 mm, are shown.

the extinct phase before the space charge dissipates too widely. It would also be of interest to note the motion of this glow, if any.

The Townsend ionization, as calculated along the axis before space charge distortion, is shown in figure 2.6. In 10^{-8} sec the ions to the left of the line A have moved in to the point. In 10^{-7} sec, if no distortion had taken place, the ions to the left of line B would have reached the point. The total ions and electrons created in the first avalanche were 2.8×10^4. If γ had been 10^{-2} in the surface field of 120 kv/cm, 170 new avalanches would have been created, so that a total of 5×10^7 positive ions would have been created in 10^{-7} sec. These calculations have been made without considering space charge distortions. However, in some 4×10^{-8} sec, a continuous sheet of positive ions would have extended from 0.4 mm up to the point surface. This would enhance the field at the surface in 10^{-7} sec to some 400 kv/cm and give a potential drop of at least 4 volts across an ion free path at the surface. Here the Morton regime sets in. Ionization is by fast electrons diffusing in the glow region.

Such a calculation is oversimplified, and data are not sufficiently accurate to warrant a more elaborate computation. It is obvious that the rate of growth will, in fact, be much faster than indicated in view of the high electron speeds if photoelectrons from the cathode lead to an effective γ_p.

The choking of the discharge by negative ion space charge could not at that time (1948) be properly accounted for. However, another feature could be studied. Rough estimates of the potentials caused by space charges at different distances from the surface permit a comparison of the existing potential distribution along the discharge axis with that in the undistorted gap. This was done by the author for three phases in the pulse (figs. 2.7, 2.8, and 2.9). It is noted that, once the discharge sets in, the potential drop at the cathode is steeper than before breakdown. This means that the potential along the axis near the point is less negative by a considerable factor relative to the region of no discharge, which is radially removed from the axis. That is, there are strong potential gradients forcing electrons toward the axis, with the *undistorted* regions more negative than the distorted region along the axis. These fields can reach the order of 200 kv/cm in a radial sense within 2.5×10^{-2} mm of the axis in the Crookes dark space region of the brightest phase of the glow. The mixture of ions and electrons ionizing in the plasma of the negative glow reduces the field in that region. In the Faraday dark space, the potential along the axis again falls. Thus, there tends to be a constrictive force on electrons in that region also. Therefore, all regions off the axis act to drive all photoelectrons in the gas toward the axis and to resist spread of the glow over the point. Hence, each Trichel pulse spot is limited in its spread over the surface and has a strong negative potential outside, or a strong virtual positive charge along its axis. Two such spots will thus *repel* each other, as has been observed. The brush region, on the contrary, has the electron and negative ion space charge accumulation. This makes this region negative to the surrounding space. Electrons that remain free are driven outward and ionize. Thus, there are stong compressive fields near the dark spaces restricting the discharge area between, causing the negative glow and a weak expansive force in the fan spreading the discharge radially. Since electron energies are high in the glow, they lead to spectral lines of NI and NII, as well as to bands of N_2, N_2^+, and NO. In the brush region, electron energies are low and most lines of NI and bands of N_2 and NO appear, giving the characteristic violet-purple glow, while the negative glow is bluish to bluish white when currents are high. It is also noted from figures 2.7, 2.8, and 2.9 that the luminosity changes during the growth and decline of the pulse. In between pulses, the region is possibly dark. As indicated, this can now be studied if desired.

In all point-to-plane coronas, expecially at higher pressures, there is an electrical wind. This results from the momentum given the air molecules by impact or drag of the ions and electrons as they move out from the high field regions. The wind given by a corona point with a fairly high current is sufficient to blow out a 1 cm diameter candle flame when placed within $\frac{1}{2}$ cm of the wick. This matter will be treated in more detail in section D. Since the electrons expend their energy from the field largely in creating positive ions and excited states and negative ions by dissociative attachment in the high field region, they do not impart as much momentum as do the positive ions created in the high field regions with positive point discharge. The wind velocities are on the whole low compared with the ion velocities, since the concentration of ions relative to neutral molecules is very low.

Following the author's study, there appeared a paper by Moore and English on point-to-plane *impulse* corona.[3.19] It was considered of interest to initiate coronas in a clean gap by the sudden application of a square impulse potential of short duration, especially as regards the positive point corona. A typical square wave pulse generator giving pulses of 1 and 2 μsec duration at potentials up to 12 kv and repeat rates from 50 to 2000 pulses per second was set up. The 1 μsec duration pulses were quite square, but the 2 μsec had a low peak near the start and a slightly higher one at the end. Rise time was shorter than 10^{-7} sec. The point was a steel needle of about 0.1 mm diameter with 2 to 3 cm gaps. These were studied by synchroscope, telemicroscope, and camera. Onset for the negative corona was around 5400 volts. On the synchroscope, the Trichel pulse occurred randomly during the high voltage pulse, although not on every pulse, as might be expected with triggering difficulties. Above 7000 volts Trichel pulses became more frequent, and two or three might occur during one impulse. At 7700 volts, the pulses started to localize at the beginning and end of each square potential pulse. At about this potential visually and photographically, a long narrow spike extended axially through the center of the brush. The spike and fan became more intense and extended further into the gap as potential increased. At 9000 volts, two side spikes appeared.

The development of the spikes is shown in figure 4.11. This was the limit of the negative pulse. By lowering the pressure, the potential increase could be pushed further above threshold. Diffusion was greater but did not materially change the appearance. Between 4000 and 5000 volts at 375 mm the development of spikes began. As the potential increased, the background luminosity increased in the gap out from the point. At about 5600 volts, the spikes became broader and difficult to distinguish. Luminous streaks or filaments occurred amid the general glow. With increase in voltage these became brighter and longer and extended beyond the general glow. They appeared to concentrate toward the sides of the chamber rather than toward the plane. At around 9000 volts, these streamers presented a spectacular appearance, something like a Roman candle. This was not a wall effect in the chamber. Only composite pictures of many streamers could be had. These are shown in figure 4.12 for various concentrations of O_2 in N_2 from 0 to 20% at 375 mm, 9100 volts, 4200 volts, and at 750 mm and 9100 volts with 100 to 500 pulses photographed in each exposure.

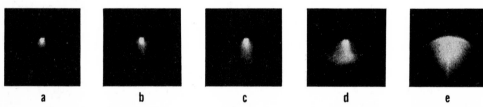

a   b   c   d   e

FIG. 4.11. Moore and English's spikes and corona in $N_2 + 5\%$ O_2 at 375 mm for 100 pulses with point of 0.1 mm diameter negative. (a) 4200 volts, (b) 5600 volts, (c) 6500 volts, (d) 7800 volts, (e) 9100 volts. In (e) the central spike can be seen clearly. This spike phase is more clearly differentiated as a pulseless mode in Miyoshi's work.

N_2 and H_2 were tried to see if the Trichel pulses appeared. Only the Townsend type glow discharges observed by Weissler[1.8] were observed up to 9100 volts at 760 mm. These are seen in the first column of figure 4.12. Addition of 1% O_2 gave Trichel pulses but no spikes. If the pressure was lowered to half an atmosphere and X/p was increased by a factor of two at 9100 volts, spikes were observed, as seen in the second column of the figure.

No simple explanation is possible. It was pointed out by the author that,

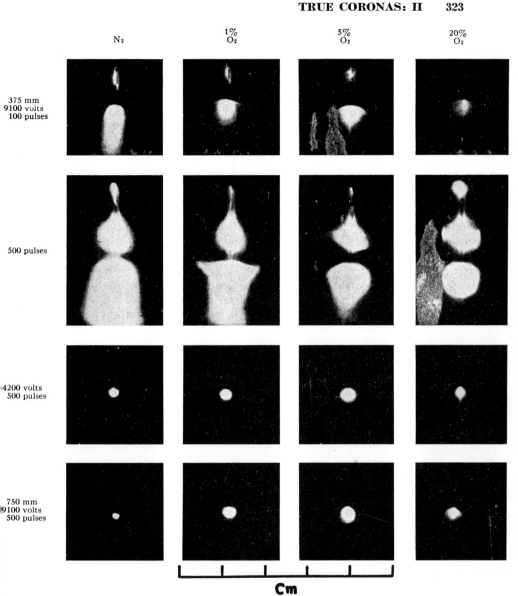

FIG. 4.12. Moore and English's negative impulse coronas in N_2 and O_2 mixtures, showing changes with potential and concentration.

with two Trichel pulses per potential pulse, the space charge cloud of negative ions of the first pulse is out in the gap when the second and final Trichel pulse occurs. It appears possible that discharges may occur between the space charge clouds of the separate pulses. The behavior is, however, too complex, and there are too few controls to permit any hypothesis.

At this point attention should be called to a short article by the author[4.7] on the role of the cathode in discharge instability. This article derived largely from the studies and theories developed on negative electrode coronas. It is based on the inherent condition of instability resulting from the space charge of positive ion build-up at the cathode and its influence on the Townsend coefficients. In principle, at least, it signifies that, unless current increase at some potential above threshold (that is, overvoltage), is controlled by external circuit resistance or internal space charge resistance in low field regions, an autocatalytic build-up of current will follow. Such a situation leads to a negative current characteristic, or apparent negative resistance, as Miller showed in coaxial geometry.[4.8] With this, a change in the nature of the discharge can be initiated which will spontaneously change a corona to a glow discharge or to an arc, or a glow to an arc. Often such a change produces a heavy transient increase in current, which, for some reason or other, space charge increase or alteration of the cathode to a low γ state restores the original condition. Given proper external circuit properties, such fluctuations, if there is feedback, can lead to oscillations. Oscillations of this nature are to be expected where current density is in a critical region between clean-up by sputtering and formation of oxide layers, or where clean-up leads to a sudden state of overvolting. In principle, the article generalizes the information derived in the preceding studies.

A paper by the author and English on the relative starting potentials of positive and negative point coronas in air, which has already been cited, appeared in July, 1949.[4.9] The first part of the paper considers the difficulties of triggering and influence of point material and ultraviolet illumination already presented. There is nothing new in that part. One point of interest discussed, however, lies in the choice of the threshold potential which must be used in the comparison of the starting potentials. Since in the past this choice has been optional, it has led to much confusion. It is thus best to recall one or two matters. The negative corona threshold in air has in the past been taken as the potential at which *the first current jump from field intensified ionization to single large Trichel pulses occurs, or else the point at which onset of a regular Trichel pulse corona with 500 to 1000 pulses per second appears.* This second region, it will be recalled, may range from 50 to 250 volts above the first, depending on circuitry, point diameter, triggering, and so on. It is not a trivial difference. Many observers choose as the threshold the onset or—what is equivalent but more reproducible—the offset of the steady sequence of pulses.

From a threshold viewpoint, onset or offset is *not a good* criterion, as it *may depend on an alteration or clean-up of the point.* With positive point corona, the situation is even more complicated. A good example is seen for a 0.38 mm diameter point with varied gap length, in which the appearance of various breakdown modes are plotted as observed at atmospheric pressure by English.[4.9] With this point, abrupt increase in current occurs with a burst pulse corona, followed at a higher potential by streamers and at a still higher potential by the onset of the Hermstein glow corona. For a 5 cm gap, the potentials are 5300, 5450, and 5850 volts, respectively. At gaps of 1 cm, all three thresholds lie close together. Actually, it has never been established that burst pulses always have a lower threshold than preonset streamers. It is probable that streamers initiate most burst pulses at the positive threshold. However, while they may trigger burst pulses, the space charge of the succeeding burst precludes further streamers until potential is raised. The point size and gap length may also alter these potentials. The glow corona *onset or offset*, which is too often used as a criterion as it is marked by a reproducibility that makes measurement easy, *is not a threshold phenomenon.* It is a new discharge condition caused by negative ion formation in the gap and depends on the current density and gap conditions favoring negative ion formation in the low field gap regions. While it may mark the end of a counting range, being current-dependent, it is a meaningless threshold to use and should *never be related to Townsend coefficients*, as has often been attempted. Here again, *the true threshold is the sudden increase of current and the appearance of the first pulses, streamers, or bursts*, signaling transition from a field-intensified current to a self-sustaining one. The appearance of luminosity is a worse criterion, since it is an ill-defined subjective phenomenon.

With this criterion, thresholds for positive and negative point-to-plane coronas, with polished hemispherically capped Pt points from 0.03 to 2.4 mm diameter and with the ratio of point radius to gap length kept constant at 1/160 (standard geometry), were determined in room air at atmospheric pressure, with the results shown in figure 4.13. It will be noted that the two thresholds are about the same at a point radius of 0.05 mm, the negative threshold lying above the positive for larger points. Other electrode materials might alter this ratio. Change of gas composition assumedly will.

A valuable feature of this study was that while gap length appears to play a minor role, its increase separates the thresholds so that they are clearly delineated, as seen in figure 4.14. This was one of the features that led to the development of the standard geometry in point-to-plane studies.

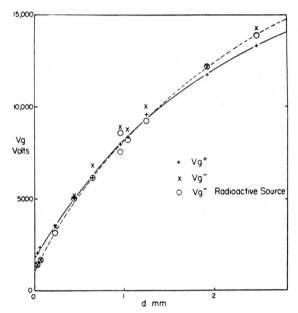

Fig. 4.13. Static thresholds of positive and negative burst pulse and Trichel pulse coronas in air at 760 mm pressure and standard geometry as a function of point diameter in millimeters with and without radioactive triggering. V_g^- lies above V_g^+ below 1 mm, but it lies below it above 1 mm.

On the other hand, the use of larger points with standard geometry shortens the potential range between appearance of breakdown streamers and a spark. To observe breakdown streamers over a large range of potentials below the spark, still smaller points and longer gaps (ratios greater than 160) are needed.

The paper which logically followed the threshold study was the paper on a choice of suitable gap forms for point-to-plane corona study by Loeb, Parker, Dodd, and English.[1.1] This has been adequately discussed in chapter 2 and in chapter 3, section 4, on the work of Bandel. It has only one important feature in regard to the negative point-to-plane corona. This is that the field along the point axis, and hence the computation of the Townsend integral avalanche size and so on, at the beginning of a Trichel pulse, is now open to much more accurate calculation.

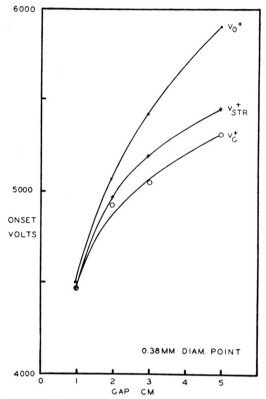

Fig. 4.14. Thresholds for positive point corona for which no triggering problem exists, with a 0.38 mm diameter point as a function of gap length, according to English. Note the increase in range between V_g^+, burst pulse threshold, streamer threshold, and onset of steady corona at longer gaps.

3. BANDEL'S COMPREHENSIVE STUDY IN STANDARD GEOMETRY

In 1951 there appeared Bandel's[1.14] quite complete collection of data on the negative and positive point-to-plane coronas in dry air using standard geometry. This study covered point diameters of 1 mm, 0.5 mm, and 0.25 mm, at pressures of 750, 400, 200, 100, 50, 25, and 10 mm. It ranged in currents from pre-field intensified currents around 10^{-13} amp and less to currents just preceding sparkover of 300 μA. Complete descriptions of the phenomena as well as the curves were given. The data were of value as they permitted calculation of the fields and integrals before field distortion sets in. By that time, all thresholds had been clearly delineated and carefully noted. The successful use of photoelectric triggering with ultraviolet light and the influence of the initial photocurrent on the course of the current potential curves before threshold were of utmost value. Reproducibility was good. The significance of observed curves under different triggering conditions was clearly revealed. All these results help to define and clarify the meaning of thresholds and ensure uniformity of study in future work if needed.

This investigation established the nature of the ultraviolet illumination needed for triggering; enough has been reported on that subject earlier (chap. 1, sec. 4, and chap. 4, sec. 2). The curves with γ ray triggering not only showed high values for thresholds but yielded values of steady corona directly at 500 volts above offset. They also showed that too much ultraviolet light would choke off pulses above onset by providing a field intensified steady current of 10^{-8} amp above that for the Trichel pulses at threshold. This is akin to the effect Kip[3.1] first observed with pre-onset streamers when too much field intensified current filled the gap with space charge. The light from a quartz Hg arc through a quartz window is sufficient without a lens to focus the light on the point.

Bandel found that clean-up by running the discharge persisted so that a small point cleaned up would start again more readily than one that was not cleaned up. A larger point, however, allowed the discharge to wander over the surface, so that cleaned-up spots did not remain clean long enough for the effect to be detected with certainty.

One effect noted by Bandel was that the current potential curves for declining potential fell below those for increasing potential shortly above Trichel pulse threshold. There is quite a difference at low pressures. That is, currents at a given potential were lower after the discharge had run for some time. This could perhaps be ascribed to chemical alteration of the air by the discharge in the confined chamber, which perhaps raises γ and alters α. It is probably largely due to photoelectric "fatigue" or alteration of photoemission of the point surface by the discharge which lowers the triggering current in the field intensified region, thus reducing the photocurrent. The current at threshold is slightly decreased, but the threshold potential is not altered.

With ultraviolet triggering at pressures below 200 mm Hg, a second type of pulse appeared on the oscilloscope screen. These pulses were very small, about 0.2 those of the Trichel pulses. Their amplitude was random. Just at threshold they appeared more frequent than Trichel pulses, but as Trichel pulses appeared and became regular, these disappeared. Their amplitude at lower pressures was greater than at higher, as with Trichel pulses. At 400 mm they were seen only rarely, and at 760 mm only on occasion with a 0.5 mm point. They could have been statistically integrated groups of single avalanches which the oscilloscope integrated. They could have been associated with dust specks that were not readily removed by ion impact, as they would have been at high pressures.

Pulses above onset of steady Trichel pulse corona were triggered by radioactive material, as these were mostly self-sustaining discharges. Ultraviolet illumination interfered with visual observations in this region.

If the negative glow button is small compared to the radius of the point, the discharge tends to wander from one active cathode spot to another. The intense ion bombardment degasses and deactivates the spot, while marginal discharge cleans up a new area to which the spot wanders. Since the area of the glow varies inversely as the pressure at the lower pressures, the wandering depends on pressure as well as on point size. It occurred mainly with a 1 mm diameter point at 200 mm and up, and with a 0.25 mm point at atmospheric pressure.

As current increases, the discharge moves faster and faster around on the point. Sometimes the fan appears to oscillate to and fro on the point as well. The point at which wandering begins is indicated on the curves of figures 4.15b, 4.16b, and 4.17b by the symbol W. If the point is small, the motion becomes more and more rapid, so that the glow appears to be at two or more places at once. With larger points, this probably leads to the ring discharge of Greenwood,[4.6] but was not noted by Bandel with his relatively small points. When many spots are active, they fall into orderly patterns because of self-repulsion and conditions of symmetrical field uniformity. They occur in one or two circles around the tip and move slowly over the surface, maintaining their spacing. The appearance of multiple glows is indicated by the symbol MG where it appears on the traces of figures 4.15b, 4.16b, and 4.17b. At around 150 or 200 mA, depending on point diameter and pressure, there is a change in the corona. This is indicated by a sudden decrease in current for the same potential and a rise of current along a different curve. Phenomenologically, this change is accompanied by a disappearance of Trichel pulses and the typical glow discharge structure. The Trichel pulse amplitude decreased to zero within a few volts increase in potential. Visually, the glow was observed to lift somewhat from the point, lengthen in the direction away from the point, and decrease in diameter.[1.14,4.3] The entire visible discharge shared in this contraction. The brush nearly disappeared at lower pressures. At higher pressures, the fan suddenly contracted to one-half its diameter or less and at times was even more intense, leaving the Faraday dark space lengthened and more sharply delineated than before. Above this point, as potential increased, it grew gradually fainter and finally disappeared. The change did not occur at a specific voltage but seemed to be associated with some conditioning of the points. It occurred at lower voltages if the discharge had been run for a longer time at lower currents. In rapidly traversed runs at lower pressures, it failed to appear.

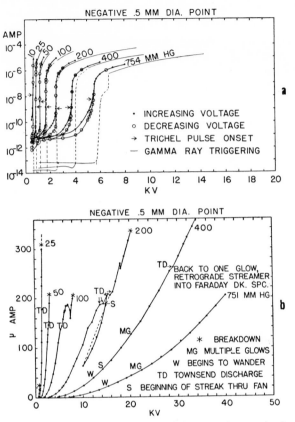

FIG. 4.15a and b. Bandel's current-potential curves in standard geometry for a negative 0.5 mm diameter point, 4 cm gap, from below threshold to nearly spark breakdown. indicating all the various phenomena for a series of pressures.

The transition to a pulseless discharge was accompanied by a jump in current of 5 to 10 μA but is not shown on the curves at lower pressure. If the potential was held constant after transition, the current decreased and the pulses reappeared. This led to the breaks in the curves. In figure 4.16b at 200 mm, the fact that the curve ran with decreasing potential after the break followed the higher voltage trend indicated that the point had been altered by the discharge. Points examined under the microscope after this change showed that the tip was highly polished because of melting, and surrounded by a sputtered area that had lost its initial polish. Cessation of pulses meant that the space charge of negative ions was no longer able to choke off the discharge. The pulse shape in the transition region shows that the pulses are drawn out in time before space charge builds up enough to quench it. The lateral constriction of the fan is reduced at this point, as

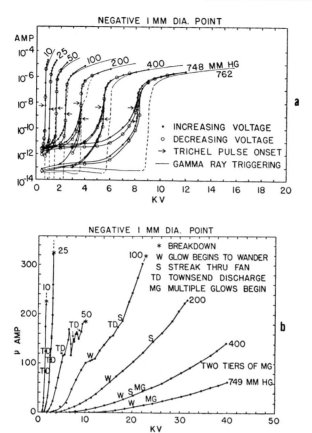

FIG. 4.16a and b. The same as figure 4.15 for a 1 mm diameter point.

the negative charge is no longer heavy enough to cause a flaring. This difference between the pulseless discharge and the pulsed discharge is similar to that observed by Weissler[1.8] for negative point coronas in H_2 and N_2 where negative ion space charge does not form and there are no pulses. Introduction by Weissler of O_2 into these discharges in the order of 1% or more restored pulses and reduced the current to about one-half.

Since the transition observed here causes only a small increase in current, it would seem likely that the space charge is being limited by the supply of electrons rather than the lack of O_2 molecules to attach to, as in N_2. The change in the point can be held responsible for the decrease in supply of secondary electrons and can also account for the subsequent decrease of current at constant voltage. If the area melted were melted all at once, it would yield many too many electrons thermionically. It is probable that only microscopic areas are melted at one time. Why the current should

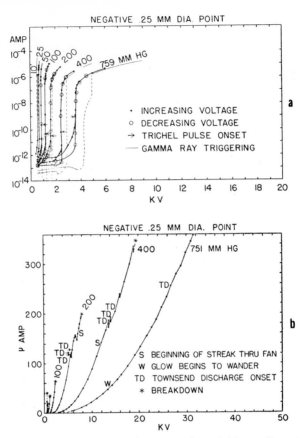

FIG. 4.17a and b. The same as figure 4.15 for a 0.25 mm diameter point.

revert to Trichel pulses at constant voltage after the current has dropped is not clear. However, by then the point is much altered. Perhaps the melting of the tip flattens that region and shifts the region of higher curvature out to unaltered portions of the tip.

At higher voltages at higher pressures the retrograde streamers previously reported by Kip[1.2] and associated with an ultimate spark breakdown appear.[4.3] This is indicated by the symbol S in figures 4.15b, 4.16b, and 4.17b. The development always started with the appearance of a streak of increased intensity through the center of the brush. In some cases it extended well out into the space beyond the fan. Only with the two larger points at higher pressure did the streak develop into the very intense sharp projection back into the Faraday dark space. The value of potential at which a spark appeared is indicated by a star; the spark was avoided, as it damaged the point.

4. The Short Time Duration of the Pulse and Negative Ion Formation

The first realization of the short duration of the Trichel pulse corona in air came when English[4.10] had access to the micro-oscillograph at the Chalk River, Canada, Atomic Energy Research Laboratory for one day. He was able to get a time resolution for both the pre-onset streamer and the Trichel pulse, using a photomultiplier. In this, the pulse duration was observed to be on the order of 2×10^{-8} sec in air at 760 mm.

The next important breakthrough occurred the next year, in 1951. At that time Geballe and Harrison,[4.11,4.12] stimulated by problems of the action of spark suppressing gases, made a study of the influence of negative ion formation on the first Townsend coefficient in such gases. This investigation indicated that at values of X/p of around 30 in air and in O_2 there was a very much increased attachment of electrons above that for the formation of the O_2^- ion discovered by the author in 1920. Compared with the curves of Bradbury and Cravath,[4.12] it corresponded to a prolongation of the second rise of their curves which was never satisfactorily explained. It was several times greater than their peak. The curve of Doehring,[4.12] going further in X/p with superior techniques, almost dovetailed into the beginning of Geballe and Harrison's curve. At first they were inclined to claim that the lower coefficient of 1920 was wrong, but it was pointed out in conference that they were dealing with a different mechanism. Whereas negative ion formation with O_2 and a free electron is a three-body impact mechanism at low energies, the process of Geballe and Harrison occurs at average swarm energies of about 1.6 ev. The O_2^- formation requires collision between an electron and two O_2 molecules. The second process is a *dissociative* impact leading to O^- ion formation with a cross section of 3×10^{-19} cm² at peak. The energy needed is around 3.6 ev. If the electron swarm has an average energy of around 1.6 or 2 ev, enough electrons can attach to be perceptible experimentally.

It occurred at once to the author[4.13] that this must solve the difficulty, which had not been clarified, as to how attachment to form O_2^- ions could occur within a few tenths of a millimeter from the point in sufficient amounts to choke off the Trichel pulse within the short time of less than 10^{-7} sec. Calculations based on Geballe and Harrison's data at once revealed that, even without the use of the enhanced energies of electrons caused by the ionic space charge, the attachment could be accounted for. The pulse rise time as observed by English in one case and later extensively studied by Amin was 1×10^{-8} sec.[4.14] The pulse of luminosity lasted 1.5×10^{-8} sec further as a peak and declined to zero in 3×10^{-8} sec. This pulse contained some 2×10^9 ions. Calculations made for available data on a 0.019 cm radius point on a 3.1 cm gap at 760 mm with a threshold potential of 5000 volts permitted the quantities to be evaluated. The first avalanche yielded 6×10^4 electrons and positive ions. If the value for γ_p is assumed to be 5×10^{-5} (which is probably low by an order of magnitude), three new avalanches are yielded in the second generation. If we neglect statistical fluctuations and alterations by space charges of positive ions and so on, there are 3.6×10^9 ions and electrons produced in ten generations. The distance for maximum ionization before space charge intervention is 4×10^{-2} cm, and the electron drift velocity is at least 2×10^7 cm/sec on the average, so that one avalanche generation takes 2×10^{-9} sec. Thus in 2×10^{-8} sec the full ionization is achieved, as observed by English and Amin. The positive ion space charge has had only 1% of the ions reach the cathode in this time. This is 3.6×10^7 ions, and if γ_i had been 10^{-3}, this arrival would have yielded a total of 3.6×10^4 new avalanches spread over an exponentially increasing curve beginning perhaps within 1×10^{-8} sec or somewhat more. Thus γ_i ionization will rapidly augment the rate of growth as soon as more than 10^3 positive ions have reached the cathode. When this occurs, ionization by avalanches, which are created from electrons liberated by ion impact, rapidly takes control of multiplication, as soon as on the order of 3×10^3 ions per second or more reach the cathode.

The electrons created in the avalanche swarm have average energies appropriate to X/p values which range from 30 to 70 volts/cm per millimeter pressure. The electrons have a *random* velocity in excess of 10^8 cm/sec. Their average energy is easily about 3 ev. The effective cross section for dissociative attachment in air is 6×10^{-20} cm^2. At 760 mm pressure there will be 1.6×10^8 ion-forming collisions per electron per second. This means that $(1 - 1/e)$ or 0.62 of the swarm of electrons will have attached in 0.62×10^{-8} sec. With an electron drift velocity of 2×10^7 cm/sec, the swarm will have attached to within 0.37 of its number within 0.124 cm; and, if in the reduced field beyond the positive space charge the electron velocities have declined to about 1×10^7 cm/sec, the space charge will have formed within 0.062 cm of the positive space charge. The diameter of the region in which the negative ion space charge forms is around 0.02 cm. Thus, the negative ion space charge density by attachment in this area is confined in a volume roughly 2×10^{-5} cm^3. This gives a negative ion space charge density of 1.2×10^{14} ions/cm^3 or, at worst, if they extend out 0.12 cm, a density on the order of 6×10^{13} ions/cm^3. This would produce a field at the cathode of 6×10^6 volts/cm were it an isolated space charge. Actually, the positive ion space charge lies between the negative ion cloud and the cathode. Its center is spread out between 4×10^{-2} cm and the cathode. It has less effective positive charge than the negative ion space charge for two reasons: (1) it has lost some ions to the cathode; (2) it has also retained quite a number of negative electrons and ions in the plasma. Considering such actions, the net reduction by a counterfield, caused by negative ion dissociative attachment, is that of a dipole field composed of a stronger, pure negative ion space charge with more ions, say at 0.12 cm from the cathode, with a 10 to 20% weaker positive ion space charge centered at around 0.04 cm from the cathode. The cathode surface field is on the order of 1.2×10^5 volts/cm at the threshold value of some 5000 volts point potential. If, at threshold, the surface field is reduced by 5%, the cumulative ionization would cease. Thus, the effective counterfield at the surface for terminating a pulse in this region would be on the order of 6×10^3 volts/cm. With dipole space charge fields of the negative ions, without disturbance, on the order of 10^6 volts/cm, the termination of ionization and pulse choking is understandable. Actually the positive region is a plasma, and only a small fraction of the total electrons create the negative ion space charge. If one had depended on the creation of negative ions by triple impact to O_2^- ions, the distances of negative space charge would have been increased by a factor of five or more, and negative ion fields would not have been effective.

The clearing of the space charge fields for resumption of the next discharge would depend on the rate of removal of the positive ion space charge field by migration to the point. This will proceed slowly at first; but at these fields, within a microsecond, the positive ion space charge should be nearly absorbed by the point; and the negative ion space charge will have moved outward some 0.06 cm and expanded the volume of space occupied, because of its internal repulsive forces. It is clear that at least 1 μsec is required before a new pulse can form. If point potential increases materially, then for reasons discovered by Amin the removal of positive space charge and the creation and dissipation of the negative space charge will consume about a microsecond. In consequence, it will not be surprising to find that the maximum repeat rate of the Trichel corona pulses is limited to the order of 10^6 cycles per second.

The analysis above, for growth of a Townsend discharge in the presence of space charge, is probably now open to studies of much greater accuracy and refinement with the use of more subtle equations using modern computer techniques. In principle, they will lead to more precise results but will differ by less than an order of magnitude from the simple example given above, which anyone can follow. With the aid of this complete interpretation of Amin's, observations are possible.

5. The Fast Oscilloscopic Analysis of Trichel Pulses by Amin

In 1953 Amin[4,14] employed the new techniques with an oscilloscope having a time resolution of 1×10^{-8} sec, using photomultipliers to study the breakdown as well as the electrical pulses, as in the past. These have been described in his study of burst pulses and streamers in chapter 3, section A.5 and 6.

Figure 4.18, upper trace, shows a 1×10^{-8} sec square wave timing pulse, while the trace B below shows the Trichel pulse photon impulse for a 2 mm diameter point with 4 cm gap at 760 mm pressure. Figures 4.19, 4.20, and 4.21 show both photon and electrical pulses as different variables are changed. Figure 4.19 shows the variation of the pulses with increase in frequency of the pulses and thus potential and current in multiples of 10^3 cycles/sec. This was for a 1 mm diameter point, 4 cm gap, 100 mm Hg pressure, RC time constant of 10^{-8} sec, and sweep rate of 0.5 μsec per division. Figure 4.20 shows the influence of pressure on the two recordings

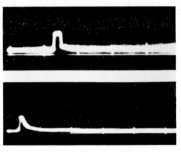

Fig. 4.18. Amin's photomultiplier analysis of a Trichel pulse in the whole gap. Above, a 1×10^{-8} sec square wave timing pulse, showing the resolution possible. Below, the Trichel pulse at the same sweep rate. Some ten pulses were superposed to give the traces shown.

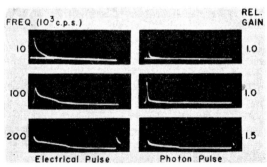

Fig. 4.19. Variation of the Trichel pulse with repeat rate. 1.0 mm diameter point, 4 cm gap, pressure 100 mm, RC time constant, 10^{-8} sec sweep rate, 0.5 μsec per scale division. Electrical and photon pulses at different frequencies and different relative gains for photon pulses indicated. The higher the frequency, the higher the acting potential.

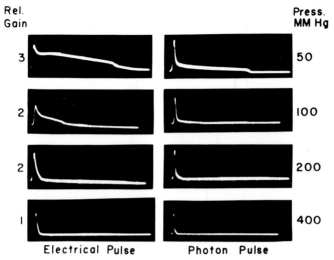

FIG. 4.20. Variation of electrical and photon pulses with pressure
at 100 kc frequency.

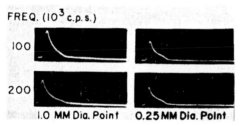

FIG. 4.21. Electrically analyzed Trichel pulses for a 1.0 mm and a 0.25 mm
diameter point with 4 cm gap at 200 mm Hg pressure. Sweep rate, 0.2 μsec per
division. Relative gain: left 1, right 2.

for the same gap as in figure 4.19 at 100 kc/sec. Figure 4.21 shows the
influence of point size, 4 cm gap at 200 mm pressure at 0.2 μsec per scale
division with 1 mm and 0.25 mm diameter points.

Certain features are quite prominent. The photon pulse is considerably
narrower and shorter than the electrical pulse. This comes from the fact
that the light pulse shows only the brightest phase of the pulse during and
very shortly after electron excitation. Thus it largely resolves the electronic
phase. The current, or electrical pulse, resolves the slower ion movement
as well. However, in both oscillograms of figures 4.19 and 4.20, it is seen
that the initial spike is followed by a plateau. The photon pulse, however,

drops sharply to zero at the end of the plateau, while the ion phase slowly tapers off. This is more clearly seen at a still lower pressure in figure 4.22. If the extrapolated slope of the tailing ion pulse is subtracted from the electrical pulse, the remainder due to electronic action is seen to resemble closely the photon pulse.

Electrical Pulse   Photon Pulse

FIG. 4.22. Electrical and photon pulses at lower pressure and higher potentials, showing the fast, sharp rise and decline in both, due to the electronic component produced by γ_p followed by a slower decline due to ionization caused by a γ_i and followed by a further decline in the ion movement phase of the electrical pulse (left), the last of which is absent from the photon pulse (right).

The crossing time of the ions is indicated by the final decline of the tail. This is shown in figure 4.23 with an R of 10^6 ohms and an RC of 10^{-4} sec, a sweep rate of 1000 μsec per scale division for a 1.0 mm diameter point, and a 4 cm gap at 760 mm. The upper trace is for a positive burst pulse at 7.6 kv, and the lower one is for a Trichel pulse at 8.2 kv. It is seen that the positive ions are slightly slower in crossing and that it takes on the order of 10^{-3} sec for the ions to go the 4 cm. Figure 4.24 shows the succession of Trichel pulses at 100 mm for a 1 mm diameter point with a 4 cm gap. The sweep speed is 5 μsec per scale division. The frequencies are given in units of 10^3 cycles/sec, that is, respectively, 10^5, 2×10^5, and 2.5×10^5 cps.

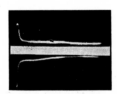

FIG. 4.23. Electrically induced pulses for a 1 mm diameter point, 4 cm gap at 760 mm, R = 10^6 ohms, RC = 10^{-4} sweep speed, 1000 μsec per scale division. Upper trace, burst pulse at 7.6 kv; lower trace, Trichel pulse at 8.2 kv. This compares the ion crossing times for positive and negative ions. They are nearly the same, as in both the ions are largely O_2^+ and O_2^- undergoing charge exchange. However, only O_2^+ ions exist. While there are in the swarm O^-, O_2^-, and O_3^- ions, the first and last have higher mobilities than O_2^- and perhaps cause the difference.

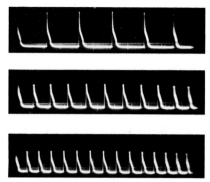

F$_{\text{IG}}$. 4.24. A succession of Trichel pulses at 100 mm for a 1 mm diameter point and 4 cm gap. Sweep speed is 5 μsec per scale division. Frequencies are 10^5, 2×10^5, and 2.5×10^5.

Figure 4.25 shows pulses as the point conditions at high current density near the transition to a pulseless discharge. This is for a 1 mm diameter point, a 4 cm gap at 100 mm pressure, an RC $\times$ 10^{-8} sec, and a sweep rate of 0.5 μsec per scale division. The left traces show a relative gain of 1.5; the right show a relative gain of 1.0. The unconditional points at 10, 100, and 200 kc/sec are to the left; conditioned ones are to the right. Condition-

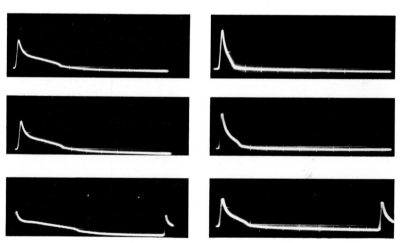

F$_{\text{IG}}$. 4.25. Pulses seen as the point "conditions" at high current density near transition to the pulseless state, for a 1 mm diameter point, 4 cm gap, 100 mm pressure, RC $= 10^{-8}$ sec, sweep rate 0.5 μsec per scale division. Left traces, relative gain 1.5; right traces, relative gain 1. The unconditioned points at 10, 100, and 200 kc/sec are to the left; the "conditioned" ones are to the right.

ing gives higher spikes and shorter pulses. The conditioning also causes an alteration of the frequency-current and frequency-potential curves, as seen in figures 4.26 and 4.27. The data on pulse frequency against current, and frequency against potential, are shown in the plots of figures 4.28 to 4.33 for various points and pressures.

Limitations were imposed on these measurements. Triggering by ultra-violet light was precluded in photomultiplier studies, as it gave too much background noise for the electrical pulses. Thus, the pulses could only be studied above the threshold for Trichel pulses. Appearance of multiple and diffusive spots at lower pressures and higher potentials, and the limitations of voltage and current output of the power supply prevented study at higher frequencies.

At lower pressures, 20 and 40 mm for a 0.5 mm diameter point, the frequency-current plots reached a maximum at about 130 μA and declined rapidly until the pulseless discharge was reached. In these cases, the negative glow disappeared and a diffuse glow covered the whole point.

An attempt was made to study the movement of the brush of purple glow, resulting from the steep potential front of the negative ion space charge, which chokes the discharge after the luminous part of the discharge is largely over. The background noise proved too high to observe this movement.

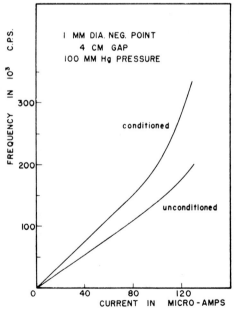

Fig. 4.26. Frequency plotted against current for Amin's conditioned and unconditioned points, showing a decrease in the number of ions per pulse on conditioning.

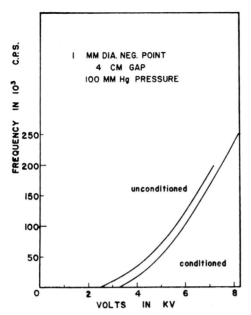

FIG. 4.27. Frequency plotted against potential for conditioned and unconditioned points.

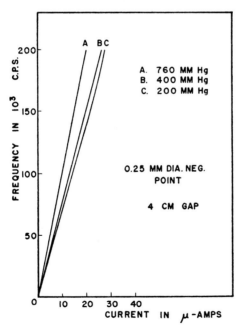

FIG. 4.28. Frequency plotted against current for a 0.25 mm diameter point, 4 cm gap at various pressures.

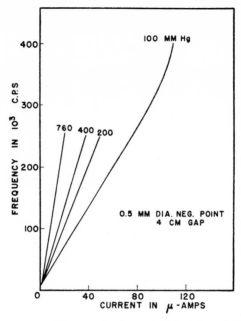

Fig. 4.29. The same conditions as for figure 4.28 but for a 0.5 mm diameter point.

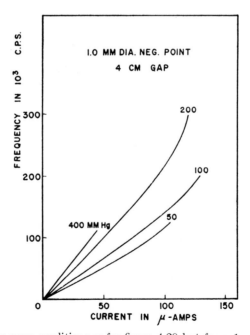

Fig. 4.30. The same conditions as for figure 4.28 but for a 1 mm diameter point.

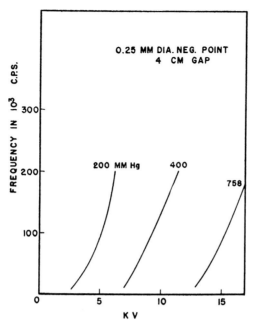

FIG. 4.31. Frequency plotted against potential for a 0.25 mm diameter point and 4 cm gap at three pressures.

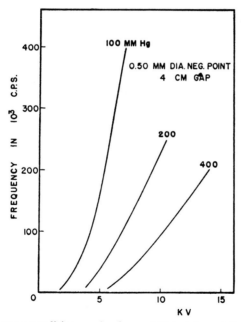

FIG. 4.32. The same conditions as for figure 4.31 but for a 0.5 mm diameter point.

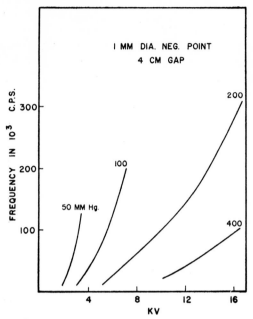

Fig. 4.33. The same conditions as for figure 4.31 but for a 1 mm diameter point.

We have described the mechanism of the pulse beginning with its rise, through a photoelectric γ_p, to the creation of the positive space charge sheath and formation of negative ions beyond the positive charge. As positive ions reach the cathode, the γ_i mechanism becomes active and predominates. When this happens, the discharge and glow are restricted to a narrow dense region along the axis, since these do not diffuse as readily as the photons. Thus, the space charge formed by γ_p would be more nearly in the form of a spherical shell, and that resulting from the γ_i action would be in a narrow dense region in the direction of the axis.

It is probable that in the longer, more clearly delineated pulses at low pressures, seen in figures 4.20, 4.21, and 4.25, the sharp rise to the spike is caused by a γ_p action, while the prolonged plateau is caused by γ_i.

The effect of the change in pressure is clearly seen in figure 4.20 and in the curves of figures 4.28 to 4.33. At high pressures, X/p at onset is slightly lower than at lower pressures. That is, at high pressures onset occurs at slightly lower fields per millimeter pressure. Here excitation is high at low X/p and high pressure. This gives dense photon liberation near the surface. Thus, the initial multiplication by γ_p is very high. In only a few generations the discharge is built up. At high pressures, the dissociative attachment is accomplished in a short distance, and the discharge is rapidly quenched. Thus, the pulses show a high initial current spike which dies out readily.

On the contrary, low pressure reduces the rate of generation by γ_p. It diffuses the photons over the point. The negative ion space charge is diffuse and created further out in the gap. Thus, more ions are needed to quench the discharge. Positive ion mobilities are increased; thus γ_i becomes active sooner and becomes large relative to γ_p. The change in shape and duration of the pulses is clearly seen in figure 4.20. On the other hand, because of increased ion mobility, the clearing time will decrease as pressure decreases. Repeat rates are thus higher for equivalent potentials, as seen in figures 4.31 to 4.33. In contrast, as seen in figures 4.28 to 4.30, the current per pulse is lower at higher pressures. Thus, for a 1 mm diameter point, the currents at 400, 200, 100, and 50 mm at 100 kc are 35 μA, 50 μA, 75 μA, and 90 μA, or a threefold increase in ion count per pulse for an eightfold decrease in pressure.

As potential is increased beyond threshold, the frequency at first rises nearly linearly with the current. Ultimately, the current grows more slowly than does the repeat rate. In the linear region, there is thus a constant number of ions created per pulse as potential increases. At higher potentials, there is a smaller charge per pulse. The linear regime is one in which the potential across the dark space does not increase as applied potential increases. That is, the ion generating conditions are the same. Increase in applied potential then merely goes to increase in potential fall across the low field region, removing the increased space charge of negative ions faster. This phenomenon is akin to that in the normal cathode fall region in a glow discharge. It is even closer to the phenomenon of current increase in a positive coaxial cylindrical steady corona where ion space charge limitations apply with constant amplification. This was explained for that corona qualitatively by the author and verified quantitatively by Colli, Facchini, Gatti, and Persona[2.17] following the author's suggestion, as seen in chapter 2, section E.6.

In this potential region, the γ_p action is largely active, as the positive ions arrive too late to be effective. Further increase in point potential begins to alter the discharge, since at that point the maximum avalanche proliferation over the limited surface available to the discharge cannot be exceeded. Increase in current to increase space charge thus requires higher point fields and further extension of the high field region into the gap in order to increase the size of each avalanche instead of the number of avalanches per square centimeter of surface, as in the constant amplification phase. This means that the ionization extends farther from the point. The negative space charge forms farther out. Positive ions reach the cathode sooner, and γ_i increases. Thus, electron liberation by ions begins sooner and lasts longer. Photons are lost to the cathode by diffusion. Pulses have a longer plateau and a smaller spike, as seen in figure 4.19. The electron liberation by positive ion impact constricts the discharge radially but lengthens it along the axis. Here, the concentrated space charge density chokes the discharge more rapidly with fewer negative ions than the photoelectric γ_p discharge, which is spread over a larger radial volume. Thus the charge per pulse decreases. The smaller charge and higher fields make removal of negative ion space charge faster. Thus, the dead time between successive pulses decreases, and the frequency of the pulses increases. This action is clearly seen in any of the curves of figures 4.31 to 4.33, especially at lower pressures where the linear regime is exceeded.

The influence of gap characteristics is readily understood. The distance to the anode, except for short gaps, is not important, since the clearing time which sets the dead time depends on regions close to the cathode, and the negative ions for gaps in excess of 2 cm do not have to cross the whole gap. Thus, gap length is not an important factor. Increase in radius, as shown in figure 4.21 and as noted by looking at the frequency for the same potential relative to the starting potential as point radius increases, causes the frequency to decrease and the pulse height and duration to increase. Thus, there is a sharper initial spike, a larger charge per pulse, and a slower repeat rate, that is, a longer clearing time the larger the point. A large point presents a larger surface for photon action. The fields extend further into the gap. Thus, avalanches are larger, photoelectric liberation is greater, and ions reach the cathode at a later time so that build-up is by photon action and therefore rapid. It takes more negative ions to form a choking space charge at the greater distance. Hence, pulses are prolonged, and, with more negative ions at greater distances from the point, surface clearing times are longer, and repeat rates are lower. The difference is clearly seen in the oscillograms of figure 4.21. The contribution from γ_i is definitely more pronounced with the smaller points.

The effect of conditioning now needs to be analyzed. The oscillograms at 100 mm show that the unconditioned point has a lower γ_p, a smaller photoelectric peak, a large γ_i ion contribution, and a great duration. The conditioned point which is bombarded for some time at high current density and high potential, showing alteration through sputtering and melting, on the other hand shows an enhanced photoelectric γ_p peak, a small short ion plateau, and a duration about a third that for the unconditioned pulse. Thus, conditioning the point, presumably through a clean-up by outgassing, increases γ_p. This builds up a discharge rapidly. The negative ions are formed near the point in such quantity as to choke off the discharge rapidly. Ions do not reach the cathode in quantity. Fewer total ions are thus needed to choke off the discharge, as they do not move far while the pulse is building up. Since the negative ions are produced in large quantity close to the point and the surface distribution of charge is greater with a photoelectric γ_p, the time to remove the negative ion space charge will be increased slightly. Thus, repeat rate is reduced. This is clearly seen in figures 4.26 and 4.27. Here at 80 μA, for example, the conditioned point has 150 kc/sec while the unconditioned point has 100 kc/sec. On the other hand, at 6 kv the unconditioned point has 120 kc/sec while the conditioned point has only 100 kc/sec.

Amin's paper compares the Trichel and burst oscilloscope traces at atmospheric pressure for a 1 mm diameter point and 4 cm gap at 1000 μsec/cm sweep rate with the burst pulse at 7.6 kv and the Trichel pulse at 8.2 kv. This permits comparison of the motion of ions across the gap, as seen in figure 4.23. The positive ions of the *burst* pulse which are generated in over 500 μsec take 5000 μsec to cross, as shown by the break in the plateau of the upper trace. Following the break, the trace approaches zero gradually in some 10^{-3} sec. This arises from the long duration of the burst pulse and diffusion of ions. The Trichel pulse lasts only some 4×10^{-8} sec, and the break appears at 2500 μsec. The mobilities of positive and negative ions, of which there are three, are given as 2.68 cm^2/volt sec for O_2^-, according to Doehring, a slower ion around 2.25 cm^2/volt sec, and a faster ion given by Burch as 3.38 cm^2/volt sec. Eiber[3.59] has made a study of these ions by a modification of the Doehring method and also finds three ions. The fastest one, 3.2 ± 0.3 cm^2/volt sec, is assumed to be an O^- ion; the slowest one has a value of 2.25 ± 0.15 cm^2/volt sec and is identified as O_2^- undergoing charge exchange with O_2 molecules. The third ion has a mobility of 2.5 ± 0.1 cm^2/volt sec; Eiber tentatively identifies it as O_3^-. It is present only in small quantities. Clearly, within the accuracy of the rather poorly defined breaks, the burst pulse corona has a crossing time of about 5000 μsec for its O_2^+ ions, and a crossing time of $\approx$2500 μsec for the Trichel pulse ions. This indicates that most of these could well be O^- ions. These are needed for the choking and are formed in such low fields that they do not transform to O_2^- ions as they do at higher X/p in Eiber's study.[3.59] The tails in the Trichel pulse could be ascribed to O_2^- ions, attachment to form O_2^- in the gap, and transformations of some O^- to O_2^-. In air, the long tail of the positive ions could be ascribed to formation of slower complex ions from impurities in the low field region in air. This crossing time is of general interest only and has no direct relation to the clearing time, except perhaps near threshold, that is, below a 10^3 per second repeat rate. No further work was done on this corona in the author's laboratory.

6. Studies of Guck and Das

In 1951 L. Wagner completed a doctor's thesis at the Technische Hochschule Karlsruhe under the direction of Dr. Ing. H. Lesch. This work was carried further by Dr. Ing. R. Guck, who made a complete and exhaustive study of the negative point-to-plane corona in air. He summarized this work in a booklet entitled "Die Negative Koronaentladung in der Spitze-Platte-Funkenstrecke." It appeared as No. 1 of a series published by Dr. Lesch from the press of C. F. Miller, Karlsruhe, in 1955.[4.15]

A rather complete study had been made, primarily from the engineering standpoint, using all the techniques previously carried out in the author's laboratory, including electrical noise studies. Hemispherically capped cylinders of Pt were used with diameters of 0.2, 0.05, 0.02, and 0.012 cm. The anode was a brass plate of 12 cm diameter with gap lengths of 0.5, 1.0, 2.0, and 4.0 cm. Pressures ranged from 75 to 760 mm, but most of the work was done at 760 mm. Room air was used. Great emphasis was placed on the current-potential curves; however, photographic and telemicroscopic studies of the discharges were also made. It will be noted that point diameters ranged from 2 mm down to 0.12 mm, the next largest point having a diameter of 0.25 mm. Gap lengths were generally shorter than those reported previously, and many data were taken with the largest point and a 2 cm gap, because this gave the greatest range of phenomena.

Since in general the longer gaps did not show all the phenomena, the work in the author's Berkeley laboratory missed one peculiar form of pulsed discharge. This was the ring discharge discovered by Greenwood in 1951.[4.6] This and a spiked form of discharge, called by Guck "Stachel entladung," and probably identified with Bandel's multiple spot corona, were investigated at length by Guck. This may be the same as Greenwood's spiked discharge.[4.3] Using room air, Guck observed the pre-Trichel pulse discharge from dust specks and failed to recognize it, or else may have missed it in reading the literature. He did not trigger with ultraviolet light and thus missed the pre-onset region below the regular pulse regime. Therefore, he was unable properly to interpret the threshold phenomena. His oscillograms left much to be desired, as they had inadequate temporal resolving power, which he should have recognized had he read Amin's work. This hampered otherwise good interpretation. He did investigate the luminous phenomena and measured the width of the various zones and the dimensions of the Trichel pulse luminosities more completely and carefully than did English.[3.17]

In typical engineering fashion, he set up a power law to describe the current-potential curves in which the current was expressed as

$$(4.1) \qquad i = c(V - V_s')^m.$$

Here, V_s' is the onset of the regular Trichel pulse corona and is a function of the point radius, the gap length, and gas density (or pressure) at constant temperature. The constants c and m, as well as V_s, were functions of radius and gap length, and c depended on pressure as well. A table of values of these constants was derived for the range of variables studied. This permitted the variation of m and c to be derived. The value of

(4.2) $m = 2.1 - 8r/a,$

where r is the point radius and a the gap length; m is constant for pressures between 760 and 300 mm. The value of c is given by

(4.3) $c = 1.1 \times 10^{-5} \sqrt{\dfrac{r}{a^3}} \left(\dfrac{1}{\delta^2}\right);$

here

$$\delta = \frac{0.386p}{273} + \theta$$

with δ fixed at 760 mm and $\theta = 20°$ C as standard air density. These relations do not apply over the whole range of the curves. Deviations appear at the beginning and end of the curves. With the appearance of the ring discharge for the 2 mm diameter point, the Trichel pulses and the power law hold, but there is a change in the exponent.

These empirical relations are of value for practical applications, that is, for engineering purposes. As long as the applied potential does not alter the dimensions of the dark space too much relative to the gap length, as elsewhere indicated, the current is governed in principle by the Townsend space charge limited current equation (eq. 2.30) $i = cV(V - V_s)$, in which c has, aside from a geometrical field factor, a dependence on the ion mobility. However, since the shape and nature of the pulses alter as current and potential change, probably no single relation gives a satisfactory fit. Thus, for practical purposes a crude empirical law such as that given may be of use. Probably an analysis in terms of space charge effects analogous to Townsend's relations could lead to a more satisfactory solution.

Recently a complete survey of equations of this nature relative to engineering applications, starting from Townsend's equation, has been made by J. H. Simpson and A. R. Morse (Laboratory Report, Electrical Engineering, EE 82, of the National Research Council of Canada).

A rather illuminating three-dimensional diagram describing the discharge forms and the regimes in which they occur in terms of potential point radius and gap length is an interesting feature of Guck's book. For smaller point diameters, the ring discharge region is narrow indeed or vanishes, and the region of steady glow discharge increases at the expense of the Trichel pulse regime below the arc stage. These regions are more properly to be expressed as functions of gap length and potential for various points rather than in terms of current and gap length. It is to be noted that most of this work was done for gaps so short that the opposite electrode and the field distribution affected by it influence the breakdown characteristics. In chapter 3 of the present book, the objective is to study primarily the behavior of the coronas from isolated electrodes, and short gaps do not permit this.

The frequency of the Trichel pulses in relation to current was next investigated. Actually, as Guck points out, the frequency of the pulses is perhaps a misnomer, since there are marked irregularities. Instead, he used the number distribution with dimensions of second^{-1}, which is the average number of pulses per unit time, best termed "repeat rate" in English. Thus, he plotted the number frequency, or repeat rate, against the galvanometrically measured current. It then appears that for each point radius there is a linear relation plotted to a log-log scale between repeat rate and current, more or less independent of gap length at 760 mm. Guck plotted *minimum* repeat rate, since there is scatter, and repeat rate h may be higher. As the position of the discharge on the point altered and the frequency shifted to off-axis positions, h was usually greater than the value h_{min} which he plotted. These changes probably correspond to the "two moding" observed by English and the "current jumps" of Bandel. With the 2 mm diameter point the range of h values, even with 10^5 impulses per second, remained broad. With other points, the range converged to h_{min}. This large range of the values for the large point is related to the vagaries of the ring discharge because of motion of the discharge over the surface. When the "Stachel entladung," or spike discharge, appears, it has its own frequency law, with a much more rapid increase in frequency with current increase. Guck again derived an empirical law for the value of h_{min} of the form

$$(4.4) \qquad \Delta h_{min} = ki^w.$$

He gives no direct relation to potential, which entered indirectly through the corresponding current increase. The external circuitry did not influence h_{min}, as had long ago been reported by Trichel. Gap length exerted little influence except for the 2 mm diameter point, in which case the value of h_{min} for the same current was observably higher in a 4 cm gap than in a 0.5 cm gap. That is, as the gap contracts, the charge per pulse is larger for the short than for the long gap. This merely indicates that in the short gap the proximity of the anode to the extensive tip field extends ionization further into the gap, produces larger pulse charges before choking, and requires longer clearing times. The data for various point radii allowed the constants k and w to be evaluated. The constant w in the exponent at once empirically appears to take the value $w = 1.21$ or closely $\sqrt{3/2}$. Thus

$$(4.5) \qquad i = k_1(h_{min})^{w_1}$$

in which

$$(4.6) \qquad k_1 = (1/k)^{w_1}$$

and

$$(4.7) \qquad w_1 = 1/w.$$

Then $k_1 = 8.1 \times 10^{-8}r$, and with $k = (1/k_1)^w$, we have

(4.8)
$$h_{min} = \left(\frac{10^8}{8.1r}\right)^{1.21} i^{1.21} \text{ per sec}$$

with r the point radius in centimeters and i in amperes. This law extends to 1000 pulses per second. There is obviously no exact or semi-quantitative basis for the empirical frequency law deduced. However, it is worthy of some attempt at analysis.

Basically, pulse duration is short compared to the clearing time except near the upper limit of repeat rates. Thus, clearing depends on the distance of displacement needed for clearing the charge and the clearing fields available. The time to clear the charge depends on the amount of charge in a pulse, its location, and the field across the clearing region. The current is merely the product of repeat rate h and the quantity Q of charge. The quantity of charge depends on point potential and field distribution, which influence the rate of breakdown and of choking negative space charge. Thus, field geometry near the point, pressure, state of the surface, and the applied potential influence this. In general, if Q is large, then clearing time will be long and h will be small. Thus, the frequency is dependent on the potential, the cathode geometry, and physical condition, and the current is merely a product of repeat rate, or frequency, and charge per pulse. Thus other than delineating a combination of charge per pulse and its associated clearing time, current repeat rate relations are not revealing. Guck should have studied the potential as the important factor in order to learn more about what happens.

The steady current contribution (that is, that due to ion movement in the gap) is only a small fraction of the charge revealed in the pulses, so that the charge per pulse Q can be derived from i/h. Since h_{min} is the regularly measured quantity, Guck uses

(4.9)
$$Q_{max} = i/h_{min} = N_{max},$$

the maximum number of electrons generated per pulse. This leads to the relation

(4.10)
$$Q_{max} = (1/k)i^{(1-w)}.$$

Putting this in numbers

(4.11)
$$Q_{max} = 1.59 \times 10^{-19},$$

$$N_{max} = \left(\frac{8.1r}{10^8}\right)^{1.21} i^{-0.21}.$$

Using his relation (eq. 4.1) that $i = c(V - V_s')^m$, one then has a relation between Q_{max} and V in the form

(4.12)
$$Q_{max} = c/k(V - V_s')^{m-0.21}$$

with

(4.3) $c = 1.1 \times 10^{-5} \sqrt{r/a^3}[0.386p/(273 + \theta)]$

(4.13) $k = \left(\dfrac{8.1r}{10^8}\right)^{1.21}$,

whence

(4.14) $Q_{max} = 1.1 \times 10^{-5} \sqrt{\dfrac{r}{a^3}} \left(\dfrac{0.386p}{273 + \theta}\right) \left(\dfrac{10^8}{8.1r}\right)^{1.21} (V - V_s')^{m-0.21}$

$= 1.1 \times 10^{-5} \sqrt{\dfrac{r}{a^3}} \left(\dfrac{0.386p}{273 + \theta}\right) \left(\dfrac{10^8}{8.1r}\right)^{1.21} (V - V_s')^{(2.1-8r/a-0.21)}$.

One interesting study was the oscillographic investigation of duration and rise time of the pulse plotted against the repeat frequency for various radii and gap lengths. The scatter of points for duration of the pulses, which range from 6×10^{-7} to 10^{-6} sec, was considerable in the frequency range 10^4 to 10^6 per sec. The times were plotted to a linear scale, and the duration to a logarithmic scale. In general, the 2 mm diameter point showed shorter duration, but this applied also to the smallest point. The rise times were more closely grouped around 2×10^{-7} sec than were the durations. It is clear from the duration and rise times shown, both in the magnitude and in the shape of the oscillograms, that Guck's oscillograph did not have the temporal resolving power necessary to portray the pulses correctly. Thus, the constancy observed was primarily instrumental.

His plot of pulse heights (that is, maximum current per pulse against frequency to a log-log scale) was relatively significant, however. The pulse heights indicated maximum currents of 6×10^{-3} amp per pulse for the 2 mm diameter point. These decreased in height as frequency increased, as previously observed by English, Bandel, and Amin. The pulse heights for the 0.012 cm diameter point were at most 3×10^{-4} amp and declined with repeat rate. No influence of gap length was noted. Also plotted were Q_{max} as a function of i, the current. Again, these varied linearly in a log-log plot, declining slowly as current increased, the value of Q_{max} being greatest for the 2 mm diameter point and least for the 0.12 mm diameter point. For the 0.50 mm diameter point, the electron number declined from about 7×10^9 electrons at 10^{-7} amp to 1.2×10^9 at 10^{-4} amp.

Measurements of current-potential curves for gap lengths of 1, 2, and 4 cm were made at 500 and 300 mm pressure for the various point radii. The longer the gap, the lower the current, although the starting potentials were approximately the same. This unquestionably must be attributed to field distortion by space charge in the low field portions of the gap, according to Townsend's law. Lowering the pressure obviously lowered the starting potential and the potential value for a given current. Current-potential curves were made from crossplots of the previous data giving the curves for the 4 and 0.5 cm gaps at 300, 500, and 760 mm. Repeat rates against current were plotted for gap lengths of 1, 2, and 4 cm for 2 mm and 0.5 mm diameter points on a log-log scale. All data for 1 and 2 cm gaps fell on a single curve of constant slope except for a change when the spiked discharge occurred at 500 mm pressure and at 300 mm pressure. There is nothing significant about these curves except that the frequencies for the same current were lower, the lower the pressure. This signifies larger pulses and greater charges at lower pressures, as observed by Amin.

Guck then analyzed the breakdown theoretically and semi-quantitatively. In that, he had the complete analyses of the author's group before him. Since the observations of that group were thorough and complete, the rather complete theory having been pioneered by the author, Guck's work added little new. He first calculated the fields involved in breakdown. He was unaware of the paper by Loeb, Parker, Dodd, and English on the fields in point-to-plane corona. He did not cite this paper, nor did he use the standard geometry which yields such fields accurately. He used a relation of a modified sphere gap due to Zeleny. The field calculations are quite crucial, and hence, except in order of magnitude, his calculations are not very good. He used his own empirical expression for the influence of air pressure. He next computed the values of the starting potential and starting field strength along the axis, for various radii and gap lengths (between 0.5 and 4 cm) used. While not accurate, they are sufficiently indicative to show the very slight dependence of the starting field strength on gap length for all the points.

Guck then used the approximate Townsend expression $\alpha/p = Ae^{-Bp/X}$ to calculate the electron yield for the initial avalanche with his field distribution. He calculated the various factors previously reported by the author—total electrons and ion distribution for the first avalanche for a 2 mm diameter point and a gap length of 2 cm at 760 mm with an onset potential of 10.6 kv. The total number of electrons in one avalanche was 7.9×10^3, which is probably too low for the extensive fields. Ionization virtually ceased within 0.13 cm of the point surface. The peak of the positive ion distribution lay at 0.048 cm from the point, and reached the point in 7×10^{-7} sec. All positive ions were removed in 2.3×10^{-6} sec. Guck underestimated the mobility of the positive ions by a factor of two, so that the times should be halved. He considered the positive ion clearing time important in his later considerations. He failed to note that the initial γ process was a γ_p and that avalanche succession is very rapid. This led him to neglect lateral spread and to assume a much greater localization of the discharge. With this concept, he calculated the influence of the positive ion space charge field. This gave an overestimate; however, it followed the present author's original reasoning on the course of the development. Guck noted the acceleration of the electrons in this high field into a Morton regime and arrived, as had the author, at the glow discharge-like excitation and ionization distribution at the height of the Trichel pulse. Thus, in the negative glow of 3 or 4×10^{-3} cm depth, ionization exceeds the α coefficient ionization.

The Faraday dark space is again a region of high field in which the electrons from the glow get enough energy to excite within another 3×10^{-3} cm in the shaving-brush–like positive column. This leads to the very steep rise in current in the pulse, which Guck showed schematically, following the author and, later, Wagner. In the low field region near the end of the Faraday dark space, negative ion accumulation begins. Guck gave a figure in which he accepted the author's region for the peak of the negative ion formation at about 0.3 cm.

His further discussion of the decline time and clearing time appears to be confused. He indicates that the rise time is connected with the positive ion movement, and the decline time with negative ion movement. This is, of course, incorrect but follows from the poor resolving power of his oscilloscope, which failed to reveal the true duration of the electrical pulse. As Amin showed, the electrical pulse parallels the luminous pulse. The process involves ion movement only in certain cases, as indicated earlier.

In his explanations, Guck failed to realize that the decline of the pulse was not related in any way to the clearing, or dead, time and hence to the repeat rate. Pulse decline depends on the clearing of the charge of negative ions, so that its field will no longer influence the field at the cathode surface. It probably would be governed in a measure by the removal of most of the positive ion space charge so as to release the negative ion charge of the dipole. With many of the excess positive ions on the cathode side of the dark space removed, the plasma of the negative glow becomes polarized in the cathode field and pushes the high field region outward from the point to some 6×10^{-3} cm or more (on Guck's scale) to help dissipate the negative ion charge. How far this high field region must move for its effect to be unimportant is the vital question. It cannot comprise the crossing time of negative ions to the anode plane. At a gap length of 4 cm this takes ≈ 2500 μsec. For a 2 cm gap it would be perhaps 1000 μsec or less. Such a clearing time would limit the repeat rate to the order of a kilocycle, which is the region where regular onset usually occurs. However, it must be noted that, in going from 10 kv to 23 kv for a 0.50 cm diameter point, the value of the repeat rate increases from 10^4 to 5×10^5, or fifty-fold; the clearing time does not involve crossing the whole gap even though it did require nearly this at threshold, for the field strength across the low field region was not more than doubled at double potential.

That the ion clouds from separate successive pulses in transit across the gap do exert some influence is seen in Guck's proper interpretation of the ring form of the corona discharge.[4.3] Thus, if the point is large enough, owing to conditioning and clean-up, the discharge point wanders over the surface as potential and current increase. Perhaps this wandering is influenced slightly by the dissipated negative ion cloud still in transit. If so, this concept is implied by Guck. If the repeat rate becomes very high, Guck computed roughly that the transit time of the ions for various potentials from 12 kv to 16 kv, depending on the gap used, ranged from 5.8×10^2 to 4.4×10^2 μsec. The repeat rate of the pulses was 2×10^3 and 1.85×10^4 per second for these potentials. Thus, there are at these potentials 1, 5, and 8 pulses in transit, respectively. When a sufficient number of such space charge clouds are in transit, they exert an influence on the next space charge. Thus, starting at the center of the point, the first discharge will, because of the space charge cloud, seek a clean spot away from the axis. The next discharge will also take place either to one side or the other away from the space charge fouled localities. Consequently, the succession of spots will follow a ring of discharges at some distance from the axis, falling in clean regions, each as close to the last discharge as space charge clouds permit. The accumulated charges, when eight or more are in transit, act to deny the axis region to a discharge, so that the discharges follow a ring pattern around the point at a distance above the center, appropriate to the repeat rate. As this rate increases, and more and smaller clouds are in transit, the ring spreads up the point away from the axis.

Guck's picture nicely accounts for the ring discharge.[4.3] The spiked discharge and the transition to the pulseless discharge follow the author's explanation, previously indicated.[1.14,4.6]

Following Guck's work, Das[3.70] carried on studies in the laboratory at Karlsruhe on breakdown in pure inert gases, mixtures thereof, and mixtures with electron attaching gases such as O_2 and CO_2. As in previous studies of Weissler with point-to-plane geometry, and Miller and Lauer in coaxial cylindrical geometry, the addition of as little as 0.1% of such gases produced the Trichel pulses. While Das's study used what *he* termed "pure inert gases," it was hampered by the fact that he was actually using Penning mixtures, since his gases were not pure to start with. This circumstance does not affect the mixture studies with O_2 and CO_2. To get a pure inert gas that is *not* a Penning mixture requires heroic measures, including gas cataphoresis. Das used commercially pure grade gases. He had before him the previous work of Guck and others in the group and used the empirical power law for current and potential curves, and followed other precedents in that laboratory. He had the advantage of somewhat better oscilloscopes than Guck had. This, together with the fact that small concentrations of O_2 and CO_2 in inert gases led to longer breakdown times, enabled him to observe the *whole Trichel pulse*. Thus, he was quite surprised at the true shapes of the pulses with photon γ_p and ion γ_i secondary emission, as well as the ionic component of the current. He did not carefully read or else he ignored Amin's papers and thus failed to recognize the significance of the shapes of the pulses.

Basically, his theoretical background and intuitive analysis of the phenomena fell far short of those of Guck. Thus, there is little except useful systematic data to be gained from his work. In what follows, some inferences will be drawn from the oscillograms which appear in his paper. With pure Ne and A, there was no corona. As potential increased, a spark materialized suddenly. There were no pulses, but a weak conditioning current cleaning up the point cathode might precede breakdown to a spark. In any event, it was less than 10^{-9} amp. Addition of 0.02% of CO_2 to pure A at 600 mm gave a weak glow discharge at 1400 mm with 2×10^{-5} amp, increasing up to a spark at 2500 volts. There were no pulses. Similar results occurred at 300 mm except that there were only 20 volts between threshold and spark. At 75 mm there was no corona, just a spark. With 0.1% CO_2 at 600 mm, irregular Trichel pulses set in at 2230 volts, becoming regular with potential increase up to nearly a sparking value of 3850 volts. In that interval before sparking, the pulses ceased. At 300 mm, irregular Trichel pulses began at 1500 volts and became regular up to sparkover. At 75 mm, pulses were observed only just below breakdown. Addition of 2.5% to 10% CO_2 did not materially alter the picture; Trichel pulses appeared and threshold increased with increase in CO_2 content.

The oscillograms closely resembled those of Amin. There was a rapid spike caused by the photoelectric γ_p build-up and a decline to a plateau in so short a time that Das's sweep speed did not permit its evaluation. The peak height was also lost because of the slow time resolution. The plateau set in, however, as the positive ions reached the cathode. Attachment was slow, and the ionically determined component lasted for $\approx 10^{-6}$ sec. It is quite possible that the photon γ_p peak was low relative to the γ_i component. Reducing the pressure to 300 mm increased the length of the pulse with 0.1% CO_2 to about 5×10^{-6} sec. With 2.5% CO_2 at 600 mm and 3150 volts, a faster sweep speed was used. Quenching was rapid. The build-up was largely by a photoelectric γ_p, and what there was of a γ_i-induced plateau occurred in 10^{-6} sec, but the pulse quenched too rapidly to allow this to develop. The chief rise and decline occurred in 5×10^{-7} sec. Reduction of A pressure decreased the γ_p peak but gave a γ_i lasting nearly 2×10^{-6} sec before it declined. Here amplitudes of the pulses were about twice those at 0.1 mm pressure CO_2 and lasted only one-half to one-third as long. At 75 mm and 2.5% CO_2, the photon pulse was very short, and the γ_i effect predominated. The current was about that with 0.1% CO_2 and lasted about as long, decline being complete in 4×10^{-6} sec. Thus, reducing pressure brought γ_i in earlier because of increased mobility, increased quenching time, and enhanced γ_i relative to γ_p. Increase of CO_2 to 10% at 600 mm at 3800 volts gave a γ_p pulse nearly twice as high as with 2.5% and was followed by a γ_i plateau of 5×10^{-7} sec duration. The pulse lasted somewhat less than for 2.5% but not much less. Probably this is a delusion caused by the heavy initial current such that ion movement makes the pulse appear longer. At 10% CO_2 and 75 mm there was virtually no γ_p, but γ_i rose to a peak in 10^{-6} sec and declined in the next 10^{-6} sec, being over in 2×10^{-6} sec. The current was about a third the peak γ_p at 600 mm. (It is clear that it was probably the A resonance photons which caused the γ_p, since the absorption in 10^{-1} mm is not great.) At low pressures the ionizing zone was further out in the gap. Many photons were lost. CO_2 might also have absorbed many photons so that the γ_p action was much reduced, say to a third, but the initial rise was by γ_p. Ions quickly reached the cathode, giving an early and heavy γ_i. Quenching was slow, as attachment occurred further out in the gap, and the negative dipole was too far away from the cathode surface.

A series of traces at potentials above threshold in 600 mm A with O_2 showed the following. With 0.01% O_2 at 2020 volts, Trichel pulses appeared with a small γ_p at rise and a delayed γ_i plateau lasting 20×10^{-6} sec. In some of these oscillograms, *there was a dip between the γ_p peak and the γ_i plateau*, or better peak, in such cases. Its cause is not clear, but it involves a decline in the γ_p discharge before the γ_i action can take place. It could be that the dense plasma created in such strong γ_p discharges absorbs the active photons sufficiently to stop the γ_p breakdown. The amplitude of the current peak was about 1.5×10^{-4} amp. With 0.5% O_2, the peak at 7250 volts was a γ_p peak with little plateau due to γ_i. Its amplitude was 5×10^{-4} amp at peak, and its duration was 5×10^{-7} sec. With 2.5% O_2 at 8800 volts, the amplitude of the γ_p peak was 8×10^{-4} amp, and the duration of the pulse about 2×10^{-7} sec. With 10% O_2, the γ_p peak had an amplitude of 1.3×10^{-3} amp. It lasted 4×10^{-8} sec and was followed by two γ_i plateaus, each about 3×10^{-8} sec long and of 1.3 and 0.7×10^{-4} amp, respectively. This represents contributions to γ_i by two ions of different mobility, presumably O_2^- in A and A_2^+ in A, or else the two A ions A^+ and A_2^+. It was not certain how much A_2^+ could form in 4×10^{-8} sec. These ions have mobilities of 1.94 and 1.67 cm²/volt sec for A_2^+ and A^+, respectively. What mobility O_2^+ has in A is not known, but it will be in excess of 2.5 cm²/volt sec. Thus, since this appeared only at 10% O_2, it seems best to assign this phenomenon to a fast O_2^+ ion and a slow A^+ ion. Reduction in pressure produced the same results as for CO_2.

Neon should not show materially different results than A. Mobilities of ions at the same pressures in Ne were about twice those in A, that is, about 4.2 and 6.5 cm²/volt sec for Ne^+ and Ne_2^+ in Ne. In very pure Ne with 0.01% and 0.1 CO_2, no corona was observed. A spark set in abruptly at an appropriate potential, characteristic of pressure and percentage impurity. With 0.5% CO_2, corona at 600 mm started at 910 volts, some 100 volts below spark breakdown. No Trichel pulses were observed. At 300 and 75 mm, no corona occurred. At 2.5% CO_2 at 600 mm, Trichel pulses appeared. These continued up to spark breakdown at 1780 volts. However, at 300 mm and 75 mm pressure, no corona occurred even with 2.5% CO_2. With 10% CO_2, Trichel pulses appeared at all pressures, including 75 mm.

As indicated by the oscillograms, there was a γ_p peak of amplitude, perhaps 3×10^{-3} amp, followed by the rise of a γ_i peak of 2×10^{-3} amp at 8×10^{-7} sec. The pulse declined in 4×10^{-6} sec. The small minima between γ_p and γ_i peaks noted above Das called "Höcker impulse," literally, hooked or humped impulses. In ignorance of the proper interpretations of Amin, Das was mystified by the plateau and humps. Ne had apparently a very effective γ_p, and electron drift velocities for Ne were apparently high, for example, larger than in A by a factor of 2 at $X/p = 0.4$. Thus, the ion space charge formed at a considerably greater distance and built up in a very short time. The ions, even though fast, reached the cathode with some delay. The hump, which Das found inexplicable, beautifully reflects the density contour of the positive ion distribution in the gap. Difficult to explain only is the rapid quenching of the γ_p pulse. Multiple ion mobilities were well smeared in CO_2. Through reduction in pressure to 75 mm and in potential to 1070 volts, the pulse was altered only by the reduction of the photon portion; there was a 30% increase in pulse length, and pulse height was reduced to one-fourth; but the area under the curve was materially increased by the increased duration.

Addition of 2.5% of O_2 to Ne at 300 mm pressure with 1640 volts gave a strong photon peak of 1.5×10^{-3} amp amplitude with virtually no γ_i plateau. The duration of the pulse was less than 2×10^{-7} sec. At 600 mm Ne with 10% O_2 at 8200 volts, the pulse duration was a photon γ_p peak of amplitude 2×10^{-3} amp followed by three distinct small plateaus, due to γ_i, to 1.6×10^{-7} sec. At 300 mm Ne with 10% O_2 and 4900 volts, the peak was a photon γ_p peak of 3×10^{-3} amp and a duration of about 4×10^{-8} sec followed by three γ_i plateaus that were less clearly defined, extending to 1.6×10^{-8} sec. Here again, it is clear that O_2 is a far more successful gas at forming negative ions and choking off the discharge than is CO_2. In consequence, it gives shorter pulses. The multiplicity of ions produced smeared the curves. O_2, on the other hand, gave much clearer results, since only three positive ions are involved, these being Ne^+, Ne_2^+, and O_2^+. The behavior is clearly explained through the earlier observations of Amin and further confirms the general theory of the phenomena outlined.

The rest of the section of Das's paper dealt with various current-potential and starting potential curves for the different mixtures, as well as repeat rate and charge per pulse, and so on, much as was found in Guck's study but for these mixtures, including the onset potential for steady pulses.

Finally, Das gave oscillograms of the Trichel pulses in 99.99% pure CO_2, 99.8% pure O_2 with A and N_2, and in 99.9% N_2 with residual A and (not stated) some O_2 and H_2. Unless further purification of tank N_2 is resorted to, there are present O_2, H_2, and some organic vapors from the iron carbides in interaction with H_2O vapor. Purity is not critical in O_2 and CO_2; in N_2 it is vital, as countless observations by several observers in the author's laboratory have shown.[1.8,1.9,2.22] The relatively pure N_2 gave a corona discharge at threshold which was pulseless and has been described under cleaner conditions by Weissler.[1.8] Miller and Loeb,[1.9] however, showed that for a clean coaxial wire cathode and really pure N_2, no corona occurred, but cathode instability led to a spark. However, if the cathode was soiled, a low order pre-breakdown current preceded flashover. Once coronas operated, enough impurity was created to give a negative ion space charge resistance, but not Trichel pulses. Coronas could also be maintained above threshold without spark if a large external series resistance limited the current. Das observed a current of 10^{-4} amp at threshold. He had a limiting resistance of some 5×10^5 ohms in series with his point, which limited the current to 10^{-4} amp with space charge resistance in the tube and allowed sparkover to occur at 10^{-3} amp when space charges were adequately removed.

With CO_2, Trichel pulses were displayed at 300 mm at 6100 volts, and at 75 mm at 3450 volts. For 300 mm, the photon γ_p pulse was 30% greater than the succeeding γ_i plateau. Its duration was less than 5×10^{-8} sec, and the γ_i plateau declined slowly, possibly for 10^{-6} sec. At 75 mm, the photon pulse was very short and about 25% greater than the ion γ_i. Its duration was possibly 2×10^{-7} sec. The arrival of the peak of the γ_i plateau region was delayed by 3×10^{-7} sec, and the pulse lasted 2×10^{-6} sec.

In O_2 at 300 mm, as expected, the Trichel pulse was entirely a photon γ_p peak of total duration 3×10^{-8} sec. There was no time for the ion γ_i plateau, as negative ion formation choked off the pulse so rapidly. The peak current was 3×10^{-3} amp. The pulse at 75 mm was likewise largely a photon pulse of 2.5×10^{-3} amp lasting 4×10^{-8} sec with some hint of a current of longer duration. In air, for comparison, at 300 mm, the pulse was also a pure γ_p pulse of amplitude 1.5×10^{-3} amp lasting 4×10^{-8} sec, quite in agreement with Amin and English's earlier measurements at 760 mm.

Again, data on repeat rate thresholds and current potential curves followed. The theory, as indicated, is not as well oriented as that of Guck and is probably not relevant.

B. FREE ELECTRON GASES

1. INTRODUCTION

The study of the Trichel pulse corona in electron attaching gases has developed a number of basic concepts concerning mechanisms and what is to be expected. These may be summarized as follows:

a. There is a difference in the rate of ionization, and thus electron and positive ion distribution in positive and negative point corona, in that the slow positive ions relative to the much faster electrons accumulate in greatest numbers at the highly stressed anode surface, in a sense extending the anode field out into the gap and by ion movement reducing it in the critical region. With the negative point, the positive ion cloud is created *at some distance from the cathode, increasing the cathode field* between the cathode and the charge. This has two consequences, of which the first is to increase ionization efficiency leading to the Morton regime and a glow discharge-like structure. The second is to create a condition of inherent instability, yielding a continuously augmenting current which will lead to a power arc unless an inhibiting negative space charge in the low field region or an external limiting resistance prevents it.

b. The second circumstance involves the very small, sensitive volume about the cathode point. Thus, initial field intensified currents will be extremely low unless electrons are liberated from the point by some agency such as ultraviolet light, heat, or x rays. Triggering such a discharge is difficult and critical. If too few electrons are triggered from the point, great statistical time lags will falsify thresholds. Too many electrons quickly lead to field intensified currents that alter the gap by space charge or exceed and mask the normal self-sustaining threshold currents. Thresholds are very difficult to fix. The role of insulating dust specks on such surfaces is important. This is especially so if positive ions charge them up, since the uncontrolled field emitted currents give localized field intensified currents and spurious discharges as well as rendering surfaces emitting and active for long time intervals.

c. The breakdown, being a cathode controlled phenomenon, is extremely sensitive to the surface properties of the metal cathode. That is, all breakdown or self-sustaining discharge has its threshold V_s governed by the Townsend condition (eq. 2.9) $\gamma e^{\int \alpha dx} = \gamma e^{\eta V_s} = 1$, where γ is the coefficient of secondary emission of the surface by photons from the discharge γ_p and by ion and metastable atom impact, γ_i or γ_m. Moreover, not only is the threshold very sensitive to the surface state and nature of the metal and surrounding gas, but what is worse, as in glow discharges, the quantity γ will also influence the current density under fixed conditions. Thus, the value of the threshold and of the current density at any time depends on *the state of the surface, which will change in time and will depend on the nature and pressure of the gas* and the *current density*, and possibly on the *cathode fall*. That is, positive ion bombardment sputters oxide films, gas films, and cathode material from the surface. Ambient gases reacting chemically or physically with the surface, as well as with ions driven into the surface by their impact energy, will alter or strive to alter the surface in various and sometimes opposing fashions. Thus, depending on current density and rates of gaseous diffusion, as well as ion impact energy and momentum, surfaces will alter one way or the other in time. Initially, clean-up of the first oxide layers will occur and may lower or raise γ_p, and will usually but not always raise γ_i. Clean surfaces may have a high γ_i. Too heavy bombardment and high current densities will melt and/or sputter the surface. They may also trap gases which can erupt; or else vapor jets from local hot spots can erupt. These can cause the discharge to wander over a clean surface. It is possible that γ_p and γ_i from a moderately cleaned surface will remain highest. Thus, discharge will start from a surface cleaned by field intensified currents or a low order pre-discharge. When current density increases so as to completely denude the surface, the discharge will wander, following the margins of the discharge spot that are optimum. Sometimes a discharge starts with a coated surface and a low γ. As it continues it cleans up, and γ suddenly becomes so large as to lead to a major current jump, by a factor of 10 to 10^3, leading to an incipient arc. A spot that has been cleaned and has a high γ_i or γ_p as a result of a discharge will maintain its properties for varying times after a discharge has ceased. Re-ignition of the discharge will thus occur at a lower threshold and yield a larger current than that at the initial breakdown. After some hours γ will revert to its former value in contact with the gas. These vagaries have some very important consequences in the study of breakdown from such points. These are:

(1) Initial threshold or onset potentials will usually be higher than the offset potentials or those after extinction and immediate reapplication of potential. The current-potential curves taken with increasing potential and after decreasing potential will not coincide but lead to a hysteresis loop. This is almost universal when the true threshold is observed, that is, the region where one goes from field intensified to self-sustaining discharge as indicated by an oscilloscope or a sensitive current detector. Such thresholds may be difficult to observe in some cases. This does *not* apply where arbitrary criteria such as onset of *regular pulses*, or a visible glow, or a current of arbitrarily chosen magnitude is used as a criterion, since such phenomena are *not* thresholds but arbitrarily chosen breakdown criteria picked for convenience or perhaps reproducibility.

(2) Currents and often actual visual appearances will alter with time after application of a steady potential. What is worse, the rate of alteration of current and arrival at a stable form, if such exists, will be higher the higher the potential.

(3) The arrival at a given state will and does vary with current density. Thus, in consequence, items 1, 2, and 3 mean that any formative time lag studies have no fundamental significance as to discharge and build-up mechanisms. They will vary from milliseconds to minutes and represent a "conditioning time" of the surface which depends on its past history, the

gases present, cathode material, and, worst of all, potential and current density.

d. If negative ion space charges accumulate in the near vicinity of the positive ion cloud in sufficient density, they will terminate the discharge in a short time interval and clear rapidly, leading to a pulsed discharge, often of high frequency. If the negative ion cloud forms slowly, it will gradually increase to a point at which it will interrupt the discharge near threshold. Clearing times will be in the millisecond range, and choking times may be longer. This action will not often be observed. The negative ion space charge will, as with the positive corona somewhat above threshold, merely give an internal space charge in the low field regions furnishing an *internal* resistance in the discharge tube and limiting the current. Thus it will act to prevent the unlimited growth of the inherently unstable cathode to a power arc.

e. When space charges cannot accumulate, as in free electron gases, breakdown from initial feeble glow currents to a power arc can be expected. Frequently after such a breakdown or even before it culminates in an arc, the very growth of the current will liberate enough gas from electrodes to give a space charge resistance and prohibit the arc. The processes of this type may show quite a formative time lag.

f. The actual formative time lags of such discharges will show an enormous range of values. Rise of low current and current density discharges such as the Trichel pulse can be exceedingly rapid with a high γ_p and a small point diameter, since distances are on the order of a few tenths of a millimeter and fields are very high. In the Trichel pulse regime alone they showed variations from 10^{-9} sec to 10^{-6} sec where ion movement was required. Where gas films and oxide layers are present on the cathode, as with pure H_2, N_2, and the inert gases, rise of the feeble threshold currents may be relatively slow because conditioning by the field intensified corona current may be very slow. Threshold in these cases is exceedingly poorly defined and very frequently irreproducible. In fact, if the potential is left on long enough at a fixed voltage, clean-up of the point may eventually, after minutes, yield a self-sustaining discharge. If the potential is increased, say some ten to twenty volts above this, the self-sustaining discharge materializes in one-tenth or even a shorter fraction of the time.

It is now of interest to describe the relatively few studies which have been made in such gases with point-to-plane geometry.

2. Weissler's Studies in Pure Gases

While some excellent earlier studies of these coronas, especially as regards the starting potentials and current-potential characteristics, were carried out in coaxial cylindrical geometry at lower pressures both by Townsend's school and by the group at Philips under Penning in Holland, with really pure and controlled gas conditions, nothing of this nature with comparable precautions had been done at higher pressures in free electron gases with point-to-plane geometry until Weissler's study in 1943.[4.16] The experimental arrangements and purity controls have been described in the section on the positive point corona in such gases (chapter 3, section E.2).

a. *Hydrogen*

At 760 mm with the 0.25 mm diameter point and 4 cm gap, negative corona appeared to yield a self-sustaining discharge at 2600 volts, which was 900 volts lower than the threshold for the positive point. It began as a continuous discharge with no Trichel or other pulses. To begin with, it appeared to be spread over the point tip. This mode was apparently unstable and sooner or later contracted to a concentrated point discharge at the region of highest field strength. This change was accompanied by a current increase of a factor of 10^3. It occurred more rapidly at lower pressures and at a potential slightly above 2600 volts. At 2800 volts the transition took but a few seconds and gave a glow discharge-like structure with no fluctuations. At lower pressures the Crookes dark space became clearly discernible. In the region from threshold or, better, from onset of the concentrated discharge to spark breakdown, the concentrated discharge appeared to move continually over the surface of the point. Furthermore, the form of the discharge changed frequently and unsystematically from the concentrated to the dispersed glow over the surface, with, however, *no change in current*. When the localized glow was stationary it revealed no pulses, but the uniform glow at high currents showed on the oscilloscope a succession of irregular inductive pulses of no definite character. Microscopic examination of the point, after it was run with a negative corona of uniform glow with high currents, showed that the whole tip over which the glow had been visible was pitted with innumerable holes. This indicates that the apparent diffuse glow was probably composed of a large number of localized discharges that rapidly moved over the point and many of which were simultaneous on a time scale of visual perception, for example, ≈ 0.1 sec. No spectacular changes were observed at lower pressures, thresholds were correspondingly lower, and current increase was more rapid because of higher mobilities. The admission of the smallest traces of O_2, $\approx 0.1\%$, raised the threshold to 4200 volts and gave Trichel pulses, currents being reduced to about half the value they had in pure H_2.

Interpretation is not difficult. Self-sustaining breakdown threshold appeared at about 0.1 μA, either as a diffuse Townsend discharge, probably due to a γ_p on a slightly oxide coated Pt point, or else as many small discharges at sensitive spots that eventually damaged the spots and caused movement. Bombardment finally cleaned the point surface to such an extent that a highly effective γ_i appeared, localizing the breakdown to one spot, lengthening the discharge, and yielding a current 10,000-fold greater— about 140 μA. Here the discharge burned for some time but may have wandered very slowly. Evidence is that Pt, which readily dissolves H_2, probably became saturated with an influx of H^+ ions, since the H_3^+ ions doubtless break up in the very high point fields. When this occurs there appears to be either an eruption quenching the discharge or a desensitization of the surface because of some hydride formation. Thus the discharge either shifts to another spot or extinguishes, and a new layer forms on the tip, except at the point of discharge, leading to another diffuse γ_p breakdown until a new spot forms. The pitting of the cathode surface indicates damage either by jet formation, gas eruption, localized melting with intensive sputtering, or even chemical sputtering owing to instantaneous high current density discharges.

One thing must particularly be noted: on clean-up the discharge underwent its 10,000-fold increase in current and vaster increase in current density, but did not go all the way to a power arc via a spark, as theory indicates it might. This current increase is in a large measure to be ascribed to a change to current continuity instead of to a succession of interrupted discharges and to the increased discharge area. In the light of Miller's work on coaxial cylinders it is to be concluded that Weissler's Pt point was not sufficiently outgassed. That is, heating to dull red heat for 3 hours is not equivalent to heating to 900° C for 8 hours. Thus, in the conditioning low current discharge phase, enough H_2O, O_2, CO, and so on, were released by the discharge to give an internal negative ion space charge in the 3.1 cm gap great enough to limit the current and prevent complete arc formation. This was true in Miller's case,[4.8] and his precautions, coming at a later date, were even more drastic than were Weissler's.

Weissler investigated another interesting phenomenon bearing on this question. Bennett[1.2] had noted a clean-up of H_2 contaminated with O_2 by running a point-to-plane corona discharge in the chamber for some time. To study this behavior in a controlled fashion Weissler used as his test point the Pt point originally used. In addition, to clean up the gas, he had, like Bennett, an auxiliary W point that could be run for hours and be damaged by sputtering. The Pt point could thus be used to test the degree of purification.

The procedure was to clean the chamber carefully and admit either pure H_2 or N_2. Then a calibrating run was made with the W point. Next 0.1% or 0.4% of O_2 were admitted and a new run was made with the Pt point. The clean-up discharge was then run from the W point for 8 to 12 hours. Finally the Pt test point was again run. The result is clearly seen in figure 4.34 for

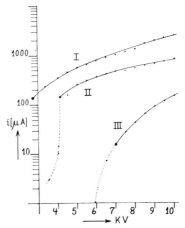

Fig. 4.34. Data on Weissler's study of the clean-up of H_2 gas contaminated with O_2 by running the discharge for a prolonged time. Here current-potential curves are given for negative point in pure H_2 (I); $H_2 + 0.7\%$ O_2 (III); and after a clean-up corona has run for some hours on an auxiliary W point in mixture (II). Dotted portions of II and III represent intermittent corona.

H_2. Here it is seen that a discharge removes the O_2 to a considerable extent. Trichel pulses, increased threshold potential, and decreased currents caused by the O_2 disappear in time with the cleansing W point discharge. The clean-up is, however, not complete, as the threshold potential is at about 3200 volts instead of 2600 volts and the higher current values at 5000 volts are one-third those for clean H_2, indicating a material negative ion space charge resistance. The removal of the O_2 was not caused by creation of W oxide directly, as a similar clean-up occurred with a Pt clean-up discharge point. With N_2 the W point produced virtually no clean-up effect or at best only a very slight one. The clean-up effect of the auxiliary discharge was thus a mass action conversion of O_2 to H_2O vapor in the H_2 discharge region. The outgassed water-starved glass walls of the chamber absorbed most of the water at 0.1 mm O_2 pressure but were incapable of absorbing all of this at higher O_2 concentrations. Enough water vapor was still present to give a good internal negative ion space charge resistance. This is a most revealing experiment on the influence of trace amounts of impurity.

b. *Pure Nitrogen*

In this gas the discharge threshold was, as in H_2, a diffuse glow at 3600 volts and ≈ 0.1 μA, which contracted to a localized corona at 3700 volts with 90 μA. No pulses were observed, and the corona went over to a power arc via a spark at 14,000 volts. Current increased again by a factor of 10^3 on contraction to a glow, so that it in this respect resembled H_2. Currents were lower because of the lower mobilities of N_2^+ ions, and threshold was about 1000 volts higher, as might be expected, as N_2 does not appear to yield much γ_p but γ_i is relatively high compared to that for H_2. The transition from low threshold current to the contracted form with its current increase was slow enough so that the visual change and current increase could be followed. In only one respect did the discharge differ from H_2. Once the discharge had contracted, it burned quietly *at one spot* and did not move over the surface. Nitrogen may form nitrides with Pt, but it does not dissolve readily in it nor react with it chemically. Thus once a spot cleaned up, the discharge anchored at the region of high field and high γ_i. Addition of O_2 to N_2 in its smallest amount of 0.1% yielded Trichel pulses and reduced the current by a factor of 12, raising the threshold potential to 4300 volts with a much longer conditioning transition to the localized discharge covering a region of some 600 volts. That is, the oxide layer was trying to re-form on the Pt when O_2 was present, and it took heavier conditioning currents to clean up an active spot. Here, as in H_2, it is probable that the conditioning discharge acted to contaminate the gas so that the current did not grow to an arc at once because of negative ion space charge in low field regions.

c. *Pure Argon*

This gas was obviously purer than the corresponding H_2 and N_2. There was no corona. Application of 3500 volts led directly to a power arc, which extended from point to plane in a continuous luminous channel at 10^3 μA. There was no change when the gap was extended from 3.1 to 4.6 cm. Once the arc was kindled, it persisted as potential was lowered to a definite offset value. Increasing the gap length had the effect of increasing the formative time lag or the build-up time of the arc. It took seconds at 3.1 cm and tens of seconds at 4.6 cm. It is possible that this breakdown partakes of a cathode conditioned glow discharge. At any rate, photons resulting from metastable molecules created at high pressures by triple impacts of metastable atoms, as well as resonance radiation so near the cathode, increase the cathode emission most effectively.[4.17] The increasing current flow from the discharge builds up a streamer generating electron space charge near the anode. The shorter the gap and the higher the anode field, the more rapidly the spark materializes.

Addition of 0.3% of N_2 to the A resulted in a change in behavior. A corona set in at 2500 volts. Once it was started it continued down to 1300 volts. The current at 2500 volts was steady and of about 200 μA. A few hundred volts above the onset, for example, at 4100 volts, there was a continuous transition to an arc. Adding larger percentages of N_2 merely shifted the threshold and arcing potential to higher values. Addition of small amounts of N_2 accomplishes the following: N_2 molecules quench A metastable states and will thus interfere with photon production in A by triple impacts yielding the radiating photoelectrically active states. However, N_2 is ionized by impact with higher states of A. It is dissociated by A metastable states. It is excited to radiating states by metastable A and excited A atoms. Thus γ_p is much increased. It probably increases Townsend's first coefficient, because the inelastic impacts of electrons with N_2 molecules radically alter the electron energy distribution at a given X/p, which in A is unique and very sensitive. Thus a glow discharge has a distinctly lower threshold. It is possible that with the N_2 some negative ion forming constituent was introduced in small amount which prevented growth of the arc until higher currents and potentials were achieved. It is also possible that metastable A atom destruction, while permitting a lower threshold, prevented ultimate increase of the current to arc-over. Increase in N_2 concentration unquestionably acted to destroy metastable A states and thus raised thresholds and potential for arc development. Addition of 0.1% of H_2 to the A produced about the same effect as the 0.3% of N_2. Otherwise, the action appeared to be about the same. However, again the presence of H_2 in the discharge caused the discharge to wander over the point, as in pure H_2. Here again H_2 deactivates the A metastables, and is both dissociated and excited by A metastables. It is ionized by higher A excited states.

Thus it removes A metastables but gives copious photons that reach the cathode and give γ_p as well as ions. This lowers threshold. It has the possible effect of introducing impurities as well as destroying metastable states of A and thus raising the spark transition threshold. Addition of small amounts of O_2 initiated irregular Trichel pulses. With 0.1% threshold was 3000 volts, with 1 μA of current and regular Trichel pulses. If O_2 exceeded 0.3% the spark threshold was raised to 14,000 volts. Addition of 1% O_2 so completely destroyed metastable states that the triggering of new pulses by photons from residual metastable atoms ceased. The pulses became quite irregular, as were Hudson's in dust-free air without triggering. Such action and the negative ion space charges with O_2 reduced the currents by a factor of 10^3.

One notable feature in A was that, once a breakdown was initiated and the glow or arc was burning, it would operate at potentials as much as 1000 volts below onset before going out. This is not true hysteresis caused by conditioning of the surface. A high potential is required to build up a mechanism leading to a self-sustaining low order discharge that goes over to an arc-like glow. Such discharges as Townsend breakdown and normal glow discharges require high cathode fields to operate. Once current densities of the initial glow reach values leading to an arc breakdown, the highly ionized gas channel of a thermal nature has carrier densities that give it a low resistance, and ionization is maintained by thermal mechanisms. Thus cathode and anode potential falls can be low as long as current density remains high. This requires a very small sustaining potential across the arc. Table 4.2 summarizes Weissler's findings for the negative point corona in pure gases, as described above.

d. *Sputtering or Etching of the Cathode*

Weissler devoted some study to this phenomenon, since it was associated with phases of corona behavior. As later discovered by Weissler,[4.1] the ion impact energy in these coronas resulting from the high positive ion space charge fields is on the order of 10 ev. Electrodes of W, Pt, Cu, Al, and Pb were tested on negative points in H_2 and N_2 gases. This discharge was run for 12 hours at 1.25 mA and a current density of about 1 A/cm^2, gap length 3.1 cm, point diameter 0.25 mm. All points were carefully polished with SnO_2 and examined under the microscope. All points except the Al point were etched to a symmetrical crater-like pit with a shiny depression in the center and heaped-up metal around the rim. The amount of sputtered material could be roughly estimated and appeared to be greater in N_2 than in H_2. No discoloration or sign of chemical change was observed. No correlation of the sputtering with any properties of the metals could be made. The etching in Al did not result in a crater-like pit. Instead, the whole surface of the point was roughened by a large number of fine pits caused by the wandering of the discharge. In Al the very highly insulating thick and strongly adhering layers of oxide present a barrier and surface vastly different from those oxide films on other metals. The sputtering off of thick layers results in a most irregular surface because of the peculiar vagaries of the Malter[1.13] effect, causing localized emission and back discharges. This undoubtedly contributes to the peculiar pitting.

3. STUDIES OF DAS

As indicated in the analysis of positive point coronas, Das[3.70] used inert gases Ne and A, mixtures thereof with CO_2, O_2, N_2, H_2, and did some little work on relatively pure N_2 and H_2. He used supposedly pure—good commercial grades of—Ne and A from Linde purified by circulating over a hot calcium and a cold trap. The other gases were commercial tank gases with perhaps 1% of impurities, no purification being undertaken. As indicated in the comparison of Weissler's and Das's results with highly stressed anodes in the inert gas A, there was no doubt that Weissler's A was cleaner than that of Das. Das's inert gases had all the earmarks of

TABLE 4.2

ESSENTIAL DATA ON THE NEGATIVE POINT-TO-PLANE CORONA

Gas	Localized Corona			Trichel Pulses	Approximate Breakdown Potential, volts	Peculiarities
	Onset Potential	Onset Current, μA	Visual Character			
H_2* pure	2800 v	140	Unsteady glow†	None	—	Alternation uniform + localized glow
H_2 + 0.1% O_2	4200 v	70	Unsteady glow	Yes	—	Same as above
H_2 + 0.6% O_2	5800 v	2	Unsteady glow	Yes	14,000	Same as above
N_2* pure	3700 v	90	Steady glow‡	None		
N_2 + 0.1% O_2	4300 v	50	Steady glow	Yes	—	
N_2 + 1% O_2	5500 v	0.1	Intermittent§	Yes	—	5,000 to 7,000 volts intermittent corona
Room air	8000 v	0.1	Intermittent	Yes	—	8,000 to 11,000 volts intermittent corona
A‖ pure	—	—	—	—	3,500	No corona
A + 0.3% N_2	2500 v	200	Steady glow	None	4,100	
A + 1% N_2	3900 v	250	Steady glow	None	—	
A + 0.1% H_2	3200 v	400	Unsteady glow	None	4,100	
A + 1% H_2	3300 v	300	Unsteady glow	None	—	
A + 0.1% O_2	3000 v	1	Steady glow	None	—	
A + 0.4% O_2	3000 v	0.1	Intermittent	Yes	24,000	3,000 to 6,000 volts intermittent corona
A + 1% O_2	3300 v	0.1	Steady glow	Yes	—	

* Gap length 3.1 cm.
† See experimental data in H_2.
‡ See experimental data in N_2.
§ Due to lack of triggering electrons.
‖ Gap length 4.6 cm.

contamination on the order of perhaps less than 0.1%, and probably about 2×10^{-2}% of other gases both in these studies and in those he made of Penning mixtures. This statement applies as well to the starting potentials for breakdown in the so-called pure gases as to the mixture studies. Das, working with what he believed to be pure gases, was really working with inert Penning gases of near minimum breakdown potential. Impurities doubtless came from his Ca and perhaps his cold trap, as well as from traces of A in the Ne, and N_2 and Kr in the A.

Das further largely studied short gaps, of 2.5 cm, using hemispherically capped Pt cylinders of 0.1 cm diameter, opposite a nickel plated brass plane 7 cm in diameter. He observed starting potentials and current-potential curves. He broke off measurements just short of the spark to arc transition to avoid changing the point. His empirical current-potential curve has been discussed under the positive point as well as in Guck's study reported in this chapter. He studied the spark breakdown transition curves, or sparking potentials, for both positive and negative corona points from half a millimeter pressure on up to 750 mm. These curves show the minimum value at pressures of around one millimeter. In his Ne the minimum for the negative point occurs at about 183 volts and 2 mm pressure and at 235 volts for the positive at 15 mm pressure. For A these occur at about 292 for the negative at 0.30 mm and at 317 volts for the positive 0.30 mm. Addition of 0.1% A to Ne *raises* the negative minimum to 216 at 1.8 mm and *lowers* the minimum for the positive point to 207 volts at 12 mm. This contrasts with results by Penning on really pure Ne and A using coaxial cylinders. Pure Ne for a certain wire, negative, gave 220 volts, and 320 volts with positive wire. Addition of 0.02% A lowered both, the negative to 200 volts at 3 mm, and the positive to 280 volts for the first minimum at 2 mm with a *second* minimum of 150 volts at 30 mm. That is, *pure* Ne had *one* minimum when positive and *one* minimum when negative, both of which occurred at around 2 mm. These were *lowered* slightly by addition of as little as 0.002% A, and a second high pressure minimum was introduced at 30 mm for the *positive point* only. This clearly indicates that Das's pure Ne was already contaminated with *more* than 0.002% A and that he failed to observe the normal minimum at 2 mm. The value of the starting potentials in the A used by Penning did not greatly differ, over all, from that used by Das. The minimum for the negative wires was 320 volts, and for positive 290 volts, at about 1 mm. However, the negative threshold was markedly lower than the positive for the really pure A, while in Das's A the two differed relatively little.

The difference in behavior between A and Ne and the mixtures can be traced back primarily to the values of the first coefficient α, or better the Stoletow-Philips efficiency function η (eq. 2.1, and as seen in fig. 2.2). In

Ne η has a very high value relative to A at an X/p = 3 exceeding A by a factor between 30 and 60. It reaches a peak at about X/p = 40. In A the efficiency function η reaches a peak at X/p = 200 and is roughly 2.5 times as great as for Ne at its peak. Thus Ne will show a minimum breakdown and/or sparking potential at low X/p and at a corresponding pd value. The *mixture* increases α enormously at low **X/p** values because Ne metastables created with little electron energy loss can ionize the A present. Thus the η curves as a function of X/p *in mixtures* show *two maxima*, one at an X/p between 2 and 3 for from 10^{-6} to $10^{-4}\%$ A, and a second one at around X/p = 30. When this appears in the sparking potential relation with γ sensibly constant (eq. 2.3) $\gamma e^{\eta V_s} = 1$ makes V_s low when η is high. There will thus be *two minima* in the sparking potential curve, one at an X/p $\approx$ 2, which is lowest, and a second at X/p $\approx$ 30, which is higher, as η at X/p $\approx$ 2 is on the order of five times as large as at X/p $\approx$ 30. If one looks at the pd curves with d constant, then V_s will have two minima in Ne in such mixed gases. Unless care is taken, the lower minimum which appears at pd $\approx$ 2 or 3 for both positive and negative corona may be overlooked because of the presence of the more prominent minimum at 30 for the positive point. This is precisely what occurred for Das, who noted one minimum at 30 mm for Ne with positive point and one minimum at $\approx$2 mm for the negative point. This definitely establishes Das's gas as a Penning mixture and not pure Ne. The quantities of A added to the already effectively contaminated Ne + 0.002% A which Das used were too large to produce an effective Penning gas and merely served to raise potentials and throw his gas into the less sensitive portion of the Penning range. All this would have been perceived by Das in his work had he mastered Penning's data and understood it. Since he did not, his data are of no significance in the sense intended by Das. They do represent mixture studies well along in the insensitive larger percentage of the Penning range. In this sense the data are of use for workers who are interested in this range of values.

Das states that in his "extremely pure" inert gases and their mixtures for a negative point, no currents in excess of 10^{-9} A were observed up to a sparking value. At a given potential the spark occurred. Changes in pressure merely changed the breakdown potential. This contrasts with the space charge controlled currents with positive point. Beyond this there is little of consequence in this study except for the data. Conclusions are drawn largely from writings of previous workers, except where they are falsified by ignorance of the Penning effect and of the gas purity which he used, and are basically correct.

In conclusion, it can be seen that the observations of Weissler and Das in general substantiate the inferences and conclusions derived from the study of the electron attaching gases and basic gas ion theory.

4. Negative Point Breakdown Streamers;
   Studies of Nasser and Adipura

It is clear from the data derived in the study of the positive streamer corona using the Lichtenberg figure technique of chapter 3, section A.8, that the method is equally applicable to the study of the negative streamer process. Thus far little attention has been paid to it in corona studies, since it does not dominate breakdown as does the positive streamer. Some little attention was paid to the traces of negative streamers from the plate cathode in the positive point work of Nasser.[2.19] Here it was noted that the high tip fields of positive streamers in the breakdown region caused emission of the negative Lichtenberg figure pattern from the cathode even if the intercepting film was as far as 1 to 2 cm from the cathode surface.

It is, however, clear that the chance of observing and studying such streamers from a negative point with plane anode under impulse conditions is far better. This is especially so as the negative point Lichtenberg figure has long been known and shows striking differences from those for the positive point.

Actually much more was known about the negative or anode-directed streamer from Raether's cloud track pictures of the midgap streamer breakdown in highly overvolted uniform field gaps with impulse potentials. Raether,[3.26] in his earlier work with the highly overvolted gaps, seemed to favor the negative anode-directed streamer as the more important factor in midgap spark breakdown. This was so, although Raether recognized certain inherent drawbacks to the negative streamer advance. Since the author and Meek[3.28] had worked more or less with positive points at threshold, the anode streamer with its charge-conservative nature had appeared to be the more important agency. This was also borne out by the high mobility of electrons in the negative space charge cloud and diffusive actions. Raether, using the cloud track method, had observed the velocities of the positive and negative streamers for three gap forms, and a summary of his findings is shown in table 4.3. It will be profitable to

TABLE 4.3

Raether's Velocities for Positive and Negative Streamers in
Three Types of Gaps, as Given by Cloud Track Measurements

Nature of Field Degree of inhomogeneity indicated by ratio of max. to min. field	Point to Plane	Two Spheres	Parallel Plates, 10–20% Overvoltage
	100	13.5	1
+	2.5×10^7 cm/sec	10^8 cm/sec	1.5×10^8 cm/sec
−	2×10^6 cm/sec	2.5×10^7 cm/sec	7.9×10^7 cm/sec

outline the inferences concerning relative behavior of positive and negative streamers, as they were essentially unchanged at the end of the pre-World War II period and remained so until clarified by Nasser's direct studies.

Both the midgap and the anode originating *positive* streamers represent a charge conservative system. When, through the positive ion space charge density at an avalanche head, the vectorially combined space charge and applied fields along the field axis are sufficient to draw in effective avalanches from photoelectrons produced by the avalanche photons in the adjacent surrounding gas, the positive space charge or streamer advances into the gas. Since the applied field is very strong, advance will generally move along its direction. In this fashion the positive ion space charge is advanced toward the cathode, maintaining the field at its tip and creating more photoelectrons ahead of itself to continue the advance. However, since the photoelectrons are created radially about the head of the avalanche as well as along the field axis, and statistics play an important role in avalanche growth and photoionization, the incoming avalanches have a considerable tendency to branch radially about this field axis. The tip of the streamer in the declining field of point-to-plane geometry becomes more and more radial. Thus branching will increase until the branches attenuate. It is to be noted that, since the new avalanches are drawn into the head of the initiating avalanche or advancing positive streamer head, concentrating near its surface, the electronic charges created in the advance will be conserved and go into a current flow up the plasma channel of relatively immobile positive ions created as the streamer advances.

In contrast, when a midgap avalanche head builds up an adequate electron concentration, or charge, on its *anode* end, or when the avalanche starts from the highly stressed cathode surface to achieve the same electron concentration, the electrons diverge outward from a high field to a lower field. Thus any electrons created by photoionization will move ahead to build up a new set of avalanches, jumping their photoionizing free path in advance of the parent head, and will move into a rapidly attenuating space charge field component. If the external guiding field is adequate, the new avalanche will advance, and, as it does so, its positive ion counterpart pole will start a *positive*, or so-called *retrograde*, streamer toward the cathode to join the original electrons in the initiating negative avalanche head preceding. In the meanwhile its negative counterpart will again after some photoionizing free path launch a new avalanche toward the parent avalanche head, as shown schematically in figure 4.35. This picture was invoked by Raether and by Loeb and Meek independently before 1941. If the ionizing free path is long, the external field will indeed have to be high, and the field of the avalanche head great, to produce such action.

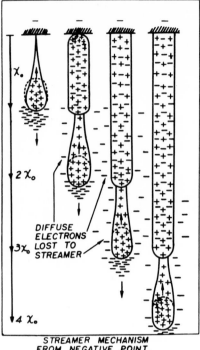

STREAMER MECHANISM
FROM NEGATIVE POINT.
AVALANCHES ADVANCE DOWNWARD
STREAMERS MOVE UPWARD BY STEPS

FIG. 4.35. Schematic diagram of the mechanism of advance of the cathode-directed streamers. The electron cloud moves out and ionizes; photons in advance will start new avalanches ahead. If the field is such that the negative streamer ahead can advance, its positive ion space charge, shown at $2x_0$, will start toward the cathode and allow the negative streamer to advance again. As a result, the positive streamer moving in a rectrograde direction to the cathode makes up for the weak ionization in the avalanche which initiates at the dipole at x_0. The high negative space charge field and high electron mobilities tend to diffuse and dissipate the negative electron streamer head, as indicated.

It is to be noted that, with the distance of the new avalanche origin an ionizing free path ahead of the initial tip, the influence of the applied or guiding field will be more important than for the positive streamer. Thus the influence of the *radial* field components of the negative avalanche head field is much reduced. Except for photoelectrons liberated in the gas close to the original negative avalanche head, the *tendency for radial branches to develop will be suppressed*. This situation in the negative streamer head is rendered, if anything, more critical, for the following reason. The negative space charge at an avalanche head, with high electron mobilities and self-repulsion among the electrons, causes the negative space charge to diffuse and dissipate radially instead of conserving it, as for the immobile positive ions in the anode streamer. This means that the electron space charge density decreases as the negative streamer advances, requiring high guiding fields to ensure advance. Thus negative streamers should show less branching than positive streamers, and require higher guiding fields. It is thus not surprising to find that the negative anode-directed streamer is not easy to create and is best observed in overvolted uniform fields, where Raether discovered it. It also follows that in highly asymmetrical fields, such as point-to-plane gaps, the negative point streamer sparking potential should be large compared to its positive point counterpart. Waidmann[2.36] has shown that for 1 mm point diameters and 2 cm gaps the impulse potential is $\approx 10\%$ above the positive point, while with static potential it is 30% higher. For longer gaps the discrepancy becomes much greater. Thus where 50 cm long impulse sparks in such gaps have been photographed, a negative streamer from the pointed cathode is met at one-third the gap length from the cathode by a positive streamer originating at the plane.[3.31] (See chapter 7, section B.1, as well as the presentation of Kritzinger's work in Appendix I.)

Thus impulse potentials, even with highly stressed cathode, are unsatisfactory for negative streamer study by means of earlier techniques. With low overvoltages in impulse breakdown it would be expected that the speed of the positive streamer would be initially greater than that of the negative streamer. As fields become more uniform, the speed of both positive and negative streamers should increase as the cathode and anode, respectively, are approached. However, at adequately high fields the negative streamer, with its ability to jump ahead by ionizing free paths (for example, by x_0 in fig. 4.35), should give much higher speeds than for the positive point, where photoionization cannot be quite as effective. The speed will be strongly influenced by the photoionizing free path. Table 4.3, taken from Raether, indicates the initial relative speeds of positive and negative streamers in midgap, where indeed the positive streamer is the faster. The technique could not be applied near the electrodes.

With this background of positive and negative streamer theory, it is possible to present the observations with the new technique in such a fashion as to be readily understood and interpreted by the reader.

Figure 3.63 shows a diagram of the apparatus, which is similar to that used for positive point except that the polarity is reversed. The points used were 0.5, 1, and 2 mm in diameter, and gap lengths will be referred to by the letter d. The distance of the photographic plate from the point will be given by the symbol s. Figure 3.65 shows a typical series of exposures for a *positive* point for an 8 cm gap at various increasing displacements s. This is to be compared with figure 4.36, representing a similar series for a negative point. Note the single point of impact at the cathode, the few

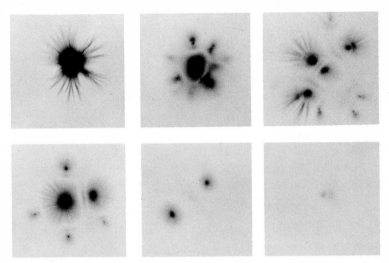

FIG. 4.36. Nasser's Lichtenberg figures for a cathode point of 1 mm diameter at various distances s from the point. Note the few branches and their straight, feather-like character.

branches, and the tendency for the branches to advance largely axially rather than radially. The dark, soft shadows are produced largely by long wave-length quartz penetrating ultraviolet light from the advancing diffuse streamer tip before impact on the film. The thin, sharp, dark centers of the longer streamers are probably high field induced, very short length discharges in the emulsion by the heavier streamers, as they consist of minute δ ray-like tracks under heavy magnification. More work is needed in the analysis of the negative figures. Important here are the feather-like branches. The light regions between figures are caused by the Clayden effect, owing to double exposure first to ultraviolet and, within a short period (10^{-7} sec) to visible light from the streamers, which reduces the blackening relatively.

From data yielded by photographs such as these, it has been possible to plot the curves to follow. Figure 4.37 shows the number of streamers as a function of distance for d = 2 and 4 cm at 30 and 36 kv, respectively. It is seen that branching is very slight and reaches a maximum in the number of branch tips striking the plate at about 1 cm. Figure 4.38 shows the radial spreading of the branches at d = 2 cm at different point potentials. It is seen to be relatively small, with maxima quite near the anode. That is, the branches persist clustered about the axis and proceed, deviating radially relatively little from their original course, until they attenuate short of the anode. This pattern contrasts markedly with that of the positive streamers, as will be shown. Figure 4.39 shows the same trend for a 4 cm gap at 36.2 kv.

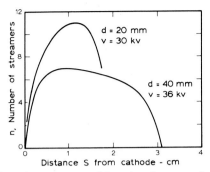

FIG. 4.37. Number of streamers striking the plate as a function of distance s
for two gap lengths, d = 2 and d = 4 cm, at 30 and 36 kv.

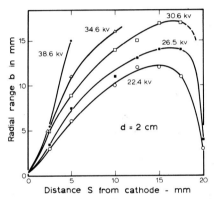

FIG. 4.38. Radial range b of negative streamers as a function of s for
d = 2 cm at various potentials.

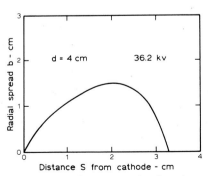

FIG. 4.39. The same as figure 4.38, for d = 4 cm at 36.2 kv.

From the unpublished data taken by Adipura in a study under Nasser's supervision at the Technical University of Berlin in 1960, the velocity of the negative streamers has been computed as a function of the distance s for 34.6 and 50.9 kv, respectively, as shown in figure 4.40. The velocities range from 2×10^8 cm/sec to 1.4×10^9 cm/sec under the conditions given. The gap here was d = 2 cm and the point diameter was 1 mm. Note the exceedingly high speeds under these conditions, and the increase in speed as the streamer tip approaches the anode. Contrast this with the positive streamer, where in general at lower fields the speeds decrease as the gap is crossed. These fields are much higher than most of those used in the study of positive streamers. Waidmann (see Appendix II) at lower potentials confirmed the short range and reported negative streamer tip velocities slightly less than corresponding positive tips.

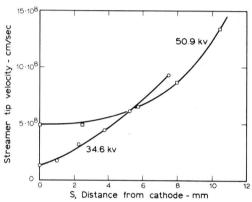

FIG. 4.40. Adipura's data on negative streamer velocities as a function of s for 34.6 and 50.9 kv. Here d = 2 cm and point diameter is 1 mm.

Figure 4.41 shows the range of negative streamers along the axis as potential increases. The increase in this range appears to be linear, with potential applied to the point, as might be expected. Figure 4.42 shows the number of branches n as a function of impulse potential. Streamer threshold here is 13 kv. The gap length is 5 cm and the point diameter 1 mm. The upper limiting number n appears to increase linearly with potential applied. However, to be noted is the excessive scatter of the value of n at any given potential. This scatter must be ascribed to the pure chance character of the branching process. With few branches, statistical fluctuations should obviously be great. Figure 4.43 clearly illustrates the great difference between the branching characteristics of positive and negative point streamers. The scale to the left of the axis of

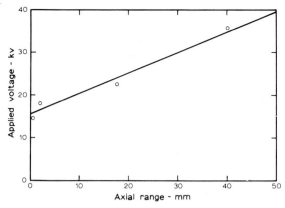

FIG. 4.41. Axial range of negative streamers as a function of potential.

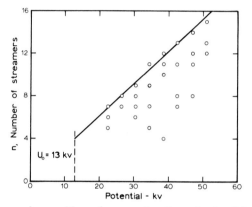

FIG. 4.42. The number n of branches as a function of potential at cathode point. The streamer threshold was 13 kv, and the gap length was 5 cm, with point diameter 1 mm. Note the scatter owing to the poor statistics with only a few streamers per film.

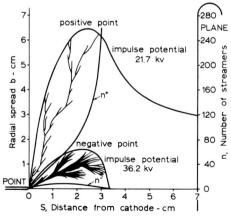

FIG. 4.43. This contrasts the radial spread, b, and number of streamers as a function of s for positive and negative streamers, the former at 21.7 kv and the latter at 36.2 kv. The difference in branching is indicated schematically. Here d = 7 cm with a 1 mm diameter point.

ordinates represents radial range in centimeters plotted against the point distance s, with d = 7 cm and a point diameter of 1 mm, while the inner scale gives n as a function of s.

It must be noted that the large n^+ curve occurs for 30 kv while the smaller n^- requires 50 kv (corrected on Nasser's scale). Note the short axial range of the negative streamers relative to the positive streamers, despite the higher potential. Note also the schematic trend of negative streamers in the field direction. This comparison appears to agree very well with the inferences drawn by Loeb and Meek and by Raether concerning the nature of the negative streamers. Finally, the axial range of the streamers for positive and negative points for different diameters of points is shown as a function of the applied potential in figure 4.44. Here again the need of very high potentials to advance the negative streamer to the same range as the positive streamer is seen. It is to be noted that above 10 cm the positive streamer range is proportional to the potential and that the slope of the line is less than half that for the negative streamer, which appears to increase linearly from its starting potential. This accounts for the great difference in breakdown voltage reported for the two cases, as well as for the much greater advance of the positive streamer once launched, relative to that of the negative streamer.

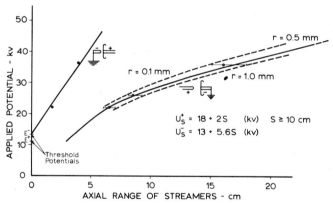

FIG. 4.44. Axial range of streamers for positive and negative points. The positive point data are given for point radii r of 0.1, 0.5, and 1.0 mm. The equations for the potential range curves in the linear portions are given below the curves. The threshold potentials are indicated by E^+ and E^- on the scale of ordinates.

The appearance of negative streamers in spark breakdown from positive points became evident in the positive point-to-plane Lichtenberg figure studies. Here, as potential was raised, with film emulsion facing the cathode, it was observed that typical negative Lichtenberg figures appeared on the cathode side. On the cathode these appeared as fine black lines following the axes of the strong positive streamer traces on the anode side. These again were caused by field induced avalanches in the emulsion, as was proven earlier (chapter 3, sec. A.8). However, near sparking potential with the film as far as 2 cm from the cathode, several, or even many, typical negative streamer figures were observed. These were invoked by the electron avalanches initiating from the cathode surface through electrons emitted from the cathode thermionically or otherwise, under the very high fields produced by impact of the positive streamer tips on the photographic film relatively close to the cathode. The potential of these tips in some instances was of the order of half or more of the applied anode potential. The character of some of these negative Lichtenberg figures was complex, and more work is needed to clarify all the details.

With a positive point, the negative streamers from the cathode are not necessarily responsible for the ultimate spark breakdown resulting from the positive streamer. The cathode streamer from the negative plane only appears here and is amenable to study because the presence of the film diverts and stops the axial breakdown streamer that, reaching the cathode directly, could liberate photoelectrons from it and send a return stroke in the form of a potential space wave of ionization up the streamer channel. The positive streamer is thus diverted over the film surface to a radial distance which so attenuates it that, if it reaches the cathode, no spark can ensue. If by chance a streamer strikes near the edge of the film, so that its diversion is so slight that it can reach the cathode, it will invite a return stroke and lead to spark. Direct evidence for this has been obtained by Waidmann.

In conclusion, it is clear that, because of the diffusive nature of the electron cloud in the negative streamer head, relative to the conservative nature of the positive ion space charge with incoming avalanches, there will be a marked difference in their character. Positive streamers advance readily in low fields, whereas negative streamers require high guiding fields. Positive streamers, by virtue of the radial character of photoionization, will develop radial branches far more readily than will the negative streamers, and will increase rapidly in number and vigor as potentials increase. Only those negative streamers moving more or less parallel to the guiding fields can develop. Thus, despite equal opportunity for branching, the negative streamers will branch less and always progress more or less in the field direction. The positive streamers, with their concentrated space charge at the head, will advance rapidly and further in low fields. With adequate photoionization in the gas and long photoionizing free paths, negative streamers in high fields show exceedingly high velocities far in excess of those possible by electron movement alone.

It is difficult at this point to assess the relative importance of the negative or anode directed streamer. It will undoubtedly play a most important role in spark breakdown of highly overvolted uniform field gaps. Here it will lead to exceedingly short formative time lags. These may even be considerably less than an electron gap crossing time, owing to the photoelectric ionization in advance. Only with very long sparks have negative streamers been observed by fast time resolution photography, as shown in the studies of Meek and Allibone in 1938.[3.31,4.18] The most striking photographs were with negative point and positive plane. Here the negative streamer was stepped, probably because of line resistance. Eventually, when it had progressed out to one-third the gap length toward the anode, it was met by a positive streamer created presumably by inductive action from the plane metal cathode that had advanced two-thirds of the gap length to meet it. This has been confirmed in detail by Kritzinger, as stated in Appendix I.

With negative point-to-plane gaps of shorter length, the nature of events is not known. Studies of Waidmann, using positive and negative point-to-plane gaps with 0.1 cm point diameter and a gap length of 2 and 4 cm, yielded the results shown in table 4.4. This indicates a 10% higher

TABLE 4.4

RELATIVE SPARKING POTENTIALS FOR POINT-TO-PLANE AND POINT-TO-POINT
GAPS OF 2 AND 4 CM UNDER IMPULSE AND STATIC POTENTIALS

Gap Length	Impulse Breakdown	Static Breakdown
2 cm	+ point − point 34 kv	13.8 kv
2 cm	+ point to plane 28 kv	28 kv
2 cm	− point to plane 31 kv	38 kv
4 cm	+ point − point 4 cm, 55 kv	38 kv
4 cm	+ point − plane 4 cm, 55 kv	50 kv

potential for the negative point for impulse where corona space charges do not intervene, and 30% higher potentials for the negative point in static breakdown with space charges which are all of one sign and exert a strong inhibiting action in both cases.

It had been the author's impression that for a point-to-point gap, the breakdown potential should be twice that for the point-to-plane gap of half the length, since the point-to-plane gap shows a mirror point image of *opposite polarity* at twice the gap length. Actually, since the field distortion of needle points enhances the point fields, the corona and breakdown thresholds should be somewhat lower than twice that for the point-to-plane gap of half length. Both impulse and static potential breakdown values were made with two equal 0.1 cm diameter points separated by 4 cm and compared with the data for a 2 cm gap. The results are of considerable interest and are seen in table 4.4. The impulse breakdown in a clean 4 cm + point-to-plane gap is on the order of 1.8 times that for the positive point potential for a 2 cm point-to-plane gap. The potential is also the same as for the positive point and negative plane separated with a 4 cm gap length. This indicates that the breakdown depends essentially on the ability of the positive point streamer to cross the 4 cm gap, and that the advance of the negative streamer plays a relatively small role. The negative point-to-plane breakdown potential at 4 cm was so far above that for the positive that the pulser could not yield the adequate potential.

In static point-to-point breakdown, where the interdiffusing space charges of opposite sign from corona from *both* electrodes have formed a somewhat conducting plasma, the breakdown potential is much lower than that for the impulse. It appears to be representative of the threshold potential of the negative streamer spark. The corona discharge plasma of positive and negative ions across the gap enhances the fields at both positive and negative points, and gives a spark threshold that is essentially the one needed to develop a negative streamer that can cross 2 cm to yield a spark, or cause a spark in point-to-plane across a 2 cm gap by combined negative and positive streamer. These data are discussed in more detail in chapter 7, section B.1. To date, this is about all that can be said of the role of the negative streamer in spark breakdown. It is seen that, under normal conditions, it will not be a prime factor, as the positive streamer usually predominates.

C. THE EXTENSIVE STUDIES OF MIYOSHI IN AIR

After the manuscript for this book was completed, the author had the pleasure of a visit from Professor Y. Miyoshi of the Nagoya Institute of Technology. Professor Miyoshi together with his staff has been conducting an intensive study of the point-to-plane corona over a very extended region of point diameters, gap lengths, point materials, and a large range of pressures in clean dry air. Static breakdown measurements and transition regions between various discharge phases were exhaustively covered. In addition, using an impulse potential consisting of the H. J. White stabilized voltage supply used by L. H. Fisher and H. W. Bandel plus a step voltage, Professor Miyoshi and his staff obtained time lag studies and oscillograms of the breakdown. To date the work has appeared only in Japanese, though publications in English are contemplated. Professor Miyoshi kindly turned over to the author his voluminous data. It is only possible to cover in this section the more important contributions to the field, as judged by the author.

While some work was done with positive point corona, where satisfactory time lags could be measured, most of the work was done on the negative point corona. In this study various point forms were used, ranging from cones to flat-ended cylindrical rods. Most of the data were taken with conical points. The point radii ranged from 0.02 mm to 3.0 mm. Gap lengths ranged from 1 to 30 mm. Pressures of clean dry air ranged from 10 to 760 mm. The static breakdown studies were made with stabilized rectifier circuit, permitting accurate potential measurements to be made. The pulses were analyzed by synchroscope across a 1 kilohm resistor. Accuracy of measurement and experimental controls were exceptionally good. The gap length was adjusted by micrometer. The plate was made of brass 7 cm in diameter. Point materials used were Pt, Ag, Al, Fe, brass, and W.

With this wide range of variables and the many combinations of measurements possible, certain generalized aspects of the corona that escaped the more limited range of studies of Bandel and others came to light. Clearly delineated were three essentially different breakdown forms. These were the corona threshold, which was probably taken as the onset of regular Trichel pulses at a potential designated as V_c, and a current designated as i_c. The Trichel pulse corona had the characteristic structure and the regular pulses discussed earlier in the chapter. It had the wide shaving-brush–like flare of the column seen in the upper inset of figure 4.45. At higher potentials in certain regions of point diameter and gap length was the pulseless corona noted by English and Bandel as well as by Guck and Greenwood. However, with the conical point, apparently the multiple spots observed by Bandel and others were precluded. Thus the so-called pulseless form was clearly noted. In this the form was that indicated in the lower inset of figure 4.45, with a narrow central spike extending far out into the gap, flanked by two lateral spikes, or perhaps vestiges of the outer edges of the shaving brush. It was a quiet pulseless discharge. Miyoshi called this the *glow* discharge form, designated by a threshold potential V_G and a current i_G.

Finally the discharge reached a limit, yielding a spark with potential V_s. As will later be noted, the oscillograms of the development of the negative impulse spark from the Trichel pulse corona indicated that it built up to a streamer spark on a time scale of 10^{-7} sec after a prolonged variable delay with an overvolted Trichel pulse or pulseless conditioning corona. Here Professor Miyoshi visually noted that, just before the spark, the spike from the cathode lengthened as if a negative streamer were advancing toward the anode. It never reached there, as it apparently was joined by a positive streamer from the anode plane, as in longer sparks. This was a most significant observation.

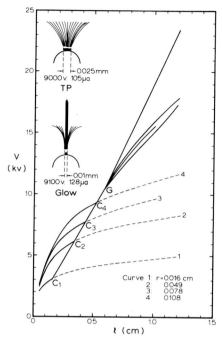

F<small>IG.</small> 4.45. Miyoshi's curves for potentials for breakdown as gap length varies for various point radii. These are for the smaller radii. The insets (left) show the visual forms of the Trichel pulse corona and the spiked glow form of corona. On the graph, the solid lines to the left, which fuse into a single curve at values of l_c characterized on the plot by C_1, C_2, and C_3, and so on, represent direct spark breakdown potentials with no corona. The dashed prolongations to the right represent the appearance potential of the TP corona for each value of 1. If the potential increases along a line at a set l, then to the right of G for a given r, the TP corona appears at the intersection with the appropriate dashed curve. At the intersection of that vertical with C_1–G, a spark appears. Above G for points 1 and 2, 3 and 4, three full curves appear. Here above $l_G \approx 0.6$, the vertical l line strikes, say, curve 2 and gives a TP corona. When it intersects the lowest G curve a glow intervenes, and when it strikes the projected line C_1–G a spark ensues.

A systematic study revealed that for points of a radius less than a critical value r_G, which varies with pressure, threshold potentials of the various breakdown forms plotted as a function of gap length showed a most interesting behavior, as seen in the curves of figure 4.45. Below a critical gap length l_G, at points C_1, C_2, C_3, C_4 for each point radius r, depicted by the dashed portions of the curves marked 1, 2, 3, 4, the spark occurred directly with no corona following along the V_S–l curves with solid lines. Beyond the gap lengths exceeding those corresponding to points C_1, C_2, C_3, C_4, the discharge started as a corona discharge with regular Trichel pulses. That is, a TP corona at a potential V_C preceded the spark. These corona thresholds are indicated by dashed lines beyond l_C. The points C at V_C and l_C lie on a straight line. That is, V_S plotted as a function of l, the gap length, in millimeters beyond the points C lies on a common straight line. This represents the maximum value of l_C below which direct transition to a spark can start without either Trichel pulses, hereafter designated TP, or glow corona. If l is less than the l_C appropriate to the radius r used and if V is plotted along the curve, say from near the origin to C_4 and then along the straight line V_S, no corona appears but only a spark.

If l exceeds the l_C value for a given r, then a Trichel pulse corona sets in along the dashed curve at a potential V_C well below V_S. If then the potential at fixed l greater than l_C is raised from below V_C in this region, a TP corona sets in at V_C and continues until the potential reaches V_S, at which point a spark will appear. In this region a TP corona goes directly to a spark when V_S is reached. If, however, l exceeds the value l_G at the point marked G in figure 4.45, then the phenomena observed as potential V is raised is as follows: The corona starts as a TP corona at V_C, where the constant l line intersects the dashed curve, and goes to a glow corona at V_G, where the constant l line intersects the V_G curve appropriate to r. The glow continues until the line at constant l intersects the common V_S–l line C_1, at which point a spark occurs. The thresholds for V_G lie along the solid lines starting at G appropriate to the value of r.

If now the point radius is chosen greater than r_G, one has the curves of figure 4.46. Here it is seen that, as before, there is a gap length linear for V_S as a function of l above a critical gap length for each value of r. Below this l the curve for short gaps has V_S increase along a parabolic path in which sparks occur across the gap with no corona preceding. It is interesting to note that, beyond the critical gap length for V_S for that value of r, the breakdown appears as a glow corona as long as r is above r_G and l is above l_G at the appropriate V. Thus if l_G exceeds the critical length for V_S only, raising the potential up to breakdown gives a glow corona on a prolongation of the V_S–l curve for that r. Thus if the potential is raised further to V_S a spark occurs. If the gap is made longer than l_S and l_G for that particular r, the first breakdown on raising the potential is a TP corona at V_C. At a somewhat higher potential this leads to a glow at V_G for that point, and at a still higher potential it reaches V_S for that point.

The most remarkable feature of the curves is that if, at a fixed r, l is increased, the general V–l breakdown curve (which at first yields a spark threshold V_S, then a glow threshold V_G, and finally a TP corona threshold) is a continuous curve. That is, for a given r the three thresholds lie on a single curve. Another interesting feature is that, when for each r value a length l_C, such as l_{C_1}, l_{C_2}, or l_{C_3}, is reached so that a TP corona can precede the spark, the points V_C–l_C delineate a straight line for l_C and the potential V_C is proportional to it. Since streamer tip potentials determine their ability to invoke a spark when crossing the gap and since the streamer tip at the point partakes of the point potential, it would appear that for these short gaps the axial range of breakdown producing streamers that cross the gap l is directly proportional to the point potential V. However, since the TP space charge is involved as well as the V_G region, this naïve explanation may be incorrect. As noted, the threshold curve for V_G–l_G is not as accurately linear and the glow transitions are not as clearly delineated.

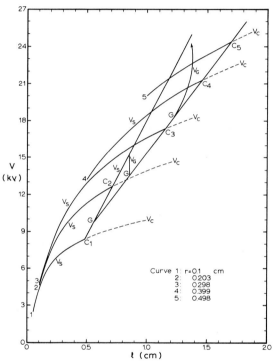

Fig. 4.46. The same as for figure 4.45 but for larger values of r near G. In this region the curves are not quite so reproducible or simple. It is noted that above G the spark line lies along C_1–G–C_2 prolonged. On curve 2 the TP corona begins at C_2, and the glow corona starts at G on that curve. Thereafter, the line G–G–C_3–G–C_4–C_5 delineates the glow–TP corona boundary. To the left of C_1–G–C_2 are prolonged sparks. Between this line and the G–C_5 line is a glow, and to the right of G–G–C_5 a TP corona. Here again, to the right of a given G for a given radius curve, the increase of V at constant l begins with a TP corona, goes to a glow, and then goes to a spark.

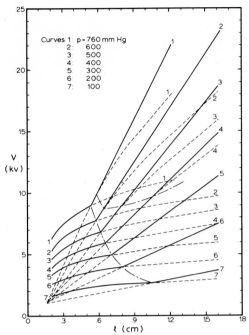

Fig. 4.47. This represents the series of V_C lines, shown solid where a spark gives way to the TP corona and then to a spark as l increases, as well as the V_G lines, which are dashed. The curves are all drawn for one value of the radius r at the various pressures of air in millimeters Hg, indicated on the legend. Both the TP curves and glow curves are dashed. The appearance lengths l for the TP corona as p varies are along the dot-dashed curve to the left, and this has l_C increase as p decreases. The appearance lengths for the glow curves are marked by the upper dot-dashed curve, which is nearly constant and lies above about 9 kv.

Figure 4.47 represents a typical V–l plot for a brass cathode r = 1.08 mm at a series of pressures from 760 to 100 mm. The solid line curves, as before, represent the V_S lines at V_C and l_C for the fixed r at each pressure. The linear V_S–l trace is extrapolated back as a dotted line. These dotted lines intercept each other at a value of V_S of about 1 kv and an l≈ 1.5 mm. This again may or may not have real significance in terms of streamer advance, since Nasser has shown that streamers cease to advance much below 1 kv. It is also possible that the intercept represents some minimum sparking potential for the brass point in air. A more complete plot down to 10 mm is too large to reproduce here. Again note that for each pressure at a fixed r there is a V_S–l line of decreasing slope as pressure is lowered and that l_C increases along the dot-dashed curve indicated. It will also be noted that for that there is a point l_G at each pressure at which the TP corona goes to a glow before the spark. The V_G–l curve is interesting in that it does *not drop much below 8 kv*, usually occurs at 10 kv, and in fact even increases as pressures become lower, while the V_S–l curve as a function of pressure continuously falls but at a decreasing rate.

Figure 4.48 is again a V–l diagram, but it omits the V_{TP}–l and V_G–l curves and merely plots the points l_C for different radii r and different pressures p, as well as the l_G and the V_S–l_C lines for different pressures and the various radii. Interpretation of the GG' dot-dashed and CC, as well as V_S, lines becomes clear by comparison with figure 4.47 at one r.

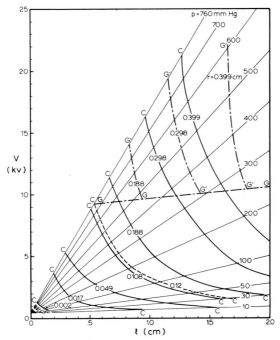

FIG. 4.48. Summarizes all the data for points from r = 0.002 cm to r = 0.399 cm and pressures from 10 to 760 mm Hg. The TP curves and the G curves are omitted for clarity and only the C point and G point intersections are shown. Note that the C points lie on straight lines of decreasing slope as pressure decreases. Note that all G curves start and lie at a potential above 9 kv. Also, the G intersections for V at fixed r as l increases fall along steep G'G' lines as pressure increases, and all C points for a given r value lie along lines that decrease in V and increase in l at a given r value. That is, the glow regime potential falls sharply as pressure decreases and space charge dissipation is more rapid, but glows do not appear below about 10 kv. The beginning of TP corona in place of a spark requires a lower potential for fixed r as pressure decreases but also requires a longer gap to give adequate space charge build-up to prevent a direct spark. These curves are most suggestive.

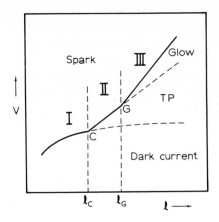

Region	Trichel pulse corona	Pulseless glow corona
I	unstable	unstable
II	stable	unstable
III	stable	stable

Fɪɢ. 4.49. Shows the various regions, spark, TP, and glow, as well as dark, or field intensified, current regions. The chart below characterizes as stable or unstable the different phenomena, TP and glow, in terms of these regions.

Region	Trichel Pulse Corona	Pulseless Glow Corona
I	Unstable	Unstable
II	Stable	Unstable
III	Stable	Stable

Figure 4.49 shows the regimes observed in *impulse* breakdown as compared to static breakdown. The designations of the stability of the phenomena in the various regions, as revealed by the oscilloscope, are indicated in the caption. It can be seen that in impulse and static breakdown the spark appears only in region I. In region II a steady series of TP can occur, but at higher potentials (overvoltage) the spark may be preceded by a succession of TP. In region III, at appropriate impulse potentials TP, glow or spark will appear. But application of an impulse potential exceeding V_S may result in a sequence of transitions of TP and glow corona breakdown before the ultimate spark.

Figure 4.50 represents the current I_c in the Trichel pulse regime and going into the glow regime, as a function of the potentials for a series of gap lengths from 6 to 15 mm. It is to be noted that at the transitions from TP to glow, that is, at V_G, there is an abrupt change of current. For a small point and short gap, the change in current Δi is an *increase* of negative current. This is plotted by Miyoshi as a negative Δi. The negative sign is a bit misleading, since the current, which happens to be from a negative point, actually increases discontinuously in negative value at transition. It is the fact of *increase* that is significant, not the sign of the current. At longer gaps and longer points this is changed, and there is a *decrease* in the negative point current, a true $-\Delta$i, plotted by Miyoshi as $+\Delta$i.

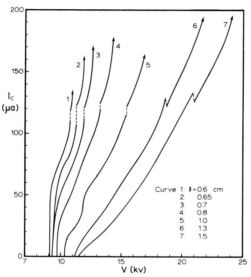

Fɪɢ. 4.50. The corona current is plotted as a function of point potential for gaps of different lengths at a fixed radius, showing the TP regime for the lower current values and the glow in the higher current values, above the dashed discontinuity regions. The arrows indicate a spark transition. It is noted that between l = 1.0 and 1.3 cm, the negative point current increases at transition from TP to glow, while at greater lengths it decreases. These changes are labeled $-\Delta$i and $+\Delta$i by Miyoshi.

This decrease was noted by Bandel and English in their longer gaps, but its significance relative to a transition to a pulseless discharge was not noted. It should here be added that, as most previous work was with longer gaps and intermediate point diameters, the beautiful range of phenomena indicated here could not be observed. The data for a large range of point diameters and gaps are depicted in figure 4.51, where Δi is plotted as a function of point radii for various gap lengths l. The $-\Delta$i values mean that there is a negative point corona current *increase*, and $+\Delta$i means an actual current decrease. This has some significance in interpretation.

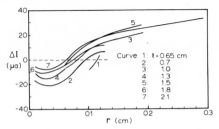

FIG. 4.51. Δi following Miyoshi's definition is plotted against r in centimeters for various l values. Negative values of Δi represent an actual increase of negative current at transition, while +Δi represents a decrease in such current.

Figure 4.52 represents a plot of the point radius r against gap length l for the various phenomena, spark, glow, and TP corona thresholds. The region marked I is the r–l area, in which only a spark occurs on raising the voltage. One goes from the pre-threshold field intensified current directly to a power arc via a spark. No corona appears. In region II there

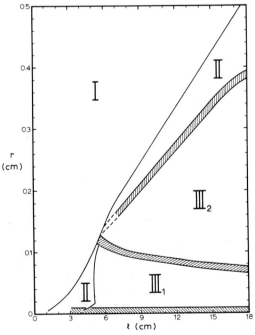

FIG. 4.52. A plot of radii against gap lengths showing region I, spark only; II, TP corona; III, glow corona regions, with III$_2$ representing +Δi and III$_1$ representing −Δi. Crosshatched regimes are transition regimes with no clear-cut results.

is a TP corona preceding either a spark or a glow as potential is raised. Regions III$_1$ and III$_2$ are both regions in which one goes from field intensified current to TP corona to a glow and then to a spark as potential is raised. Region III$_2$ is the region in which the negative point current decreases in transition from the TP; that is, Δi is $+$ in figure 4.51. III$_1$ is the region where the negative current Δi increases in going from the TP to glow; that is, Δi is $-$ in figure 4.51. Note that the crosshatched regions are regions of ill-defined transition.

Figure 4.53 gives comparative values of potentials for threshold of sparks and coronas with positive and negative points as a function of the product pl, pressure in millimeters Hg times centimeters of gap length. The point radius is not given, but from other data it would appear to be a 3 mm radius point. Note the fact that the curves for onset of burst pulse or pre-onset streamer corona and the TP corona cross each other three times in the range of pl given. The thresholds for the pulseless negative point corona are given. The sparking potential of the negative point lies below that for the positive point up to about pl = 650, where they cross.

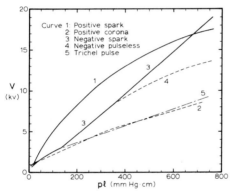

FIG. 4.53. Potential plotted against pressure times gap lengths for various aspects of positive and negative point corona. The solid curves 1 and 3 are sparking potentials for positive and negative corona. At high pl values the positive streamer corona has a threshold lower than the negative spark transition. Curve 2 is for the threshold of the burst pulse corona, while curve 5 is that for the negative TP corona. Note the frequent crossing of these curves as a function of pl. This follows from the influence of photoionizing free path and space charge removal in the gap on the positive corona, and the influence of space charge on choking of the TP corona as p and l increase. The threshold for the glow corona is shown by curve 4. Again it starts above about 9 kv.

Measurements were made of the formative time lag for the threshold of the positive point streamer corona. Here reproducible results were obtained. These results, displayed as time lag in microseconds against the relative voltage $\Delta = (\Delta V + V)/V_s$ applied at $\Delta < 1$ or applied at $\Delta > 1$, are shown in figure 4.54. At $\Delta < 1$, one is below V_S. With $\Delta > 1$, one has an overvoltage. ΔV is the actual step voltage. The data represent the transition from streamer pulse to a spark for a 1 cm gap with r = 0.7 mm at 100 mm pressure. The circles give the transition time from pre-onset streamers to a spark through the Hermstein glow corona region at 200 mm. The crosses give the time for transition from a pre-onset streamer pulse corona to a Hermstein glow corona at 300 mm. The transitions to sparks from the time of appearance of streamer pulses, either pre-breakdown or pre-onset, appear to be about the same and on the order of 1 μsec. The time decreases to about 0.3 μsec as Δ reaches the threshold for sparking. These findings are in keeping with Hudson's photomultiplier observations as well as with Menes and Fisher's data. The lag decreases when the number of abortive streamer pulses in the series required to cause the spark decreases. The greater time for the transition to the Hermstein glow and thence to the spark is to be expected. Below $\Delta = 1$ the streamer sparks pass through the Hermstein glow region, but the time lag of spark build-up follows that for the direct transition from a breakdown streamer.

A great deal of effort went into attempts to obtain time lags for the negative point breakdowns. The results could have been anticipated from what has earlier been said of the meaningless nature of negative point lags that are governed by conditioning of the point surface. They will be func-

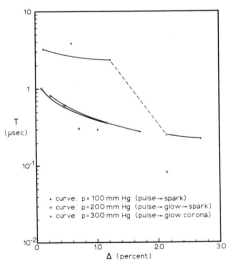

FIG. 4.54. This depicts the formative time lag for the transition burst pulse or Hermstein glow to a spark as a function of $(V - V_s)/V$ or $\Delta V/V$ in percentages at 100 and 200 mm pressure, respectively. Here the points are "conditioned" and the formative time lag is the time for a streamer to grow and close the gap. It decreases slowly with overvoltage. The transition from the burst to glow corona is slow at 300 mm at low overvoltages. It drops sharply when Δ is equal to 1% and then roughly follows the prolongation of the other curves.

tions of the point material, state of the surface, its past history, pressure, and so on. There is an enormous range of formative lags for the TP, the glow, and the spark breakdown. Statistical time lags seem to have been reduced by use of ultraviolet light, judging from the Laue plots. As an example of the sort of fluctuations that were observed, one may cite the following: The time lag for TP corona was determined as a function of gap lengths from 2 to 25 mm, for a W point in dry air at 760 mm with $r = 0.7$ mm, at an overvoltage for threshold of 6%. These lags ranged from 2 to 150 μsec at $l = 2$ mm. The fewer points seemed higher, lying between 30 and 500 μsec for $l = 10$ and $l = 15$ mm. At $l = 20$ mm, the thirteen points observed ranged from 2 to 100 μsec.

Similar data were obtained for the appearance of the pulseless corona where the step potential was added above the spark breakdown potential, that is, spark breakdown overvoltage Δ_s, which ranged from 0 to 12%. The lags appeared to decrease along a parabola, and the scattered points at each Δ_s were bracketed by two parallel curves that ranged from 11 to 800 μsec at $\Delta_s = 1\%$, and lay between 2 and 40 μsec at $\Delta_s = 12\%$. Two sets of curves were plotted for $r = 0.7$ mm and $l = 0.4$ and $l = 1.5$ cm, respectively, for the formative time lag of the TP corona against Δ_c, ranging from 1 to 12% for pressures of 760, 500, 300, and 100 mm of Hg. The scatter was so extreme that no orderly trends could be derived. There appeared to be a sort of parabolic decline as Δ_c increased, but the curve for $p = 753$ mm had the highest lag, with the 100 mm next, the 300 mm next, and the 500 mm the lowest for $l = 1.5$ cm. With a gap of $l = 4$ mm, the curve at 761 mm was highest, that at 500 mm next, that at 100 mm next, and that at 300 mm lowest. Possibly if averaged the shorter gap might have had the shorter lags, but this is not significant.

It is interesting to note that the values of T_C for the TP corona had the same wide spread at constant current as at constant overvoltage. Here again, perhaps more clearly, the decrease of T_C with increase in Δ_c was apparent. For the 4 mm gap with $r = 0.7$ mm, the total electron transit time across the gap was computed at various values of Δ_c in percentages and for pressures of 100, 300, 500, and 761 mm Hg. These varied only slightly with Δ_c and were 2×10^{-8} sec, 2.5×10^{-8} sec, 3×10^{-8} sec, and 3.2×10^{-8} sec, respectively. They were thus 0.1 to 0.01 of the formative lags. Ion transit times would be comparable with the minimum values of the observed lags.

The oscillograms, which are not readily available, are, however, very interesting. First is seen the rapid rise due to the current charging the point. This in the TP regime is followed after some delay by the appearance of the TP, which are spaced at regular intervals if the value of $V_C < V < V_G$ or V_S. If $V > V_S$ and one is in region II where TP precede the spark, then the TP, since they are highly overvolted, begin as soon as the point is charged and continue until the spark suddenly materializes and develops on a time scale of tenths of a microsecond. If one is in region III, then after some time the TP corona goes over to a pulseless corona and there is either an increase, $-\Delta i$, in current when the new mode comes on, or a $+\Delta i$. The pulseless regime may not last a long time if V is such that ΔV_S exceeds V_S appreciably. But the *formative times* of the transitions which occur at various intervals of TP corona are exceedingly short.

Finally, some data were obtained showing the influence of point material on currents as a function of frequency, that is, charge quantity per TP pulse for 1 cm gaps with r = 0.72 mm at 100 mm pressure. Here, for example, the Ag point gave more charge per pulse than did the Pt point by a factor of 3 to 2. This in terms of Amin's study would indicate that γ_i for the Ag was greater than that on Pt, prolonging the pulse. The Pt point pulses doubtless are dependent more on a γ_p. Note that the differences came to light at lower pressures. Generally speaking, it is very difficult to find reproducible and comparable data on the influence of point material in such studies.

It is possible now to make an attempt at physical interpretation in the light of these revealing studies. Probably the most significant data are as follows:

a. The visual appearance of the negative point spike in the glow phase projecting out into the gap just before spark-over.

b. The very short time of breakdown of the spark from either the TP phase or the glow phase, falling into the 0.1 μsec time scale.

c. The very nearly linear character of the V_S curve from the point V_C onward at any given r as l increases.

d. The fact that the TP curve as l increases at a given r in both the sparking region I and the glow region III lies along the same curve up to the critical gap lengths l_C and l_G.

e. The significant fact that the glow phase with its spike and negative streamer does not appear below 9 kv. This limits these glows to pressures above 200 mm, except for much longer gaps.

f. The fact that, as pressure decreases, l_C for any point seeks longer gaps.

g. The convergence of the V_S lines projected back to the axis to a point of $V \approx 1$ kv.

h. The $-\Delta i$ and $+\Delta i$ at TP to glow transitions.

i. The only suggestive formative time lag data on the negative point discharge are the minimum times observed, as longer times represent some sort of clean-up times. The formative lag of spark growth is again significant.

j. It should also be noted that the gaps used are much shorter than in standard geometry and that much of the data come from points of diameter around 2 mm with gaps rarely over 20 mm. Thus some of these gaps are nearer uniform field gaps. The influence of the other electrode, here the anode, is not to be neglected. High fields persist across the gap in most cases.

The least complicated observation to account for is the continuity of the V–l curve for sparking, that for the TP, and that for the glow corona at a given r as l increases in the proper relation to the point G for that r and pressure. It is clear that the TP curve marks the threshold of self-sustaining discharge. Below values of V appropriate to this, there is only field intensified current. For very short gaps relative to r, the only break-down is a spark. Thus, as observed and discussed elsewhere, if the gap is short enough the threshold is marked by a spark. If l is in the right region and above G, then glows will form if they can precede the spark in thresh-old, and the threshold will be a glow. For a cathode of a given metal conditioned by field intensified current or previous use to a given γ, the discharge threshold given by $\gamma e^{\int \alpha dx}$ will lie along a single curve and is that for some form of Townsend threshold at V_T. This curve will have the shape of the curves seen for any one of the different r values of figure 4.45. At short gaps V will be small and a spark will result, but as l in-creases relative to r, the gap length plays a smaller and smaller role in the threshold value. Hence the value of V_T increases rapidly with l at first and eventually flattens out where V_T is a threshold, perhaps the Townsend threshold for the point. Note that all these measurements were conducted in such a fashion that the hysteresis produced was lost because of the quite thorough conditioning of the points at the start or because of the method of evaluating V in measurement. Hysteresis effects appear in time lag studies but not in breakdown threshold measurements.

The next feature to explain is how, or why, V_T should lead to a spark for short gaps at smaller r, TP at longer gaps below r_G, and to glows at still longer l. The answer is associated with the fact that beyond l_G, as V is raised at one l, one has in succession TP, glow, and spark. To account for this, one must state that certainly at the higher pressures—perhaps even at the lower—the spark from its formative time lag, that is, break-down time of $\approx 10^{-7}$ sec, is presumably a streamer spark. Now a good deal is known about such sparks. The positive or anode streamer is favored over the negative or cathode streamer because of its charge conservative nature. The positive streamer in longer point-to-plane sparks is known to leave the plane anode and meet a negative point streamer within about one-third the gap length from the cathode point. The axial range of neg-ative streamers in air relative to positive streamers has been shown by Nasser to be less than one-quarter of the latter at the same point poten-tial. The starting potential for Nasser's streamers for 0.1 mm and 0.5 mm radius points with 10 cm gaps lies around 10 kv for the positive streamer and 13 kv for the negative streamer. For shorter gaps these values could well be around 9 kv. It is known that, when the point potential of a streamer tip falls to ≈ 1 kv in low gap fields, it ceases to advance. It is further known that the conducting streamer carries a rather large fraction of the point potential out into the gap. Near breakdown, if the tip potential of the positive streamer tip is on the order of 25 kv, it will lead to a spark as it approaches the low field region of the cathode. For a 3 cm gap near but below spark breakdown, the positive streamer tip has within 80% of the anode potential out to 2 cm from the anode.

In shorter gaps with more nearly uniform and stronger fields at the plane, the streamer tip potential need not be 25 kv to cause a spark. It is only when the field is low in the gap, as at the cathode of an 8 cm gap with a 1 mm diameter point at 60 kv, that the tip potential need be so high. This projection of the high point potential into the gap by a streamer results in a strong distortion of the field at the plane along its axis. This becomes stronger as the streamer nears the plane. Here the image force in the plane permits calculation of the fields in the absence of space charge. In air at all pressures the thresholds for TP pulse corona and/or pre-onset streamers and burst pulses is not much different, as is clearly seen in figure 4.53. Thus induced pre-onset streamers can begin at a plane anode when the field reaches the threshold for breakdown for the geometry of a point, giving the same field configuration as induced in the plane by the point, the negative streamer from the point, or the negative ion space charge of the anode plane. Where ion space charges do not interfere, these streamers can cross the gap and lead to a spark.

In the negative point-to-plane geometry here used there is no reduction in the field in midgap relative to that at the plane surface as affecting an induced positive pre-onset type of streamer. Positive streamers leaving the anode plane progress into a continually increasing field as they approach the cathode point. As Waidmann has shown, they increase in speed as they near the cathode in increasing fields, because of the distortion of their tip potential, even in the case of positive point to plane. Increase in speed is proportional to the increased tip potential of the streamer. Thus in the high field of the point cathode, the streamer from the anode plane acquires such vigor that a return stroke and spark are assured if the gap is short. It is therefore possible that the anode spark yielding streamer may initiate at a *pre-onset potential at the plane anode* and reach its spark, producing 25 kv tip potential as it reaches the cathode point.

If now the gap is short relative to the cathode radius (this ratio of r/l for the appearance of TP rises rapidly from about .09 for a .17 mm radius gradually to approach 0.32 at 760 mm pressure), the field is more uniform across the gap and fields are high. A Townsend type breakdown arises at the cathode, which with inherent instability rapidly grows, leading to the Morton regime. Negative ions form but are swept out. The breakdown at the point, throwing negative ions and electrons across the relatively short gap, enhances the anode field, and a pre-onset type of streamer corona starts from the anode, which accelerates and leads to spark breakdown. Here the short gap and field conditions militate against too much negative ion accumulation *near the point,* and these ions by their space charge near the anode plane favor streamer formation. Thus the streamer spark rises as soon as threshold for self-sustaining discharge is reached. The negative ions do not choke the discharge, and there are no TP. The short gap with negative point and little space charge resistance is unstable.

Once the gap increases beyond l_C for the given r, then the gap is too long to clear the negative ion space charge near the cathode before it chokes off the breakdown at the cathode, and the TP corona intervenes. If in this region, however, the potential is raised, the TP current increases, and a greater negative ion density occurs at the anode because of the increased i of the TP corona. As will be seen, eventually a streamer from the anode plane can again cross the gap and a spark ensues. This action will later be justified in more detail.

If now two conditions are met—that r exceeds a value r_G at which V_G at a given l_G is above ≈ 10 kv—negative streamers can start from the cathode point. These streamers project out into the gap a considerable distance. The negative ion formation at the point is no longer adequate to quench the discharge and a pulseless corona current arises through a very rapid succession of negative streamers, or better by a streamer-like discharge triggered by a strong γ_i and avalanches that propagate outward by photoionization in the gas. These streamers may at onset consist of a continuous succession of γ_i triggered avalanches that advance by photoionization in the gas rather than by regenerative cathode γ mechanisms close to the cathode, as in TP corona. They are initially not individually of such size as to approach negative breakdown streamers in magnitude and so give no intermittence such as the fully developed cathode streamers show. But they are avalanches of a form initiating the cathode streamers. As potential increases, these advance further and further into the gap and enhance the anode field until the anode streamer spark is launched. It is possible that just before spark breakdown the cathode streamer pulses could be detected by fluctuations on the oscilloscope against the background of the steady glow corona current. When the fluctuation occurs, and/or the quiescent streamers from the cathode sufficiently shorten the gap, the anode streamer spark occurs.

It is of interest to note that for small points and shorter gaps, where the TP changes to the glow, the negative current abruptly increases. With space charge no longer choking and with a short gap, spark transition will follow and the cathode-like streamer pulses increase current by alterations in gap field, as seen at the ends of the curves of figure 4.50. If the gap is long the transition to pulseless corona instantaneously increases the current *near the cathode point*, but this increased current piles up the space charge near the anode, which at this juncture cannot produce an anode streamer. A small current decrease thus follows, so that higher potentials are needed to remove the space charges near the anode and increase cathode streamers. This gives the ultimate current decrease $+\Delta i$. However, not far above this the increase in current and potential lead to a spark.

It will be noted that generally the glow region is relatively narrow in terms of potential compared to the TP region for the shorter gaps, where $-\Delta i$ is observed. Where the region is more extensive a $+\Delta i$ occurs. Thus the reason for the existence of such a discharge and a breakdown by this means stems from the fact that, when gaps become too long and radii of such value that above some 10 kv the negative streamer can materialize at the cathode, the glow which represents the growth of the incipient negative streamer appears. Once the potential launches a stronger negative streamer from the cathode, the anode streamer is triggered by image forces and a spark ensues. Below this potential and point radius, if l is adequate, that is, above l_C, the TP pulse will burn until the potential and space charge fields enable the anode streamer to start.

There remains one regularity which is so far unaccounted for. That is the proportionality between V_S and l once the threshold value of l_C of the TP corona is exceeded. This means that, once the negative ion space charge chokes the autocatalytic rise of current at the cathode, thus producing TP corona, lengthening the gap increases V_S in direct proportion. The explanation becomes obvious: the potential drop across the ionizing zone, as long as TP corona operates, has more or less constant value V_C, and the remainder of the potential drop across the gap is taken up in moving the space charge across the gap. Hence the consequences of this must be considered.

The best approach to an explanation is to consider the Townsend theory for the space charge in a coaxial cylindrical gap. The author has contended that this theory, except for modifications suggested by geometrical changes which alter the constants, can in principle be applied to point-to-plane-like gaps as well. The Townsend theory is derived from the constancy of the ion space charge density created by the current as the charge crosses the gap. This follows the fact that the volume of the charge increases with r across the gap, but that the ion drift velocity is proportional to the fields in the low field regions and these decrease in the same proportion as the volume increases. The Townsend equation (eq. 2.29) for a space charge controlled corona says that $V - V_C = \dfrac{piA^2}{2KV} \log A/a$. Here V is the applied point potential and V_C is the TP corona threshold potential. The quantity

$V/A \log A/a = X_A$, the field at the anode cylinder, a centimeter distant from the wire of radius r, and $K(X_A/p)$ is v_A, the drift velocity of the positive ions in transit at the anode surface at a pressure p. Since A is equivalent to l, this makes

(4.15)
$$V - V_C = \frac{iA}{2v_A} = \frac{il}{2v_A}.$$

Now $il/2v_A$ is in principle the density of ions per cubic centimeter across the gap, for the charge density of positive ions is constant across these gaps and A is the gap length l. It is seen that, when this density reaches some critical value, because of increase of i with the applied potential, a spark occurs by anode plane streamer. In consequence, if there is a limiting $(i/v_A)_s$ value at which sparks occur, the spark will occur when this value is reached.

Now observationally the curves for i as a $f(V)$ at higher V are nearly proportional to V^2, while v_a varies linearly with V. Thus i/v_a will increase with V and, when the density of positive space charge at the anode is such as to launch a pre-onset type anode streamer, the spark can follow.

Thus from equation 4.15, $V_S - V_C = \frac{1}{2}\left(\frac{i}{v_A}\right)_s$ which defines the spark

threshold. This states that as long as i/v_A is independent of l, $V_S - V_C$ will vary linearly with l. Thus $\frac{V_S - V_C}{l} = \frac{1}{2}\left(\frac{i}{v_A}\right)_s.$ If further V_C in the region above l_C also varies approximately linearly with l, that is, if $V_C = Bl$, then

(4.16)
$$\frac{V_S}{l} - \frac{V_C}{l} = \frac{V_S}{l} - B = \frac{1}{2}\left(\frac{i}{v_A}\right)_s,$$

such that
$$\frac{V_S}{l} = \frac{1}{2}\left(\frac{i}{v_A}\right)_s + B,$$

and V_S is also proportional to l, as noted from the curves of figure 4.45.

The curves of figures 4.47 and 4.48 allow one to evaluate the slopes of the sundry curves for V_S against l in centimeters and for V_C against p at each of the pressures from 10 to 700 mm Hg pressure. It turns out that V_S/l is remarkably closely proportional to pressure with a value 22.2 volts/cm per mm gap length. Likewise, while the V_C–l curves are not exactly linear beyond V_C, their slopes are very small and are proportional to pressure at about 3.15 volts/cm gap length per mm pressure. Thus within the accuracy of the data, the quantity $1/2(i/v_A)_s + B = 22.2$ volts/cm l per mm and $B = 3.15$ volts/cm l per mm, so that to a sufficient precision $1/2(i/v_A)_s = 19$ volts/cm l per mm Hg. This represents a sort of average X/p equivalent if the average field across the gap were the field uniform. As the field is not uniform, it is only an analogy. It is nonetheless a condition derived directly from observed data and as such represents an experimental fact, irrespective of the naïve representation. In consequence, at any pressure p, by multiplying 19 volts/cm by p in mm, one has the value of the $(i/2v_A)$ at that pressure required for a spark. If this is further multiplied by l, the term gives the potential equivalent to $il/2v_A$ due to the space charge accumulated in the gap to cause the streamer spark. If one could translate the equivalence of the space charge at the anode surface to the field strength, he could determine the starting potential at the anode to cause the streamer. By adding B at the appropriate pressure to $i/2v_A$ and multiplying both by l for the gap, one has the value of V_S beyond l_C.

It is seen that, while the important quantities needed cannot at this time be derived, analysis of the data in terms of the analogy of Townsend's relation to the problem has led to unexpected and remarkably interesting relations between the variables and a sort of intuitive interpretation of the significance in relation to the narrow discharge regimes. The fact that the data used embraced the region where the glow and cathode streamers begin and still holds as valid as for TP corona alone makes one suspect that the space charge in transit may even play a role here, though the spark may involve the cathode streamer as well. It is possible that its propagation as well as that of the anode streamer may depend on the space charge field, which depends on the space charge density at all times.

D. THE ELECTRICAL WIND

It is perhaps of importance to say a word or two about the electrical wind, which is a characteristic feature of the asymmetrical field corona discharge. Since ions of one sign are in principle created and repelled from the highly stressed electrode, they gain momentum from the electrical field and in consequence of the resistance to their motion imposed by the gas molecules this momentum is passed on to the gas. This creates a pressure difference in any section of the gas and produces a current of gas in motion, or an electrical wind. This wind has been known since the earliest electrostatic studies in the late 1600's. A sharp point yielding a current in the hundreds of microamperes from a good static generator will blow out a small candle flame at 1 cm distance. The wind leads also to the electrical pinwheel or whirl, in which a set of radial spokes on a movable shaft having sharp points fixed at right angles pointing, say, in the clockwise direction, will on electrification cause counterclockwise rotation.

While there are some misconceptions about the rotational mechanism, regarding the phenomena as being caused by the reaction to the creation of the jet of air by the discharge, a careful analysis in terms of the action and reaction of the system indicates that the motion results from the difference in electrostatic attractions, that is, the fields between the stationary reference system and the pinwheel in virtue of the screening of the charged point by the space charge in the gap.[1]

The gas velocities resulting from the wind are small compared to the ionic drift velocities and are of no great consequence in the breakdown mechanism. They do lead to circulation of the gases. Perhaps the only excuse for a detailed discussion of the phenomena, other than giving the basic equations, lies in the reported and ill-advised attempts at utilizing the phenomenon for levitation aircraft, for reducing skin friction of aircraft in flight, and similar devices. These have cost the investors, some speculators, and, sad to say, the government unnecessarily large expenditures in research funds because of the enthusiasm of inventors who "rediscovered" the classical phenomena.

Consider a point-to-plane corona with the plane a gauze permitting the gas flow. Choose two planes normal to the electrical field X and thus to the flow separated a distance d. The n ions per centimeter[3] of charge are in motion in the field. From the third law of motion it is clear that the force exerted on the ions by the field F_X must just balance the force F_W exerted on the gas molecules creating the wind. Thus the force on the ions between the two planes

$$(4.17) \qquad F_X = AXend = -F_W.$$

Here A is the area of the planes normal to the field enclosing all the ions traversing the planes, e is the ionic charge, n the ion density, and X the electrical field.

$$(4.18) \qquad F_X/A = Xend = -F_{W/A} = \Delta p,$$

where Δp is the pressure difference across d. Since the drift velocity of the ions is v = kX with k the mobility, X = v/k,

$$(4.19) \qquad \Delta p = \frac{vend}{k}.$$

The total current measured is i = Aven with ven = i/A. Thus,

$$(4.20) \qquad \Delta p = \frac{i}{Ak} d$$

across the distance d. More formally, this can be expressed by calling x the direction of flow, and y and z the coördinates normal to the flow along the planes. Then

$$(4.21) \qquad F_X = \int\int\int Xendydzdx = \frac{i}{k}\int\int\int vendxdydz.$$

[1] This is discussed in the author's *Fundamentals of Electricity and Magnetism*, 3d ed. (John Wiley, New York, 1947), p. 193.

The current

(4.22)
$$i = \int\int vnedydz$$

such that

(4.23)
$$F_X = \frac{i}{k}\int_a^b dx = \frac{i}{k}(x_b - x_a),$$

where x_a and x_b are the coördinates of the planes between which the force is exerted. Now $F_X/A = P_b - P_a$, where P_b and P_a are the pressures producing the wind at x_b and x_a. Whence again

(4.24)
$$P_b - P_a = \frac{i}{Ak}(x_b - x_a), \quad \text{or} \quad \frac{\Delta P}{\Delta x} = \frac{i}{Ak}.$$

An estimate can be made of the value of the pressure differences experienced across a distance of 1 cm between planes if the area $A = 1$ cm², as in a tube that confines the flow of gas and current to this area. Let the current $i = 100$ μA or 3×10^5 esu. The value of k is 600 cm²/esu sec for ions in air. Hence Δp for 1 cm² is 500 dynes/cm². This represents a pressure of 0.38 mm Hg. Using an inclined oil manometer with a point in a cylindrical tube and a ring collector, Chattock and later Chattock and Tyndall[4.19] tested the law by measuring k from $\Delta p/\Delta x$, correcting for the effect of pressure borne by the ring by measuring Δp for two values of Δx and thus getting $\Delta p/\Delta x$. Later Ratner[4.20] used a negative point and a gauze plane, measuring the pressure with a moving vane suspended from a calibrated torsion head. He discovered that, as pressures in the gas were lowered so that electrons remained free for most of the path, Δp fell and consequently the average k increased. Free electron gases should give Δp on the order of 0.0005 that for ions in air.

To estimate the sort of gas velocities which can be achieved, one may assume that the work of moving the ions through the gas goes almost entirely to kinetic energy of motion with little loss to heating the gas. Consider the work done on the gas mass of area A by the pressure difference Δp in moving the column of gas of length S the distance S cm down the tube.

(4.25)
$$\Delta p A S = FS = 1/2Mv^2.$$

But the mass M is M $= \rho AS$, where ρ is the density of the gas. Accordingly,

(4.26)
$$\Delta p = 1/2\rho v^2.$$

Thus under the conditions assumed above, Δp is 5×10^{-4} at atmospheric pressure, so that $v^2 = \dfrac{2 \times 5 \times 10^{-4}}{1.3 \times 10^{-3}} = 0.78$ and $v = 0.88$ cm/sec. In the case of the candle flame, as one approaches the point, the area A over which the constant current flows varies as r^2 with r the distance from the point. Hence at 3 mm from the point the value of Δp is nine times as great as at 9 mm and the velocity is roughly 8 cm/sec.

We can now consider the problem of a typical proposed electrical wind levitator. Assume a balsa wood frame 10 cm by 10 cm on a side. Assume ten wires 10 cm long arranged in parallel and attached by a light copper wire to a source of potential giving some 25 kv and the current needed. Assume that the wires have downward deflecting baffles above them such that the total air current is downward. If the wire radii are 1.5×10^{-2} cm, the starting potential V_s is some 5 kv, and if breakdown to a positive streamer spark to ground does not occur below 25 kv, the current given per centimeter of length of each wire at that potential is around 10^{-6} amp. The ten wires 10 cm long give a total current of 10^{-4} amp. Then

$$(4.27) \qquad\qquad \Delta p = \frac{i}{Ak} X \Delta x$$

with Δx taken as 10 cm. If one converts to esu for the area A of 100 cm², the value of Δp is 50 dynes/cm². The total lifting force is then 5 grams. If the frame is made of balsa wood pieces 0.5 cm $\times$ 0.5 cm, the weight would be between 1.5 and 2 grams. With baffles and connecting wire it could well be less than 3 grams, and it would levitate.

The velocity of the air mass set in motion would be $v = \sqrt{2p/\rho} = \sqrt{2 \times 50/1.3 \times 10^{-3}} = 270$ cm/sec. This velocity is perhaps somewhat high, but it is what the value of Δp ideally would give. Accurate details of one such levitator which was actually demonstrated have not been given. All that is known is that the frame had an area of 500 cm², and the velocity of the air was about 60 cm/sec. The power in the exemplary calculation was put in 20 kv $\times$ 10^{-4} amp = 2 watts. The grams lifted per watt of power expended were approximately 2.5 g/watt. This is equivalent to about 4 pounds lift per horsepower. The modern light airplane with a 50 hp motor weighs between 1500 and 2000 lb. Thus the airplane actually flies and does so with a lift of 40 pounds per horsepower; that is, it has 10 times the lift of the levitator.

The comparison is strengthened when it is realized that the model exhibited and calculated was near the most efficient range. If power is to be increased, then the only way to do it is to increase the current i. Since the current per unit length of wire cannot much exceed the value given, the only way to increase lift is to increase the length of wires by increasing the area of the frame, because crowding wires too closely limits current flow and efficiency. Thus the lift increases with area of the frame, that is, as the square of the average linear dimension of the frame. But the mass of the frame will increase as the cube of the linear dimension of the frame. Hence by increasing the average linear dimension of the plane by fourfold, the lift would be increased to 80 grams, but the weight would increase to 96 grams even if a rigid frame could be constructed of 0.5 × 0.5 cm balsa rods.

Thus the device would no longer levitate. It should also be pointed out that the generator, electrostatic or otherwise, that gave the 10^{-4} amp at 25 kv was not supported by the frame but reposed on the ground. The mass of such a generator, and especially as power needs increase, if added to the weight of the frame, could not be airborne. From an engineering viewpoint, the efficiency of converting heat to mechanical energy and air thrust directly is so much greater than that of the inefficient conversion of mechanical energy to static electrical energy that the procedure might a priori have been seen to be futile. Similar considerations will be found to apply generally to attempts to utilize electrical wind for reducing surface drag and similar effects.

The law for the corona current, that i is proportional to $V(V - V_g)$, holds in still air. In connection with the study of potential gradients in thunderstorms, Chapman[4.21] has shown that in a transverse wind of velocity v the current is much increased at constant voltage as space charges are removed. Under these conditions the current varies more as $(V - V_s)$ and with the wind velocity w. As was stated in section A.6, a survey of these equations with engineering applications was recently made by J. H. Simpson and A. R. Morse.

REFERENCES

4.1   G. L. Weissler and M. Schindler, J. Appl. Phys. *23*, 844 (1952).

4.2   L. B. Loeb, Phys. Rev. *86*, 256 (1952).

4.3   A. N. Greenwood, Phys. Rev. *88*, 91 (1952).

4.4   L. B. Loeb, Phys. Rev. *71*, 712 (1947).

4.5   G. L. Weissler, J. Appl. Phys. *24*, 472 (1953).

4.6   A. N. Greenwood, Nature *168*, 41 (1951).

4.7   L. B. Loeb, Phys. Rev. *76*, 255 (1949).

4.8   C. G. Miller and L. B. Loeb, J. Appl. Phys. *22*, 614 (1951).

4.9   W. N. English and L. B. Loeb, J. Appl. Phys. *20*, 707 (1949).

4.10 W. N. English, Phys. Rev. *77*, 850 (1950).

4.11 M. A. Harrison and R. Geballe, Phys. Rev. *84*, 1072 (1951).

4.12 L. B. Loeb, *Basic Processes in Gaseous Electronics* (University of California Press, Berkeley and Los Angeles, 1955), pp. 399 ff., 416 ff.

4.13 L. B. Loeb, Phys. Rev. *86*, 256 (1952).

4.14 M. R. Amin, J. Appl. Phys. *25*, 627 (1954).

4.15 R. Guck, *Die Negative Koronaentladung in der Spitze-Platte-Funkenstrecke*, ed. by H. Lesch (C. F. Miller Press, Karlsruhe, 1955).

4.16 G. L. Weissler, Phys. Rev. *63*, 101 (1943).

4.17 L. Colli, Phys. Rev. *95*, 892 (1954).

4.18 I. S. Stekolnikov and A. V. Shkilev, Dokl. Akad. Nauk SSSR *145*, 781 (1962).

4.19 A. P. Chattock, Phil. Mag. *48*, 401 (1899); *1*, 79 (1901). A. P. Chattock and A. M. Tyndall, Phil. Mag. *19*, 543 (1909).

4.20 S. Ratner, Phil. Mag. *32*, 442 (1916).

4.21 S. Chapman, *Recent Advances in Atmospheric Electricity*, ed. by L. G. Smith (Proceedings of the Second Conference on Atmospheric Electricity, Pergamon Press, New York, 1958), pp. 277 ff.

5

Asymmetrical Gaps with Both Electrodes Participating; Pseudo Coronas; Coaxial Cylindrical Geometry; Highly Stressed Anode

A. FREE ELECTRON GASES

1. INTRODUCTION

The first basic study of the breakdown in coaxial cylindrical geometry with free electron gases was carried out in Townsend's laboratory in 1914.[2.23] The current-potential curves at higher pressures led to the relation between these two variables given in chapter 2. It was so accurately obeyed that Townsend, Boulind[2.14] and Huxley[5.1] extended it to lower pressures to evaluate positive ion mobilities at higher X/p. Unfortunately in that region one of the basic assumptions underlying the theory—that the ionizing zone is small compared to the gap length—no longer applies and ion mobilities inferred were erroneously high at the lower pressures. The starting potential for such coronas with highly stressed anode was ascribed to the Townsend condition (eq. 2.9) $\gamma e^{\int \alpha dx} = 1$, since in this geometry the encircling cathode cylinder is an efficient source of secondary electrons yielding a γ. However, for such tubes Townsend also considered ionization by positive ions in the gas as a factor. The starting potential thresholds for positive and negative coronas were determined by Boulind,

408

Huxley, and Bruce for the He, Ne, O_2, N_2, and H_2. None of these studies made from 1928 to 1930 were in really pure gases as judged by modern standards, and the data were shown to be in error by Penning,[5.2] who did use quite pure inert gases in 1931. In the 1931 studies the highly stressed positive electrode breakdown threshold was invariably higher than that for the negative, whereas the curves of the Townsend co-workers crossed each other at low pressures. These peculiarities were shown by Penning[5.2] to be caused by the Penning effect or its modification by unsuspected amounts of trace impurities.

Interest in a study of these coronas was stimulated around 1928 in consequence of the counting properties of such tubes for ionizing events initiated by Geiger and Mueller. These studies were largely empirical and invariably made in connection with some sort of a counting circuit with triggering by ionizing radiations. They led from the high series resistance controlled counter of Geiger and Mueller (slow counters) to internal ionic space charge quenched fast counters using inert gas-organic quench gas mixtures. In the process of development and study of the latter counters, Greiner[2.27] and others showed the role of photoelectric ionization of the gas to be a vital factor in the operation of such counters. This concept was developing almost contemporaneously with Trichel and Kip's studies of streamers and burst pulses. A great step forward in the study of the basic behavior of the coaxial system tubes came when Lauer,[2.21] stimulated by a study of Colli and Facchini in Italy,[3.71] conceived of the device of triggering a sequence of ionizing events, open to oscilloscopic study, by shooting α particles parallel to the axis very close to the outside cathode cylinder. These gave a sufficient number of virtually simultaneous electron avalanches to yield a good oscilloscope signal. The very sharp, clearly defined pulses permitted transit times of electrons and ions to be measured. They permitted evaluation of amplification and led to estimates of γ_p and γ_i; this study was made by Lauer in the author's laboratory following more conventional oscilloscopic studies of the author and Miller[1.9] in pure N_2 and air, for both positive and negative wire.

In what follows it will be best to present Miller's[1.9] results in N_2 first and then extend these to the analyses of Lauer[2.21] and Huber.[2.22] After these studies have been described and the negative wire studies have been presented, it will be possible to consider the lower pressure studies and the related matter of relative starting potentials.

2. MILLER'S STUDIES IN PURE N_2

In this study fairly modern techniques were used except that the Alpert valves had not been perfected so that greased stopcocks were still used.

Miller used a Ni collecting cylinder of 28.5 mm diameter and 5.55 cm length accurately placed coaxially with cylindrical Ni guard cylinders of the same diameters at each end spaced 1 mm distant. The central electrode had two viewing slots 1 × 45 mm milled into it at 90 deg angular separation for viewing the axial wire electrode. The tube was mounted with axis vertical. The slits could be viewed with a pair of telemicroscopes. There was, in addition, a hole 1 cm in diameter screened with 60-mesh Ni screen through which triggering ultraviolet light could be focused on the cylinder or wire through a side arm quartz tube.

By means of two W filaments inside the glass envelope but outside the Ni cylinders, the cylinders could be raised to red heat by electron bombardment for outgassing purposes. The W vapor was also thought to act as a getter in removing the last traces of O_2 from N_2. The central Pt filament had a diameter of 0.0174 cm. It was mounted accurately coaxial with the cylinder and kept from sagging under heat by the tension of a spring at one end. It was outgassed by running a heating current through it. At one side of the tube, there was a side arm connected to the top and bottom ends of the glass envelope of 1 cm diameter to permit circulating the gases in the tube to facilitate mixing and removing or homogenizing the impurities created in discharge. The glass envelope was so arranged that it could be baked out at 425° C in six to twelve hours while the metal parts were glowing, so that pressures under the pump were better than 10^{-5} mm while hot.

The tank gas, for example, N_2 was purified by passing in sequence over tubes of NaOH, hot CuO, and hot Cu, and then through three liquid air traps in succession at less than 100 cm³ flow per minute.

Power was supplied by two 5000 volt reversible polarity, well regulated, dc power supplies. Corona currents were measured by a Dolazalek electrometer using a set of internally consistent but not absolutely calibrated S. S. White resistors ranging from 10^8 to 2×10^{12} ohms. The electrometer had its highest stable sensitivity at 3150 mm per volt at 1.25 m distance from the scale. Higher currents were measured by shunting the electrometer with suitable resistances or by microammeter. A separate microammeter measured currents in the guard ring cylinders to evaluate spread of discharge along the wire. An oscilloscope pickup resistor ranging in value from 100 to 0.5 megohms going from cylinder to ground was used for registering the current pulses. The applied high potential was measured with an L and N, type K, potentiometer using a 1000 ohm wire-wound resistor in series with 12 megohm wire-wound resistors. The Pt central wire was protected by 5 and 10 mA Littel fuses rated to open in 0.1 msec. This was very important for the negative wire studies. Miller's tube is shown in figure 5.1.

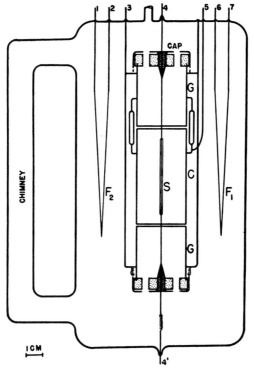

Fig. 5.1. Miller's coaxial cylindrical tube arrangement for study of coronas in pure gases. Notice electrode outgassing filaments F_1 and F_2. Also the recirculating chimney.

Using pure N_2, the current-potential curves observed are shown in figure 5.2 for pressures ranging from 27 to 617 mm. Here a reasonably large volume of gas is ionized by cosmic radiation or other ultraviolet light if needed. In consequence, currents increase steeply with potential from field intensified currents of 10^{-12} amp to a critical value lying between 10^{-9} and 10^{-8} amp. At these critical current values, which are mentioned as being more reproducible than the potentials causing them, there is an even more abrupt increase in current to the order of about 10^{-5} amp, after which it increases relatively slowly. The rate of rise of the field intensified current appears steeper at lower pressure on this scale of plotting. However, at 27 mm the rise takes place in some 300 volts at 1000 volts, and at 617 mm it occurs in some 1200 volts out of 4000. Thus the rate of increase in current in relation to the percentage increase in potential is not very different.

It is to be noted in the figure that there are data points marked "C" with additional branch lines that extend to the smooth curves at the top

of the rise. These side branch "C" curves are obtained by reducing potential after the current jump is established and continuing the reduction to
an offset potential. The curves thus represent a hysteresis phenomenon
that is particularly pronounced at the higher pressures but does not appear
at 27 and 50 mm. The presence of such hysteresis in running the curves
clearly indicates the influence of the outer cathode cylinder in the breakdown, for it always represents a cathode conditioning in the breakdown.
This is absent in the studies with positive point and distant plane.

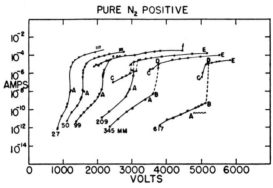

FIG. 5.2. Miller's current-potential curves for highly stressed anode in pure N_2
at several pressures. Note the steep current rise at points marked A, the more
abrupt transitions at points marked B. At points DC note the lack of reproducibility of the curves on decreasing potential after the first current jump. This is a
typical hysteresis curve, indicating a conditioning of the cathode by a low order
pre-discharge. The vertical lines above the curves at 50 and 27 mm pressure indicate breakdown streamers preceding the spark.

Below the large current jump the oscilloscope showed a steady current
without pulses. As soon as the jump occurred the oscilloscope showed
very irregular current oscillations on a millisecond time scale. These were
not burst pulses or streamer pulses. They were much too slow and too
large. The course of the oscillations at once became clear on visual observation. Below the jump no luminosity was seen on the wire. Coincident
with the jump to 10 μA there appeared multiple bright spots on the anode
wire, as sketched in figure 5.3. These spots were in continual motion. Increasing potential multiplied the number of spots, decreased the fluctuations, and smoothly increased the average current about which the
fluctuations occurred. As the current decreased below 10 μA, the current
became very unsteady and the spots extinguished one by one except for
one luminous spot at a current of $\approx$2–5 μA. Further decrease in potential
extinguished the one spot, and the discharge current decreased by a factor
of 5 or more. At 617 mm the current at 5050 volts was 10 μA with multiple

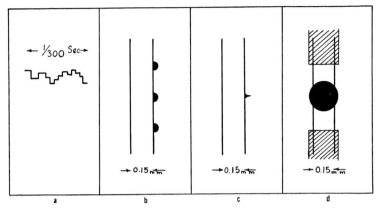

Fig. 5.3. Miller's oscilloscope pattern after the current jump to the left in a, caused by multiple spots. (b) Sketches of the wire with moving luminous spots observed with current jump and oscilloscopic pattern. (c) The single spot just before offset at low current at high pressure. (d) The same as c but at 99 mm; note the velvety glow, indicated by the shaded portion, separated from the spot by a dark space.

spots. A single spot appeared at 8.5 μA and persisted until the current was reduced to 0.5 μA at 5040 volts with 0.1 μA to the guard rings. Below this the spot went out and the current dropped to 10^{-10} amp with a very slight ripple on the oscilloscope. If the potential was raised immediately, the spot reappeared at a slightly lower potential than before. It usually required the same overshoot to 5140 volts that it did the first time. With one spot near 0.5 μA the current was steady, indicating a glow discharge concentrated at one spot.

The length of the guard rings relative to the cylinder was 0.85. Above 10 μA the current was nearly evenly divided between guard ring and central cylinder. This meant that the one spot seen at 0.5 μA carried most of the current, but with multiple spots owing to self-repulsion they were spread evenly along the whole wire and into the guard rings as well. The spots were akin to the localized concentrated glow discharge observed by Weissler[1.8] with a point.

The sequence was the same at lower pressures except that overshoot was absent at 27 and 50 mm when potential was raised *very slowly*. It is possible that, had potential been raised slowly at 99 mm, there would have been no overshoot. At 99 mm and 7 μA the whole wire showed a uniform glow instead of a spot. At 15 μA a single spot appeared superposed on the glow, although there was a dark space between spot and glow, as seen in figure 5.2. At 50 mm and below there were no spots to 100 μA. Both 27 mm and 50 mm curves show characteristic breakdown streamers before filamentary spark breakdowns at 10^{-4} amp.

It is clear that with the coaxial cylinders, in contrast to the point-to-plane geometry, the phenomena are in no way related to the regular positive wire coronas and to the Geiger counter action associated with photoionization in the gas. This change must be ascribed to the strong influence of the surrounding cathode cylinder, which conserves photons and yields a γ_p which enters into the increase of the field intensified currents. These, in turn, bombard the cathode surface with positive ions leading to a local clean-up in N_2. This clean-up creates spots of high γ_i which in turn lead to localized filamentary glow discharges between spots on the cathode and on the anode wire opposite. The glow on the anode wire is the familiar anode spot-like glow seen in lower pressure glow discharge tubes. The discharge in this case is localized at the cathode and leads to a localized spot at the anode. This comes from the rectilinear motion of the γ_i, producing *ions* in the fields at higher pressures. The spot-like constrictions of the discharges at cathode and anode arise from field configurations between the distorted breakdown region and that of no breakdown, as seen in chapter 2. As pressures decrease, the diffusion of ions, γ liberated electrons, and photons over the whole cylinder surface leads to the general glow discharges. That this interpretation is correct was dramatically proven by Lauer with his α particle triggered ionization sequences in H_2.

It is clear from all the observations, including current distribution, that there is a relatively smooth transition from a field intensified current to a low-order, self-sustaining Townsend discharge at potentials well below the large current jump. The transition appears to result from a clean-up of the cathode by this current, which changes the feebler γ_p determined discharge in N_2 to a very much more intensive γ_i conditioned one with localization at the surface. The accompanying hysteresis in the current potential curves results from this gradual transition and makes onset differ from offset potential, especially where the discharge is confined by lack of diffusion at the higher pressures. The conditioning does not last long after current interruption in the N_2 used, for the high value of γ_i appears to depend on positive ion impact cleansing of the surface. This peculiar temporary character of the conditioning is also attested by the wandering of the spots over the surface. This comes from overloading it at one point, which somehow deactivates the surface and leads the spot to seek out a freshly cleaned surface.

The spots, as indicated by the author, repel each other as well as repelling the photoelectric conditioned glow when it is contemporary with the spots, as at 99 mm pressure. Obviously single spots are confined to small areas by the nature of the equipotential surfaces, as in point-to-plane discharge, and can carry only a certain amount of current. Increasing potential overloads the single spots by excessive ion bombardment and leads to wandering, and the extra current is achieved by new spots. The wandering, deactivation of spots by excessive bombardment, spot extinction, and the starting of new spots produce the fluctuating current pattern. Apparently a single spot reduced to nearly its extinction potential and with low current density can settle down at one point and burn quietly with constant current.

It was repeatedly observed that the current was a much more reliable criterion for the transition than was the applied potential, indicating that clean-up is conditioned by current density.

As in all glow discharge and Townsend discharge phenomena, near threshold the potential fall across the ionizing zone next to the wire remains sensibly constant as current increases. The same applies to its depth from the wire. All that increase in potential does below threshold is to *increase current density* or the number of spots along the wire above threshold, the increase of potential being taken up by increased potential fall across the positive ion space charge, as indicated by the author[2.17] and proven by Colli, Facchini, Gatti, and Persona.[2.17]

If there are no spots by which the area of the discharge can be known, then, as at low pressures or in the pre-spot regime at higher pressures, the Townsend current potential (eq. 2.30) appears in the form

$$V(V - V_s) = \frac{piR^2}{2K_1} \log R/r,$$

where V_s is the threshold potential for the *threshold of self-sustaining discharge, not of onset of spots*; p is the pressure; i, the current per unit length of cylinder; R, the radius of the outer cylinder; and r, the wire radius; with k_i the mobility of the positive ions. The mobility calculated by Miller at 50, 27, and 12 mm pressure reduced to standard conditions was close to 3 cm²/volt sec. The ion in N_2 at low fields is predominately the N_4^+ ion, and its low field mobility as given by Varney is around 2.5. Within limits of error, which arise from the exact estimate of the threshold potential, the mobility calculated from theory gives results that are in good agreement with the accepted value of 2.5. An overestimate of V_s, by choosing the onset with spots instead of the true threshold, would make k_i appear too large. Miller used the beginning of the large potential jump at the low pressures, which is undoubtedly too high, as the conditioning Townsend discharge sets in below this, so that his values of k_i are high.

To show how sensitive to impurity this N_2 was, it suffices to state that on occasion one of the W filaments that was not outgassed was flashed by accident. This released enough adsorbed gas to alter the phenomena. The *negative* corona thresholds and current values were not affected. The positive corona current started from a fairly high background, but there was no overshoot or hysteresis. With an abrupt increase of current a Geiger counter-like burst pulse discharge set in. The glow adhered closely to the anode wire. This phenomenon was caused by a discharge initiated by photoelectric action in the gas through trace contamination. It persisted up to 20 μA at 34.5 mm. Above this a clean-up of the cathode and spots appeared as with pure N_2 at onset. The flashed filament had contaminated the cathode, reducing γ_i and leading to a burst pulse threshold that cleaned the cathode.

With 1% O_2, currents did not rise to 10^{-8} amp at 765 mm as they had with pure N_2 before the jump. They rose to 10^{-10} or 10^{-9} amp and abruptly jumped to 0.1 μA, at which point the *burst pulse* pattern appeared on the oscilloscope screen. Because of a spread of the burst pulse discharge along the wire, the duration of the burst pulses was 10 to 20 times those observed with the point-to-plane geometry. The spots, however, no longer appeared. This finding is in agreement with those of Huber, that photoionization of pure N_2 by N_2 photons is very low, but if a trace of O_2 or H_2O is added, it becomes large enough to lead to spread along the wire. The impurities if adequate also reduce γ_p and γ_i and so raise the threshold for the Townsend discharges.

3. Lauer's Study with Controlled Electronic Triggering

a. *Techniques and Principles*

Lauer[2.21] had been assigned the task of carrying over Miller's study to pure H_2 and A. In doing this he read of the work of two young Italian physicists, L. Colli and U. Facchini,[3.71] who were studying the basic physics of inert gas Geiger counters, for example, coaxial cylindrical geometry in which the discharge was triggered by α particle tracks projected near one end of the tube. The nature of their data inspired Lauer to make a very simple modification of their technique. Without informing the author, under whom he was working, he introduced an α particle gun into the tube of Miller. This consisted of a capillary of 1 mm diameter just outside the guard ring outer cylinder at one end. The axis of this capillary was parallel to the wire and cylinder axis, so that the trajectory of the particles was a millimeter or two from the cathode cylinder wall and parallel to it for some distance inside the cylinders. The range of α particles outside the capillary was about 2 to 3 cm at 700 mm and covered the whole length of the cylinders at around 100 mm. At the end of the capillary a small wire coated with Po was inserted. The Po wire tip was 1 to 2 cm from the upper end of the capillary tube. This allowed about one α particle per second to emerge from the capillary and launch its 10^4 electrons per centimeter of path at 760 mm within 10^{-8} sec along a line closely parallel and close to the cathode. This linearly strewn group of electrons formed avalanches of $\approx e^{\int_r^R \alpha dr}$ electrons in average number, simultaneously moving radially toward the anode wire. The liberation of positive charge in the gap as the electrons entered the anode wire yielded a sharp inductive kick on the oscilloscope placed across a resistor of some hundreds of ohms between the cathode cylinder and ground. If each avalanche yielded 10^4 electrons and 10^4 electrons were started, then the 10^8 electrons at the anode in 10^{-8} sec gave a readily detected pulse. The rise of this impulse was of the order of the crossing time of the electron avalanche, that is, $\approx 10^{-7}$ sec. As the positive ion cloud, created largely at the wire, crossed to the cathode, the current peak declined exponentially, or perhaps parabolically, and fell to zero as the last of the positive ion cloud was picked up by the cathode. This declining current was due to the motion of the positive ions in the gap. As 75% of these ions are created in $2/\alpha$ cm from the anode, the effect of the ion movement on the cathode is greatest at first while they are near the wire, but continues until all the ions reach the cathode.

If on creation of the avalanche of M electrons and ions many photons are liberated within $2/\alpha$ cm from the anode wire, accompanying the ionization, most of these will reach the cathode with the speed of light and will traverse the few centimeters between anode and cathode, even with several reflections, within some 10^{-9} sec. A certain fraction of γ_p secondary electrons will be released from the cathode by these photons for each positive ion of the avalanche. Of the total $M\gamma_p$ photons, a fraction ν will leave the cathode against back diffusion and *send a second wave of electrons* from cathode to anode. The bunching of ionization in the α particle pulse, together with the speed of photoliberation at the cathode, will make the duration of liberation and crossing of the second group of avalanches yield an exceedingly sharp pulse. Thus, when the second generation of electron avalanches reaches the anode, a new pulse is generated on the oscilloscope.

If potential is close to but below that of self-sustaining corona, a succession of secondary peaks will be displayed on the oscilloscope screen spaced by about an electron crossing time, provided there is no delay in creation and emission of the photons. The successive secondary peaks will decline in amplitude in the measure that $\gamma_p\nu M$ is less than unity, each peak being a power of $\gamma_p\nu M$ less than the one preceding it. Here M represents $e^{\int_r^R \alpha dr}$, the amplification of the initial electron by the avalanche group.

If there is no photoelectric liberation of the secondary electrons from the cathode, then, when the ions reach the cathode after the first avalanche, these will liberate γ_i secondary electrons from it for each ion of the avalanche. Here again a fraction ν will escape back diffusion and lead to a new sequence of avalanches. Again, as potential reaches values near that of a self-sustaining current with $\nu \gamma_i e^{\int_r^R \alpha dr} = 1$, there will be a sequence of pulses of diminishing amplitude of $\nu M \gamma_i$ for each successive pulse spaced at intervals of an *ion* crossing time.

The value of ν at higher pressures and at the low cathode surface fields will decrease as X/p decreases, so that the actual values of decrease in pulse height for successive pulses must be divided by the M and the appropriate ν to yield the true γ_i. Theobald[5.3] has given relations that now permit such calculations of ν and resulting corrections to be made.

In principle, at threshold for a self-sustaining corona when $\gamma_p \nu$ or $\gamma_i \nu$ multiplied into $e^{\int \alpha dx} = M$ equals unity, a continuous sequence of such pulses should be observed which increase in amplitude as potential is increased, except as the space charge of positive ions may influence the field in the cylinders. The space charge will tend to decrease multiplication and reduce pulse size. Such a temporary reduction of multiplication and current in time lowers the space charge so that pulse heights can again increase. The succession of pulses will then have fluctuating amplitude, which may die out in time, superposed on them by the action of space charge.

Despite the sharpness of the pulses, there is diffusion of electrons and ions in crossing the cylinders. This diffusion will gradually increase pulse duration and decrease amplitudes so that after perhaps some hundred pulses the current becomes nearly continuous and the sequences become more and more blurred. In addition, all ionization initially starts as a sheet along a plane between α particle track and filament. However, diffusion of photons around the cylinders, as these are emitted in all directions radially, will spread the discharge coaxially about the central anode in some hundred or so electron crossing times. The γ_i conditioned discharge will spread radially around the anode wire much more slowly, as diffusion of ions is a hundred or so times less. Eventually the space charge and ionization will uniformly fill the tube, as is indicated by the validity of Townsend's equation for steady current.

It is seen that, given sensitive oscilloscopes of adequate time resolution, a most complete diagnosis of events in the tube can be made, yielding invaluable quantitative data. These are as follows:

1. The duration of the pulses from rise to decline to zero, or the interval between peaks in successive γ_i pulses yields at once the ion crossing time. Ion mobilities responsible can be calculated even in the presence of some space charge (which, however, for the first few pulses is nearly negligible) through the relation

$$(5.1) \qquad k_i = \frac{R^2 \log R/a}{2\tau_i(V_0 + 2Q_0 \log R/a)}.$$

Here R is the radius of the outer cylinder, a that of the anode wire, V is the potential across the gap, and Q_0 is the charge per unit length of ion sheath. The value of Q_0 is determined by the pulse height in volts on the oscilloscope multiplied by the capacity per unit length of the cylinder of the discharge tube and the amplifier input. In Lauer's study space charge represented a correction of 0.2 to 3%. The quantity τ_i is the ion crossing time.

2. If the interval between pulses with photoelectric action at the cathode γ_p, as observed with fast oscilloscope, is ≈ 0.1 μsec, then the average electron drift velocity across the tube can be determined. If the field across the tube is not seriously distorted by space charge, the field across the gap being known, it is possible to integrate the known electron drift velocities across the tube as a function of X/p and compute the expected crossing time τ_e for electrons for comparison with the observed τ_e from time between pulses. If these two computed and observed values agree, the γ_p is caused by photons created simultaneously with the ions in the avalanches. If the active photons are indirectly created by some secondary interaction, then the τ_e observed will exceed τ_e calculated, and pulses will be broad and diffuse. That is,

$$(5.2) \qquad\qquad \tau_e = \int_a^R v_r dr,$$

in which v_r is the drift velocity at any point. Here $v_r = f(X_r)$ where

(eq. 1.4) $X_r = \dfrac{V_0}{2r \log (R/a)}$, and v_r can be obtained from tables of constants

appropriate to the gas.

3. In addition, as indicated, the ratio of the number of ions in the nth secondary pulse relative to the $n - 1$, or succeeding secondary pulses, is given by $\gamma \nu M$, where $\nu \gamma$ is either $\nu \gamma_p$ or $\nu \gamma_i$ and M is the multiplication factor $e^{\int_a^R \alpha dr}$ in the avalanche. What one must do to get the value of γ is to measure the areas under successive peaks, since the oscilloscope deflection is proportional to current, and divide the area by M. Here there is some difficulty, since M is not accurately known. In theory, the gaseous purity is that under which α has been determined in previous studies and is given in tables as $M = e^{\int_a^R \alpha \, dr}$. However, since for example in Ne or He traces of impurity $\approx 10^{-4}\%$ of a more ionizable gas are present, values of α can unknowingly be altered by a factor of 2 or more at the same X/p. Hence the values of M cannot be determined in this fashion more accurately than is permitted by the difference between existing and assumed purity.

In A, where the gas multiplication is not calculable through α, Lauer resorted to attempts at direct evaluation. Where pressures were high enough so that the n_0 ions created without noticeable multiplication by the α particle path could be determined, the ratio between n_0 and n_1, the number of ions in the initial ion pulse, gives M through the relation $M = n_1/n_0$. In Lauer's measurements his oscilloscope could measure n_0 at higher pressures. At lower pressures n_0 could not be observed with his existing techniques and the oscilloscope; today it may be possible. Thus he got significant deflections with M = 20 in H_2 and M = 2 in A. He used an electrometer to measure the current due to a γ ray source as a function of voltage, and the n_1/n_0 curve from the pulse was joined to the i/i_1 curve at the lowest gap voltage for which pulses could be evaluated. There was no significant rise of the $\log i/i_0$ curves above the $\log n/n_0$ curves until M = 50. At the lower pressures the n_0 currents could not be measured with an electrometer, so that the value of n_0 was calculated from the known ionization-range curve of the α particles and the path length in the cylinder at that pressure. This gives a value of n_0 which is possibly too great, since it is not certain that all electrons generated in the α particle track are recovered and reach the anode in the low X/p region near the cathode. The error here would make M appear to be lower than it really is and thus raise the computed value of γ. In H_2 calculations of M from Rose's[5.4] good values of α in H_2 are probably quite satisfactory, since traces of impurity in H_2 do not as seriously affect α as they do in A, where metastable atoms have 11 eV of energy.

b. H_2 Gas

Lauer purified his H_2 by the standard procedure with tank H_2, by passing it over drying agents and then hot Cu powder to remove O_2, and finally through two liquid N_2 traps. He outgassed effectively, but still employed stopcocks, since Alpert valves were just then being developed. His apparatus is shown in figure 5.4. He arrived at the following results. In figure 5.5 are shown the particle pulses in H_2 at 200 μsec sweep for various pressures and corresponding potentials below breakdown. The initial rise of current is not seen at this sweep speed, but the plateau and decline due to ion movement are seen. From such data he was able to evaluate the mobility of the positive ions in H_2. Figure 5.6 shows a plot of the reciprocal of the mobility against pressure in millimeters of Hg. The plot is linear, leading to an evaluation of the mobility of the dominant ion in H_2 as 13.4 ± 0.4 cm²/volt sec at 0° and 760 mm. This agrees very well with other values and represents the mobility of the dominant H_3^+ ion at this temperature and pressure.

At a pressure of 400 mm, with 10 μsec for a full-scale deflection for the top three traces and 20 μsec for full-scale for the last two traces, figure 5.7 shows the initial rise (dashed) and a succession of secondary peaks which constitute the decline to the plateau of figure 5.5 at slower sweep. The successive traces (from top to bottom) are at 2200, 2400, 2500, 2550, 2600,

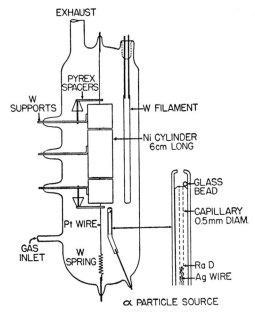

FIG. 5.4. Lauer's tube for α particle-triggered avalanches to study the breakdown in coaxial cylinders. His tube is closely similar to Miller's in dimensions except for the addition of the α particle gun shown on the right.

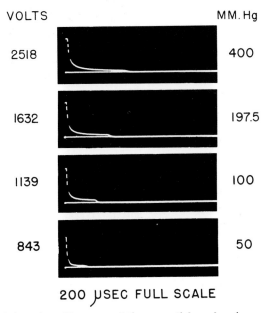

FIG. 5.5. Reinforced oscillograms of the α particle pulses in pure H_2 at a sweep rate of 200 μsec for the full length, at various pressures with corresponding anode wire potentials indicated. The rise due to the electron component is not shown, but the decline and full traces are due to the motion of the positive ions.

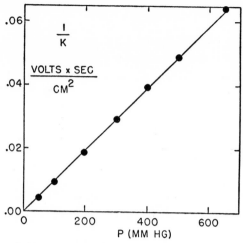

FIG. 5.6. Plot of the reciprocal of the ion mobilities in pure H_2 obtained from the time of ion transit shown in figure 5.5 against pressure in millimeters. This gives a reduced mobility in excellent agreement with other values for the H_3^+ ion.

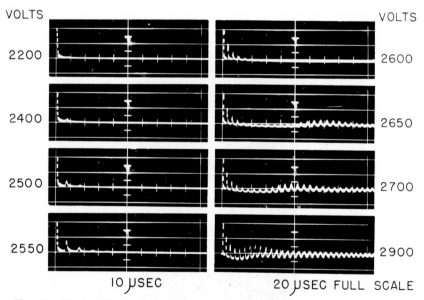

FIG. 5.7. The initial rise of the electronic component of the α particle avalanches, dashed, and the subsequent electronic events on a time scale of microseconds. These are concealed in the rise of the traces in figure 5.5. The secondary pulses are due to electron liberation by photons from the cathode. As potential increases, first one, then two and more pips, due to second, third, and later, generations of avalanches, are seen. Between 2600 and 2650 volts the discharge is a self-sustaining Townsend threshold. At 2900 volts the amplitude oscillations due to space charge choking are obvious.

2650, 2700, and 2900 volts. Here the interval between peaks represents about 1 μsec. The decline in amplitude of the peaks in successive generations is clearly seen. Between 2600 and 2700 volts the discharge becomes self-sustaining; that is, a corona has set in. Note after the initial decline the rise and subsequent fall of pulse amplitudes, presumably as a result of space charge action. The electron transit times as determined from τ_e, the time between successive peaks in traces such as those of figure 5.7, yielded values as shown in table 5.1 for four pressures lying between 650 and 100 mm. To ascertain if there was any time lag in the emission of photons, the drift velocities for electrons across the tube at existing values of X/p, neglecting space charge, were taken from Bradbury and Nielsen's data on H_2 of comparable purity. These were integrated across the tube to yield the calculated values of τ_e shown in the last column. Agreement between observed and calculated values of τ_e is seen to be easily within the limits of error. Thus the photons in the avalanche are created in large measure at the same rate as the ions.

TABLE 5.1

ELECTRON TRANSIT TIME, HYDROGEN

| Pressure, mm | Gap, volts | Transit time, μsec | |
		Measured	Calculated
650	3400	1.12	1.13
400	2500	1.02	1.02
200	1650	0.90	0.88
100	1135	0.77	0.75

In H_2 the presence of traces of impurity are not as serious in their influence on α as in A. Thus M could be computed from Hale's data to within $\pm 20\%$. The ratio of the *amplitudes* between the nth and (n − 1)th peaks are clearly proportional to the areas under the triangular pips. From this ratio $\nu\gamma_p M$ was estimated. These led to values for $\nu\gamma_p$ in H_2 shown in the third column of table 5.2 for the pressures given. Using values of ν as later evaluated by Theobald for the values of X/p existing at the cathode surface, Lauer corrected the back diffusion, yielding the true values of γ_p for avalanche photons in the Ni cylinder, shown as γ_p in the fourth column of table 5.2. It is seen that γ_p is a function of the pressure and potential in the tube at the wire surface. It varies from 0.014 at 100 mm down to 0.0015 at 650 mm. That is, it decreases as pressure increases and as X/p at the anode decreases. No data are available for γ_p on Ni under these conditions. It is probable that H_2 absorbs the photons active; this would account largely for the decline in values. The values are comparable in magnitude to the γ_i values of Parker[5.5] at lower ion energies, and Rose has shown that the values of γ_p are $\approx 10^{-2}$ in the range of lower X/p values in glow discharges at low pressures.

TABLE 5.2

LAUER'S γ_p ON H_2

Pressure, mm	X/p at cathode	γ_p	γ_p/ν
100	1.43	6.9 $\times 10^{-4}$	0.014
200	1.04	1.5 $\times 10^{-4}$	0.004
400	0.79	0.74 $\times 10^{-4}$	0.002
650	0.66	0.49 $\times 10^{-4}$	0.0015
		Argon	
25	3.12	7.8 $\times 10^{-4}$	0.035
50	1.92	7.0 $\times 10^{-4}$	0.051
100	1.18	4.0 $\times 10^{-4}$	0.040
200	0.79	0.6 $\times 10^{-4}$	0.006
400	0.52	0.2 $\times 10^{-4}$	0.002

No γ_i was detected in this gas, since γ_p sufficed to yield a discharge. Rose[5.6] has observed that γ_i is strongly dependent on X/p, aside from the influence of X/p on ν at very low values. In fact, at X/p values where ν was already low, Rose observed γ_i on Mo to be $\approx 10^{-4}$. It is presumed that with the very low X/p at the cathode, γ_i was at best 0.1 of γ_p in this study. The work of Hale[5.7] also showed a pronounced γ_p peak at low X/p in the gas on Pt in a uniform field gap and a very low γ_i. It is difficult to compare the two quantities with Hale's data, as X/p in the region of ionization is high and at the cathode is low. Parker's results for γ_i on Pt show that in vacuum γ_i falls off very rapidly as ion impact energy on the cathode decreases. Thus Lauer's data are the only ones applicable to this situation and must be accepted as correct.

c. A Gas

The data obtained on Linde's spectroscopically pure grade A, cooled over liquid N_2—which contains traces of Kr and N_2—are as follows. Figure 5.8 shows the pulses in A for a sweep speed of 1000 μsec for the full deflection. It is seen that at 400 mm and 1655 volts, there is just an ion peak, the duration of which gives the ion crossing time. At pressures below 200 mm this is followed by a succession of peaks of decreasing amplitude separated by just an ion crossing time. Thus there is strong evidence of an active γ_i, and at higher potentials this process leads to a self-sustaining discharge at which $\gamma_i \nu M = 1$. It is noted that the traces of figure 5.8 at higher pressures in A have a flat top of some duration. Analysis with faster sweep in A indicated that there were two successive rather fuzzy peaks that could have been due to a γ_p. These peaks were not clearly separated by an electron crossing time in A, nor are they sharply delineated as in H_2. For this reason Lauer ignored them. He was not able to work much above 400 mm pressure with his A source. He was disturbed to note that Colli, Facchini, and Gatti,[5.8] who used pressures of A above 200 mm and up to nearly two atmospheres, got a succession of γ_p pulses with a brass cathode which led to a self-sustaining γ_p corona discharge in A at higher pressures. They *did not observe a* γ_i.

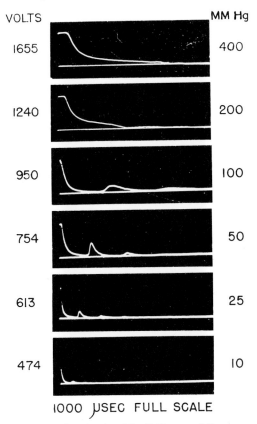

VOLTS

MM Hg

1655 — 400

1240 — 200

950 — 100

754 — 50

613 — 25

474 — 10

1000 μSEC FULL SCALE

FIG. 5.8. Oscilloscope pulses in A with 1000 μsec full sweep, showing the ion movement across the gap at various pressures and the corresponding potentials indicated. The electronic rise is not seen, but the ion pulses with arrival of ions at the cathode are clearly seen. There is a succession of secondary peaks, indicating that a γ_i is creating new avalanches on ion impact on the cathode. The flat top of the higher pressure traces conceals a low order prolonged γ_p.

d. *Colli and Facchini's Clarification of Discrepancies*

This discrepancy prompted the author to have Colli and Facchini[3.71] work in his laboratory for some weeks under the generous auspices of the Office of Naval Research. The results of this collaboration at once clarified all the discrepancies. The A used by Colli and Facchini had initially been impure when introduced into the tube. Their measurements were made after the A had been purified by circulation over a Ca arc. Thus their A was as pure as that of Lauer when finally studied. From 1000 mm to about 150 mm the γ action predominating was a γ_p from either a brass or a clean Ni cathode cylinder. There is, however, something peculiar about the γ_p peaks. The peaks were diffused in time and quickly became more diffuse, unlike the sharply defined γ_p peaks of Lauer in H_2.

The time between peaks was longer than just the electron crossing time, such that there appeared to be a delay on the order of some 3.5 μsec in the emission of the photons and the creation of the avalanche. There was another difficulty with a photoelectric γ_p for A photons on Ni. The Ni cathode used had a work function of about 5 ev. The spectrum of A has only one wave length corresponding to about 11 volts electron energy. This is its resonance line; all other lines correspond to energies less than 4.1 ev. The resonance line is very heavily absorbed at higher pressures, and it also degrades to the neighboring metastable levels and is lost in its delay of some 10^{-4} sec in reaching the walls by repeated absorption and remission. Thus it gives little γ_p and that is a very diffuse one which is delayed by 100 μsec beyond the avalanche time.

In keeping with results of Phelps and Molnar[5.9] on the destruction of the metastable states of A, the metastable A atoms in one out of about 10^5 collisions are destroyed and create excited A_2 molecules by triple impact with A atoms. These are slightly metastable and have a lifetime of about 3.5 μsec, emitting a band of about 1250 Å or about 10 ev energy.[3.71] These lines are not absorbed by the A atoms, and the A_2 molecules are too few to absorb them. The value of γ_p inferred from this work was $3-5 \times 10^{-3}$. It was not very pressure-dependent in the range of pressures used, and was about the same for Ni and brass. Colli[5.10] had earlier shown that in a Townsend avalanche lasting 5×10^{-7} sec the photons are emitted according to the equation $n(t) = \exp -(9p^2t) - \exp -(3 \times 10^5 t)$. This is in conformity with the law for the rate of destruction of metastable A atoms in pure A by triple impact with A atoms, discovered by Phelps and Molnar. The photon burst has a rise time of 1 μsec and a delay time of 3.4 μsec at pressures above 300 mm Hg. This readily accounts for the diffusion and smearing of the succession of photon pulses.

The tube used by Lauer had a diameter of 2.9 cm so that the transit time of electrons in A was 2–3 μsec, while the one used by Colli and Facchini had a crossing time of 6–8 μsec. Thus at 400 mm in Lauer's apparatus the overlapping of the pulses almost obscured the γ_p effect. The pulses obtained by Colli and Facchini with Lauer's tube are shown in figures 5.9 and 5.10 and confirm Lauer's observation. The values of γ_p obtained in this study were close to 1.5 to 2.5×10^{-3}, in agreement with their earlier values.

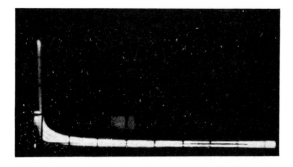

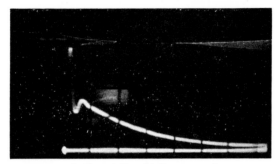

FIG. 5.9. Colli and Facchini's curves for a γ_p action in A at 400 mm in Lauer's tube at 1660 volts, 100 μsec. The upper pulse at 100 volts *below* corona threshold shows the primary avalanche with no secondary effect. The lower pulse at threshold shows the peak and a decline due to a diffuse γ_p.

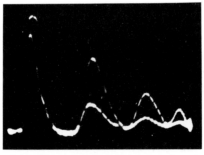

FIG. 5.10. Two superposed traces of γ_p in A at higher pressures taken by Colli and Facchini. At onset these pulses become continuous. Note the breadth and smearing at intervals that exceed the electron crossing time.

The question of the appearance of a γ_i observed by Lauer below 200 mm pressure was a more difficult one to settle. Colli and Facchini used Lauer's outgassing procedures of heating the cathode to 900° C for many hours under the pumps, and then admitting the Linde A over liquid N_2 into the well outgassed tube. They then obtained the same results as Lauer with $\nu\gamma_i$ ranging from 10^{-3} at ≈ 100 mm Hg and going down to 6×10^{-5} at 200 mm and 2×10^{-5} at 600 mm. Here M at threshold lay between 400 and 600. All these values of γ_p and γ_i were *not* corrected for back diffusion by division by ν.

Colli and Facchini had, it was noted, not outgassed their brass cathode in Milan above 350° C. They also had initially impure A with some N_2 and O_2 undoubtedly present before using the Ca arc. Thus it appeared that they had desensitized their cathode so that γ_i was too small to measure. They also did not work below 150 mm. To test for possible desensitization, the tube was filled as it had been in Lauer's work and good γ_i readings were obtained at 200 mm and below. Then either the cathode was exposed to air for a few minutes in the cold or else 99.5% pure A was admitted after the pure A was removed, and then either the air or the 99.5% pure A was pumped out. Admission of spectroscopically clean A then permitted a repetition of the study. Here *no γ_i effect was observed*. Apparently the trace of O_2 that adsorbed on the clean Ni surface sufficed to suppress the γ_i. The γ_p effect was little altered by this treatment. Only after the cathode was again outgassed at 900° C for several hours and pure A readmitted was the γ_i again observed. In their γ_i study Colli and Facchini obtained the oscillogram of figure 5.11. Here A pressure was 50 mm at 765 volts using clean and outgassed Ni. In the upper trace the 2000 μsec sweep time shows the γ_i pulses in succession with diminishing amplitude. The secondary γ_i peaks show two peaks at each generation. The lower trace resolves these two peaks. These indicate that the γ_i is composed by secondary emission of A ions of two velocities. At higher pressures the two peaks are not observed because they are not clearly separated, as seen in figure 5.8 at 100 mm. These peaks correspond to the A^+ plus the A_2^+ ions which are created by the Hornbeck-Molnar process from highly excited A atoms in collision with unexcited A atoms. The reduced mobilities of these ions are, respectively, 1.63 and 1.94 cm^2/volt sec. In fact, this geometry was utilized as early as 1951 by Hornbeck and Molnar to resolve the two mobilities of the A^+ and A_2^+ ions by using pressures of 5.3 mm of A gas.[5.11] Here the avalanche succession for several generations was so clearly resolved that the relative abundance of the two types of ions could easily be determined. The γ_i and γ_p phenomena in breakdown at different pressures were studied by Menes[5.12] in pure A in uniform field geometry. The γ_p process was found active at higher pressures. At lower pressures but not the lowest, data could not nicely be accounted for by the γ_i

process. The problem here is quite complex. Lauer's values for γ_i for A are shown in table 5.2, corrected for ν using Theobold's data. Here values remain high until high pressures are reached. This could arise from a lower value of γ_i for the A_2^+ at low X/p which predominates at high pressures. Higher pressure A may have desensitized his cathode by trace impurities. Lauer's ion crossing data gave a value of 1.94 ± 0.08 cm²/volt sec for the A_2^+ ion in A. This agrees well with other values for this ion. It is not clear why he did not observe A^+ as well. Colli and Facchini did observe both at 50 mm with Lauer's apparatus.

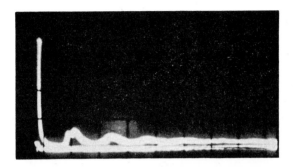

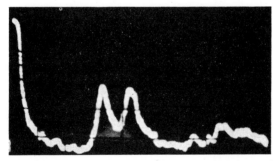

FIG. 5.11. Colli and Facchini's curves for γ_i in A at 50 mm pressure, at corona threshold at 765 volts. The upper pulses were taken at 2000 μsec sweep time. The first secondary pulse is composed of two distinct peaks. The lower trace shows the pulse enlarged. There are two A ions, A^+ and A_2^+, of different mobility.

e. *Effects of O_2*

Lauer added small amounts of O_2, for example, 0.1% and 1.0%, to the H_2 gas. This had the effect of increasing the photoionization in the gas, probably decreasing γ_p somewhat, and creating negative ions from avalanche electrons. This at once decreased α slightly as far as the immediate avalanche size, that is, that in one crossing time, was concerned. It probably in the end gave back the electrons by dissociating the negative ions at values of X/p in excess of 90 near the anode. Multiplication was reduced, since the negative ions do not ionize until they shed their electrons. The delayed arrival of these ions after the avalanche has reached the anode should yield secondary photoelectrons from the cathode at times between avalanche crossing and the arrival of most of the positive ions at the cathode. If $\gamma e^{\int \alpha dx}$ is nearly unity, these secondary delayed electron avalanches will superpose small pips on the declining positive ion pulse. Figure 5.12 shows three traces with no O_2 and with 0.1% and 1% O_2 in H_2 under the same conditions. While the top trace in H_2 indicates a self-sustaining discharge, the pulses decline much more rapidly with 0.1% and very rapidly with 1% of O_2. This indicates that $\nu\gamma_p M$ is decreasing, and, since $\nu\gamma_p$ with 0.1% and 1% O_2 is probably closely the same, the decrease is largely in M because of reduction in α.

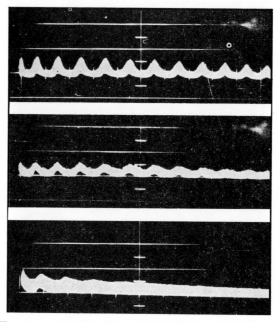

Fig. 5.12. Three traces for H_2 with no O_2, 0.1% O_2, and 1% O_2, under the same conditions near threshold for steady corona. The sweep speed is 10 μsec for the full scale. The top trace shows a self-sustaining discharge in the pure H_2. With 0.1% O_2 the pulses are seen to decline. With 1% O_2 there are only three avalanche sequences possible before extinction, indicating the decline of $\nu\gamma_p M$ due to attachment and loss of electrons in O_2.

VOLTS

VOLTS

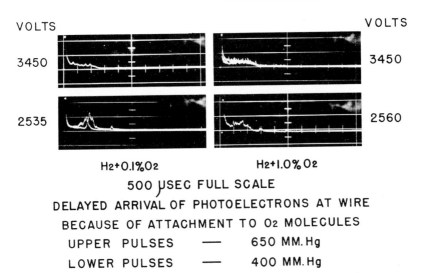

3450

3450

2535

2560

H2+0.1%O2

H2+1.0%O2

500 USEC FULL SCALE

DELAYED ARRIVAL OF PHOTOELECTRONS AT WIRE

BECAUSE OF ATTACHMENT TO O2 MOLECULES

UPPER PULSES   —   650 MM. Hg

LOWER PULSES   —   400 MM. Hg

FIG. 5.13. Traces in H_2 plus 0.1% O_2 with slow sweep, indicating the presence of delayed pulses during the crossing of ions. These come from groups of negative ions, created in mid-gap, that on reaching the anode liberate electrons and start new avalanches of small size. These are shown at two different pressures and the corresponding voltages. The sweep speed is 500 μsec for the full scale.

Figure 5.13 presents the traces at slow sweep, showing the ion pulses with, respectively, 0.1 and 1.0% O_2 at 500 μsec for full sweep and indicating the delayed γ_p induced pulses on the declining phase of positive ion current. The effect is small with 0.1% O_2, but becomes very prominent when more O_2 ions are created. These effects are not pulses due to single negative ions but are statistical fluctuations due to nearly simultaneous arrival of a number of negative ions at a region of $X/p = 90$ near the anode. The situation is complicated by the fact that electrons attach to O_2 molecules at various points in the gap. Most electrons that form negative ions are created nearer the anode, and some of the ions are O^- ions created by dissociative attachment where the electron avalanche has an average energy in the field of around 1 ev. These ions break up at $X/p = 20$ to form O_2^- ions, which have mobilities on the order of 2.3 cm^2/volt sec in O_2 and perhaps 9 in H_2. They are thus much slower than the H_3^+ ions. However, it is seen from the traces that quite a few O_2^- ions may be formed within about 0.2 of the interelectrode distance from the anode to give the delayed pulses.

In order to complete his study of the coaxial cylindrical discharge in H_2 and A, following Miller's studies, Lauer took the current-potential

curves for pressures from 25 to 650 mm in H_2 (fig. 5.14a), and between 25
and 400 mm in A (fig. 5.14b). The curves are seen to be similar to Miller's
for pure N_2. The self-sustaining corona set in, as in H_2, at currents around
0.1 to 1.0 μA. The initial current could only be measured at the highest
pressures with no amplification present where it was around 10^{-13} or
10^{-14} amp. The discharges become visibly luminous at currents above
threshold for self-sustaining corona with 10 μA of current. There was no
spark in H_2 for a considerable range of potentials above threshold, even
though potentials were increased to relatively higher values above threshold than were Weissler's for point-to-plane corona.[1.8] On the other hand,
A broke down to a spark from the glow at around 50% increase in potential above self-sustaining threshold, as seen in figure 5.14b. In Weissler's
point-to-plane corona gaps, A gave no corona, but went directly to a
power arc via a filamentary spark.

This leads one to the conclusion that the influence of an effective γ

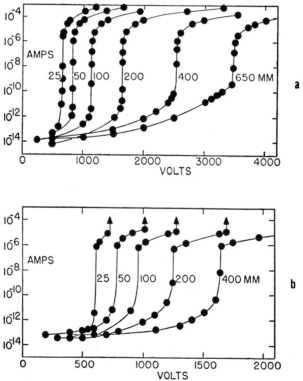

Fig. 5.14. (a) Current-potential curves for pure H_2 at various pressures, taken by
Lauer. (b) Current-potential curves for relatively pure A at various pressures.
Note the early breakdown to a spark, indicated by the arrows.

source from the surrounding cylinder with coaxial cylinder geometry leads to a ready Townsend breakdown. Once this is achieved, there is quite a range of potentials with corresponding increase in current which can be accommodated by the effective cathode space charge collection before the streamer spark occurs. This is probably due to the nature of the positive ion space charge distribution in the relatively confined gap. The space charge distorted gap field conditions are much more conducive to streamer action with point-to-plane than with coaxial cylinders.

That A leads to a streamer spark more readily is now understandable. The spectroscopically pure grade A has sufficient Kr and N_2 to lead to streamer sparks quite readily. Though H_2 can yield a two-component gas of H atoms and H_2 molecules with ionization potentials of 10 and 15 volts, respectively, so that photoionization by its own photons should be effective, it is easier to obtain H_2 in a pure state. There are undoubted elements in the H_2 bands that can photoionize the gas, but this is not particularly efficient. H_2 shows considerable reluctance to photoionize itself at low X/p and yield streamer sparks—in fact far more reluctance than N_2, O_2, and the impure A. Pure A would show the same tendency as H_2, if not a stronger one.

4. Huber's Study in N_2 and O_2

Lauer's investigation was carried further by Huber[2.22] on pure N_2, O_2, and mixtures thereof. Only the study of N_2 will be given here. The technique was the same as that of Lauer and of Colli and Facchini[3.71] in the author's laboratory. The apparatus is shown in figure 5.15. Here there were two α particle guns, one firing α particles parallel to the axis, the other one firing α particles at right angles across the tube normal to the axis at one end, one-third the distance from anode wire to the cylinder. Both sources could be cut off by magnetically operated gates at the exit end of the capillary. About one α particle was emitted per second to trigger pulses. Both Airco purified N_2 with less than 0.002% O_2 and ordinary tank grade N_2 passed over the tubes of KOH pellets, hot Cu shot, and liquid N_2 traps gave the same results. Use of the flashed W filament as a getter introduced more impurity than it removed, as the mass spectrograph revealed.

Pure N_2 behaved in a uniform fashion for all pressures from 25 to 600 mm of Hg. The primary pulse following the avalanches was followed successively by one, two, three, and so on, secondary pulses as the potential was raised from below upward to the threshold. The pulses were spaced at an ion crossing time. Figure 5.16 shows a sequence at 400 mm with the lowest trace 10 volts below self-sustaining corona. This definitely indicates

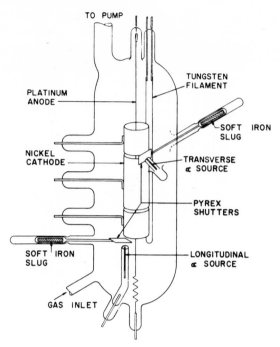

FIG. 5.15. Huber's modification of the Lauer technique, including the transverse α particle gun as well as the parallel gun.

that the principal active mechanism is a γ_i on outgassed Ni with pure N_2. Others, including Fisher and Kachickas, Llewellyn Jones's group, and Raether's students, have reported the mechanism to be largely a γ_p in N_2, mostly at higher N_2 pressures. Since Huber's study was made on definitely purer N_2 and really outgassed systems using Alpert vacuum techniques, with atomically clean Ni except for N_2 adsorption, the difference must be ascribed to the insensitivity of Ni to the N_2 photons, whereas O_2 coated Ni is a good emitter with the photons.

There appeared to be little photoelectric ionization in the gas, as indicated by the smooth decline of the ion pulses. The results for the mobility of the ions, as inferred from τ_i, was linear when plotted against $1/p$, yielding a reduced mobility of the positive ion in N_2 of 2.53 ± 0.08 cm^2/volt sec, at 22° C and 760 mm. This is in agreement with Varney's value at very low X/p, which indicates this ion to be largely N_4^+.

The evaluation of γ_i was achieved by measuring the values of n_0 and n. The latter was derived directly by measuring the area under the primary pulse. The decrease in amplitude under successive pulses $n_1/(n_1 - 1) = \nu M \gamma_i$, then $M = n/n_0$. The value of n_1 was estimated from the known

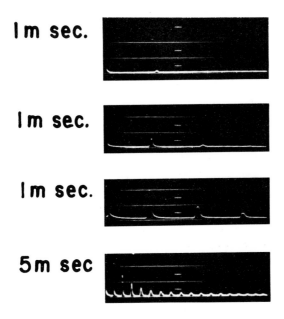

I m sec.

I m sec.

I m sec.

5m sec

NITROGEN - 400 mm.

FIG. 5.16. Huber's γ_i pulses in pure N_2 at a succession of potentials at 400 mm pressure. The sweep speed is indicated at the left. The bottom trace is at 10 volts below self-sustaining corona.

ionization of the α particle. The methods of estimating M used by Lauer and by Colli and Facchini were also used as checks. $\nu\gamma_i$ was not evaluated to better than $\pm 20\%$. Theobald's data for back diffusion loss in N_2, ν, were used to reduce the $\nu\gamma_i$ readings to the true γ_i for vacuum. These data are given in table 5.3. The value of γ_i ranged from $2-3 \times 10^{-2}$, with a mean value of about 2.5×10^{-2}. The error here could be by a factor of 2, since Theobald's corrections are only approximations. No data for γ_i on pure Ni exist for comparison. Parker's data for N_2 coated Pt, which is similar to Ni, was for N^+ and N_2^+. At very low impact energies, values were on the order of magnitude of 1×10^{-2}. In Huber's work the ions were N_4^+ in transit, but N_4^+ is very unstable, so that probably the actual ions effective were N_2^+ ions.

Addition of a small percentage of O_2 gave very interesting results, which are essential to indicate because of their bearing on the data for pure N_2 and those of other workers. The results are summarized as follows:

1. Secondary pulses from γ_i were smaller and fewer in number at given potential and pressure than in pure N_2.

2. The γ_i pulses began to lose their smooth shape and become rather ragged.

3. Onset of steady self-sustaining corona began at a potential approximately 10% higher than in pure N_2.

4. With faster sweep and shorter RC time constant, a new type of secondary pulse was observed, spaced at intervals of about 1 μsec from the primary.

TABLE 5.3

HUBER'S γ/ν

Pressure, mm	γ_i M, Measured	γ_i M, Calculated	X/p Maximum of Influence on M	γ_i/ν
600	0.6×10^{-3}	7×10^{-3}	220	2×10^{-2}
500	0.7×10^{-3}	1×10^{-3}	230	2.3×10^{-2}
400	1.1×10^{-3}	1.2×10^{-3}	250	3.2×10^{-2}
300	1.0×10^{-3}	1.3×10^{-3}	290	2.7×10^{-2}
200	1.2×10^{-3}	1.5×10^{-3}	360	2.9×10^{-2}
100	1.5×10^{-3}	2.4×10^{-3}	530	2.5×10^{-2}
50	1.4×10^{-3}	3.4×10^{-3}	780	2×10^{-2}

Figure 5.17 is for pure N_2 at 200 mm Hg with 1000 μsec full sweep, RC = 3 μsec. The upper trace is a sequence of several repeated sweeps, showing the beautiful reproducibility of each sweep. This was at 2640 volts

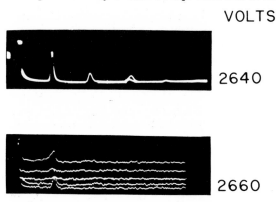

VOLTS

2640

2660

FIG. 5.17. Pure N_2 at 200 mm at 1000 μsec full sweep with RC = 3 μsec. The upper trace is below self-sustaining corona; and several repeated sweeps are shown. The lower trace represents the small fluctuations on the steady corona current just above onset. The separate sweeps are displaced vertically. It is seen that, while the first and second pulses appear, due to α particle triggering, the fluctuations of the continuing current furnish the background as a result of diffusion of the discharge over the wire. Only occasionally do the α particle pulse and the next generation appear against the background.

below self-sustaining corona. At 2660 volts the self-sustaining corona appeared. Here there is a repeated sequence of sweeps displaced vertically to separate the sweeps. The secondary sweeps can be seen, but here it is impossible to get more than the self-sustaining corona, which has spread around the tube. Only an occasional sweep shows one or two of the sequence of initial pulses. Figure 5.18 shows secondary ion pulses and self-sustaining discharge in N_2 plus 5% O_2 at 200 mm Hg. Sweep speed is 1000 μsec full scale; RC = 3 μsec. Traces such as c and d are single sweeps. Figure 5.19 shows photoelectric pulses in N_2 plus 5% O_2 at 200 mm Hg, with RC = 0.03 μsec. The sweeps are no longer reproducible. Trace e at 2960 volts shows that the photon γ_p does not here produce the self-sustaining corona as a discharge. The diffusion of the peaks suggests a delay of photon emission for which there is other evidence.[2.4]

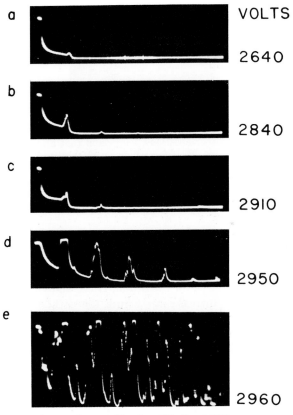

FIG. 5.18. The secondary pulses at N_2 plus 5% O_2 at 200 mm pressure. Sweep speed 1000 μsec full scale, RC = 3 μsec. Traces c and d are single sweeps. Note the appearance of photoelectrically triggered pulses, burst pulses, in between the γ_i triggered pulses.

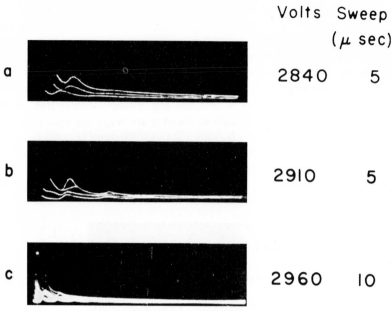

Volts   Sweep
(μ sec)

a   2840   5

b   2910   5

c   2960   10

Fig. 5.19. Photoelectric pulses in N_2 plus 5% O_2 at 200 mm Hg with RC = 0.03 μsec. Sweeps are no longer reproducible. Trace c would indicate that the photon γ_p does not give a self-sustaining corona at this pressure. This observation appears to differ from that of other observers, who claim that a photon γ_p leads to a self-sustaining glow discharge in uniform field geometry in relatively impure N_2 and N_2–O_2 mixtures. The γ_p in this geometry is low because of the low ν at low X/p at the cathode. In uniform field geometry X/p is high at the cathode and ν is negligible.

It is seen that addition of O_2 at once complicates the phenomena. It reduces γ_i by a factor of 5 with 5% O_2. It probably reduces γ_i by this amount, even in trace amounts. This is in keeping with Parker's findings about the influence of O_2 on Pt. Burch, Irick, and Geballe[5.13] observed an order of magnitude increase in γ_p from the cathode with 4% O_2 in N_2.

At the same time, as was suspected from Kachickas and Fisher's[3.40] results, it brings in a γ_p. Here γ_p is reduced by the small ν at the low X/p and is not as large as it is in the uniform field gaps with very high cathode X/p. To test the reduction of γ_i by O_2 directly, the O_2 in Huber's experiment was removed and pure N_2 admitted. The γ_i remained low. It was only restored to its value of 10^{-2} on glowing the Ni at 900° C for hours and using only pure N_2.

The smearing of the avalanches due to γ_p in time indicates a probable delay in the photoeffect over that of the avalanche ionization. This phenomenon was first reported by Bandel[2.4] in uniform field geometry in air and was also observed by Raether's[2,4] group. It is the O_2 that appears to be the active agent in this emission, according to studies by Raether's group. Ward,[5,14] using a computer, has reproduced Bandel's curves by judicious combination of γ_i and γ_p in air. The question of the time delay is still open. However, computed agreement by juggling adjustable constants is misleading if contradictory to directly observed smeared γ_p pulses.

That the short pulses were indeed photon pulses is shown by comparison of the calculated electron crossing time τ_e in microseconds, as calculated from Nielsen and Bradbury's data in N_2 and as observed for the interval between avalanche and the first γ_p pulse. However, these data are for only the first peak of the pulse, which is quite well smeared in longer time intervals.

It is clear that, in addition to γ_i and γ_p, there is a third type of action present that becomes more prominent as potential is raised with 5% O_2, as is seen in the large rough bumps on the curves of figure 5.18. This is not due to electron attachment and detachment as for O_2 in A, where roughness is a statistically conditioned minor manifestation. The bumps indicate that a very active photoionization in the gas actually sets the threshold for steady self-sustaining corona in this mixture, for which the γ_i sequence can be roughly noted in trace e at 2960 volts. The secondary pulses at much shorter intervals are comparable in size to the γ_i created pulses. These are caused by photoionization in the gas and a Geiger counter-like action. Such a phenomenon, with spread of the discharge along the wire, does not occur in uniform field geometry. It is characteristic of the asymmetrical field conditions and is brought out most strongly in the coaxial cylinders, since spread over the point in point-to-plane geometry is limited. With a feeble γ_i and a feeble γ_p, because of low cathode fields in coaxial geometry, as μ increases, the photoionization in the gaps in N_2–O_2 mixtures becomes strong enough to give the self-sustaining corona at lower M than either γ_p or γ_i.

B. ELECTRON ATTACHING GASES, NOTABLY AIR AND O_2

1. INTRODUCTION

Unfortunately, all there is to know about processes here at work is not yet known. Accompanying the electron attaching property, photoionization in the gas also becomes a strong feature. This leads to the Geiger counter-like spread of the discharge along the wire anode surface, contributing very heavy self-quenching discharges below the threshold for the Townsend type of breakdown involving the cathode action. Thus discussion of this type of discharge properly belongs in the Geiger counter section (sec. C) of this chapter.

The negative ion formation, however, acts to give ready self-counting and delayed pulses, so that it does not make a good counter. It does exhibit all the properties of a counter, and some basic studies made with it are most valuable in the interpretation of Geiger counter behavior. The role of the change of photoionizing mean free paths in the gas relative to the diameter of the wire unfortunately is also not well known. It is a field currently under study. However, some work subsequent to that of Miller and Huber is now available.

2. MILLER'S STUDY IN N_2–O_2
   MIXTURES AND IN O_2

Some intimation has been given of the change produced in the corona in pure N_2 on addition of 1% O_2. This produced the curves shown in figure 5.20. Below threshold the currents were weaker, with only a small Townsend contribution to the field intensified currents. However, the currents were stronger than in air. There was no current limiting resistor so

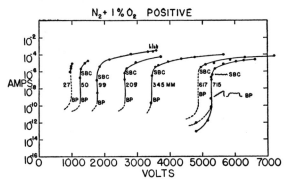

FIG. 5.20. Current-potential curves in N_2 with 1% O_2. Note the abrupt current jump to a burst pulse corona, marked BP, and onset of steady corona, SBC. Note the appearance of breakdown streamers at 99 mm, indicated by vertical lines above the curve.

that, unlike those in pure N_2, the ionic space charges largely controlled the current. Currents rose to 10^{-10} amp only and jumped abruptly to 0.1 or 1 μA when burst pulses appeared. In pure N_2 currents rose to 10^{-8} amp with relatively little jump when the glow began. There was no overshoot with 1% O_2. The curve at 715 mm is branched at its lower part, the larger current branch being caused by increasing ionization in the tube by an external γ ray source. The burst pulses began coincident with the sharp rise in current and lasted 10^{-3} sec. This is in contrast to the duration of 10 to 100 μsec for point-to-plane corona because the pulses can spread along the whole wire and thus take longer to build up a quenching space charge, especially in the more efficient coaxial clearing field relative to point to plane. Decreasing pressure to 50 mm increased the duration of the burst pulse from 1 to 6 msec. This change is due to diffusion and the decreased density of positive ions.

The intermittent regime ranged from 10^{-9} amp for occasional single pulses to 10^{-7} amp. This change is produced by a very small range of potential of some tens of volts. In Geiger counter parlance this is the region of the counting plateau. Here the bursts depend on the number of external ionizing events that can trigger a self-quenching discharge. The symbol BP indicates the appearance of the burst pulses, and the symbol SBC indicates onset of the steady glow corona that terminates the Geiger counting regime. This begins around 0.1 to 1.0 μA. The outer cylinder probably plays some role in ensuring the very narrow counting regime, as it can liberate photoelectrons to trigger discharges from the cathode and help spread the discharge along the tube.

One may ask whether the formation of negative O_2^- ions and O^- ions in the dilutely contaminated N_2 is responsible for the steady glow corona as is the case with Hermstein's glow corona in air.[1.16] It is less likely that this occurs in N_2 with 1% O_2, but when O_2 reaches 5% or more, negative O_2^- ions could yield a Hermstein sheath. However, burst pulses and their continuity at the end of the counting plateau here are probably not caused by negative ions but by increase in duration and frequency of the bursts so that the current is not interrupted by positive ion space charge. Negative ion formation may, however, help in this regard. In pure N_2 the positive ion space charge intermittence is not noticed. In the region below the onset of the heavy spot corona in pure N_2, the small field intensified currents of the Townsend discharge are too feeble to choke themselves off. Once the fields are great enough, as required by the clean-up of the surface, the potentials are so great that positive ion space charge can no longer interrupt the discharge. Actually, clean N_2 Geiger counters of the slow type worked, but they had such a narrow space charge interrupted plateau that series resistors were needed to quench the discharge.

Where the cathode plays a large role in yielding a discharge, as in pure N_2, the electrons multiply well across the gap and somewhat aid in alleviating the positive ion space charge action, as they come from outside the zone of intense ionization. With the photoelectric ionization in the gas, the ionization is produced so close to the anode that the electrons are quickly absorbed and the positive ion space charge is little diluted with electrons.

Thus at the critical region near threshold, the ionization efficiency gives a quenching positive space charge close to the anode in the burst pulse regime. This is, however, very narrow in terms of applied potential.

Above 1 μA the glow of closely adhering blue color is seen about the anode. It generally coincides in appearance with the burst pulses. Because of the decreased current density of the glow spread along the wire, the luminosity is less than with point to plane. Above the onset of steady corona, the earlier linear current rise and the rise following the form $V(V - V_s)$ occurs in accordance with Townsend's law (eq. 2.30) for the space charge limited current. Where the potential source at 99 mm allows a sufficient potential increase, the curve is followed to the breakdown streamer regime. Breakdown streamers are observed as low as 27 mm pressure in this mixture. At no time are pre-onset streamers observed near threshold as with point-to-plane corona. It is possible that, because of secondary cathode emission, Townsend-like currents below burst pulse threshold build up space charges that inhibit pre-corona streamers. It is more probable that the photoionizing free path in 1% O_2–N_2 mixtures is too great for the effective generation of streamers from the fine anode wire used. The photoionizing free paths relative to the length of the ionizing X/p zone and relative to $1/\alpha$ are a critical factor in pre-onset streamer propagation. This matter is being investigated in the author's laboratory using Nasser's Lichtenberg figure technique, as seen in chapter 3, sections A.7 and A.9.

As earlier indicated with steady corona, Townsend's law (eq. 2.30) for i as a function of V appears and yields a value for the reduced mobility of ions in 1% O_2–N_2 mixtures of about 2.8 cm^2/volt sec. This is a bit higher than the observed mobilities for N_2^+ ions in N_2, of 2.5. It could be ascribed to an assigned value of V_s which is too high.

As the value of p goes down, the threshold for breakdown is reduced, but not as rapidly as p is reduced. This follows from the condition for steady burst pulse corona. This may be set as (eq. 2.35)

$$\beta^1 \nu f^1 \exp \int_a^{r_0} \alpha dx = 1.$$

Here f^1 is the fraction of the electrons in the last avalanche of a dying burst pulse (choked off by space charge which is clearing elsewhere along the wire) which liberate photons capable of being absorbed by the cathode to emit new electrons. β^1 is the chance that they strike the cathode, and ν is the correction for back diffusion. f^1 and β^1 are more or less independent of pressure, since the starting current for onset is not varying widely with pressure. ν is a function of X/p at the cathode only. Thus the variation of threshold will depend primarily on the variation of $\left(\int_a^r \alpha dx \right)$ with pressure. Since the field $X = kV_r/x$ so that $dx = \left(\dfrac{kV_r}{X^2} dX \right)$,

$$(2.16) \qquad \int_a^r \alpha dx = kV_r \int \alpha \frac{dX}{X^2} = kV_r \int_0^{X_{r/p}} [(\alpha/p)/(X/p)^2] \frac{dX}{p}$$

as shown more generally by Dodd, as seen in chapter 2, section E.2. With the electrodes used

$$X_r = kV_r/r = 25V_r$$

$$(5.3) \qquad \int_a^r \alpha \, dx = 0.19V_r \int_0^{25V_r/p} [(\alpha/p)/(X/p)^2] \, d(X/p).$$

Now α/p is a $f(X/p)$ only and is independent of p. The integral then depends on V_r and p only as they appear outside the integral or in the upper limit. Here, as equation 2.35 has $\left(\int_a^r \alpha dx \right)$ essentially constant, that is, $= \left(\frac{1}{\beta^1 f^1 \nu} \right)$, then the value of the integral remains constant as p changes. As p decreases in the upper limit of the integral of equation 5.3, V_r must decrease so that the upper limit may strive to render the $\int \alpha dr$ constant. However, as V_r decreases in the limit, the multiplier V_r also decreases. Thus the upper limit to the integral V_r/p must increase to compensate for the decrease in V_r. Hence V_r will not decrease in proportion to p but as demanded by the constancy condition.

That is, it will decrease more slowly than p, as observed. Were α/p proportional to $(X/p)^2$, as it is over some of its effective range, then $0.19V_r^2/p$ would have to remain constant and $V_r \propto \sqrt{p}$. In figure 5.20 it is seen that the variation is very nearly that of $\sqrt{p}$. The change in p is 26.5-fold, and its square root is 5.15. The change in V_s is by a factor of 5.4. This indicates that indeed β^1 is nearly constant, as assumed. According to equation 2.35, β is given roughly by $\beta = 0.5 \exp(-1.86 \, \mu_1 ap)$, where μ_1 is the absorption coefficient of the active component for burst pulse creation, assuming only a single one active. Then if $\mu_1 ap$ is very small, β approaches 0.5 and varies very slowly. Thus the values of a and $\mu_1 p$ are important factors in the variation only if $\mu_1 p$ and a are such that $e^{-\mu_1 ap}$ varies rapidly with p. For propagation of the counter pulse down the wire, $\mu_1 p$ cannot be too large, as will be seen from the theory and equations 2.50 and 2.51. Thus if $a = 0.01$ cm and $\mu_1 p$ is on the order of 0.1 cm, β^1 will be nearly constant and the relation as derived will hold. Actually the $(X/p)^2$ variation of α/p with X/p is probably the dominating variation in the region of X/p that is important. This will not be generally true under all conditions.

Increasing the O_2 concentration to 21% that for clean dry air alters the phenomena relatively little over those with 1% O_2 in N_2. Pre-onset currents are lower and increase more slowly as potential increases, as seen in figure 5.21. O_2 has absorbed many of the active photons and γ_p and γ_i are reduced. Electrons from the cathode have a good chance of attaching to yield negative ions. Photoelectric ionization in the gas near the anode is

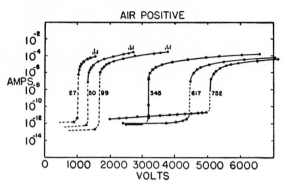

Fig. 5.21. Current-potential curves in clean, dry air with coaxial cylindrical geometry. Note the very slight slopes of the pre-threshold currents. Note also the appearance of breakdown streamers at lower pressures.

much increased. Thus the low order Townsend discharges are suppressed. Field intensified ionization currents are small. At threshold around 10^{-11} amp, depending on radioactive triggering, the current abruptly jumps to 0.1 μA. Here the oscilloscope shows occasional burst pulses. Concerning the growth of current as revealed by modern amplifying techniques in the air Geiger counter, more will be related when α particle triggering studies are described[5.15] later in this chapter (sec. B.3).

From the burst pulse threshold on, the bursts get longer and more frequent with increase in potential until the discharge becomes continuous with only fluctuations but no interruptions of the steady corona at onset. Here undoubtedly Hermstein's negative ion glow corona sets in near but above onset. The *onset* of *steady corona* was poorly delineated at 752 and 615 mm, but from 345 mm on down it was more clearly noted, and set in at 0.1 μA. Above 0.1 μA the current was irregular, but there were no periods of extinction at this lower pressure. This was caused by frequently overlapping bursts which started at different points of the wire because of strong absorption and more limited spread along the wire. The whole regime at 200 mm comprises some 40 volts in 2000. With greater photon spread at lower pressures conditions reverted to those with diluted O_2. Even then the burst pulses were two to four times longer than for point-to-plane corona in air and seemed to be of larger amplitude. The increase in amplitude may reflect more sensitive pickup with coaxial cylinders, as pointed out by English.

No pre-onset streamers were noted with certainty. Although such streamers were observed for some points in moist air, these are very sensitive to geometry and with the great length of anode for gathering ionizing events and the longer currents with spread, it is possible that they were missed.

Breakdown streamers could only be observed below 99 mm as the potential source was not adequate. They are indicated by vertical lines at the top of the curves in figure 5.21.

The observations with pure O_2 showed interesting deviations from air. The data are summarized in figure 5.22. The currents start at 10^{-12} amp but show an appreciable rise and slope well before the steep rise chosen at the threshold—perhaps even greater than in pure N_2. This characteristic differs from air and indicates an ion multiplication which did not occur in air and which leads to a steep rise nearly 500 volts higher than in air. Curves are retraceable and show no hysteresis or overshoot. The slow rise and the steep rise must thus be independent of cathode action and rise in the gas. With the steep rise at 10^{-9} amp, the oscilloscope showed *definite streamer pulses* and *no burst pulse corona*. These phenomena were really not clarified until Huber's α particle trigger study. The streamers appear up and down the whole length of the wire, as indicated by the ratio of currents to guard rings and central cylinder. The glow when it appears does not adhere to the wire but extends a millimeter or two outward around the wire. Streamers are of short range. This indicates some diffusion of triggering electrons by photoelectric emission from the anode cylinder.

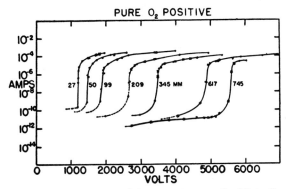

Fɪɢ. 5.22. Miller's current-potential curves in pure O_2. Note the gradual current rise below the sharp jump, in contrast to air. Above 200 mm the corona consisted of streamers along the whole wire, while below 200 mm streamers and burst pulses appeared simultaneously. Below 50 mm only burst pulses were observed.

If the peculiar behavior of O_2 is due to the short range of photoionizing photons in the gas, then reduction of pressure to 150 mm, or $\frac{1}{5}$, as for air at 750 mm, should restore burst pulses and yield the phenomena observed in air. In fact, burst pulses were noted around 200 mm in pure O_2 but not above this.

When potential was raised above the streamer producing threshold.

streamers became more frequent, yielding a proportionate increase in current. The streamers appeared to come in groups, as if one triggered a sequence of streamers. The groupings may have been purely statistical fluctuations in the random appearance of streamers. When currents reached 5 μA the amplitude was constant and the streamers came in groups of 3 to 10 on the screen; the individual streamers were separated by about 280 μsec. The groups were separated by much longer intervals of random lengths. As current continued to increase, the intervals shortened and groups were more closely spaced. At 617 mm and 9 μA the interval was 245 μsec, while at 20 μA it was 172 μsec. What determines this interval between streamers is hard to ascertain. The ion crossing time for complete clearing of space charge is 1000 μsec, and not the 250 observed. Thus triggering by electrons from the cathode by the avalanche is *not* involved.

However, if the negative ions created earlier in the avalanche can reach the anode just as the space charge clears sufficiently to permit another streamer from the same locality, the shorter interval would be accounted for.

The pale blue luminosity of O_2, in contrast to the bright blue second positive group band spectrum in N_2, and the diffusion of light to 1 mm about the wire make visual observation of the streamer glow difficult and possible only at the higher currents. With the oscilloscopic observation of streamers at 200 μsec intervals or 2×10^{-4} sec, there will be some 5×10^3 streamers per second along the wire, with possibly some 8×10^{-6} amp of current. Streamers become visible at well above 10 μA. The glow is then an integration of the light from some 10^4 or more streamers per second liberated over a length of some 6 cm of anode. It is important to note that the *Hermstein glow* corona did *not* appear *in pure O_2*.

At 345 mm threshold showed occasional random streamers. At 8 μA the heights became uniform and repetition regular. At 13 μA groups appeared at 160 μsec intervals and separation between streamers appeared. At 18 μA the streamers became so frequent and continuous that the current registered as continuous with fluctuations. Visual glow appeared much above 12 μA.

At 209 mm the threshold current began at 10^{-9} amp, rising to 0.1 μA with mixed streamers and burst pulses. With potential increase burst pulses continued, and large fluctuations set in which reached a maximum at 2 μA. Further increase in current resulted in space charge fouling which suppressed streamers and led to the Hermstein glow corona. The glow was visible only at 100 μA. Below 200 mm only a burst pulse corona, probably a Hermstein glow, was observed and, as with air at 50 to 27 mm, lasted 1 to 5 μsec before extinction.

3. Huber's Study with α Particle
   Triggering; Measurements in O_2

Using Lauer's techniques but with modifications indicated for pure N_2 (see sec. A.4), Huber studied N_2 and O_2 mixtures as well as O_2. Her data on adding 1 to 5% O_2 to N_2 have already been detailed in connection with the N_2 study. The deactivation of the cathode for γ_i, the appearance of a γ_p that was smeared and did not seem to yield a Townsend discharge, and the appearance of burst pulse phenomena associated with photoelectric ionization in the gas near the cathode have been discussed. At 5% O_2 in N_2 at 200 mm, the number density of O_2 molecules corresponds about to that of a 1% O_2–N_2 mixture at 760 mm. Even here it was clear that a threshold suddenly set in with the γ_i sequence interspersed with equally large burst pulses at about half an ion crossing time. These no doubt came from burst pulses triggered by negative ions from the avalanche.

Going over to air, Huber attempted to study the propagation of the burst pulse along the wire. In air the γ_i and γ_p phenomena should be completely submerged in the photoelectric anode process, with the pulse spread along the wire by triggering electrons. To attempt to observe the spread along the wire, recourse was had to triggering the burst pulses in two ways. In the first, the α ray track was parallel to the axis. Then along the wire the α ray triggered avalanches which started and arrived essentially simultaneously at the anode surface. This produced a contiguous series of burst pulses along the wire which could only propagate through each other and lead to a single burst pulse that rose very rapidly and quenched itself locally by space charge.

In the second method the transverse source of α particles triggered a pulse at one end of the wire. This must propagate itself over the length of the wire before it can quench itself as a whole. Thus the pulse triggered at one end should rise more slowly and last longer than the one triggered by the particle parallel to the axis. Figure 5.23 shows the oscillograms obtained in the two cases. The traces represent longitudinal and transverse α particle triggering at three sweep speeds in air at 300 mm pressure and 2720 volts and with the same gain. It is clearly seen that the longitudinal pulse rises more rapidly, is of greater amplitude at its peak, and dies out somewhat more rapidly. The areas under the two pulses are about equal. The difference in rate of rise to a peak in the two cases is particularly clear at 50 mμsec/cm and is still clear at 200 mμsec/cm.

To ensure that the differences are real, pure N_2 was used at 300 mm and 3000 volts with 20 mμsec/cm sweep speed, as shown in figure 5.24. Here

LONGITUDINAL
α SOURCE

TRANSVERSE
α SOURCE

mµ sec/cm

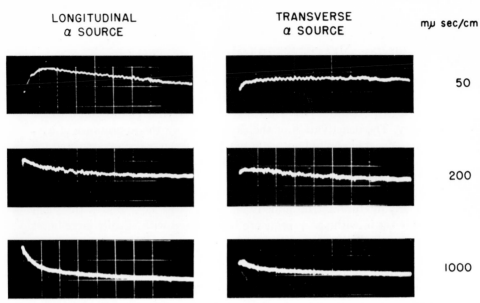

50

200

1000

FIG. 5.23. Huber's oscillograms of α particle triggered pulses in air, parallel to axis or longitudinal and transverse to axis, at three sweep speeds. Note the rapid rise of the longitudinal pulse relative to that of the transverse pulse.

α SOURCE

REL. GAIN

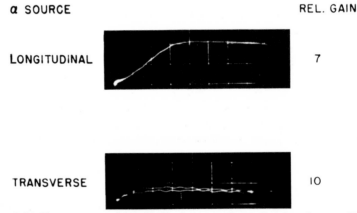

LONGITUDINAL

7

TRANSVERSE

10

FIG. 5.24. Transverse and longitudinal α ray triggered pulses in pure N_2. Note that the longitudinal pulse rise is here slower and the current much greater even with an unfavorable relative gain in comparison with the transverse pulse. This is very different from that in air, where burst spread occurs, while in N_2 it is a field intensified corona current that is proportional to the number of ions in the α particle tracks.

there is no photoelectric ionization in the gas, and all one notes is the rise and decline of the fast electronic component of the avalanches. In this case the efficiency of collection of longitudinal avalanches is much greater than for the transverse trigger. The pulse height and area for the former are thus greater than for the latter, and yet both pulses reach their peak at the same time. The rise of the transverse pulse is if anything faster, as it should be, because α particles are created nearer the central anode, while the longitudinal pulse probably suffered from poor collimation as well as from delay in generating all its electrons at a greater distance. The total ionization in N_2 was that expected from the number of avalanches that could reach the wire from the tracks in the two cases. In the air pulses, the total ionization for longitudinal and transverse pulses was the same and larger than in N_2, showing that in air there was a self-sustaining but self-quenching discharge in which ionization was independent of the triggering track.

From the shape of the pulses in air for the two cases, the velocity of spread was estimated from the rate of rise of the transverse pulse to a peak relative to that for the longitudinal one. The values at various potentials and pressures are shown in table 5.4. The velocity compares favorably with

TABLE 5.4

Huber's Values for the Velocity of Burst
Pulse or Geiger Pulse Spread in Air

Pressure, mm Hg	Potential, volts	Velocity, cm/sec
100	1580	3×10^7
200	2190	3×10^7
300	2720	4×10^7
400	3240	6×10^7

that recorded for fast self-quenching Geiger counters by Alder, Baldinger, Huber, and Metzger.[5.16] These observations prompted the author to analyze the question from the viewpoint of photoionizing absorption coefficients and led to the theory (eqs. 2.50 and 2.51) given in chapter 2, section G.3. This says that there are upper and lower limits to the velocity of spread of the ionization down the wire; it cannot exceed the time to create the full avalanche. It must not be so slow that in the time of n generations the pulse has not spread an ionizing free path down the wire, for with n generations it is locally quenched by space charge. In general in the propagating range, decrease in absorption coefficient (increase in ionizing free path) increases the speed of propagation. The free path depends on the photons involved, on the concentration of the photoionizable gas, on the sensitive zone for ionizing avalanches about the anode, and thus on potential. Where the photoionizing free path becomes comparable in magnitude

with the length of the wire or not less than one-tenth of it, the spread of
the pulse down the wire is nearly instantaneous. Isolated pulses are triggered nearly simultaneously. These rise at the build-up rate of each local
photoelectron produced avalanche sequence. Further proliferation comes
from the later avalanches. The rise will not be as rapid as for the axial
α ray track, but will be much faster than the one propagating more slowly
along the wire.

In table 5.4, *increase* in air pressure, that is, *increase* in μ, increased
the velocity, contrary to expectations. But as V the potential changed, the
effective ionizing avalanche zone r_{0g} also increased, as did the drift velocities of the electrons. Thus the failure of the theory cannot be regarded as
critical. Furthermore, as will be noted, the velocities may be miscalculated,
as spread down the whole wire was assumed and probably did not occur.

Huber then studied the influence of the percentage of O_2 at constant
pressure and at the same potential. The data at 300 mm and 2720 volts for
changes in O_2 concentration from 0 to 40% are shown in table 5.5. Here

TABLE 5.5

Huber's Data on Velocity of Geiger Counter Spread
in Air as a Function of the Percentage of O_2 in N_2
Potential = 2720 volts, p = 300 mm Hg

Percentage of O_2	Velocity, cm/sec
0	no spread
5	9×10^7
10	6×10^7
21	4×10^7
40	no α pulses

conditions are more constant. This led to an increase in velocity as μ
decreased, as theory indicates. However, the range in velocities and O_2
concentration for which spread is observed was small. There was spread
at 1% O_2, but its velocity was not established. It was undoubtedly faster
than at higher percentages, but less than the rise with a single α ray track
parallel to the wire, and spread down the wire varied with O_2 concentration.
Actually, too little is yet known about the relation between μ and r_{0g}, and
a and V. Contemporaneously, as shown in chapter 2, section G.3, Tamura[2.33]
has used the author's theory to evaluate μ and f_1f in relation to radius a
of a point in point-to-plane corona, for very fine points.

Pure O_2 gave interesting results. Figure 5.25 shows the pulses observed
below the threshold for self-sustaining corona at 300 mm pressure. The
time constant RC was 3 μsec, and the time scale 1000 μsec full sweep at
2775 volts. Townsend avalanches were small with little secondary mech-

anism. The pulse duration, as in N_2, yielded the mobility of the ion in O_2 at the low average fields, which, reduced to 20° C and 760 mm, was 2.2 ± 0.1 cm²/volt sec. This agrees within the limits of error with Varney's low field value for the O_2^+ ion of 2.25 ± 0.1 at 0° C. In figure 5.26 at 2900 volts under the same conditions as before, one sees a secondary mechanism superposed at the later portion of the decline of the positive

FIG. 5.25. Oscillogram of Huber's pulses in pure O_2 below threshold at 300 mm pressure, RC = 3 μsec, at 2775 volts with 1000 μsec full sweep.

FIG. 5.26. Oscillogram of a single sweep at 2900 volts, with other conditions as in figure 5.25. Here are seen the irregularities superposed on the ion movement curve. As these last for the whole crossing time and beyond, unlike Lauer's data in air, which are due to negative ions near the anode, these are a series of small, very localized burst pulses that do not spread and that extinguish themselves by space charge.

ion pulse. It is really difficult to state whether these are delayed secondary pulses due to fluctuations in the arrival of the slower negative ions at the anode or to essentially small localized burst pulses that extinguish themselves rapidly. Since only a few negative ions are formed near the cathode by the avalanches, the prominence of the sustained currents near the end of the positive ion crossing might point to small gas photoelectric conditioned local discharges at the anode. These pulses were, moreover, characteristically potential-dependent at all pressures in pure O_2. That is, the fields had to be great enough to cause a field intensification of photoelectrons created very near the wire. The statistical attachment behavior would not be so characteristically potential-dependent. However, these burst pulses do not spread down the wire, and they last on the order of about 0.2 the duration of those in air, which are shown in figure 5.27 for 200 mm pressure at 2120 and 2145 volts in air at 1000 μsec full sweep. These also occur with natural ionization triggering in the gas as well as with α particles.

VOLTS

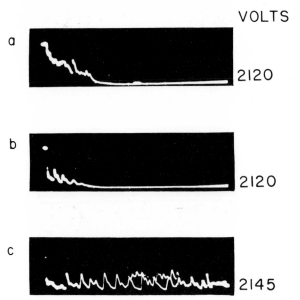

a

2120

b

2120

c

2145

Fig. 5.27. Oscillograms at 200 mm pressure in air at 2120 and 2145 volts at 1000 μsec full sweep time. These are burst pulses that propagate down the wire, that is, Geiger counter pulses, and go to self-sustaining corona at 2145 volts.

If the potential is raised to 3050 volts under the same conditions as in figures 5.25 and 5.26 but with gain reduced 200-fold, the pulses of figure 5.28 are seen. With fast sweep at RC 0.003 μsec and 1 μsec full scale, figure 5.29 shows a single sweep trace of one of these pulses. The amplitude is more than 200 times as great as that of the preceding ones. They are spaced at uniform time intervals of around 200 μsec. Their duration is 10^{-7} sec. Instantaneous currents in these pulses are as high at 10^{-4} amp, corresponding to $\approx 10^8$ electrons. These are identified with the streamers reported by Miller. The streamer spacing is closely related to ion transit time, and corresponds to two-thirds of the duration of a single ion pulse at lower potentials. It appears that the ions from the first streamer must be swept out of the gap before the next one comes. Huber did not carry these observations on to breakdown and higher potentials, as did Miller. The observations of Miller indicate that these are not a continuous sequence of γ_i conditioned pulses leading to a steady corona, for his occurred without α pulse triggering and represented random groups of streamers. There was no suggestion of a continuous succession, as might be implied by the sequence of five regularly spaced pulses triggered by particles in Huber's study. The visual length of the streamers in Miller's study was from 0.1 to 1 mm out from the wire. In Huber's work they were not much in excess of 2 mm.

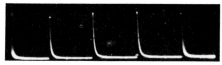

F<small>IG</small>. 5.28. Pulses seen for pure O_2 under conditions analogous to figures 5.25 and 5.26 at 3050 volts, but with amplifier gain reduced 200-fold. These regular pulses are a sequence of streamer pulses.

F<small>IG</small>. 5.29. High-speed oscillograms of several superposed streamer pulses triggered on the electrical rise, RC = 0.003 μsec and 1 μsec full scale. Amplitudes here again are 200 times those observed in figures 5.25 and 5.26.

4. P<small>HOTOMULTIPLIER</small> S<small>TUDIES OF THE</small> B<small>URST</small>
   P<small>ULSE</small> S<small>PREAD</small> D<small>OWN THE</small> W<small>IRE AND</small> I<small>TS</small>
   R<small>ELATION TO</small> G<small>EIGER</small> C<small>OUNTER</small> A<small>CTION</small>

As Huber terminated her study, it was suggested that she confirm the spread of the Geiger counter pulse by using a screen for a cathode cylinder and triggering the pulse by an α particle gun at one end, employing the two-photomultiplier technique of Amin to record the passage of the luminous phase of the pulse along the anode wire. It was suggested that, in view of its luminosity, air be used as a filling gas and that speed be measured at various pressures. Huber attempted to detect the pulses by photomultiplier with slit transverse to the axis triggering on the electrical pulse rise. She got no results. It was also suggested that she coat the glass walls of the tube with one of the inert substances, such as Octoil S, used in nuclear studies, which fluoresce strongly in the visible when exposed to ultraviolet light. She was not successful.

The study was resumed by Condas[5.17] in the author's laboratory. Condas did succeed in getting the photomultipliers to respond and record the Geiger counter-like light pulses. He triggered on the electrical rise but could use only one photomultiplier. Even then the cells had to be carefully chosen and the signal was so weak that a high resistance with an RC time constant of 100 μsec was needed. It was impossible because of noise and other problems to get two sufficiently balanced photomultipliers to register. This precluded very accurate or careful study. Only once did he succeed in recording the light pulse without time distortion when there was no resistor. This pulse had a duration of about 10^{-7} sec, and its width was estimated to be on the order of 1 mm, with a speed of propagation of $\approx 10^7$ cm/sec down the tube. What did emerge from the study was that, over the very narrow Geiger counting plateau, covering a range of some 20 volts in 1200 to 1600 for pressures of 50 and 100 mm, there was an enormous statistical fluctuation in the pulses observed. That is, under *fixed* conditions the succession of pulses viewed were found to range in distance of propagation from 1 to 6 cm from the gun along the anode.

Very many readings had to be taken at a given distance from the gun to get the number of pulses that exceeded that distance, or range, at a given potential and pressure. By taking a large number of readings, one had the fraction of light pulses that traveled x cm or more from the gun. The number of oscillograms required for this was unbelievably great, because not every trigger yielded a pulse at a given setting, so that progress was slow at best. Also, the amplitude of the pulses was never the same. There were unquestionably spurious pulses caused by cosmic ray events that started elsewhere along the axis than at the gun.

Figure 5.30 shows Condas' tube, and figure 5.31 shows the results of this study. Here the settings of the photomultiplier slit from the gun are given, and the time of arrival of signals is given. The full lines show the ranges in time at each distance observed. A line drawn through the midpoints gives the velocity of the pulses as about 2×10^6 cm/sec. This value is low compared to Huber's values. Numbers to the right give the number of pulses observed in the interval. The velocity spread is from 5×10^6 to 1.4×10^6 cm/sec.

It appeared that, between threshold for propagating pulses and self-sustaining burst pulse corona, the *average distance of propagation of the pulses increased with potential*. Not enough data at different pressures could be obtained to get a pressure variation of velocity. The work was reported at the American Physical Society meetings at Stanford in the summer of 1957.[5.17]

After Condas observed such fluctuations in his study, a careful review of the Geiger counter literature revealed that indeed, even when the insensitive counting techniques were used, there were very large and disturbing fluctuations. Thus the nicely plotted curves of Alder, Baldinger, Huber, and Metzger[5.16] really were average curves drawn from a mass of statistically distributed data.

Condas was unable to continue his work owing to ill health. The study was carried to a successful conclusion by H. M. Herreman and is here reported for the first time. Herreman first confirmed Condas' observations, but was not satisfied. Accordingly, improved oscillographic techniques were developed, thanks to the technical advice of Dr. Julius Muray.

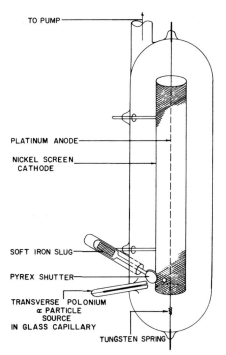

FIG. 5.30. Condas and Herreman's tube for observing the passage of Geiger counter pulses in air when triggered by an α particle using photomultipliers.

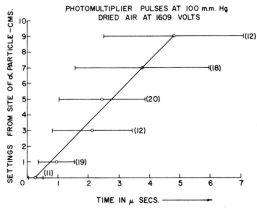

FIG. 5.31. Summary of Condas' preliminary study of the time-distance data for luminous Geiger counter pulses propagating down the anode wire in the counting region.

The counter consisted of a 0.13 mm diameter Pt anode wire mounted along the axis of a cylindrical glass tube and held taut by a spring at one end. The cathode cylinder was a 1 mm² mesh Ni screen of 3.5 cm diameter and 20 cm long, as shown in figure 5.30. The anode wire extended 4.8 cm beyond the Ni cylinder so that the gun was 5.3 cm from one end of the anode wire. The other end of the anode projected 7 cm beyond the Ni cylinder. At 1.5 cm from one end of the cylinder was an α particle gun. This consisted of a 0.3 mm inner diameter capillary tube. Into one end of this was inserted a wire coated at first with Po and later with Pu^{238}. The tube extended 9 mm beyond the end of the wire and could be closed by a shutter to prevent the emission of α particles. The axis of the gun traversed the tube at a distance of 8 mm from the wire. It liberated a burst of α particle-created electrons along a trajectory such as to give rise to a large number of nearly simultaneous avalanches to initiate a Geiger counter pulse at one end of the system. The measured activity of the source and the geometry indicated that there should be about 2 α particle pulses per minute. The gas was room air dried over silica gel.

A block diagram of the apparatus is shown in figure 5.32. The power supply consisted of a bank of dry cells giving a total of 5 kv with taps and a potentiometer that permitted control of the positive potential on the

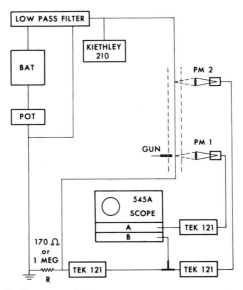

FIG. 5.32. Block diagram of Herreman's arrangements for observing Geiger counter pulse phenomena, including velocity of propagation, with two photo-multipliers.

anode wire to within 1 volt. The gauze was grounded. The most difficult problem in measurements of this sort lies in providing adequate electrical shielding. In the block diagram of the measuring system, the shielding is not indicated; however, everything from the filter, including tube photomultipliers and wiring, was carefully screened. Leads used carefully shielded coaxial lines. The power from the extensive area of the batteries went through a low frequency pass filter to eliminate high frequency noise. Potential was measured by a 100 to 1 potential divider and a Keithley model 210 electrometer. For rapid time resolution analysis, the current from the cathode went to ground through a 170 ohm resistor, labeled R in the figure. Otherwise a 1 megohm resistor R was used where amplification of weak electrical pulses was needed. The improvement in technique with adequate instrumentation came through the suggestion of Dr. Julius Muray that the 545A Tektronix Synchroscope could be triggered on the rapid rise of the electrical pulse and that two photomultiplier signals could be displayed in proper time sequence on a single trace, using type CA plug in dual preamplifiers. It was observed by Herreman that, if a simple T connector was used, the electrical current pulse could also be displayed. The arrangements used then follow.

The Tektronix 545A Synchroscope had two inputs, A and B, leading to the single sweep. The dual inputs A and B each went to two Tektronix 121 preamplifiers, which in turn went to the two photomultipliers. Input A went to PM 1, which viewed the end of the tube proximate to the plane of the gun. Input B went to PM 2, which was movable and could view the cell at a plane at any distance x from the source along the anode. The light from the anode at the gun and at x was picked up by means of lenses and focused on the PM's. The current due to the rapid rise of the electrical pulse taken from R was fed through a preamplifier and a T connection to input B. Thus the steep rise of the electrical pulse triggered the sweep of the scope. Recorded on the screen were the algebraic sum of the rise and decline of the electrical pulse, together with the light pulses from PM 1 and PM 2 as the light of the pulse passed by each PM. These light pulses were sharp and of short duration, ≈ 0.2 μsec. Since PM 1 had a characteristic signature, because of a slight ringing, its pulse was clearly differentiated from that of PM 2, which had none. This proved to be most valuable in cases where the pulses did not originate at the gun, so that sometimes PM 2 registered before PM 1, indicating a spurious count. Furthermore, the electrical rise of the pulse was the same, regardless of the part of the tube in which the ionizing agent appeared, though the shape of the pulse differed.[5,16] Differences were noted in that at 300 mm the light pulse at PM 1 appeared slightly earlier relative to the peak of the electrical pulse than it did at 200 mm, while at 100 mm it occurred later. This would indicate different rates of rise of the electrical pulse relative to the appearance of light at different pressures at the same location. In actual counting at a given potential, the gain of the preamplifiers was set so that the pulses of maximum amplitude just filled the screen. Then only those pulses that registered above one-third the maximum height were counted as significant. Such pulses usually corresponded to the pulses that spread down the tube and were thus true Geiger counter pulses. As assorted sizes of smaller pulses were present, this precaution was essential, as will be seen.

With these improved techniques, the statistical fluctuations in size, range, and velocity of pulses was large, but much less than in Condas' observations.

Two serious problems, however, were encountered that required adequate control measures. One of these was the increased number of pulses counted as sensitivity was increased at a given potential setting. The large range of pulse sizes was related to the potential applied relative to its location on the counting plateau. This at once suggests that the second problem involved the establishment of a reference point on the counting plateau to which applied potentials could be related. In consequence, a method of discriminating between valid and spurious counts, which was capable of being assigned to the potential relative to the counting plateau, was needed. The solution to these problems required a critical study of electrical pulse amplitudes at increasing sensitivity below and along the counting plateau, and the determination of some fixed plateau potential.

Since the range of the phenomena at 200 Torr was the greatest, many of the data were taken at this pressure. Figure 5.33 shows two photographs taken to investigate electrical pulse size at a high sensitivity at two different potentials. Each photograph shows ten horizontal sweep lines. Each line represents an exposure at 2 msec/cm sweep speed with 60 successive sweeps superposed during 1 sec. The length of each sweep line was $\approx$10 cm, thus totaling counts for 20 msec/cm of time. With a total time of 12 sec per photograph, the number of pulses per photograph multiplied by five gives the approximate number of pulses of any type per minute. To observe all pulses, great and small, the oscilloscopes' matched impedance dropping resistor R of 170 ohms was replaced by a 1 megohm resistor, since amplitude, not detailed time resolution of the pulses, was required. The photographs of figure 5.33 were taken on the counting plateau at 18 and 28 volts below onset threshold, and have been retouched for purposes of reproduction. It will be noted that the pulses fall roughly into three amplitude groups, although there were always scattered pulses of intermediate sizes. These groups were class a, smallest (barely perceptible); b, intermediate; and c, large.

Calibration of the oscilloscope pulse height in terms of electrical quantity involved indicated that the a group had less than 10^6 and perhaps as few

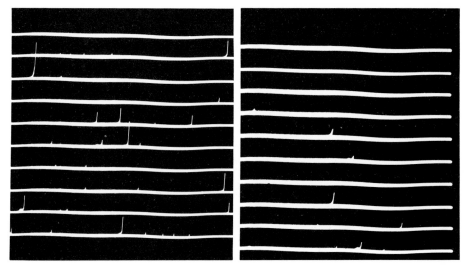

FIG. 5.33. Reinforced oscillograms of sweeps over a total time of 12 sec per photograph, of which two are displayed with 10 lines each, each line having 60 successive sweeps recorded. The number of pips of each size group per photograph multiplied by 5 gives the number of those counts per minute. Note the three principle size groups: large pulses about the height of each trace; intermediate pulses; and pulses so small that they are just perceptible.

as 10^5 ions and electrons. These could have been avalanches or groups of fewer than ten nearly simultaneous avalanches consisting of $\approx 10^5$ or fewer ions and electrons. Alternately, they could have been caused by a succession of avalanches lasting on the order of ten generations, covering perhaps 10^{-7} sec total. The termination of such avalanche sequences could be caused by statistical avalanche fluctuations and/or by local positive ion space charge choking. The second, or b, group had between 10^6 and 10^7 ions. These probably corresponded to Amin's primary and secondary sequence of burst pulses that spread very little, < 1 mm along the anode, and extinguish locally through space charge choking. The last, or c, group contained more than 10^7 ions and ranged up to several times 10^8 ions per pulse. These were identified with the Geiger counter pulses that propagated along the anode wire. Their amplitude is limited by the distance traversed along the anode, as will be seen. They were triggered by α particles at the gun or by γ and cosmic ray events elsewhere.

The gathering of statistics from the record, as shown, required care and experience gained by visual observations. There was some persistence of luminosity of pips on the screen. The presence of pulses of intermediate size increased the difficulty. Thus only clear and sharp pulses were counted, leading to the data summarized in table 5.6.

TABLE 5.6

Tabulation of Count Rates for Various Pulse Sizes

Sweep rate, 2 msec/cm; pressure, 200 Torr;
print scale factor, 0.7; cathode resistor, 10^6 ohms

Voltage (Below Onset)	Sensitivity, millivolts/cm	$< 10^6$ ions/pulse, counts/min	$10^6 - 10^7 >$ ions/pulse, counts/min	10^7 ions/pulse, counts/min
−18	80	15	35	15
−28	80	15	15	5
−28	50	35	25	15
−38	50	25	25	15
−48	50	5	35	5
−58	50	20	15	5
−68	17	70	50	5

Before discussing these results, one must consider the question of a reference potential. Small changes of conditions from day to day, as in all such measurements, lead to variations in absolute values of the onset or other threshold potentials. More significant than the averaged numerical threshold potential value is the determination of the potential for onset of self-counting, or steady corona, at the time each measurement is made. This fixes the upper limit of the counting plateau at which individual significant ionizing events may be differentiated from spurious or self-induced pulses. The extent of the counting plateau in air at best is very limited, falling in a range of perhaps 60 volts below onset and extending to within 5 volts of onset. A few larger nonpropagating pulses could at times be observed at as low as 200 volts below onset. Onset was checked for each series of observations by a quick analysis based on a change in form of the pulse traces at onset as revealed on high gain, using R = 170 ohms. The more exact determination consisted in a tedious search observing the onset of self-triggering discharge with the α particle gun inactive, a practice among cosmic ray investigators at Berkeley. This was used to check the more rapid method of measurement, which agreed within about 3 volts.

Table 5.6 assembles the data from seven such photographs, giving the potential in volts below onset at which they were taken, the sensitivity of the recording in millivolts per centimeter, and the number of counts per minute in the a, b, and c groups. It must be noted that the data were meager and observed counts were multiplied by five. Since fluctuations are large in such small counts, the data are symptomatic only. The sweep rate was 2 msec/cm, pressure p = 200 mm, print scale factor 0.7, R = 10^6 ohms. It is seen that at low sensitivity at −18 volts, only the larger pulses are observed and counted. There are but few a pulses and the b group dominates, although there are 15 c type counts. Not all c type counts in the 12 sec periods were α particle triggered. In fact, many larger counts came from cosmic ray bursts. The tube was mounted vertically, so that on the order of 8 or 10 counts per minute could have been due to cosmic rays. The oscilloscopic velocity studies indicated that about 2 counts per minute came from the α particle gun as estimated. At this same sensitivity, potential was reduced to -28 volts. The c or Geiger counter pulses dropped to 5, the b pulses dropped to 15, and the small pulses remained the same. Obviously at lower potentials a number of events that previously grew, spread, and propagated at the higher potential could no longer grow.

Thus potential is a critical factor influencing pulse size. However, since these counts are small in number, taken over only 12 sec, the influence of statistical fluctuations on the data must not be ignored. Increase in sensitivity at -28 volts doubled the a count, nearly doubled b, and trebled pulses previously noted as c. This means that, when sensitivity is low, pulses that should have been counted as a, b, and c passed unrecognized. Progressive reduction in potential to -38, -48, and -58 volts, at constant sensitivity, produced a progressive decrease in the b and c groups insofar as statistical fluctuations permit conclusions to be drawn. The smaller a and b pulses gained as c pulses declined. A further increase in sensitivity to four times its original value greatly increased the a and b pulses at the lowest potential of -68 volts, but left the c pulses unchanged. The five c pulses were probably caused by the α particle and largest cosmic ray triggers, and thus launched Geiger counter pulses well below the counting plateau at -68 volts.

One may conclude that there are a large number of triggering ionizing events due to single, small, and large nearly simultaneously generated localized groups of electrons that give rise to pulses. The smallest of these groups gives rise to on the order of 10 avalanches or short avalanche sequences that die out, largely through statistical fluctuations and space charge effects. The next size group are either initiated by larger ionizing events or at higher anode potentials, and are able to sustain themselves for a longer sequence of generations. They do not produce enough photoelectric ionization in the gas to spread or advance very far over the surface, and thus they choke themselves off by local space charge accumulations. There are finally ionizing events comprising some 10^3 or more nearly simultaneous avalanches. These create enough photons in the high field region to propagate along the anode wire and give large Geiger counter pulses. At low potential all groups can be detected at high sensitivity, and probably in more or less proper proportion to the activating events.

As potential is raised, all pulses increase in intensity, but, what is worse, smaller pulses increase in relative magnitude faster than do the large c group pulses, which cannot increase in size beyond a certain limit. Increase in potential can then introduce smaller ionizing events which should not be counted into the counting regime. It is at this point that careful discrimination is necessary. It is for this reason that in the study of the pulses, the criterion to limit counting to pulses only in excess of one-third the maximum height was in keeping with recommended practice in the use of Geiger counters.[5.18] With this criterion, only those events which caused pulses to propagate down the tube as proper Geiger counter pulses were recorded.

It is now of interest to report the findings, which were most complete for the 200 mm air pressure. The data assembled are represented in the composite of graphs of figure 5.34. Represented are, first, the pulse count rate for the Geiger counter-like pulses of maximum amplitude, group c, plotted against the potential applied to the anode below the self-counting onset threshold. The potential at onset was about 2150 volts. Here pulses counted

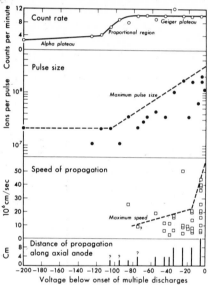

FIG. 5.34. Composite of Herreman's graphs of Geiger counter pulse counts per minute, ions per pulse, speed of pulse propagation, and distance of propagation along the anode, all plotted against potential in volts below the onset threshold for steady corona.

were those which, at the sensitivity used, filled the screen and down to one-third this amplitude. Thus at a potential at which the largest pulse amplitude covered the screen at the gain used, which might represent say 1.2×10^8 ions per pulse, only pulses from 4×10^7 to 1.2×10^8 ions per pulse were counted. This excluded the smaller pulses. In figure 5.35, center trace, at 170 ohms resistance and sweep rate of 0.2 μsec/cm, every pulse was recorded for 30 sec. There are just 5 pulses that are greater than one-

third the peak amplitude. This corresponds to 10 pulses per minute at
-5 volts below onset, with 4×10^7 to 1.2×10^8 ions per pulse.

It is noted in figure 5.34 that the count rate rose from 2 to 4 up to -100
volts. Thereafter the count increased linearly with the potential to a plateau
value of 10, which was reached at -75 volts. These counts were electrically
recorded, using the amplitude criterion adopted above. They continued
constant to onset and then rapidly increased. The points in the next lower
trace represent the magnitude of the electrical pulses, which ranged from
7×10^7 ions to around 2×10^8 ions per pulse. The pulse size at the same

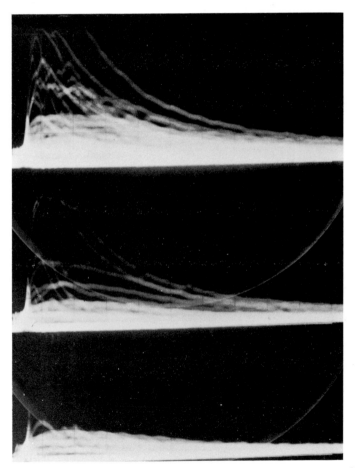

FIG. 5.35. Center trace, oscillograph traces at 0.2 μsec/cm sweep speed, with
170 ohms, showing all pulses observed for 30 sec. Here just 5 pulses are greater
than one-third the peak amplitude. Note the many smaller spurious pulses.

potential varied considerably, the extreme range at -20 volts being from 3×10^7 to 2×10^8. The maximum pulse size started at 2×10^7 before the linear rise, and increased thereafter to 3×10^8 as threshold was neared. The reason for the increase in size is clear from the increased maximum distance of propagation of the light pulses in this potential range, as will be noted.

The lowest plot shows the relative maximum distance of propagation of the luminous pulses down the wire at various potentials. It is noted that in the linear rise region of the counting range, at -100 to -75 volts, the pulses propagate no more than 2 cm, but that maximum propagation increases up to 12 cm down the tube at -5 volts. The distance of propagation was determined by setting the second photomultiplier at various distances along the wire from the gun and noting how far down the wire the light pips leaving the gun region were detectable with the second photomultiplier. From the oscillograms of the two pulses, the time of departure of a given pulse and its arrival at the second photomultiplier could be observed, and hence the speed could be calculated. There were, as stated, many spurious counts, and α particles did not always trigger the sweeps.

Examples of the photomultiplier traces yielding velocities are shown in the four sets of sweeps in figure 5.36. Trace 3 in set a shows only one clear-cut case where the sweep was triggered at the gun. Here the sharp electrical rise at the origin at the left was followed by a spike from PM 1 with its characteristic ringing. At 5 cm, or 1 μsec later, the peak from PM 2 gave a signal. The distance covered between PM 1 and PM 2 was 4 cm. The speed of the pulse was 4×10^6 cm/sec. Note in trace 1 in the same set that about 1 μsec after an electrical pulse was registered triggering the sweep, PM 1 gave a light signal, and 0.6 μsec later it reached PM 2, 4 cm distant. This can only mean that a pulse started somewhere in the tube but that the light did not reach PM 1 for 1 μsec and PM 2 for another 0.6 μsec later. This pulse probably started on the anode well before the gun, which is located 5.3 cm beyond the end of the anode. Set b again has only one clear sequence at trace 3. Note that PM 1 was activated by the electrical pulse on 3 sweeps and that only one of these had a range that reached the 4 cm to PM 2. There were also two electrical pulses that gave no signal at either PM. Set c shows one fast pulse in trace 4 at the bottom, which covers 4 cm at a speed of 1.3×10^7 cm/sec. Trace 2 is delayed some time beyond the pulse rise. It could have been triggered from behind the gun and represents a slower pulse. It was not included in the summarized data. Trace 1 shows an electrical pulse which triggered PM 2 but did not appear to have reached PM 1. Set d, trace 3, gives a record of a pulse at the sweep speed of 0.2 μsec/cm that propagated 12 cm between PM 1 and PM 2. The speed here was 1×10^7 cm/sec.

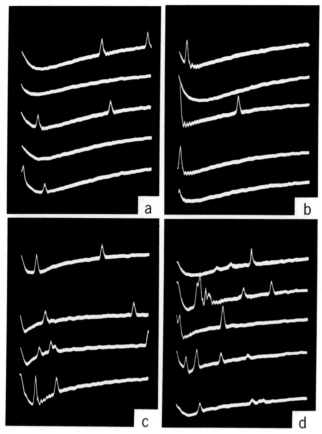

FIG. 5.36. Oscilloscope traces of the electrical rise pulse, the light signal arriving at PM 1 at the α particle gun, and then later the light signal arriving at the signal PM 2 placed at various points along the anode. Note the rise of the electrical pulse at the extreme right. PM 1 has small oscillations on its decline that identify it. PM 2 has a signal that *usually* follows PM 1. Time goes from left to right in all cases.

Such illustrations indicate that to get good data hundreds of traces must be photographed. The data on the speeds are plotted for traces taken on the counting plateau when propagation of the pulses exceeded 4 cm. The speeds in general remained low and gave much scatter, even at -5 volts. They rose to values of around 4×10^7 at -5 volts when pulse size exceeded 10^8 ions per pulse and propagation went 12 cm down the tube. It is seen that Condas' average speed of 2×10^6 was a bit on the low side, but could well have corresponded to pulses taken at potentials as low as -20 volts below onset. His highest values were on the order of 5×10^6 cm/sec.

Unfortunately Herreman was only able to devote a limited time during vacations to taking data. This precluded an extensive study for other pressures. At 100 mm the data are displayed in figure 5.37. The plateau was more limited and the counting rate was somewhat less. The ions per pulse increased linearly as the potential neared the onset of 1550 volts over a range of 40 volts. These did not differ too much from those at 200 mm. The speed of propagation in the counting range was low and did not exceed 2×10^7 cm/sec. The pulses began to propagate about 4 cm at -30 volts, reached 12 cm as before, and are commensurate with the large ion yield.

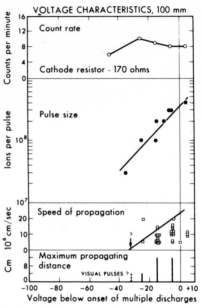

Fig. 5.37. Herreman's graphs at 100 mm.

Data at 300 mm are shown in figure 5.38. Here the counting plateau remained around 10 or 11 to as low as -110 volts. Onset occurred at 2750 volts. The pulse size increased linearly from around -80 volts, but did not exceed 1.5×10^8 ions per pulse. Speeds were mostly under 1.5×10^7 cm/sec, and some as low as those of Condas at 2×10^6 cm/sec were observed. Pulse lengths, as might be inferred from the low ion count, were under 4 cm, even at -5 volts.

At 400 mm the counting plateau was limited indeed, onset occurring at 3340 volts. The plateau was not reached until about -40 volts. The counts ranged, as before, around 10. The ion pulse size remained small and ranged from 3 to 5×10^7 ions per pulse. Pulse lengths remained too short to permit velocities or pulse lengths to be observed with any accuracy.

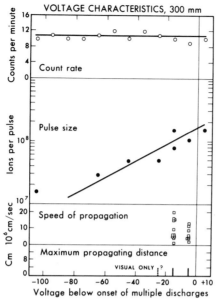

Fig. 5.38. Herreman's graphs at 300 mm.

The conclusions to be drawn from this study are that the Geiger counter spread regime is indeed very limited in air; that the appearance of pulses that propagate along the wire is restricted to pressures between 50 and 400 mm; and that the optimum pressure appears to lie around 200 mm. The plateau and range of counting are probably greatest at 200 mm and decline at 100 and 300 mm. The maximum range of travel of the pulses lies at 200 and 100 mm. The plateau at 100 mm is, however, decreased. The ions per pulse are proportional to the range of propagation of the luminous pulse, which was observed to go as far as 12 cm. The speeds are highest at 200 mm pressure. Pulse size, spread down the wire, and speed increase as the onset of self-sustaining corona is approached.

It appears that the range of the photoionizing radiations depending on the partial pressure of O_2 are an important factor in the spread. Two radiations, one of $\mu_i \approx 38$ cm^{-1} from O_2 and one of $\mu_i \approx 25$ cm^{-1} from N_2, are active. The absorption coefficients at 200 mm air pressure are thus roughly 2 cm^{-1} and 1.3 cm^{-1}, so that the photoionizing free paths for optimum Geiger counter propagation in air are around 5 to 7 mm. It is clear that not only is the photoionizing range of importance but the total ionization, that is, the quantity $Ne^{\int \alpha dx}$ of initiating ions in the pulse, is important for propagation. Here N is dependent on the triggering source, and $e^{\int \alpha dx}$ is the multiplication by the field. The latter is highly sensitive to potential. Unless, then, there are enough photoionizing photons in the triggering event, and the photoelectric absorption is adequate, the pulse will not move down the wire. At 400 mm the distance of propagation is very limited.

It is also to be noted that, in the background of any such tube, there is a whole family of smaller ionizing events in the a and b pulses that, given enough amplification by the field, can grow to propagating pulses leading to spurious counts. How many of these are due to external events, and how many are caused by secondary emission and γ effects from the cylinder is not known. These are the ones that on amplification yield the self-counting regime at onset. This study helps to clarify the basic physics underlying such counters.

Perhaps a brief discussion of the speeds recorded is appropriate. Condas observed a mean speed of 2×10^6 cm/sec at 100 mm with a range of values from 1.4×10^6 cm/sec to 5×10^6 cm/sec. Herreman observed speeds ranging from 2×10^6 to 4×10^7 cm/sec at 200 mm, and not more than 2×10^7 at 100 mm. Huber observed speeds ranging from 3×10^7 cm/sec at 100 mm and 1580 volts, the same at 200 mm and 2190 volts, 4×10^7 at 300 mm and 2720 volts, and 6×10^7 at 400 mm and 3240 volts. The anode cylinder used was about 6 cm in length.

It must first be noted that there were in Herreman's clearly defined time-distance measurements a great range in speeds; that the speeds increased as the onset was neared; and that speeds, pulse size, and distance of propagation down the anode were closely correlated. High speeds went with longer ranges. In Condas' study the data were not taken with reference to a carefully observed onset potential, as were Herreman's. Thus Condas worked at the lower portion of the speed range. Both these sets of measurements observed the distance traversed by the luminous pulse and the time required, though Condas' method included spurious pulses. Huber's speeds were computed by comparison of the rise time of the transverse and the longitudinal pulses. The calculation was predicated on the assumption that each of her pulses propagated down the 6 cm of the wire. Huber also worked at slightly higher potentials, that is, even nearer onset, than did Herreman. Thus her very few observations at 100 and 200 mm are not seriously out of line with Herreman's. They would probably conform closely if she had been able to know the true length of propagation of her pulses along the wire. Thus, for example, had her pulses propagated 3 cm instead of the 6 cm assumed, her speeds would have been 1.5×10^7 instead of 3×10^7. The great decrease in range of the pulses at 300 and 400 mm in Herreman's work unquestionably accounts for the *apparent increase* in speed of the pulses at these values noted by Huber. In fact, at 400 mm Herreman got no pulses exceeding 4 cm at this pressure, and ion counts per pulse, despite the increased pressure, were less than at 200 mm, indicating that pulses could have propagated perhaps 1 or 2 cm instead of the 6 cm used by Huber in computation.

It is probably improper to compare the speeds in the air counter with those in the A-Organic, or conventional fast ion counter mixtures. Most speeds reported ranged around 10^7 cm/sec. Here again most data came from assumed anode wire lengths and rate of rise, as did Huber's. The only observations that were more explicit and gave true time-distance curves were those of Alder, Baldinger, Huber, and Metzger.[5.16] They used a sort of a Fizeau principle. If the cathode cylinder was constricted at two points along the anode, separated by a known distance, the single long counter could be divided into three separate counters. If proper pulsed potentials were applied to the two cathode constrictions, the propagating discharge triggered at the first of the three tubes could be passed to the center section and on to the third section. Otherwise the passage was prevented by a bias potential. Then by applying pulsed potentials that opened the gates for passage of the pulses and by varying the frequency of the pulsing, it was possible to determine the time elapsed for the pulse passing through the first gate to reach the second over the length of the second chamber. Thus when pulses from section 1 reached section 3, the length of section 2 had been traversed in a known time.

The speeds there recorded ranged from 8×10^6 to 1.7×10^7 cm/sec. These, although in a different gas system, are commensurate with those observed by Herreman. In their study, average instead of individual pulse velocities were measured by determining the frequency at which the two end counters most often coincided. In Herreman's study individual pulse velocities were measured.

C. THE PHYSICS OF THE
GEIGER COUNTER

1. INTRODUCTION

In the preceding discussions concerning the burst pulse type of corona in air, and in the preceding section, in which pulse size was discussed, frequent reference has been made to the Geiger counter.

In the discussion to follow, which will be developed largely in a historical sequence, it will be noted that photoelectric ionization in the gas, which is responsible for the burst pulse corona in air, was in the period 1935–1937 independently recognized by several investigators as the mechanism causing propagation of the pulses in the *fast* Geiger counter.[2.27] Thus very early the author[2.28,3.2] identified the burst pulse corona spread as being the same as the spread in the Geiger counter action. At that time the differentiation between *slow* and *fast* counter action was just beginning to be recognized. In view of the fact that the corona studies have done a great deal to help clarify and explain details of the Geiger counter action, and since most Geiger counter investigations, being developmental, bypassed a study of the basic phenomena or oversimplified them, it is of importance to devote some space to this useful instrument.

In about 1908, when α particle counting became of importance, Geiger in Rutherford's laboratory, among others, began to search for faster, more objective counters to replace the tedious and tricky visual scintillation counting techniques. He first devised a small hemispherical, positive corona point opposite a large concentric hemispherical cathode. The α particles entered a hole in the outer hemisphere and moved along the axis of the system so that they nearly struck or even grazed the positive point. The positive point was at a potential where it just failed to give corona. The intense burst of α particle ionization generally caused the point to give a heavy corona discharge. This gave a pulse capable of being recorded in those days before electronic amplifiers were known. The discharge, however, was not certain in its action, as its potential range of response was limited to near or slightly above threshold potentials and spurious discharges were hard to prevent. The point, as well, had a limited sensitive volume. Some attempts were made to use negative points for counters, but these were even more capricious in their action. Later analysis of negative counters whose sensitive volume was even more confined that the positive counter was carried out by L. F. Curtis.[5.19] It indicated such counters to be value-less, as then used, since they depended for their action on surface conditions of the cathode which never remained constant.

Around 1928, however, Geiger and Mueller,[5.20] experimenting with posi-tive wire coaxial cylindrical coronas, discovered that a Cu wire oxidized lightly in a Bunsen flame gave quite reliable counts. Improved amplifying and recording circuits were then available, and the need for counters for γ rays and the newly recognized cosmic rays, as well as for α particles, was becoming acute. Investigation of the oxidized wire counter soon revealed that the function of the oxidized Cu wire was merely to place a fairly high resistance in series with the high potential positive wire. This resistance, once the corona ignited and drew a current i, gave an iR drop which, added to the space charge about the positive wire, reduced the potential below V_0, the onset value, and extinguished the corona. Thus, it was soon recognized that any clean wire could be used if it had sufficient resistance in series with it and the high potential source. This represents the *slow* type of Geiger counter action.

At this point in the study of Geiger counter action, probably the clearest analysis followed from the investigations of Sven Werner,[5.21] who made very extensive studies of the action. This slow type of counter, of the period from 1935 on, was limited in that the time to discharge and recharge the circuit after it fired depended on the RC time constant. Thus the period

$\tau \approx 1/RC$ became long, since the values of R needed to quench ranged from 10^9 to 10^{10} ohms. The counter could be applied either to the Townsend type of discharge with a cathode γ or to the burst pulse type of discharge in coaxial cylindrical geometry. The self-quenching properties of the positive ion sheath in the burst pulse type of corona and its motion across the gap, independent of the circuit-reduced iR drop of the earlier counters, was perhaps most clearly first delineated by the Montgomerys in 1941,[5.22] although Werner also recognized this action. This foreshadowed the development of the fast self-quenching counter. The positive ion space charge sheath can by itself be used to quench, but the potential range of operation —the plateau—is very narrow, as is seen in Herreman's study in air. Various other devices, such as external quenching circuits of short response time, were resorted to and gave somewhat better results.

In general, the large number of intensive researches from 1934 on by numerous workers increased knowledge of the mechanisms active. In these studies it was shown that most of the *secondary actions at the cathode cylinder* by photons, metastable atoms, and positive ions, while yielding usefully heavy discharges, tended to prolong discharge and/or make the counter yield spurious counts because of delayed secondary triggering. Further trouble was encountered by the formation of negative ions near the cathode. Thus, final resort was had to using insensitive cathode surfaces (oxidized Cu) and gases which did not yield metastable states, or else mixing inert gases with small amounts of "quenching" gases. These acted to destroy metastable states, altered the ions to slow types yielding a low γ_i at the cathode, perhaps by predissociation, resulted in heavier space charges, and, finally, absorbed photons giving ions in the gas and a high βf. With this development the counters became the fast self-quenching type depending solely on photoionization in the gas.

Thus, today, while other types of counters and mechanisms may on occasion be employed, most counters now in use contain quenching gases such as organic vapors (alcohols, ethyl ether, acetone, CH_4) or the halogen gases. In these, the secondary mechanism is chiefly through photoionization in the gas of the organic component with a maximum of space charge action, owing to low ion mobility. Background action owing to negative ion formation is never entirely absent in such gases, except perhaps with halogen gases that yield very stable negative ions. Thus, successful Geiger counters now operate largely by the burst pulse corona, which is essentially the true corona mechanism. Greiner and others as early as 1933 recognized the role of photoionization in the gas as one mode of Geiger counter action.[2.27] The Montgomerys indicated the function of the space charge motion.[5.22] Stever, in 1941–1942,[2.27] was one of the first to demonstrate clearly that the propagation in the more modern counters was by photoeffect in the gas. He also defined the dead time and repeat rate. A nearly contemporaneous analysis in more detail was made by van Gemert, den Hartog, and Müller.[5.23] Probably the most complete analysis was that by Alder, Baldinger, Huber, and Metzger in 1947.[5.16] Obviously, depending on gas filling, electrode material, and so forth, other secondary actions may also occur, as was shown by Craggs and Jaffe in 1947[5.24] for the conventional Geiger counters, even when the principal action was of the burst pulse corona type.

2. THE QUANTITATIVE THEORY OF GEIGER COUNTER ACTION AND SPREAD

Aside from Greiner's early recognition of the role of photoelectric ionization, the significant studies were first the general basic studies of Werner.[5.21] He worked with the corona currents in coaxial cylinders and assumed largely the classical Townsend-like mechanisms. Since he worked with many pure gases, he actually was correct in this assumption. The basic relations for the magnitude of the space charge fields and the principles of the action of the iR drop-quenched counters are to be ascribed to him. The various gases and metals were also studied by Werner. This material up to 1939 can be found summarized in the author's book, *Fundamental Processes of Electrical Discharge in Gases*.[5.25] As the low counting rate of such systems drove the instrumentalists to the eventual use of the inert gas organically quenched counters, certain investigators studied the basic mechanisms active in such counters. Partly because adequate high-speed oscilloscopes had not been developed and partly because of tradition, many of these studies were complicated and made difficult by the employment of counting circuits for observation. The data from these were rendered more obscure by the statistics of triggering, which necessitated statistical analysis of complicated data to sort out fundamental variables. In many of the studies, not only were absolute values uncertain but also the reaction of the counting circuit on the discharge served to obscure the issue. Thus, for example, no one has ever made a proper study of the physics of the counting plateau. In consequence, of the many researches on counters of the more modern type, only those studies leading to major advance will be cited.

Probably D. D. and C. G. Montgomery in 1940[5.22] were the first to recognize more intimately the principles of the self-quenched modern counter and the inductive nature of the current impulse recorded as a result of space charge movement. The shape of the current pulse as a function of time was first experimentally delineated by A. Trost,[5.26] and later by W. E. Ramsey.[5.27] In 1942 H. G. Stever[2.27] used the iR drop on a resistance across the counter and an oscilloscope for studying the pulses. He established certain important facts. (1) The quenching is entirely within the tube.

(2) The amount of charge involved is independent of circuit constants other than the applied potential difference. (3) The potential difference across the tube does not drop to threshold value in a pulse. The change in potential difference may vary anywhere from zero to considerably more than V − Vg. Here three potentials must be differentiated: V_s, the threshold of the burst pulse, or localized self-sustaining discharge, quenched by its own space charge; V_g, the threshold for a propagating discharge along the anode wire, also quenched by space charge; and V_0, the onset potential for steady corona of the self-counting regime. (4) The organic vapor is the important agent.

The pulse shape Stever observed directly is shown in figure 5.39. The rise time from the origin t_{rise} was short compared to the rest and on the order of 10^{-5} sec. The counter then could not fire again until a dead time t_d, as measured from the origin, had elapsed. After that, electrons triggered new pulses of an amplitude which increased the further the pulse voltage had fallen beyond t_d. These are shown beyond t_d on figure 5.39. At some clearing time t_{cl} measured from the time origin and shown beyond t_d of figure 5.39, the pulses can again start with the maximum amplitude of the first pulse shown at t_{rise}. The dead time t_d was first noted by Stever and defined by him. He also noted the recovery time, t_r. He set it equal to $t_{cl} − t_d = t_r$.

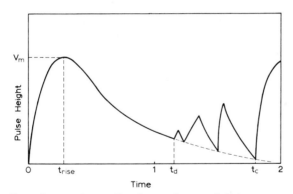

FIG. 5.39. Stever's experimentally observed type of Geiger counter pulse showing rise time, t_{rise}; peak height, V_m; dead time, t_d; and clearing time, t_c. Here the primary pulse followed by smaller secondary pulses can be seen beyond t_d, and a new full amplitude pulse at t_c.

Stever derived the basic relations for the charges involved. The wire of radius a, shown in figure 5.40, has a charge Q per unit length at any potential V. The positive ion space charge is considered to have q charges per unit length at a distance r_s from the axis, while the outer cylinder has a radius b. Let V be the potential across the counter. In region I the field is

$X_I = 2Q/r$. In region II the field is $X_{II} = 2(Q + q)/r$. Then

(5.4) $$-V = \int_a^{r_s} X_I \, dr + \int_{r_s}^b X_{II} \, dr.$$

This yields

(5.5) $$-V = 2Q \log b/a + 2q \log b/r_s$$

with Q evaluated by

(5.6) $$Q = \frac{-1}{2 \log b/a} (V + 2q \log b/r_s).$$

Now the integral of X_I across the counter, negative in the absence of space charge, is defined as an effective voltage V_e. It is the *apparent* voltage across the counter for counting action. Evaluated in terms of the potential V and space charge q, it can be written as

(5.7) $$-V_e = \int_a^b \frac{2Q}{r} = -\frac{1}{\log b/a} \int_a^b \frac{V + 2q \log b/r_s}{r} \, dr.$$

Thus

(5.8) $$V_e = V + 2q \log b/r_s.$$

When $V_e = V_g$ the counter threshold is reached. At this point $r_s = r_0$, the critical distance for a self-propagating avalanche sequence, and $q = 0$, as no space charge exists. At higher values of V, r_s extends out into the gap and can be identified with a quantity R. This is the *critical distance for*

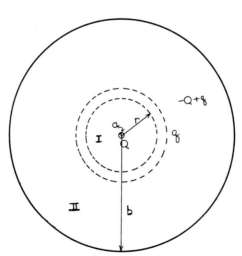

Fig. 5.40. Stever's schematic of a Geiger counter tube with space charge in transit, from which he derived his theory.

dead time, T_d. If $R = b$, then the time associated with it is T_i, the time of ion crossing for the space charge sheath. When V is just greater than V_g, then space charge formation occurs and it is only permitted to write that

$$(5.9) \qquad V_g = V + 2q \log b/R,$$

with r_0 replaced by R, the position of the sheath at the end of T_d. It follows that

$$(5.10) \qquad V_g - V = 2q \log b/r$$

and that both V_g and V are negative. Thus it is permitted to write

$$(5.11) \qquad V - V_g = -2q \log b/R,$$

so that the critical radius involved in the dead time R is evaluated as

$$(5.12) \qquad R = b \exp - [(V - V_g)/2q].$$

To evaluate the dead time and clearing time, it must be recalled that $kX = \dfrac{dr}{dt}$. Thus

$$(5.13) \qquad dt = dr/kX_i,$$

with k the ion mobility and X_i the acting field. Now the field acting on ion sheaths is

$$(5.14) \qquad X_i = X_I + q/r.$$

The q/r comes from the fact that the force on a surface charge density by a field X is $F = x\sigma/2$. Thus

$$(5.15) \qquad X_i = \frac{2Q}{r} + \frac{2q}{r} \left[\frac{1}{2} - \frac{\log b/r_s}{\log b/a} \right].$$

Here Q is the charge per unit length of counter when the potential V is placed across the counter. Actually at *clearing* time $r_s = R = ub$, where $\frac{1}{3} < u < \frac{2}{3}$. Thus the $\log b/r/\log b/a$ is negligible. In consequence

$$(5.16) \qquad X_i = \frac{2Q}{r} + q/r.$$

The positive space charge sheath therefore moves in an X equivalent to the potential V placed on the wire plus the added potential equivalent to $q/2$; that is, it is caused by charges $Q + q/2$ on the wire. In consequence, the effective potential acting on the ions causing their movement is

$$(5.17)$$
$$V_i = \int_a^b \left(\frac{2Q}{r} + \frac{q}{r} \right) dr = 2Q \log b/a + q \log b/a = V + q \log b/a.$$

Again, since

$$(5.18) \qquad X_i = \frac{V_i}{r \log b/a},$$

$$(5.19) \qquad dt = \frac{dr}{kX_i} = \frac{rdr \log b/a}{kV_i}.$$

Now, since

$$(5.20) \qquad V_i = V + q \log b/a = 2(Q + q/2) \log b/a,$$

these expressions can be written in an alternate form. Call

$$(5.21) \qquad m = q/Q;$$

that is, m is the ratio of space charge generated per centimeter of wire over the impressed charge per centimeter of wire Q to give the impressed voltage V, which for counting must be greater than V_g. Then

$$(5.22) \qquad \frac{Q + q/2}{Q} = 1 + \frac{m}{2},$$

so that

$$(5.23) \qquad \begin{aligned} V_i &= 2Q \log b/a(1 + m/2) \quad \text{and} \\ V_i &= V(1 + m/2). \end{aligned}$$

As it is convenient at times to use m, it will be used in what follows. Thus

$$(5.24) \qquad dt = \frac{rdr \log b/a}{kV(1 + m/2)}.$$

If k is constant, as is practically the case, then integration from a to R of dt gives T_d, the dead time of the counter.

$$(5.25) \qquad T_d = \frac{(R^2 - a^2) \log b/a}{2kV(1 + m/2)}.$$

Practically, the clearing time T_{cl} could be set equal to T_i, the time for ion crossing. Actually this is inexact, as a new pulse could start when the inner edge of the cloud is short of the cathode. If one lets $T_{cl} = T_i$, then

$$(5.26) \qquad T_{cl} = T_i = \frac{(b^2 - a^2) \log b/a}{2kV(1 + m/2)},$$

and the recovery time is $T_r = T_{cl} - T_d$, so that

$$(5.27) \qquad T_r = \frac{(b^2 - R^2) \log b/a}{2kV(1 + m/2)}.$$

Actually $k = Kp_0/p$, where p is the pressure at 760 mm and K is the *reduced* mobility. Again a^2 can be neglected relative to b^2 and R^2. Thus the various times become

$$(5.25) \qquad T_d = \frac{R^2 \log b/a}{(2p_0/p)KV(1 + m/2)}$$

$$(5.27) \qquad T_r = \frac{(b^2 - R^2) \log b/a}{(2p_0/p)KV(1 + m/2)},$$

and if $T_{el} = T_i$, which may not actually be the case,

$$(5.26) \qquad T_i = \frac{b^2 \log b/a}{(2p_0/p)KV(1 + m/2)}.$$

In many counters between V_g and V_0, $0 < m < 1$. In some cases $m > 1$. At the point of transition $m = q/Q$ undergoes a sharp decline in increase as $V - V_g$ increases. Here V is the applied potential.

Going one step further, since

$$(5.12) \qquad R = b \exp - [(V - V_g)/2q],$$

then

$$(5.28) \qquad T_d = \frac{b^2 \left[\exp - 2 \left(\dfrac{V - V_g}{2q} \right) \log b/a \right]}{(2p_0/p)KV(1 + m/2)}.$$

With

$$(5.29) \qquad m = q/Q \quad \text{and} \quad mQ = mV/2 \log b/a,$$

$$(5.30) \qquad T_d = \frac{b^2 \log b/a \exp - 2\{[(V - V_g)/mV] \log b/a\}}{(2p_0/p)KV(1 + m/2)}.$$

Immediate interest focuses on Stever's measurements. He used a counter with $a = 0.01$ cm, and $b = 1.11$ cm, or later with $b = 1.43$ cm. Values of V and V_g were measured, as of course was p. Now q is unknown. The total quantity of charge per pulse can be measured and divided by l, the wire length, to give q per pulse per centimeter. Measurements showed q to vary linearly with overvoltage $(V - V_g)$ for a given counter and gas mixture. The linear relation held for different pressures and different values of V_g. This linear relation at once yields R for a given counter. Since $R = b \exp - [(V - V_g)/2q]$ (eq. 5.12), $R = ub$ is then a constant for a given counter with $0 < u < 1$. Since, then, R is a constant for a counter at $R = ub$, the linearity of q with $V - V_g$ is at once clear. For $-q = (V - V_g) \log 1/u$.

The choking off by space charge q depends on its total value. It does not depend on how fast it moves nor how badly it is dispersed as long as the sheath remains clear of the radius b at the time in question. Thus pressure changes may shift the value of R, but there is no reason to expect a change in the linearity between q and $V - V_g$ at any pressure though the constant and R are changed.

Argon-xylol counters with ratio of gases 9 to 1 and at 10 to 160 mm total pressure were used by Stever. R was calculated as $R = 0.65$ cm, and pressures ranged from 132 mm to 50 mm with V_g of 1410 to 985 volts, respectively. Recovery times from 1 to 2×10^{-4} sec were calculated for a value of $K = 0.8$ cm^2/volt sec and agreed with observed recovery times within some 10%. Use of a larger b = 1.43 cm resulted in a smaller q and smaller R = 0.47 cm.

The measurements of T_d were not in agreement with the theory given. Stever set K = 1.0, b = 1.43 cm, a = 0.01 cm, V = 1035 to 1370 volts. The dead times observed were T_d = 2.6 × 10⁻⁴ sec with T_r = 4.5 × 10⁻⁴, while calculated T_d = 0.65 × 10⁻⁴, T_r = 4.3 × 10⁻⁴, and R = 0.47 cm. At V = 1035, T_d observed was 1.8 × 10⁻⁴ sec, T_r observed was 3.1 × 10⁻⁴, T_d calculated was 0.43 × 10⁻⁴ sec, and T_r calculated was 3.0 × 10⁻⁴ sec.

It is clear that the calculated values of T_d are one-fourth or thereabouts of the observed values. This raises questions as to the values of the mobilities of the ions in the *high field* region, for calculated values of T_r agreed fairly well with the measured values. Measured T_r were about twice the values of T_d, which seems reasonable, and T_r with K chosen as 0.8 or 1 agreed with the observed values. The mobilities involved in T_r all lay in the low field region. The value of R chosen seemed consistent with other data and with the theory proposed. This theory is approximate and relatively crude. Assuming R to be correctly chosen, then the calculation of T_d can be in error only because the value of K chosen is too high. This means that at values of X/p ranging from 255 at r = a, 51 at R = 0.56, and 1.8 at r = b, K must have been quite low. Not too much is known about mobilities at high X/p.

Direct measurements by Hershey and others indicate that above X/p on the order of 20 to 40 for most gases, ion mobilities either first increase by some tens of per cent and then ultimately decrease. The ultimate decrease has v varying as $\sqrt{X/p}$ instead of as X/p; that is, k *decreases* in proportion to $\sqrt{X/p}$ as X/p increases. In some gases there is no rise and the decrease begins at once. In the range of X/p from 50 to 200, if the ions were A^+ in A the decrease would begin at once after passing the critical X/p, while for A_2^+ in A there would at first be a rise and then a decline. However, there is *no* evidence on hand that in the range from X/p = 50 to 250, the value of k decreases by a factor of 4. Known decreases are on the order of values of 3 to 2 or at most reduction by one-half in this range. Stever was thus rightly uneasy in attributing the erroneous calculation to reduction in k, since T_r gave k correctly. He thus tested the rate of photoelectric propagation of the pulse along the wire; that is, he measured T_s, the spreading time, and investigated the mechanism of propagation to ensure that it was photoelectric. He succeeded in satisfying himself that the propagation was photoelectric and that spread plus burning time, $T_s + T_b$ took on the order of 10⁻⁶ sec or on the order of the 10⁻⁵ sec duration of T_{rise}. Thus T_d was not increased by build-up of the pulse, its spread along the wire, or burning time.

The only way to account for the erroneous values calculated for T_d is to consider that the theory gave results more consistent than were justified on the basis of the simplified assumptions. A very small extension of R from 0.5 cm to more nearly 1 cm, with b = 1.43, a decrease of 30% in K in high fields, and a real decrease in K in the low field region where ions can change mobilities by complex formation would lead to consistent results.

Stever in his study of photoelectric propagation down the wire placed a glass bead in the middle of the wire. Then he showed that if pressure and voltage were such that r_0 lay closer to the wire than r_{bead}, the counter triggered at either end acted like two independent counters, since the discharge could not propagate down the wire. If potential was raised or the bead made smaller so that r_0 extended beyond r_{bead}, then the impulse spread down the wire and it acted as a single counter.

Nearly simultaneously with this pioneering work of Stever, A. G. M. van Gemert, H. den Hartog, and F. A. Müller [5.23] developed a complicated circuit for analyzing the shape of pulse forms in self-quenching Geiger counters as the counter was triggered by γ rays either from one end or from the middle. The current-time characteristics differed from those of Stever when pulses were triggered from the ends of the wire instead of the middle and when care was taken to confine triggering to a narrow band. The sequence of forms are similar to those shown in the curves of figure 5.41, which came from a later, more complete study by other workers. [5.16]

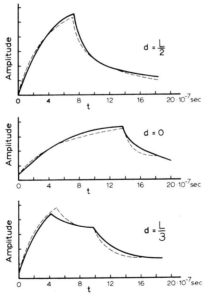

FIG. 5.41. Pulse forms for Geiger counters triggered at the middle, d = L/2; at one end, d = 0; and at L/3, which is at a distance d = L/3. These pulses were calculated by Alder, Baldinger, Huber, and Metzger and compared with *averaged* experimental data (dashed). The agreement is very good.

From these curves they were able to derive basic empirical relations for the current as a function of time with triggering at different points as influenced by the spread on propagation of the discharge down the wire.

These workers proceeded, using the theory of the Montgomerys.[5.22] In their derivation they calculated the field strength X acting on the space charge, the forces of the space charge itself, the *charge induced by the wire on the cylinder*, and the *charge induced on the wire by the space charge with proper signs*. This more elaborate expression reads

$$(5.31) \qquad X = \frac{q}{r} + \frac{2CV}{r} - \frac{2q}{r} \frac{1/(\log r/a)}{1/(\log r/a)1/(\log b/r)}$$

with the capacity $C = 1/2 \log (b/a)$ and V the applied potential. This relation is reduced to

$$(5.31a) \qquad X = \frac{V + q \log r^2/ab}{r \log b/a}.$$

Then setting $X = (1/k)dr/dt$ (eq. 5.13) and integrating, they arrive at

$$(5.32) \qquad t = ab \log b/a \frac{e^{-V/q}}{2kq} [E_i(V/q) + \log (r^2/ab)].$$

They set

$$(5.33) \qquad E_i(x) = \frac{1}{x} e^x$$

and obtain

$$(5.34) \qquad t = \frac{1}{2k} \log b/a \frac{r^2}{V + q \log r^2/ab}.$$

The space charge induces a charge per unit length on the cathode, which is

$$(5.35) \qquad q = \frac{1/\log b/r}{1/\log b/r + 1/\log r/a} = q \frac{\log r/a}{\log b/a}.$$

If equation 5.35 is differentiated, the current becomes

$$(5.36) \qquad i = \frac{q}{\log b/a} \left(\frac{dr}{r}\right).$$

Finally, eliminating X, r, and dr, the relation gives

$$(5.37) \qquad li = \frac{c}{t} lq.$$

In order to understand the current forms obtained as the discharge moved down the wire if triggered at different points, let I(t) be the total current measured as a function of time. Let the current passing through 1 cm of counter at x cm from the point x = 0 where the discharge was triggered be called i(x,t). If its velocity of propagation down the wire be v, then

(5.38) $$i(x,t) = i[0, t - (x/v)]$$

and for short we will write $i[t - (x/v)]$ for this current. Then we have

$$I(t) = \int_0^{vt} i[t - (x/v)] \, dx = v \int_0^{vt} i[t - (x/v)] d[(x/v) - t]$$
$$= v \int_{\tau=0}^{\tau=t} i(\tau) d\tau,$$

or, more simply,

(5.39) $$\frac{dI(t)}{dt} = vi(t).$$

If the propagation along the length of l takes t_1, $v = l/t_1$.

(5.40) $$li = t\left[\frac{dI(t)}{dt}\right]_1.$$

From the experimental data t_1 can be observed, and thus $i(t)$ can be derived from $I(t)$ observed. The $li - t$ characteristic so observed can be fitted by a hyperbola of form

(5.41) $$li = \frac{A}{t + t_0}.$$

For their counter $a = 0.005$ cm, $b = 3.5$ cm, using 45 mm of A, 45 mm of Ne, and 10 mm C_2H_5OH, $V_g = 1285$, $V = 1500$, the $A = 242 \times 10^{-12}$ coulomb, $t_0 = 5 \times 10^{-3}$ microsecond. This law was first established by Trost[5.26] in 1937 and later by W. E. Ramsey in 1940.[5.27] It is seen that it agrees with the law derived above if t_0 is omitted and $A/C = lq = 3170 \times 10^{-12}$ coulomb.

Van Gemert, den Hartog, and Müller then applied their equations to the calculations of the dead time. To do this they started from

(5.31) $$t = ba \log b/a \frac{e^{-V/q}}{2kq} [E_i(V/q) + \log r^2/ab]$$

and set the dead time as being achieved when X_w at the wire surface was just that when V_g was applied to the counter with zero pulse size. For limits of integration they set $r = a$ and $X_w = 2CV_g/a$. This yields

(5.42) $$\frac{2CV_g}{a} = \frac{2CV}{a} - \frac{2q}{a}\left[\frac{1/\log (r/a)}{1/\log (r/a) + 1/\log (b/r)}\right] = \frac{V - 2q \log b/r}{a \log (b/a)}.$$

Thus

(5.43) $$V/q + \log r^2/ab = V_g/q + \log b/a.$$

Inserting the limits into the integration and setting

(5.33) $$E_i(x) = \frac{1}{x} e^x = \left(1 + \frac{1}{x} + \frac{2}{x^2}\right)\frac{1}{x}$$

for large values of x, they arrived at

$$(5.44) \qquad t_d = \frac{b^2}{4kCq} \exp\left(\frac{V - V_g}{q}\right)\left[\frac{1}{x}\left(1 + \frac{1}{x} + \frac{2}{x^2}\right)\right]$$

where $x = V_g/q + \log b/a$ and $C = \frac{1}{2} \log b/a$. It is seen that this expression is far more elaborate and probably more accurate than the one deduced by Stever. The direct measurement of q as a function of V gives values to be used in the solution. Calculations were made of T_d as a function of V from 1100 to 1600. It ranged from 6×10^{-4} sec at 1100 volts, to 1.5×10^{-4} sec at 1630 volts. Here $l = 37.5$ cm, $a = 0.005$ cm, $b = 2.9$ cm, $C = 0.079$ with 90 mm A and 10 mm C_2H_5OH. The mean ion mobility was 4.8 $cm^2/volt$ sec. The reduced mobility k is 0.567 $cm^2/volt$ sec, which is reasonable.

The conclusion is unescapable that Stever's failure to calculate correctly the dead time stemmed from the incomplete theory, reflected most probably in the choice of R. It is likely that the value of K is more nearly 0.5, as indicated here. Thus it was an oversimplified theory that with choice of a higher K led to the assignments of R, leading to a very short calculated T_d.

Further studies were carried out by Hill and Dunworth,[5.28] who in 1946 measured the ion velocity v for counters with pressures varying from 25 to 95 mm of inert gas and 5 mm alcohol, observing v ranging from 4 to 16×10^6 cm/sec depending on $V - V_g$. Craggs and Jaffe[5.24] also carried out some studies in which they showed among other things that, while the discharge was propagated down the wire largely by photoionization of the gas, this was not strictly correct. Some secondary action in their counters was shown to come from the cathode. The presence or absence of cathode conditioned secondary mechanisms will depend on the cathode metal, its state, and on the filling gases, especially the concentration of organic vapor. The action of negative ion formation near the cathode is one that can also cause trouble, simulating a secondary cathode mechanism, as S. C. Brown and Maroni[5.29] have shown. This could have been the causative agent in Craggs and Jaffe's study.

Probably the most complete theory to date is that developed and tested experimentally by F. Alder, E. Baldinger, P. Huber, and F. Metzger,[5.16] also in 1947. This theory is unquestionably the most complete to date. The theory has been reformulated and generalized by D. R. Wilkinson,[5.30] who, however, has so confused and garbled it in presentation that its basic physical significance is no longer recognizable.

The Swiss authors start their analysis from an avalanche *triggered by a single electron in coaxial geometry*. This avalanche liberates photoelectrons in the gas, which then propagate new avalanches down the wire until the end of the wire is reached. As the avalanches continue a sequence in any region of the wire reached, a space charge builds up which quenches the discharge. In order to ensure adequate photoionization, gases like the inerts—for example, A—are used as the vehicular gas whose photon quanta of radiation are able to ionize the organic vapor present photoelectrically.

To calculate the velocity of propagation, the process is separated into many uniform steps. In the first step the last of the light quanta of the first avalanche are produced on the average at a distance z_0 from it, in which z_0 is governed by the mean free path for photoionization $1/\mu$. This electron starts a new avalanche, thus initiating the next step. The mean time of such an elementary process will be designated as θ. The projection of the step z_0 along the axis of the system on the wire is x_0. The velocity of propagation then is given by $v = x_0/\theta$. While the derivation is given in chapter 2, section G.3, a portion will be recapitulated here for the purpose of continuity. To calculate x_0, the component of the steplike distance of propagation along the axis, it is clear that the photons are produced in the gas about the wire. One must then take into account the angular distribution of the photons about the wire. One takes the number dN which penetrate a plane normal to the axis of x between angles ϕ and $\phi + d\phi$. Here N_0 is the number of electrons created in the one electron avalanche that is assumed to initiate the pulse. The number at an angle ϕ is

$$(5.45) \qquad dN = \frac{N_0}{2} e^{-\mu x/\cos \phi} \sin \phi d\phi,$$

with μ the absorption coefficient of the gas for the photoionizing photons. If x is small compared to the diameter of the tube, the number of photons which penetrate the plane are

$$(5.46) \qquad N(x) = \int_0^{\pi/2} \frac{N_0}{2} e^{-\mu x/\cos \phi} \sin \phi d\phi.$$

Partial integration yields

$$(5.47) \qquad N(x) = \frac{N_0}{2} [e^{-\mu x} + \mu x E_i(-\mu x)].$$

Here

$$(5.48) \qquad E_i(-\mu x) = -\int_{\mu x}^{\infty} \frac{e^{-t}}{r} dt.$$

Let f_1 be the chance that a photon liberates an electron, and let f be the chance that an electron avalanche creates a photon. Then the N quanta yield Nf_1 photoelectrons. The condition for propagation is then that $ff_1N(x_0) = 1$. This defines the propagating distance (x_0) in terms of the values of μ and N_0 and f_1. Such a definition then leads to the expansion for propagation of the light pulse by one initiating electron as

$$(5.49) \qquad f_1 f \frac{N_0}{2} [e^{-\mu x_0} + \mu x_0 E_i(-\mu x_0)] = 1.$$

Now N_0, the number of photons from the avalanche in question, is

(5.50)
$$N_0 = fe^{\int_a^{r_0} \alpha dx},$$

where the integral gives the size of the avalanche, that is, the number of electrons which at any point can yield photons.

Again, the photons that ionize are those created by the vehicular gas, in this case, for example, A or some inert gas. Let its partial pressure be p_A, while the total gas pressure is p. Thus the quantity f to be used with a mixed gas is not f but $f(p_A/p)$. Let the alcohol or ionizable gas pressure be p_i, so that the total pressure p is given by $p = p_A + p_i$. Then the absorption coefficient for photoionization is $\mu = \mu_0 p_i/p_0$, where μ_0 is the absorption coefficient at $p_0 = 760$ mm alcohol and p_i is the partial pressure of the alcohol. Again the authors have expressed the quantity $x_0 = \theta v$, assuming that propagation at a velocity v follows in the creation time of an avalanche. This would be strictly correct in an ideal single avalanche theory. It is not necessarily true in practice. It has been considered elsewhere by the author in the relation of v and x_0 more generally and will not be considered further at this point.

One may insert θv for x_0, for the sake of conformity, as did Alder, Baldinger, Huber, and Metzger. In consequence, the expression deduced above becomes

(5.51)
$$\frac{f_1 f}{2} \frac{p_A}{p} e^{\int_a^{r_0} \alpha dr} \left[e^{-\mu_0 \theta v p_i/p_0} + \mu_0 \frac{p}{p_i} \theta v E_i \left(-\mu_0 \frac{p_1}{p_0} \theta v \right) \right] = 1.$$

The above-named writers expressed the ionization in the avalanche by a very crude approximation as $2^{(r_0-a)/\lambda_i}$, where λ_i is the free path for ionization, $\alpha = 1/\lambda_i$. The value of $r_0 = V/X_i \log b/a$, where X_i is the critical field to produce ionization. Since *there is no such field*, α decreasing indefinitely exponentially as X decreases, this is indeed a poor approximation. Actually, the only practical way in which r_0 could be estimated is arbitrarily to take a value of r_0 at which the $\int_a^{r_0} \alpha dr$ is less than 1% or 2% of the total area. That is, a value in which further increase in $\int_a^{r_0} \alpha dr$ would not significantly alter $e^{\int_a^{r} \alpha dr}$. With this criterion, the value of X_i comes from the value of r_0 chosen, the geometry, and V. The better way, of course, is to carry out the Dodd treatment for $\int_a^{r_0} \alpha dr$ as indicated in chapter 2. The use of the very naïve elementary approximation for the Townsend integral $2^{(r_0-a)/\lambda_i}$ is inexcusable. It happened to be convenient because by setting $r_0 = V/(X_i \log b/a)$, the expression for the avalanche size can be approximately set as 2^{BV} as $r_0 \gg a$ when V is large.

This makes the propagation condition vary exponentially with the applied potential. Vaguely this is a very crude semi-empirical equivalent of the more accurate Townsend relation

$$\exp \int_a^{r_0} \alpha \, dr = \exp \int_0^{V_0} \eta \, dr$$

where η is the Philips-Stoletow function α/X (a function of X/p) and V_0 is the potential equivalent to the limit fixing r_0. Since η is not a constant independent of V but depends on the field geometry and its own variation with X/p, it does not follow that one can write an expression in which one has an equivalent of an e^{BV} as indicated. If η were constant with V over a narrow range, such an approximation might hold. It is unfortunate that such an otherwise fruitful theory should have been burdened with poor approximations. The worst feature, however, is that others less critical have accepted the theory as published and *tried to make the data conform to this theory*.

To consider the time constants involved in dead times and the current, Alder, Baldinger, Huber, and Metzger resorted to equation 5.15 of Stever, which if a variable r instead of r_s is used reads

$$(5.15) \qquad X_{ri} = \frac{2Q}{r} + \frac{2q}{r}\left[\frac{1}{2} - \frac{\log b/r}{\log b/a}\right],$$

with Q the charge per centimeter of wire before discharge.

For purposes of integration, it is desirable to avoid the troublesome term $\log b/r$ in the expression above. It is to be noted that X_{ri} and the current rapidly decrease as r increases. Thus only smaller values of r are of consequence. The troublesome $\log b/r$ may thus be replaced[1] by the quantity $\log b/a$, which yields

$$(5.52) \qquad X_{ri} = \frac{2Q}{r} - \frac{1}{2}\frac{2q}{r},$$

so that the integral of $\dfrac{dr}{dt} = kXr$ (eq. 5.13) permits an evaluation of t of

the form $dt = \dfrac{dr}{kX/r}$ with

$$(5.53) \qquad t = \int_a^r \frac{r \, dr}{2k(Q - q/2)} = \frac{r^2 - a^2}{k(2Q - q)}.$$

The current I^* at constant potential comes from the movement of the space charge per centimeter length of counter. Since

$$Q = \frac{V}{\log b/a} - q\frac{\log b/r}{\log b/a}$$

---

[1] It is to be noted that for the calculations to be made, which apply largely to ions in the high field regions, there is no serious error in approximating the various forces on the ion space charge sheath. When the movement concerns the ions in the weak field regions, the forces must be more carefully computed to give the dead times, and so on. For this reason the simple approximate theory of Stever may be used here.

(eq. 5.6), the current I* is

$$(5.54) \qquad I^* = \frac{dQ}{dt} = \frac{q}{\log b/a} \frac{dr/dt}{r}.$$

If

$$(5.55) \qquad t = \frac{(r^2 - a^2)}{k(2Q - q)}$$

is solved and we differentiate with respect to t, the value of I* becomes

$$(5.56) \qquad I^* = \frac{A}{t + \tau}$$

with

$$(5.57) \qquad A = \frac{q}{\log b/a}$$

and

$$(5.58) \qquad \tau = \frac{a^2}{k(2Q_0 - q)}.$$

This is the current per centimeter *due to the movement of space charge at any point*. This is not the current observed, because of the spread of the discharge down the tube. The observed current is given by

$$(5.59) \qquad I(t) = \int_0^t I^* \left(t - \frac{x}{v}\right) dx.$$

One must now consider the boundary conditions.

If now L is the length of the anode and if the discharge is triggered at a distance d from the end of the wire, the spread of the discharge will lead to different current forms depending on the relation of d and L. It is then possible to get I(t) for the various times up to d/v, from d/v to $(L - d)/v$ and beyond. The solutions are

a. For the time $t \le \dfrac{d}{v}, \dfrac{d}{v}$ = time to *near* end

$$(5.60) \qquad I = 2Av \log \frac{t + \tau}{\tau}.$$

b. For the time $\dfrac{d}{v} \le t \le \dfrac{L - d}{v}$ to the distant end

$$(5.61) \qquad I = Av \left(\log \frac{t + \tau}{\tau} + \log \frac{t + \tau}{t - \dfrac{d}{v} + \tau}\right).$$

c. $t \ge \dfrac{L - d}{v}$

$$(5.62) \qquad I = Av \left(\log \frac{t + \tau}{t - \dfrac{d}{v} + \tau} + \log \frac{t + \tau}{t - \dfrac{d}{v} + \tau}\right).$$

Two special cases follow. (1) The discharge starts at one end, d = 0:

$$(5.63) \qquad t < \frac{L}{v} \quad I = Av \log \frac{t + \tau}{\tau} \quad \text{while spreading;}$$

$$(5.64) \quad t > \frac{L}{v} \quad I = Av \log \frac{t + \tau}{t - \frac{L}{v} + \tau} \quad \text{after spreading, burn-out time.}$$

(2) The discharge starts in the center, $d = \frac{L}{2}$:

$$(5.65) \qquad t \leq \frac{L}{2v} \quad I = 2Av \log \frac{t + \tau}{\tau} \quad \text{while spreading;}$$

$$(5.66) \qquad t \geq \frac{L}{2v} \quad I = 2Av \log \frac{t + \tau}{t - \frac{L}{v} + \tau} \quad \text{after spreading.}$$

This theory is now open to direct experimental test if the constants needed are known.

The absorption coefficient was evaluated by the method of Greiner, using two tubes and coincidence studies. The measurements reveal that μ_0, the absorption coefficient for alcohol at 760 mm pressure, is $\mu_0 = 640$ cm^{-1}. At 1 mm pressure it is $\mu_0 = 0.84$ cm^{-1}. Thus for 15 mm alcohol $\mu_0 = 12.6$ cm^{-1} or $1/\mu_0 = 0.08$ cm. This yields a mean free path for absorption close to 0.08 cm, or 0.8 mm. This value of μ is just about that observed by Cravath for photons leading to burst pulses in air at 760 mm. The more exact data of Przybylski gives the value of 2 mm for N_2 photons ionizing O_2 in air with ff$_1 \approx 1 \times 10^{-4}$. Thus the argon counter with 15 mm C_2H_5OH pressure would let few photons escape more than 3 mm from the wire, in good agreement with Stever's glass bead studies.

Elaborate studies based on the principle of the Fizeau measurements for the velocity of light were undertaken to measure the velocity of propagation. For a tube with 16 mm alcohol, 64 mm A at 1100 volts with a = 0.015 cm and b = 1.8 cm, v was accurately determined as v = 8.35 ± 0.05 × 10^6 cm/sec. The current pulse form I(t) was determined by means of cathode ray oscilloscope.

The characteristic curves observed are compared with theory in figure 5.41. There the curves correspond (a) to propagation from the center of the wire, (b) propagation from one end of the wire, (c) propagation with d = L/3. The oscillographs were calibrated for amplitude and time scales. Direct calculation of these quantities is impossible, since k and q are not very well known. However, τ and Av can be determined from the curves such as figure 5.41. The time derivative of I(t) can be evaluated from dI/dt = 2Av 1/(t + τ).

Evaluation of dI/dt at two points leads to a value of $\tau = 0.72 \times 10^{-7}$ sec. Fitting the theoretical curve to the observed curve at the peak yields $Av = 298 \mu A$. The time of spread from the center to the ends permits an evaluation of v. For a value of $L = 8.5$ cm, the value of $t = 6.84 \times 10^{-7}$ sec, which gives $v = 12.44 \times 10^{6}$ cm/sec. The semi-empirical evaluation of the constants permits comparison of the observed and the theoretically derived form of the curves. Agreement is excellent, as is seen in figure 5.41. As noted by van Gemert, den Hartog, and Müller and in the earlier studies of Trost and others, it is not surprising that the theoretically derived expressions, the theory of which was guided by previous empirical data as to curve shapes, should give good agreement when constants are evaluated from the observations. It is, however, gratifying to note that a relation derived from basic theory does agree with observation.

More important for general discussion is the information which evaluation of Av and τ yield. From the relation $V = 2Q \log b/a$,

(5.57) $A = q/\log b/a$,

and

(5.58) $\tau = a^{2}/2k(2Q - q)$,

with $V = 1150$ volts, $a = 7.5 \times 10^{-3}$ and $b = 1.8$ cm, the value of q was 2.29×10^{-10} coulomb per centimeter length of wire and that of k was 11.2 cm²/volt sec. At 80 mm A gas pressure this makes $K = 1.3$ cm²/volt sec. This value of K is probably a bit high for A with 10–15 mm of alcohol.

A direct comparison of v as determined on the same counter by the Fizeau-like method and by the oscilloscopic observation of T_s was made and yielded values of v of 8.8×10^{6} and 8.75×10^{6} cm/sec, respectively, with 67 mm A + 13 mm C_2H_5OH at $V = 1050$ volts, which is very good. The velocity of spread of the pulse v was next determined as a function of V, for different pressures of alcohol filling, for it is the value of p of alcohol that yields μ and alters the spread. It would be expected that, as p decreases for a fixed value of V, the velocity would increase, since the mean free path for absorption is increased. Increase in potential at fixed alcohol pressure would increase v, since the number of electrons in an avalanche increase as well as the photons. Thus the spread of photons and of the pulse should be accelerated. That is, with the same or nearly the same value of θ the increased avalanches will increase v. The result of the measurements showed that v increases linearly with V at a given p of alcohol.

In closing, the Swiss authors indicate that their oscillograms were composites of many pulses. Thus the smooth curves averaged for comparison with theory were not really typical. Individual pulse traces visually observed on the oscilloscope were quite wavy. These fluctuations were not instrumental. They must be ascribed to statistical fluctuations of current density in the space charge sheath as it built up. There is also a clue in that the velocity depends on the number of ions initiating pulses. Thus, depending on the triggering event, the propagation, velocities, and amplitudes will vary.

It is now possible to compare the general statements concerning such corona discharges with predictions. It is clear that the time for the discharge to spread along the wire T_s is finite and that T_s is comparable with the time for the current to burn out and to dissipate. The larger $1/\mu$, the larger the burning length and the higher v and lower T_s. For a tube triggered in the center, thus having 4 cm to burn in each direction, the time of spread in the example was $\approx 7 \times 10^{-7}$ sec and the velocity of spread was on the order of 8.8×10^6 cm/sec, varying with V and pressure of alcohol. The less alcohol, the larger $1/\mu$ and the faster the spreading at constant V. Too little alcohol, however, fails to quench metastables and so on, and counting is poor. The higher the potential, the higher the velocity at constant p of alcohol.

The rise time, t_{rlse}, now designated T_R, which is the time for the current to rise from zero and reach the peak, is seen to be about 7×10^{-7} sec for the avalanche started at the center of an 8 cm wire. Thus T_s, the spreading time, is clearly designated and is nearly synonymous with T_R. If during 1 cm advance the ionization at the origin of the 1 cm has quenched itself, the duration of the pulse would be prolonged by the time to advance 1 cm at the end of the anode. Since this continued ion generation would contribute perhaps 1/8 of the pulse current at the end, it would not too seriously round the break on the curves before the decline. Thus T_b can be set at about the time of linear spread for 1 cm. The time of the avalanche assumes that θ is on the order of 1.2×10^{-9} sec. In the duration of the linear spread over 1 cm along the wire, the time is $1/8.8 \times 10^6 \approx 1.2 \times 10^{-7}$ sec. Thus the spread down 1 cm of wire lasts for about 100 successive avalanches, propagating along a line on the cylindrical anode wire.

Stated more concisely, assuming that each avalanche lasts θ seconds and that each avalanche triggers the next photoelectrically so that 1 cm is traversed at a speed v centimeters per second, the avalanches are triggered at an average spacing along the wire of x_0 centimeters where $x_0/\theta = v$, or $x_0 = v\theta$. Thus x_0 is an imaginary distance derived from the theoretical concept of an avalanche creation and radiating time and an observed velocity of propagation. If this concept and the data given are used, $x_0 = 10^{-2}$ cm or 0.1 mm.

That there is something wrong with this concept appears at once. Alder, Baldinger, Huber, and Metzger directly measured the absorption coefficient of the photoionizing and propagating photons in the gas in question, and found that at 16 mm of alcohol $1/\mu$ was 0.8 mm. This is just 8 times the value of x_0. From the viewpoint of physics, one might expect the photoionizing free path to be less than x_0. That the photoionizing free path, the distance in which $1 - 1/e$ or 67% of all the photoelectrons exceed x_0 by eight-fold, is indeed surprising, since it would be expected that it determines the average advance of the new photoelectrons along the wire which, multiplied by the time θ of each avalanche, would determine the velocity v.

Physically then one should set $v = 1\mu\theta$ and $x_0 = 1/\mu$. If one sets $x_0 = 1/\mu$ then the value of v fixes θ as 9.1×10^{-9} sec instead of 1.2×10^{-9} sec. This means that there is a delay on the order of 8 single ideal avalanche times θ before the spread can take place. This could come either from a delay in photon emission in the avalanche process caused by the needed secondary reactions, or it could mean that the yield of avalanche photons, owing to a low ff_1, would require a sequence of about 8 avalanches *in situ* before the next step in propagation.

Leaving this out of consideration for the time being, one must consider the time T_b, or more properly the quenching time of each avalanche sequence, and relate this to the charge q per unit length observed. This was set as equal to 1.2×10^{-7} sec and comprises 100 successive theoretical avalanche generations of θ sec duration each. This ideal time θ for avalanche growth of 10^{-9} sec is correct for the electron multiplicative process $e^{\int \alpha dr}$ It is not necessarily identical with the photon production time. It is not unlikely that x_0 would be slightly less than $1/\mu$, since 67% of the ions are created within $1/\mu$ and the rate of photoionization is greater to begin with. For convenience in calculations to follow, it is not very inaccurate to set $x_0 = 0.5$ mm. Then in 1 cm advance there will be, on the average, 20 new successive photoelectrically triggered electron avalanches in a line along the axis. The value of $1/\mu$ also ensures a rather even radial spread of the avalanche points around the anode wire.

How many avalanches starting from the initial avalanche can be accommodated around the wire is not known. Certainly there could be five and probably ten or more times as many as the linear separation allows. This permits on the order of 200 new avalanches to originate along and around the 1 cm of anode considered. As these are initiated in 1.2×10^{-7} sec and the ideal single avalanche time amounts to 1.2×10^{-9} sec, there will follow in T_b a sequence of 100 avalanches for each of the 200 points. This gives a total of 2×10^4 avalanches generated in the time T_b associated with 1 cm of advance.

The value of q measured in one pulse was 2×10^{-10} coulomb per centimeter. With 2×10^4 avalanches, this indicates that each avalanche amounts to 10^{-14} coulomb or roughly 10^5 electrons per avalanche. This value is about 10-fold higher than one would expect for an avalanche in such a corona. For pure A and for air at the pre-onset streamer level, avalanche size is around 10^4 electrons and ions. It must be noted that this work was being carried out in a mixture A and C_2H_5OH, for which the Townsend coefficients are *not* known. Thus $\int \alpha dr$ could be larger.

Again, these studies were carried out in a region at the high end of the counting plateau, that is, near onset of steady or self-counting corona. Thus the $\int \alpha dr$ could be somewhat larger than at burst pulse threshold. Finally, the assumption that the pulses are triggered by one electron is most doubtful. The studies of Herreman in air and other studies indicate that counter pulses are usually triggered by 10 to 100 electrons, which need not all be located in one point. Such generation can certainly multiply the number of initiating avalanche points per centimeter and will also introduce a considerable spread in time in the avalanche crossing. These initiating points could easily be as great as the 8θ avalanche times needed to give the observed speed, using $1/\mu = x_0$. The multiple initiation will permit more avalanches per centimeter and thus reduce the size of each avalanche. Thus there is no serious discrepancy in the observations.

One may now consider the photoionizing efficiency needed for propagation of the pulses. In air there are two photons, one from O_2 and the other from N_2, which have values of $1/\mu$ on the order of 2 and 1 mm with $f_1 f$ values on the order of 10^{-5}. These are associated with Geiger counter-like behavior in air at lower pressures, although some more absorbable photons of comparable efficiency may be active. If the photons in the A–C_2H_5OH counter had an $f_1 f$ of 10^{-3}, then there would be 10 photons per avalanche which, were they liberated in the right region, could propagate the discharge. Because of statistics, this number will represent a lower limit, as break-off of succession with spreading in all directions would be most likely for fewer than 10.

Since nothing is known about the nature of the photons in this mixture, not much more can be said. With this small yield a succession of avalanches of time duration θ of 10^{-9} sec might be needed, and the average advance along the wire might be in steps less than the $1/\mu$ here used. Again, with 10 to 100 pulse-initiating electrons, the photon problem would be alleviated. There are, however, some physical facts which must be clearly stated. A photons that penetrate A gas at 80 mm pressure in a reasonable time are all under 4.1 ev energy. These probably do *not* photoionize the alcohol. The A resonance radiation with 11 ev can photoionize.

However, it is very heavily resonance-absorbed by A atoms and will traverse space at only $\approx 10^4$ cm/sec. Thus for 0.5 mm it will take 2×10^{-6} sec to propagate the ionizing free path. Even were it 10 times as fast at the pressures used, each new ionizing act would be delayed by 10^{-7} sec, which is incompatible with the observations. Thus the ionizing photons coming from the excitation of A atoms can only come through interaction of A metastable atoms and resonance-excited atoms with alcohol or its fragments in the gas, probably the alcohol itself. They could also arise from excitation of alcohol by the 11 volt A photons to a radiative state that will radiate and ionize alcohol, a process that is not too probable.

This means that between excitation by electron impact of the A atoms and the radiation of the active photons, collisions must take place. Assuming the large cross section for excitation of 10^{-14} cm², the time needed for such an exciting impact at the concentrations of alcohol used is 10^{-8} sec. This is 10 times the avalanche time θ. If the cross section for excitation is 10^{-15} cm², which is also quite large, the value is 10^{-7} sec. It is thus probable —in fact, almost certain—that the theoretical estimate of 1.2×10^{-9} sec for θ_p, as regards the radiation of photons, is in error, though this does apply to avalanche growth rate. If θ_p is on the order of 9.1×10^{-9} sec, which is compatible with such a process, then x_0 is actually the observed mean free path for photoionization and these values are consistent with the velocity.

In an avalanche one could perhaps assume that the number of A resonance-excited atoms and A metastables equals the number of avalanche ions and electrons, that is, that $f \approx 1$. In any event it will not be very much different in order of magnitude. Since there are 5 impacts with A atoms to 1 with alcohol, and many impacts with alcohol will result in deactivation without photon production, one could expect perhaps only 10^3 photons per avalanche of 10^4 electrons at most. If f_1 were 10^{-2}, then one could have 10 photons per avalanche to propagate the discharge. If avalanches were larger by a factor of 10, things might be more realistic, since the achievement of 10^{-3} efficiency assumed above for $f_1 f$ is very optimistic.

However, the situation is much improved if the pulses are initiated by 10 or more nearly simultaneous electrons. It is clear that a more detailed basic physical analysis along the lines here indicated is essential before one can realistically fix θ_p for photon emission and arrive at a proper concept of x_0 other than identifying it with $1/\mu$.

From the foregoing it may be concluded that with existing data there is enough latitude that, within the existing knowledge of fundamental processes, a consistent and realistic picture of what is occurring may be obtained without recourse to naïve artificial concepts. It would of course be most interesting to obtain values for ff_1 and for the Townsend coefficients as well as for avalanche size, to make possible a more exact picture of the process.

In the Swiss counter T_R is followed at once by a sharp break indicating extinction of most of the discharge, and the ensuing curve represents movement of ions across the gap. With ion mobilities inferred, the values of T_d and T_r, as well as T_{cl}, will be in the range of several times 10^{-4} sec. The actual correctly calculated values of T_d in the Dutch counter work using xylol instead of alcohol lay between on the order of 1.5 to 6×10^{-4} sec with a large b or 2.9 cm. Here it is possible that T_d was around 10^{-4} sec and T_r was about 2×10^{-4} sec. For the faster pure A tube of Lauer, with little space charge and dimensions similar to that used by the Swiss, Lauer directly observed T_i and found it to be 2.5×10^{-4} sec. With C_2H_5OH present it could well have been less, since such gases yield inelastic collisions.

3. THE COUNTER AS A COUNTING MECHANISM

Thus far only the physical aspect of the mechanism of the Geiger counter has been presented, with no consideration of its function as an instrument for measurement. In its function as a counting tube what is desired is a tube with definite volume which can be exposed to ionization of one sort or another in which ionizing *events* in time are spaced at random according to a Gaussian curve about an average value. These events may be caused by portions of particle tracks, α ray tracks, fast electrons liberated by x rays and γ and β rays, or a few photoelectrons emitted by light from a surface. These agencies vary widely in the number of electrons they set free. The ideal in many cases would be to have a counter tube sensitive to very few initial electrons liberated in the volume by some agency whose ionizing events are to be counted.[2] In any case, ionizing events which are commonly counted may be expected to liberate from perhaps 10 to 10,000 electrons each in the counter tube. These electrons may be closely confined in a group or spread out over some distance. Again the events come at random.

Assume that n events occur per second. Then on the average the time between uniformly spaced counts is $T = 1/n$ second. However, sometimes two counts will come within a much shorter time, and others within a longer interval. If now two counting events occur within less than t_d, the dead time, they will not be recorded. A Gaussian distribution of time intervals about an average time T has the form $AT^{-1/2}e^{-t^2/T^2}$. If one set $n = 1/T$ and set t as t_d, the dead time, the fraction of counts that could be recorded would be given by $An^{1/2}e^{-t_d^2 n^2}$. If $n^2 t_d^2$ is small—that is, if $1/n > t_d$ as with $n = 10^3$ and $t_d = 10^{-4}$—then, with A properly normalized, virtually all events will be counted. If on the other hand $1/n \approx t_d$, a considerable fraction of the counts will be missed. Thus it is essential that t_d be as short as possible and that t_d set the limit to the average counting rate. It should also be noted that, in such statistical distribution of counting rates, if one is interested in accuracy one counts some $N = nt$ events in t seconds. Since n varies, or fluctuates, there is thus an inherent error in N. The average probable error in N events counted is then $N \pm \sqrt{N}$. Thus for 1% accuracy N must lie around 10^4.

The next question that arises is of far greater significance and involves

---

[2] Of course, it is possible to conceive of situations where discrimination would make it desirable to count only the events leading to heavier ionization. Such situations, however, present *no* counting problem.

the question of the system of counter and the physics of the process. A counting circuit will respond to a given magnitude of signal. This signal develops from the fast electronic component of the current as it rises in the breakdown pulse process, in contrast to the ion movement in the gap. If the rise is fast enough relative to the time constants of the circuit, such that most of the charge q is liberated within the response time, the current will be proportional to q. Now the recording and amplifying circuit has limitations on its sensitivity determined by its response time and its background noise. If the dropping resistor used in the pickup has a high value, the time resolution is inadequate, that is RC is too large. If RC is reduced by use of more stages of amplification, then background noise levels of the circuit interfere, yielding spurious counts. Thus a given circuit can detect and reliably count pulses of the magnitude of $q \geq A$.

In a true Geiger counter, as distinguished from the proportional counter, it is clear that q depends only on the discharge created in the counter. It does *not* depend on the *size* of the *triggering pulse*. Ideally q depends on the spread of discharge down the whole anode wire and its extinction by the space charge.

It may be added that larger overvoltages and large triggering signals may lead to such a fast build-up and spread of the discharge that the space charge q before choking takes place will be greater than required by theory. This comes from a rate of growth that creates a dense plasma sheath near the anode such that the electrons reach the anode slowly. This phenomenon was first discovered by the author and El-Bakkal[3.67] in all glass cells, where at about $(V - V_g)/V \gtrapprox 0.2$, or $V/V_g = 1.2$, q was no longer proportional to $V - V_g$ but proportional to V. Thus it is possible that under such conditions q may exceed theory with large pulses and fairly sharp quenching. Thus all full-length-of-wire pulses may not be equal.

Again, as shown by Stever, all discharges are not ideal. There is the primary discharge of size q_1, often followed after a dead time t_d or more by a series of discharges q_2, q_3, and so on, depending on counting events which occur between t_d and t_{cl}. These vary in size with the values of q_2 or q_3; the greater the value, the nearer the event occurs toward the end of the cleaning interval t_{cl}. Again, as Alder, Baldinger, Huber, and Metzger state, the individual pulses q_1, triggered uniformly at one end or in the middle, differ between themselves and fluctuate heavily in magnitude so that averages must be used. This was also strikingly brought out by Herreman. In fact, he found that, by increasing the sensitivity of his amplifying circuit, he could observe and count oscilloscopically avalanches from $\approx 10^5$ electrons up to the Geiger counter discharges of 10^8 or more electrons. The former were almost continuous and very frequent in his air counter even as low as 150 to 200 volts below onset of steady corona. This can be expected with delayed counts due to negative ions and some γ_p from the cathode cylinder. It is clear then that one has to consider the following situation:

1. One wishes to count all discharges triggered by the events to be counted, α ray, or γ ray, or x ray, or photon bursts, and so on.

2. These yield counter pulses of varying size, such as primary pulses and secondary pulses of varying amplitudes, but all of an order of magnitude such as produced by a self-sustaining discharge and not a highly amplified group of avalanches.

3. Even the q_1 charges differ greatly among themselves, depending on their location of generation, overvoltage, and probably on space charge conditions and the extent of the triggering event.

4. These true counting events must lead to such values of q_1, q_2, and so on, that they *exceed* the q^+ produced by secondary ionization mechanisms, or fortuitous noncounting events or agencies.

Since true counter action depends on the propagation of the discharge along the anode, then not only must the filling be appropriate but the whole process is critically dependent on the potential V applied. It is therefore essential to trace events as V is raised from below V_s and V_g up to V_0.

As potential increases from below V_s, different triggering events ranging from groups of thermionic or photoelectrically liberated electrons from the cathode, which could number perhaps up to 10 but are usually less in a 10^{-8} sec or somewhat longer interval, will give avalanches of on the order of 10^3 to 10^4 electrons each, that is, of amplification factor $M \approx 10^3$ or 10^4 electrons, depending on how near V is to V_s. Such events as α particles, γ rays, and so on will liberate 10 to 10^4 electrons, usually less than 10^3. Thus a fluctuating background noise current of 10^4 to 10^6 or 10^7 electrons in nearly simultaneous avalanches will appear, those of 10^4 electrons being quite frequent, perhaps separated by random intervals in the tens of microsecond to 10^{-4} sec range. As potential increases, the amplification factor M increases rapidly, so that background noise also increases. Reducing the spontaneous cathode emission and γ_p factor by heavy coats of oxide, carbon, or graphite will reduce such noise except for stray γ ray and cosmic ray bursts. When the potential reaches V_s, the starting potential of *localized* burst pulse coronas that cannot spread but that extinguish by space charge, occasional larger pulses of 10^5 to 10^6 electrons are triggered. These are not quite those that the counter of threshold A will pick up, but occasional ones of 10^6 may register.

As regards this threshold, much more must be said, and this applies equally to the triggering of pulses above V_g that spread. The burst pulse threshold equation of the author, now confirmed by Tamura, is given by equation 2.35. In form it represents a relation of the Townsend type $\gamma M = \gamma e^{\int \alpha dx} = 1$, with, however, a somewhat different equivalent to γ. Such a coefficient as γ represents the average value for events governed entirely by chance; that is, γ or its equivalent in equation 2.35 represents a probability. If the number of secondary electrons per avalanche are counted in individual successive avalanches, the number will fluctuate widely about the average value γ. Furthermore, α is an average value depending on a series of ionizing events in each avalanche which are determined by pure chance. Thus individual avalanches will fluctuate about the average measured α. In consequence $e^{\int \alpha dx} = M$ will fluctuate very widely between different successive avalanches.

The author requested R. A. Wijsman[2.2] to investigate the result of the fluctuations of γ and M on the value of the product γM, for the influence of γ only had previously been studied by Laue and Zuber. Assuming that space charge did not intervene to alter the situation, Wijsman showed that the chance of a breakdown set by $\gamma M = 1$ for a single triggering electron was 0 below the potential V_T at which $\gamma M = 1$. It began to rise above the zero value at $\gamma M = 1$ according to the law $P_T = 1 - 1/y$ where $y = \gamma M$ above the value unity. That is, $P_T = 1 - 1/\gamma M$, and $\gamma M > 1$. Thus P_T is 0.09 at $\gamma M = 1.1$ and 0.9 at $\gamma M = 10$. It is seen that the chance that one electron would trigger a sequence of discharges is very remote at V_s set by $\gamma M = 1$, and γM must become large indeed before a single electron is sure to start a sequence of avalanches. Again, while in theory an avalanche *sequence* should be self-sustaining at $\gamma M = 1$, once it is started fluctuations among successive avalanches are sure to terminate it sooner or later. The theory of this situation for a sequence of avalanches in uniform field geometry has been extensively developed by Legler and by Frommhold[2.6] in Raether's laboratory.

Thus at the theoretical V_s given by equation 2.35, it is very unlikely that a self-sustaining burst pulse will originate or complete itself triggered by *one* electron from the low field region. This is the more so since the positive ion space charge near the anode exerts a quenching action. If, however, a burst of 10 avalanches reaches a region of the cathode nearly simultaneously, the chance of a self-sustaining sequence starting is increased. Thus for larger electron bursts, the chance of a localized pulse starting is much greater. Small increases in potential increase avalanche size and γM as well. Thus *above*, but *fairly near*, the *theoretical* V_s, an actual occasional pulse due to a large electron burst will initiate an observable pulse, register a current increase, and give an *observational* V_s. As V increases above V_s, the number of triggering electrons needed for a localized burst decreases, so that pulses are initiated not by the very few exceptionally larger bursts of electrons from radiations but by fewer and fewer electrons. Thus it is seen that, near and above the observational V_s, more and more localized burst pulses will be measured by a detector sensitive enough to register the 10^3 to 10^6 electrons involved.

The number of such triggering electrons per event again in principle should not influence the value of q even in the localized bursts, since in theory they quench themselves by space charge and are thus self-sustaining and quenching. However, one is not dealing here with a point but with a wire, and a track of a γ ray or other device liberating 10 electrons in a region at a fairly constant distance from the anode will lead to perhaps 2 or 3 *simultaneous* localized burst pulses. These will register as a single current pulse if spaced less than T_R apart. Then the q registered will be two to three times the normal burst pulse size. Thus certainly until V_g, the potential for the spreading of the burst pulses along the wire, is reached, *it cannot be said that counts taken will be independent of the triggering event either in charge size or in certainty of being recorded.* Thus possible counting pulses will vary very much in this potential range, both in magnitude and certainty of detection, with the size of the triggering event. However, the pulse sizes will, though with no certainty, be proportional to the magnitude of the triggering event. They will depend on the location and orientation of the track.

Again, once the potential reaches V_g, where the pulses do not remain localized but can spread so that q applies in theory to a spread down the whole wire and is theoretically independent of n_0 in the triggering event, the idealized situation is never realized. Herreman's results in the air counter as well as the Swiss on a fast counter clearly show that, irrespective of theory, *supposedly equal triggering events at one end yield a wide variety of distances of spread down the wire, of velocities of spread, and of individual pulse sizes.* How these fluctuations depend on n_0, on the space charge situation in the gap, on the geometry of initiating ionizing events, is not clear.

Thus even above V_g one must consider that pulse sizes q or self-propagating discharges will vary widely depending on various triggering factors, but in addition one has primary pulse sizes q_1 as well as secondary pulses q_2 that range down to perhaps $0.05\ q_1$ in magnitude, which must all be counted. Since the extent of spread, the size of q_1 and of q_2, and t_d and t_{c1} are all benefited by increasing V above V_g, the effect of this factor must be further considered.

Now one must assume that an amplifying circuit of the counter is able to record a current due to a signal of amplitude A or above this. The pulse height depends on q/T_R. In the region of V, of value V_s, and V_g, q is roughly proportional to $\Delta V = V - V_g$, from Stever's relation (eq. 5.10) $q = (V - V_g)/2 \log b/R$. How far this extends above a $V/V_g = 1.2$ or $(V - V_g)/V_g \approx 0.2$ is not certain. Thus the rise of the pulse height as V rises above V_g should be roughly linear in theory. Hence V_g or perhaps V_s could roughly be fixed by measuring q as a function of V. The intercept of the q–V curve with a count n_0 on a line parallel to the V axis approximately sets V_g. Actually, on account of what has gone before, the rise will probably be more nearly asymptotic because of the fluctuations, even among the larger pulses.

Since there is a background of cosmic ray and/or γ ray bursts that come from outside which are not caused by the particular agency under study, counts below V_g will not be zero but will have the constant background value n_0 indicated. This is determined with certainty by counting in the absence of the agency to be counted.

Thus above V_g, if A is such that it can detect $q - A$, the Geiger counter pulse for the whole wire at threshold begins to be recorded at V_g and increases its number of counts at first slowly and then nearly linearly as V increases above V_g. Thus it first records all the primary pulses and then more and more of the secondary pulses. When it is able to detect most of these, the counter has reached its counting *plateau*, V_p, at which *nearly all true counting events exceed A* in amplitude, but none of the spurious events triggered by a single or few electrons are able to reach a q which exceeds A.

This plateau may have a large or short potential range, depending on the effect of the thermionic, photoelectric, or other background disturbances on the relative magnitude of the triggered ionizing events. Assume that the triggering needed to ensure pulses must exceed 50 electrons. When V is increased sufficiently above V_g to increase the avalanches due to a few, say 3, electrons in 10^{-8} sec, to such size as to yield pulses of size q = A (since amplification increases very rapidly as V increases), then spurious events appear. What probably happens is that avalanches grow to sufficient size that $\gamma e^{\int \alpha d}$ becomes great enough so that photons or ions from each counting pulse trigger enough new electrons to give a charge of $\approx$A. This leads to a condition of self-counting in which each count ensures a new count after t_d, and shortly thereafter there is an onset of steady corona at a potential V_0. Thus as V increases too far beyond V_g and V_p, the counts from spurious causes begin to increase the counting rate and the counts per second rapidly increase. This delimits the counting plateau at its upper end V_0. The rise near the region of self-counting is usually extremely rapid because the background ionizing fluctuations are more or less uniform in magnitude, and as soon as V reaches the critical value for one disturbance it takes very little potential increase to go to self-counting at V_0.

The reliability of a counter thus depends on the flatness and potential range of the counting plateau. Since spurious counts in a total count are so troublesome, even if the background is deducted in the absence of the radiation to be counted, it is the custom wherever possible to resort to coincidence counts on separate counters for the same event. It has been seen how important this was in Herreman's observations.

In regard to the question of a broad plateau, one must consider the sources of spurious counts and indicate the properties of a good counter.

4. REQUIREMENTS FOR A GOOD COUNTER

In the early days of counter development, before the advantages of the photoelectric burst pulse mechanism in the gas were recognized, counters consisted of ordinary coaxial cylindrical tubes using Townsend discharges with secondary actions at the cathode. The trouble with these discharges was that, even for those with photoelectric action at the cathode, in order to get satisfactory pulses the potential had to be raised very near the threshold of self-sustaining discharge (as in Herreman's air counter) and secondary actions at the cathode had to be efficient. The range between adequate amplification of pulses of a few triggering electrons and self-sustained discharge thus was very narrow, and such discharges were very *hard to quench* under those conditions.

Thus the counting regime fell between *very* narrow ranges of potential, as in Lauer's curves in H_2 and A. These discharges could only be used for counting if either *larger series resistances were used to increase the iR drop* to the anode, or quenching circuits were used. However, large resistances and quenching circuits introduce long time constants, large dead times, and low counting rates. The difficulty leading to narrow counting ranges of potential stems from the requirement of a high γ_i or γ_p on the outer cylinder or cathode. If these quantities are large the surfaces are also excellent emitters of spurious secondary electrons not associated with counting events. In any case, good secondary electron emitting surfaces are subject to spurious emission, so that counts will be registered by internally triggered discharges. Spurious counts are caused by the following agencies.

1. Thermionic liberation from large sensitive cathode surfaces of low work function, for example, the alkalis, pure Cu, brass, and Ta. In the early counters oxidized surfaces such as CuO or surfaces of carbon were usually used to reduce this effect.

2. Liberation of electrons by light and other sources outside the tube not connected with the counting source. Some tubes were placed in light-tight containers to avoid this trouble.

3. Delayed spurious counts from imprisoned resonance radiation, meta-stable atoms, and late arriving positive ions, incident on the cathode. This was so serious that it precluded use of the inert gases unless quenching agents were mixed with them.

4. Fine specks of *insulating* dusts such as SiO_2, SnO_2, Al_2O_3, Fe_3O_4, and other materials on the cathode become charged by positive ions after the first discharge. The fields across these thin insulators are such as to cause field emission of electrons from the surfaces of metal. These produce very disturbing spurious counts at such a rate as to ruin a counter. The effect

was first noted by L. Malter[1.13] in America and later investigated by H. Paetow[1.13] in an intensive study in connection with Geiger counters.

5. One of the most serious causes of spurious counts, irrespective of cathode, consists of those created by negative ions. Improperly outgassed tubes, glass walls, or metal electrodes liberate CO, H_2O, and O_2. O_2 in particular gives O_2^- ions by electron attachment, and O^- is formed from impact between fast electrons and O_2 molecules. Electron attachment to O_2 takes place in the low field regions for electrons coming from the cathode. Impact attachment takes place only in the high field region, and such ions are rapidly removed. The O_2^- ions, however, created near the cathode are in a position to arrive near the anode, shed their electrons, and start a count *just in the recovery time t_{cl} interval*. Oscillograms of Lauer in H_2 with 0.1 to 1% O_2 show the spurious discharges or counts produced by the O_2^- ions. This source of delayed counts is perhaps the most seriously disturbing factor in counting action today. It is confined to negative ion forming impurities with electron affinities generally less than 3 ev. The impurities are created in the counter in the course of its life, or by diffusion of such impurities from the walls. Impurities that form ions in regions such that they have arrival times lying between dead and clearing times are most troublesome. They must be formed near the cathode, from cathode or photoelectrons whose transit time to the anode is short. Thus new triggering electrons are created in the high field region by deionization of the negative ions just as the space charge has cleared, thus producing a spurious count.

With the true burst pulse type of photoelectric corona in the self-quenching Geiger counters, virtually all the electrons are created near the wire. Thus, disturbing ion formation seems unlikely, since photoelectrically active photons are absorbed near r_0. However, even then if the cathode is sensitive to photons and the gas is transparent to photons capable of producing a γ_p, or if the positive ions on impact at the cathode produce a γ_i, then electrons are available for negative ion formation. The stability of negative halogen ions permits their use as quenching gases *in small amounts*. These gases, however, cause other difficulties in corroding electrodes unless the cathodes are made of $SnCl_2$ plated on the glass. They also cause the growth of metal whiskers with certain metal electrodes.

In principle, therefore, the idealized fast self-quenching Geiger counter is desired to operate under the following conditions:

1. Pressure, gas filling, and potential are to be such as to propagate the discharge photoelectrically along the anode at as high a speed as possible. This requires a mixture of gases with appropriate photoionizing free paths long enough for rapid spread but too short to extend far across the gap. A mixture of an organic vapor with A or Ne is quite suitable.

2. A rapid space charge formation, as close as possible to the anode,

which requires a short but intense burning time, is needed to choke off the discharge.

3. This space charge must be one which is readily removed from the high field region to give a short dead time.

4. A gaseous mixture is needed that will destroy all photons capable of photoelectric action at the cathode, destroy all metastable states, and by charge exchange alter positive ions of high energy to ions of ionization potential less than the work function of the cathode. This sort of action can, again, be done by organic vapors. The organic ions of complex molecules are not effective secondary emitters from surfaces. Organic molecules with their absorption bands can remove most photons. Organic vapors also readily absorb energetic photons and destroy metastable atoms.

5. Cathode cylinder surfaces are needed that have a high work function and do not emit either secondary electrons or gases like oxygen on standing.

6. It is desired to have as high an ion mobility as possible, in order to give quick quenching but no spurious counts. This is not readily achieved by ions that are inert against the cathode emission. Bulky organic molecules or clustered halogen ions have low mobility. Thus clearing times are long with such counters.

7. It is desired generally to have a counter that operates at a low potential. In this connection, the inert gases Ne and A, or Penning mixtures of these, as vehicular gases mixed with 5–15% organic vapor, are best.

8. Electron drift velocities are low in the inert gases, but are higher in inert gases mixed with organic vapors giving inelastic electron impacts. However, electron drift velocities play a relatively unimportant role compared to that of the slow ions.

9. It is desired to have a counter that has a long operating life and that does not alter with time. It is obvious that electrical discharges break down the organic molecules and in time deteriorate the counters. In some cases negative ion-forming radicals are created. Actually, breakdown causes deterioration after some 10^7 or more discharges in most counters, some counters lasting longer than others. It would be best to avoid oxygen-containing vapors such as acetone or ether, since these can produce negative ion formers.

10. The short dead times and large pulses desired are definitely fixed in all these counters by limitations imposed by the counting task, which fixes design (radii, a and b, length, and so on) as well as the nature of the gases.

11. In theory a proper counter should be triggered by a single electron created outside the ionizing zone r_0 about the wire. At adequately high applied potentials, this could happen. Reliable counting usually requires quite a number of nearly simultaneously triggering electrons.

12. The counter should be insensitive to temperature changes. Thus

many organic substances are precluded because of low vapor pressure. Most counters have a temperature coefficient that is sometimes troublesome.

13. Last but not least, it is desired that the counter have a fairly long and flat so-called counting plateau. This means that there should be a fairly extended operating range of potentials over which the number of counts per unit time registered for a given source of events is constant and represents the time rate of events when corrected for statistical fluctuations and dead times. A long plateau is desirable, since the higher the potential in a given counting regime, the shorter the dead times, the more rapid the pulse generation, and hence the faster the counter. It is also desirable to have a comfortable operating potential range to count in. Higher relative potentials also ensure counting of all events, as will be seen.

5. HALOGEN-FILLED COUNTERS

Unfortunately, little is known about the basic physics of halogen-filled counters. While the point-to-plane studies of Weissler and Mohr[3.57] on air-Freon mixtures give some data concerning the action of halogens on coronas, they are not helpful in studying counters, as these use inert gases where secondary mechanisms are quite different from those in air.

It had frequently been considered that the halogens might serve as useful agencies in the inert gas counters because of their action in destroying metastable states, and, in fact, in small amounts in combination with He and Ne they yield Penning mixtures. Problems arose because of their chemical reactions with electrode materials and the consequent changes in concentration, as well as the deterioration of the electrodes. In fact, in 1937 Geiger and Haxel[5.31] obtained a patent on halogen and inert gas halogen mixture-filled counters. This patent closely followed Trost's announcement of the use of organic quenching agents. None of the halogen counters appeared to be very satisfactory, and so the organic vapor-inert gas counters became more or less universally used as self-quenching fast counters. In these early halogen counters, however, the difficulty appears to have resided in the use of too large a concentration of halogen gas (greater than 2%), which among other things caused such negative ion formation as to vitiate any other advantages that may have been gained.

It is much to the credit of S. H. Liebson[5.32] that he recognized the difficulty and developed counters using small percentages of the halogens. These counters, especially those using Ne–A mixtures, appeared to give about the same service that the organically quenched counters gave but had longer life and little temperature coefficient. Liebson was led to try small concentrations by consideration of the concentrations of halogens found by F. M. Penning to give the optimum quenching of metastable states—of the order of 0.1% or less relative to the vehicular gas. Later studies convinced Liebson that somewhat higher concentrations were needed. For Liebson's counters, however, chemically passive electrodes were required, which made good counters hard to achieve. The problem has now been nicely solved by L. B. Clark, Sr.,[5.33] of the Naval Research Laboratory, through the use of transparent stannous chloride coatings on the glass envelope for use as a cathode. This appears to have sufficient conductivity (resistance about 400 ohms) and is stable, giving very weak secondary emission if any. It may be that these counters will thus become the ultimate in fast counter development.

In order properly to assess the action of halogen counters, their fore-runners, the organically quenched counters, must be discussed. The fast self-quenched organic counter depends on the use of inert gases, or mixed inert gases, for example, Ne + A, which ensures relatively low voltage thresholds. These gases are mixed with some percentage (up to 10%) of an organic quenching vapor. This ranges from CO_2 through aliphatic and aromatic hydrocarbons like methane or benzene, the alcohols, and some ketones. The function of these appears to be (a) to yield the needed photo-electric ionization in the gas and so to speed build-up; (b) to destroy the metastable atoms, quench the resonant radiation, and absorb other photons active at the cathode; (c) to yield ions which on reaching the cathode have too low an ionization potential to liberate secondary electrons or which on neutralization near the cathode lose their energy in predissociation. Such ions as a rule have large collision cross sections in the low cathode fields near the cathode, and so have a large dead time. Acting in this fashion, the organically quenched counters are relatively successful.

However, all these organics will eventually be destroyed or broken down by the repeated discharges or occasional heavy currents. Thus such counters even at moderate current will show deterioration after some 10^7 counts. Destruction alters the quenching properties and counter characteristics. Many of these organic vapors also alter their partial pressure with temperature and lead to variation of threshold with temperature. All those containing oxygen eventually yield enough O_2 to give negative ions in the low field region which increase spurious delayed counting, thus shortening the plateau, for the ionization potential of O_2^- is less than 0.5 ev. Not enough is known of the free radicals, like CH_2, CH, and so on, which are found in discharges, to know whether these also can give negative ions that cause spurious counting at higher potentials. It should be added that one difficulty encountered with *all* counters having *massive metal cathodes* lies in the eventual introduction of negative ion-forming O_2, for it is impossible to remove all the O_2, H_2O, CO, CO_2, and so on, from massive metal cathodes.

The use of halogens as quenching agents with inert gases is an attempt to accomplish the same objectives as with the organic compounds without the deterioration, temperature coefficients, and perhaps spurious counts at lower potentials. Table 5.7 gives an idea of the behavior of the halogens.

TABLE 5.7

HALOGEN GASES

Gas	First Excitation Potential	Ionization Potential	Dissociation Potential	Energy of Negative Ion Formation
Cl	9	12.96		3.78
Cl_2	2.23	13.2	2.475	*
Br		11.84		3.52
Br_2	1.72	13.0 ± 0.5	1.971	*
I		10.44		3.12
I_2	1.47	9 ± 0.5	1.54	*

INERT GASES

Gas	Energy, Lowest Resonance Level	Energy, Metastable Levels	Ionization Potential	Low Field Positive Ion Mobilities, N.T.P., cm²/volt sec	
He	19.75	19.73	24.48	He^+	11
		20.55		He_2^+	16.5
Ne	16.60	16.54	21.47	Ne^+	4.2
		17.096		Ne_2^+	6.5
A	11.57	11.49	15.86	A^+	1.6
		11.64		A_2^+	2.4 or 1.93
Xe	8.3	8.1	11.5	Xe^+	0.771
		8.4		Xe_2^+	0.74

* Molecular negative ions do not form. Impact of electrons on molecules give negative ions by dissociative attachment with cross sections of the order of 10^{-16} cm² at thermal energies, decreasing with energy.

It shows the first excitation, ionization, and dissociation potentials of the halogens as well as the energy of negative ion formation, together with similar information about He, Ne, A, and Xe used as vehicular gases.

The ionization potentials of Cl and Cl_2 are fairly high, 13 ev, with the other halogens slightly lower. These potentials are lower than those for the inert gases, except for Xe. In the presence of halogens, the work functions of the cathodes are high, which means that they are probably well above the 6 ev for the highest clean metal Pt and thus are on the order of 7 or 8 ev. Since molecules or atoms with an ionization potential of 13 volts or less at lower kinetic energies liberate most of their secondary electrons at cathodes with low X/p through capture of electrons to excited states and ejection by inelastic impacts of the second class within the metal, an effective γ_i will require ions of energy about twice the work function. Thus halogens are effective in reducing secondary actions at the cathode, first through chemically rendering it inactive and secondly because of their low ionizing energies. They also capture electrons readily near the cathode.

The halogens effectively remove metastable states and photons because of the range of their excitation levels from a volt or two upward. This is particularly true of the molecular forms with their band spectra. This ready absorption of all higher energy photons makes the halogen-inert gas mixture an ideal source of short-range photoionization. It leads to the desirable burst pulse corona that spreads rapidly along the wire, quenches itself by positive space charge, and does *not create electrons near the cathode*, nor liberate photoelectrons from it. He and Ne metastable levels will ionize all the halogens. A will ionize I_2 and I. Xe will ionize none. A is used to produce ionization at the expense of Ne metastables. Thus A plus a halogen will quench He and especially Ne metastables very effectively. All inert ions except Xe will transfer their charge to all the halogens, and Xe ions will transfer its ions to I_2 and I only. The positive halogen ions are, as indicated, fairly inert toward secondary liberation.

The negative ions formed by the halogens are all quite stable, Cl^- being the most stable at 3.7 ev. They are more stable than O^- at 1.45 or 2.0 ev, and much more so than O_2^- at < 0.5 ev. O^- ions at an $X/p > 20$ go over to O_2^- ions on impact with O_2 molecules. The O_2^- ions can liberate electrons in fields at an $X/p \approx 90$. I^- ions would take much higher values of X/p and so are unlikely to yield the spurious counts that O_2^- does. The halogens readily form negative ions by dissociative electron attachment with large cross sections in the low X/p regions only. O^- forms under similar circumstances but with smaller cross sections. O^- ions form by dissociative attachment at 3 ev only in the high field regions. Thus if electrons originate near the cathode, these will form copious halogen atomic ions and on reaching the high field region near the anode wire could give rise to spurious counts. If, however, the electrons are not detached from the halogen ions at a sufficient distance within the ionizing zone of radius r_0, they may be ineffective in initiating avalanches of sufficient size to start a new burst pulse. Thus until high values of potential are reached, with geometrically produced large r_0, such cathode region-produced halogen ions will not cause spurious self-counts.

O_2^- ions formed in the same region will provide spurious counts, as the electrons detach as soon as they reach r_0. The O^- ions will only form within r_0 by dissociative attachment of electrons of about 3 ev of energy. After being re-ionized they thus traverse too little of the ionizing zone from r_0 to a to be very effective, even if many are converted to O_2^- close to a. In any case they do not contribute to possible delayed spurious counts after a negative ion transit time from near the cathode, as do the O_2^- ions. Actually dissociative attachment *within* r_0 to either *halogens* or O_2 will act to remove avalanche electrons and thus reduce effective electron multiplication. However, cross sections are low in r_0 where electron energies are not too high for the halogens but high for O_2. O_2 is present only in traces. If halogen concentrations are high, then even with small cross sections of negative ion formation within r_0, *a considerable electron loss will occur* and lead to increased threshold potentials. Above all, capture of the triggering electrons from the event to be counted in low field regions will decrease counting efficiency and raise V. If much negative ion formation took place outside r_0, the negative space charge would reduce the magnitude of the positive ion space charge. This increases fields inside r_0 and prolongs discharge, as it makes positive ion space charge less effective in quenching. There is one more consideration: ionization produced by some event to be counted will in general be created outside r_0. If halogen concentration is high, these electrons stand a good chance of attaching before reaching r_0. Thus, unless potential is high to reduce attachment or r_0 large to cause detachment, counting efficiency may be very low with much halogen.

It is seen that the problem of design of the halogen counter is indeed a delicate one. Enough halogen must be present to quench, yet not enough must be present to form ions in low field regions. On the other hand, the optimum quenching ratios of below 0.1% halogen observed by Penning are too low for practical Geiger counters, because they are at the border of insufficient quenching action in the counter. If quenching is not adequate, the counter oscillates and does not count. This is quite analogous to the observations of H. W. Bandel[1.14] in Trichel pulse corona, in which, when negative ion space charge was inadequate to quench, the pulse went into an oscillating discharge superposed on a background of steady current.

In considering table 5.7, some remarks should be made about the choice of gases. He is not desirable, as it has a fairly high breakdown voltage, very energetic ions, metastables, and high energy photons. Electrons lose more energy in He than in A because of relative inelasticity of collision, but the mobilities of ions are quite high. However, quenching is not so effective in He and it is not used. Ne gives the lowest breakdown potential of any gas. It again is too energetic to be used alone with halogen for quenching. A has a higher breakdown potential than Ne but is readily quenched, and its ion mobilities are low. The best is a mixture of Ne with an ideal quenching quantity, or even more of A. This has a low ionization potential and when mixed with halogen is very effective. Usually the ratio of halogen to A is 1 to 4, with Ne as the vehicular gas. Mobilities in Ne are higher than in A, but about $\frac{1}{5}$ to $\frac{1}{3}$ that in He. Xe would not be considered useful because it is quenched most effectively by I, but it is useful in a counter for x rays or γ rays where the high atomic number of Xe is useful. Of the halogens, except for special circumstances, it is clear that Cl_2 in general is most suitable. It has the highest vapor pressure and is thus less affected by temperature.

In the development of the more successful halogen counter, some of the advantages and disadvantages of the inert gas–halogen filling listed above were recognized by Liebson. These were the photoionization and rapid photoelectric spread along the anode, the rapid quenching of metastable states and absorption of high energy photons before they reached the cathode, and the creation of low energy positive ions with passive cathode surfaces. Liebson also recognized the disadvantages of heavy negative ion formation. Thus he considered the use of low halogen concentrations that would reduce the negative ion formation and temperature variation while achieving the maximum of quenching. Strangely enough, investigation revealed that concentrations of 0.01%, such as Penning advocated from other considerations, were not suitable in counters, since they led to oscillations instead of proper quenching, and higher concentrations were considered. Removal of halogen by the metal electrodes was another difficulty which was exaggerated at low concentrations.

Studies of metals suitable for cathodes and for passivizing were investigated, but there was no guarantee that a passivized surface would stay so over the life of the counter. Such passivized surfaces also gave minute islands of nonconducting chloride that led to spurious counts by the Malter effect. Liebson did find that concentrations of halogens ranging from above 0.1% to 2%, depending on the geometry of the tube, gave the most suitable mixture. While engaged in this study, Liebson conceived of eliminating the passivizing problem through the use of semi-conductor coatings containing halogen on the walls of the outer cylinder. He was unable to pursue this study, through assignment to other duties.

The discovery of the procedure for making successful cathodes of this type properly came from a man versed in the arts of glass techniques. This was L. B. Clark, Sr.,[5.33] who had originally worked with Liebson in his development of the halogen-filled counters. Clark was familiar with the ancient art of decorating glass with iridescent coatings by exposure while hot to certain salt vapors. For counters the coating is achieved by dropping stannic chloride into ethyl alcohol, using a slow stream of clean, dry air to blow the fine fog which rises from the alcohol surface through the glass tube to be coated while the tube is heated to 300 to 500° C. The thickness of the film can be judged by the colors of the thin film. The coated glass tube is then made the cathode *cylinder* and *wall* of the counter, using an appropriate well-centered coaxial wire held taut along the glass tube axis for anode. The resistance of the glass film can be made to suit by varying the thickness of the deposit. Its resistance ranges from 400 ohms upward.

The exact composition of the coating is not known. Clark believes it to be stannous chloride, but it could be a partially reduced stannic chloride, that is, a mixture of $SnCl_2$ and $SnCl_4$. The counting plateau shown by this counter seems to indicate that secondary actions liberating any number of electrons from it are weak indeed, if not absent. Visible light does not affect the counting rate of these cathodes as it does of other counters. The only metal in the counter is the anode wire and its supports. These are thin metal (in 1 to 10 mils), they can be effectively outgassed, and passivizing the wire that will be the anode is not difficult. It does not alter much by electron bombardment. The metal anode then liberates a minimum of oxygen, the cathode liberates no oxygen, and the anode removes relatively little halogen.

Thus a counter with relatively low threshold using Ne with, say, 0.4% A and 0.1% Cl_2 can be made with insensitive electrodes. Actually, Clark did not study such counters with his coatings. He used a 1.0% Cl_2 filling, which is appropriate to his geometry, according to Liebson's experience. With such a counter the operating life appeared to be indefinite. The plateau was definitely quite flat for 300 or more volts in 1200, with no marked change after 10^9 counts. There was no marked change in characteristics from 0° to 77° C. Near $-10°$ C the plateau was no longer as flat as at higher temperatures.

As for the counting rate, there is not much advantage over the organically quenched tubes. Since in most counters build-up and burning times are short compared to the ion transit times involved in dead and recovery times, it is clear that carrier mobilities, especially those of positive ions, will be very important. Electron drift velocities are relatively low in the inert gases, but the inelastic impacts with molecular gases will raise these velocities. The effect will be larger with the organically quenched counters than with the halogens, since more organic molecules are present. Electron drift velocities will, however, only alter the short build-up times by a factor of 2 or 3.

Drift velocities of the positive ions, on the other hand, present a different story. He^+ and He_2^+ ions have high mobilities of around 11 and 16 cm^2/volt sec, respectively, at 0° C and 760 mm. Ne^+ and Ne_2^+ have values of 4 and 6, while A^+ and A_2^+ have values of 1.6 and 1.9, respectively. Introduction of organic vapors in these gases reduces the mobilities to those of large, bulky organic molecules in, say, Ne or A gas. These mobilities are usually below 1 cm^2/volt sec and may be as low as 0.5 or less. The important parameter in dead and recovery times is the value of the low field mobility of the ions, corresponding to the pressure of the gas and the resultant field acting on the ions across the coaxial cylindrical gap beyond

r_0. The higher the overvoltage $V - V_g$, the higher the drift velocities and the shorter the recovery time. Lower gas pressures also shorten the transit times, but may prevent destruction of all active agents. If the plateau is flat and long, then the overvoltage can be increased and the times shortened.

The mobilities of the halogen ions in the inert gases have never been measured under proper conditions. Whether the Cl^- formed manages to cluster or form complex molecules at low energies in transit is not known. Experimental study in this area is definitely indicated. At present all that can be said is that Clark claims that in his counter the dead time-recovery envelope took 300 to 500 μsec at threshold and reduced 75 to 100 μsec at a higher potential on the plateau. This tube had 550 mm of Ne and 6 mm of Cl_2, that is, about 1%. Its starting potential was 900 volts, with a plateau of over 360 volts. With passivized massive metal cathodes, Liebson and Friedman found that, as predicted, the build-up times for a 4 to 1 $A-Cl_2$ mixture at 0.25 mm pressure in 200 to 500 mm Ne were several times slower in halogen mixtures at lowest voltages for operation than in the organically quenched tubes. At higher potential the build-up times became comparable. The dead and recovery times in the halogen-filled counters above were about the same as for conventional organically filled counters.

Liebson states that dead times vary from 400 μsec at $X/p = 7$ at the anode to 20 μsec at $X/p = 70$, with a = 15 mils and b = 10 mm. Threshold potentials at 100 mm Ne for 0.25 mm $A-Cl_2$ mixture were as low as 250 volts and rose to 350 volts when 500 mm Ne was used. With 1 mm of $A-Cl_2$ mixture the thresholds were 350 and 475 volts at 100 and 500 mm Ne, respectively. The anode had a = 5 mils and the cathode had b = 10 mm. Here no temperature effect was observed from 100° C to −80° C. In any case, the temperature effect was small because of the high vapor pressures of the halogen used. In Clark's counter 1% Cl_2 at 6 mm was used. The amount of quenching agent, such as the halogen, depends on wire radius, cathode radius, and total pressure, according to Liebson. Thus, for some counters, up to 2% Cl_2 is proper, and Clark's counters fall within the Liebson range. Liebson and Friedman's plateaus were flat, but, except at 1 mm of $A-Cl_2$, which extended 100 volts for a 550 volt threshold, the counting range was not much more than 50 volts. Here contamination of O_2 from the cathode could have occurred.

The use of A with Cl_2 in Ne helped quench the discharge when the wire radius was large and also lowered the threshold. Liebson found that $A-Cl_2$ mixture does not make a low voltage counter but one approximating that for A alone. Liebson states that this comes about as follows: Cl_2 ionizes at 13.2, and Cl at 13 volts. Argon ions have 16 volts and thus transfer charges to Cl_2 or Cl, but the metastable A^m at 11.54 ev cannot ionize Cl_2

or Cl. However, Cl and Cl_2 have many states of excitation around and below 11.4 volts which destroy A^m. Since A^m does not produce Cl^+ or Cl_2^+ ions, the threshold is set by the A itself. Cl_2 contributes little unless much is present, in which case it raises the threshold. Inert gas photons, and so on, except those above 13 ev in A, do not cause a lowered threshold by enough effective photoionization in the gas until higher potentials are reached and more highly excited states of A are created. On the other hand, Ne, A, and Cl_2 make a generally effective mixture, as one might expect from Penning's work. The cathodes used by Liebson were Ta, chrome-Fe, and stainless steel. Ti and Zn were successfully used at the University of Maryland at Liebson's suggestion.

The only way to decrease dead times would be to use a gas with faster ions. Ne should be faster than A, and He still faster. In this connection H_2 gas might be the best. Its mobilities are high, and it has no effective ions for secondary emission at low cathode X/p. The use of small amounts of HCl or Cl in an H_2 counter might prove interesting, now that Clark's anodes are available. However, the polar nature of HCl could lower mobilities of ions even in H_2 by cluster formation.

REFERENCES

5.1   L. G. H. Huxley, Phil. Mag. *5*, 721 (1928); *10*, 185 (1930). J. H. Bruce, Phil. Mag. *10*, 476 (1930).
5.2   F. M. Penning, Phil. Mag. *11*, 961 (1931).
5.3   J. K. Theobald, J. Appl. Phys. *24*, 123 (1953).
5.4   D. J. Rose, Phys. Rev. *104*, 273 (1956).
5.5   J. H. Parker, Phys. Rev. *93*, 1148 (1954).
5.6   D. J. Rose, Bull. Am. Phys. Soc. II, *2*, 84 (Jan. 30, 1957). Bell Laboratory Report, *ca.* 1958.
5.7   D. H. Hale, Phys. Rev. *54*, 241 (1938); *55*, 815 (1939).
5.8   R. Colli, U. Facchini, and E. Gatti, Phys. Rev. *80*, 92 (1950).
5.9   J. P. Molnar and A. V. Phelps, Phys. Rev. *89*, 1202 (1953).
5.10 L. Colli, Phys. Rev. *95*, 892 (1954).
5.11 J. A. Hornbeck, Phys. Rev. *84*, 615 (1951). J. P. Molnar, Phys. Rev. *84*, 621 (1951).
5.12 M. Menes, Phys. Rev. *116*, 481 (1959).
5.13 D. S. Burch, R. C. Irick, and R. Geballe, Sixth Annual Gaseous Electronics Conference, Washington, D.C., Oct. 22–24, 1953.
5.14 A. L. Ward, Fifteenth Annual Gaseous Electronics Conference, Boulder, Colo., Oct. 10–13, 1962, paper D-6.
5.15 H. M. Herreman and L. B. Loeb, Jour. Appl. Phys. *34*, 3160 (1963).
5.16 F. Alder, E. Baldinger, P. Huber, and F. Metzger, Helv. Phys. Acta *19*, 207 (1946); *20*, 73 (1947).

5.17 G. Condas, Bull. Am. Phys. Soc. *2*, 375 (1957).

5.18 S. A. Korff, *Electron and Nuclear Counters* (Van Nostrand, New York, 1955).

5.19 L. F. Curtis, Phys. Rev. *31*, 1066, 1127 (1928).

5.20 H. Geiger and W. Mueller, Physik Z. *29*, 705 (1928).

5.21 S. Werner, Z. Physik *90*, 384 (1934); *92*, 705 (1934).

5.22 C. G. Montgomery and D. D. Montgomery, J. Franklin Inst. *229*, 585 (1940); *231*, 447 (1941). C. G. Montgomery and D. D. Montgomery, Phys. Rev. *57*, 1022 (1940); *59*, 1045 (1941).

5.23 A. G. M. van Gemert, H. den Hartog, and F. A. Müller, Physica *9*, 551, 659 (1942).

5.24 J. D. Craggs and G. Jaffe, Phys. Rev. *74*, 410 (1918). J. D. Craggs, C. Balakrishnan, and G. Jaffe, Phys. Rev. *72*, 784 (1947).

5.25 L. B. Loeb, *Fundamentals of Electrical Discharge in Gases* (Berkeley and Los Angeles, Calif.: University of California Press, 1955), pp. 485 ff.

5.26 A. Trost, Z. Physik *105*, 339 (1937).

5.27 W. E. Ramsey, Phys. Rev. *57*, 1022 (1940); *58*, 466 (1940). W. E. Ramsey and W. L. Lees, Phys. Rev. *60*, 411 (1941). W. E. Ramsey and E. L. Hudspeth, Phys. Rev. *61*, 95 (1942).

5.28 J. M. Hill and J. V. Dunworth, Nature *158*, 833 (1946).

5.29 S. C. Brown and C. Maroni, Rev. Sci. Instr. *21*, 241 (1950).

5.30 D. R. Wilkinson, *Ionization Chambers and Counters* (Cambridge University Press, London, 1950).

5.31 German Patent 682657 issued to H. Geiger and O. Haxel, 1937.

5.32 S. H. Liebson and H. Friedman, Rev. Sci. Instr. *19*, 303 (1948). S. H. Liebson, Rev. Sci. Instr. *20*, 483 (1949). Also private communication, 1949.

5.33 L. B. Clark, Sr., U.S. Naval Research Laboratory Memoranda, Washington, D.C., Report No. 72 (Oct., 1952).

6

The Highly Stressed Cathode, Pseudo Coronas, in Coaxial Cylindrical Geometry and Relative Threshold Potentials for Positive and Negative Polarity

A. THE HIGHLY STRESSED CATHODE
   IN FREE ELECTRON GASES

1. INTRODUCTION

As has been indicated in chapter 2 under theory and as was clear for the negative point-to-plane corona, the highly stressed cathode leads to quite an unstable condition. Although it suffers from the drawback that it virtually requires triggering electrons to be liberated at the cathode surface, once these are liberated at adequate potential, the breakdown occurs rapidly and nearly catastrophically under many conditions. The instability is caused by the high surface X/p at the cathode, which increases γ_i directly, and both γ_i and γ_p by reducing the back diffusion loss coefficient to nearly zero.

At lower pressures this lowered diffusion loss is facilitated by the small dimensions of the cathode. Again, electron drift velocities are high and ionizing zones r_0 are small. Thus successive avalanches materialize rapidly

512

because electrons and positive ions have short distances to travel. In addition, the electrons create a positive ion space charge close to the cathode. This enhances the cathode surface field. Under some conditions it can enhance it so much that the electrons from the cathode enter the Morton-Johnson regime, where they gain the energy of the cathode fall in a few free paths. Once they have this energy, they ionize more efficiently, increasing the number of ions produced by a factor of 4 to 6 times that to be expected from $e^{\int \alpha dx}$. Finally, low-order self-sustaining discharges with a $\gamma \approx 10^{-4}$ to 10^{-3} will set in at low values of X/p. These quickly lead to a bombardment of the cathode surface with ten volts or more of energetic positive ions.

Thus sputtering or clean-up of the cathode surface is rapid. In time intervals from 10^{-3} to tens of seconds, a surface can go from one with a γ_i on the order of 10^{-4} to one with a γ_i of 10^{-2}. Since potential is constant, the value γM is rapidly increased and the point is overvolted. Unless there is much space charge resistance due to slow negative ions in the low field region, there is little to prevent the breakdown, once begun, from going into a power arc if the external circuit permits. However, the ion formation through impurities introduced by the discharge, or introduced previously in the gas, will limit the current to a corona current short of an arc-over. In the absence of space charge an arc-over can be prevented by an external series resistor. The heavy ion bombardment, if confined to a local spot, can also damage the surface so that the discharge wanders ceaselessly over the surface, choosing the areas of lowest work function at the periphery of the damaged zone but not on the surface areas on which oxide, or other, films have re-formed.

All such breakdowns in their intense phases will by virtue of the positive space charge and plasma near the cathode show the characteristic luminous and dark zones found in glow discharges, with symmetrical but non-uniform fields at the electrodes. At higher pressures the regions of the Crookes dark space, the negative glow, the Faraday dark space, and the positive column will be compressed to dimensions of the order of 10^{-2} cm.

Such discharges where diffusion is limited, as at higher pressures, will have the Crookes and Faraday dark space region as well as the negative glow constricted to a surface area of about 0.1 mm in radius, while the column will be flared. This characteristic comes from radial potential gradients between regions of breakdown which distort the field relative to the adjoining regions where there is no discharge. The discharge spots will thus also repel each other over a surface, and on a symmetrical cathode surface they will take on regular patterns of discharge spots.

In this geometry as with the point-to-plane geometry, the anode plays a very minor role. It serves merely to collect the negative carriers. Its surface fields and X/p are generally entirely too low to initiate a photoelectrically conditioned discharge *in* the gas. The only function of the cylinder is to collect the negative carriers more completely than does the plane; that is, the cylinder has less edge or end loss of carriers. The cylinder also helps to *conserve photons* which do not reach the cathode and would otherwise escape by reflecting them back to the cathode, in the measure that the anode surface is a good reflector. It is only when the induced surface field at the anode surface, through negative ion space charge or otherwise, becomes locally sufficiently highly stressed to yield a streamer that a spark can initiate from the anode surface.

In consequence of the conditioning of the cathode surface by low-order Townsend breakdowns of low threshold, some further consequences must be indicated.

First, great care must be used in such studies to ascertain the true breakdown threshold. This may not be evidenced by the appearance of luminosity or by an impressive current increase. Field intensified currents may start at some value of, say, 10^{-13} amp, changing to a self-sustaining current of perhaps 10^{-11} to around 10^{-9} amp. The real test is whether the current is a field intensified current because an external ionizing agency has given a current of electrons from the cathode which is proportional to i_0, or whether the current is a self-sustaining one independent of i_0 except in a narrow intermittent counting regime.

These threshold currents may be transient, for, in time intervals which may extend to minutes at the same low potential, the current may jump by a factor of 10^3. Unless sensitive instruments are used and the potential is raised very slowly, what is generally observed will be the more impressive transition current jump of 10^3-fold, hereafter termed the onset, which is often mistaken for the true threshold. Very little work has been done on this phase in gases sufficiently pure that these conditions can be observed and the true initial self-sustaining thresholds ascertained. Again, the degassing and alteration of metal surfaces under bakeout, ion bombardment, and exposure to reactive gases is just now coming under study. Effects are very complicated and are as yet little understood.

The degree of cleanliness of the surface in such experiments must be controlled and certain. In some cases if the applied potential is high enough, there will be no low-order breakdown but the system will go directly to a power arc. One effect that has never been carefully studied oscilloscopically is the growth of the current to a power arc under such conditions. Since it is unlikely that the power arc develops through a streamer-conditioned filamentary spark transition, its growth must consume several hundred microseconds or more. Recent evidence in air with point to plane by Miyoshi seems to point to a transition from a negative streamer to a positive-negative streamer rising in units of a microsecond (chap. 4, sec. C).

Associated with many cathode-determined threshold breakdowns is the commonly observed potential overshoot or hysteresis phenomenon. As potential is increased, the current jump occurs at a certain high potential and a current-potential curve is traced out. Once the glow discharge is burning, the curve obtained with increasing potential is not retraced as the potential decreases. In consequence, the potential may be decreased many tens or even hundreds of volts below the onset before the current ceases.

It is also clear that time lag studies of the growth of such currents can have no significance for the breakdown mechanism. They merely indicate the time required to clean up the surface, which in turn depends on the initial current density. The study of the growth of the current, once the initial overvolted current begins to rise in arc development, is not a time lag measurement but a study of the rate of growth of current which is much needed, and only recently has been made for the negative point corona in air, as seen in chapter 4, section C.

It is clear that the peculiar characteristics of the phenomena indicated above required that certain instrumental developments were needed before a proper analysis of the processes could be made. First needed was the development of atomic physics and a more detailed knowledge of the basic processes of gaseous electronics, which was not sufficiently clearly delineated until the 1930's. Secondly, as contrasted with engineering studies, there had to be an interest in and an incentive to the study of basic processes themselves. Thirdly, pure or clean gas techniques with proper outgassing procedures were essential. These developed only toward the middle of the 1930's and were not perfected until bakeable metal valves and Alpert gauges were introduced after 1950. Finally, measuring techniques required the use of fast oscilloscopes, photomultipliers, and modern electronics. In consequence, it is only in the studies of Miller and Lauer, and those in the Philips laboratory at Eindhoven that the basic elements of these discharges with coaxial cylinders began to be delineated. It is of course true that many isolated observations of single aspects had earlier been observed without, however, basic understanding. In what follows, therefore, only the basic studies will be presented.

2. MILLER AND LAUER'S OBSERVATIONS ON
   FREE ELECTRON GASES

a. *Clean N_2 Gas*

Miller[1,9] first observed that it was impossible to make any studies of coronas in pure clean N_2 with outgassed electrodes without using external current-limiting resistors or soiling the gas through a heavy discharge.

With clean wire and pure N_2 at 617 mm, it was repeatedly observed that, as the applied potential was gradually raised to around 4100 volts with no limiting resistor, a heavy discharge occurred which either melted the cathode wire or blew a 10 mA protective fuse. In contrast, at 750 mm with 4100 volts and 200 μA, a 3.5 megohm limiting resistor permitted observation of a threshold with the appearance of a luminous spot. There was no preliminary glow comparable to that observed by Weissler with point-to-plane gap, or as will later be described. This means that, because of the inherent instability of the system showing a negative current voltage (negative resistance) characteristic, corona phenomena in pure N_2 can only be observed and studied with a current-limiting external resistance. To get the true current-potential curves, the iR drop through the current-limiting resistor may be deducted from the applied potential. This procedure is permissible and leads to reasonable values only above threshold.

Above threshold the current can be measured, is steady, and does not alter with time when the breakdown potential is approached from *below for the first time with a clean gas*. No pre-threshold currents were recorded in pure N_2, nor could the *exact potential initiating breakdown* be established. Analogous results were later reported in relatively pure N_2 for uniform fields by Kachickas and Fisher.[3.40] Presumably the currents here were on the order of 10^{-13} amp before the Townsend discharge set in. Once it set in, it relatively quickly developed into an incipient arc.

Table 6.1 indicates the nature of the difficulties encountered. Thus at 607 mm pressure with 10.7 megohms, breakdown occurred at 3565 volts. The post-threshold current was 140 μA, requiring a sustaining potential of 2065 volts across the gap. The offset potential of the discharge on declining potential was 2000 volts at 80 μA, indicating an initial overshoot in potential of 1565 volts. In each case, after the *first* breakdown, which slightly soiled the gas, the discharge began as did Weissler's point-to-plane discharge. That is, at much lower threshold a faint diffuse glow appeared down the length of the wire. This in time contracted to a single luminous spot. The higher the potential, the shorter the time for this contraction. The contraction was in this case not accompanied by a great change in current, as it was in Weissler's work. This study differed from Weissler's point-to-plane study in that, while the gases were nearly if not equally pure, Weissler could not outgas his cathode by flashing at a high temperature as did Miller. It is also possible that Weissler contaminated his N_2 by assumedly "purifying" by flashing a W filament getter. Thus Weissler's point required a conditioning discharge which soiled the gas, or the gas was soiled by the W filament. Weissler may also have had a protective resistance in series, which Miller did not have. In consequence, when the arc-over first burned out his filament, Miller stumbled on the phenomenon. An added difference between Miller and Weissler's equipment was

TABLE 6.1

1 Limiting Resistor, megohms	2 Onset Voltage	3 Post-Onset Current, μA	4 Sustaining Volt- age for Currents Given in Col. 3	5 Offset Current, μA
10.7	3565	140	2065	80
6.0	3600	200	2400	80
3.5	3700	320	2580	80
2.5	3650	410	2625	80
1.5	3650	600	2750	80
0.90	3600	800	2880	97
0.45	3550	1300	2965	90
0.29	3575	1600	3110	90
0.10	3438	2000	3238	100
0	3200	2300	3200	100
0*	3100	3000	3100	
0†	3278	1900	3278	
0‡	3778	300		
0§	3180	1600		

* Wait 24 hours, glow out filament.
† Wait 24 hours more.
‡ Refill to check with purified gas above.
§ Run 19 minutes at 800 μA.

Miller's relatively large cathode surface, so that his faint glow phase covered 100 to 1000 times the area of cathode that Weissler's did. Thus contraction to the luminous spot implied no great change in total current on spot formation, but only a corresponding increase in current density.

In Weissler's case the diffuse discharge over the confined point surface had a low current density and a low total current which increased 10^3-fold on transition to a luminous spot discharge. In Miller's case, with *clean wire surface and clean gas*, the first discharge, on raising the current, needed no surface clean-up by diffuse discharge, and so went directly to a breakdown with the spot if the limiting resistor prevented the arc. If a 0.1 megohm limiting resistor was used, the currents increased to 2000 μA. A 24-hour wait and outgassing the filament gave onset at 3100 volts and 3000 μA, and after 24 hours more onset was at 3278 volts and 1900 μA. Here again the discharge contracted from an extended glow to the form indicated in figure 6.1. It is clear that running the discharge for some time with limiting resistor, so that no arc-over occurred, altered the system. This was proven by removing the limiting resistor after such a sequence and then applying the potential needed for a discharge. In this case the sequence of phenomena observed by Weissler with points was observed,

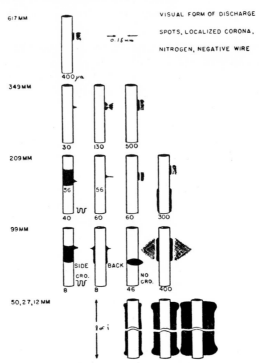

FIG. 6.1. Miller's typical forms of negative wire corona in pure N_2 in coaxial cylindrical geometry at various pressures and currents in microamperes. The scale is drawn at the top of row 3. These were all observed with limiting resistors or else in slightly soiled gas with negative ion space charge resistance.

that is, the diffuse glow contracted to a luminous spot, but no arc-over followed. In the initially clean tube this procedure would have led to an arc.

The discharge was steady, and at no time were any pulses observed. To check these conclusions, pure gas was again admitted at 609 mm under conditions which had led to an arc-over with no limiting resistor. A discharge was run with a 3.5 megohm resistor at 800 μA for 19 min. When the resistor was removed, the discharge set in at 3880 volts with 1000 μA as a glow that quickly contracted to a spot. That the changes caused came from the gas and not from changed conditions on the wire was shown by the observation that standing after the discharge had ceased did not restore the system. Had the breakdown limiting action been caused by the formation of a film on the cathode, it would have re-formed in a few hours. Such action of a gas film was observed where positive wire and negative Ni cylinders recovered their high work function on a few minutes of standing after the cleansing discharge ceased. Glowing the wire or electron bombardment did not change it. Leaving the wire as it was, but pumping out the contaminated gas and filling again with pure N_2, led to arc-over, but only when the potential was raised to above threshold so that the local clean-up of a spot did not have time to liberate enough soiling gas. It should also be noted that in this case the surface of the cathode had been fairly well cleaned in spots by the initial discharge so that with clean gas the arc-over situation was at hand.

Thus it is clear that, despite a flashing of the filament to clean it, once the discharge started and bombardment produced a luminous spot, enough gas was liberated from the metal in 20 minutes to yield a negative ion space charge resistance.

Miller's data permit a rough estimate of the number density of negative ions in the space charge needed to cause this current inhibition. In the initially clean tube, when 10.7 megohms permitted 140 μA, the space charge potential needed to prevent arc-over before contamination added internal space charge was equivalent to about 1500 volts across the tube, or 5 esu of potential. According to the theory of the Geiger counter system, this leads to a quantity of charge q per unit length of cylinder with b = 1.5 cm, a radius with an a of about 7.5×10^{-3} cm of (eq. 5.5) q = $2V/\log b/a$, and for a 6 cm long cylinder q = $12V/\log b/a$. By inserting the values this will lead to roughly 2×10^{10} electrons in 42 cm^3 of coaxial cylinder. This means that the average negative ion density in transit consisted of some 5×10^8 ions/cm^3. At around 600 mm pressure with about 2.2×10^{19} molecules/cm^3, it means that there is in transit one negative ion for 4×10^{10} molecules of N_2. If one hazards a guess that perhaps 10^{-4} or more of the oxygen molecules liberated from the Pt created negative ions by dissociation or three-body electron attachment, then the number of O_2 molecules liberated to cause the ion space charge was on the order of 2×10^{14} O_2 molecules total, with 5×10^{12} molecules/cm^3. This is equivalent to $2.2 \times 10^{-5}\%$ of O_2 which was presumably liberated by the filament.

The amount of O_2 impurity needed could well have been more if efficiency of attachment had been low. Even had it been 10^{-6} there would have been needed only 5×10^{14} O_2 molecules/cm^3 or $2.8 \times 10^{-4}\%$ O_2 present. That O_2 was chosen as the probable negative ion form comes from the fact that it is a contaminant of Pt which is dissolved in it and can only be removed from the surface layer by flashing but is readily removed by ion bombardment and sputtering to greater depths below the surface. Since the discharges used did not raise the anode and gas temperature, and only the cathode wire was subject to bombardment, the conclusion is inescapable that the O_2 or perhaps some H_2 and H_2O were liberated from the filament. Appreciable amounts of O_2 and gas can be evolved by flashing a Pt filament even after it has been flashed previously, as Miller discovered with his supposed W "getter" filaments. That perhaps as much as the order of 10^{14} molecules were liberated by the filament bombardment for 20 minutes or more is more probable, since it is hardly conceivable that Miller's N_2 was initially so devoid of O_2 as to have less than $10^{-5}\%$ present. Thus perhaps as much as $2.2 \times 10^{-4}\%$ O_2 was needed.

Miller observed that, below 99 mm, 1% O_2 did not furnish enough negative ion space charge resistance to choke off the arc breakdown with 3000 volts overvoltage instead of 1000. This means that the partial pressure of O_2 in N_2 was evidently $2 \times 10^{-3}\%$ of O_2, or 0.02 mm partial pressure. This is ten times that estimated above. However, if the overvoltage decreases the chance of attachment in the low field region, this observation need not be contrary to the estimate of $2.2 \times 10^{-4}\%$.

The fact that after flashing in the pure N_2, the breakdown to an arc, or a resistance controlled glow, proceeded without a diffuse clean-up current, while after contamination by running the glow discharge the diffuse prespot clean-up breakdown occurred, points to a surface nitration or oxidation of the Pt filament by the N_2 present or the O_2 liberated. This layer was not a monolayer, as it required quite a high value of the product of current and time to remove it. Since the surface area of the filament is on the order of 10^{-3} cm^2, there are on the order of perhaps 10^{12} molecules to be found on a monolayer. This means that, if flashing or bombardment removed 10^{15} molecules from the contaminated layer, the layer was about 10^3 molecules thick. It is probable that the layer was not that thick. Perhaps 100 molecules could be taken as a conservative estimate. However, if sputtering occurs for 19 min, presumably 5×10^{14} molecules could be freed. A layer of 100 molecules would not form in seconds, but would take hours. The first monolayer can form in 8 hours at 10^{-10} mm pressure with no foreign gas to hamper diffusion. The building up of a thick layer on the cathode could well consume hours at a partial pressure of 2×10^{-2} in a foreign gas at 750 mm. These considerations indicate that perhaps the amount of O_2 liberated from the filament corresponded to 10^{14} to 10^{15} molecules of O_2, which sufficed to give the observed effects, and that possibly some of these came from well below the surface in the course of sputtering.

It is probable that the soiled gas also facilitates the spread of the localized breakdown at spots on the cathode wire surface along the whole surface. The pure N_2 at 750 mm is perhaps too opaque to its own photoionizing photons to propagate the photoionization in the gas down the cathode. However, the photons from N_2 have now been shown to be most effective in photoionizing O_2. The spread of the discharge over the cathode is accordingly facilitated by the fouling. Thus, in contrast to pure N_2, the breakdown begins and remains at the spot where the first self-sustaining Townsend avalanche system initiates. In the soiled gas the local discharge spreads all around the wire by photons created in the gas near the wire and scattered by reflection from the anode cylinder. Once a spot cleans up completely, so that it is highly overvolted on the oxygen-coated wire, the discharge centers at that point. The spot draws all the current the resistors allow and reduces the potential of the cathode below the value to sustain the faint glow along the rest of the wire.

It was observed that, while the discharge run on the cathode wire quickly pollutes the N_2, running a discharge in pure N_2 with the wire *positive* does not contaminate the gas in as great a measure. The very low energy positive ions striking the Ni cylinder in this case, with 250 volts/cm and an $X/p \approx 0.4$ do not liberate gas, nor does electron bombardment by the diffusing electrons into the anode. The wire temperature is only slightly increased by the current flow when it is anode or cathode.

In relatively pure N_2 even with the contamination from bombardment of the cathode alluded to, the clean-up occurs at the threshold value for a γ_p conditioned Townsend discharge if given time. If 1% O_2 is added to the N_2, the chemically active impurities O, NO_2, O_2, and others re-form the coated surface as fast as the weak current density of the Townsend discharge can clean it up. There must be therefore a considerably *greater current density* to clean up a spot for a localized discharge *and to keep it clean*. Thus with 1% O_2 the threshold for the large current jump and spot formation occurs at potentials well above the Townsend threshold in slightly soiled N_2. It is conceivable that the chemical reaction consists of a nitration or an oxidation of the Pt in the presence of N_2, O_2, and NO_2, since the films formed on Pt in pure O_2 do not present the same formidable clean-up conditions as when N_2 is present.

From the foregoing, it is seen that the current-potential curves of figure 6.2 for the negative wire in originally pure N_2 are not too significant. In the plotted curves the potentials are the applied potentials across the external resistance and tube, less the iR drop across the external resistor. The slightly negative slopes of some curves are to be disregarded. They come from inaccuracies in evaluation of i and R.

The visual appearance of spot discharges at various pressures and under various circumstances are shown in figure 6.1. Too little is known about the details to speculate about the forms. This is largely because the pressures are sufficiently high to conceal much of the structure of the discharge in the limited high field region.

The diffuse broadening and spread of the discharge beginning below 99 mm should be noted. What this signifies is that, owing to electron diffusion, the current density and charge densities are not great enough to prevent radial diffusion of the discharge. The normal high pressure diameter of the discharge of around 0.1 to 0.2 mm should be noted and is comparable with that in point-to-plane geometry. At higher pressures the

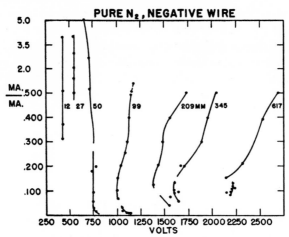

FIG. 6.2. Miller's current-potential curves in pure N_2 with negative wire. Different limiting resistors were used for the broken sections of the curves. The negative slope is not significant, coming from inaccuracies in observations of the iR drop in the external resistor which has been deducted from the observed potentials applied across the system.

Crookes dark space is hinted at, but negative glow and Faraday dark space as well as positive column are more clearly delineated. The glow appears to spread around the wire as current increases.

When a spot is overloaded, multiple spots appear. With multiple spots the current loses its steady character and fluctuations appear on the oscilloscope. These are of large amplitude and long duration, in contrast to Trichel pulses. At 209 mm and at other pressures the discharge seemed to oscillate from the general glow to the point discharge and back again, producing long period oscillations on the oscilloscope screen. Below 50 mm bombardment of the surface does not yield a spot, and a dispersed uniform glow takes its place. Increase in current spreads the glow as in the normal mode of the low-pressure glow discharge.

b. H_2 and A

Lauer[2.21] observed, but not extensively, the negative wire behavior in H_2 and A which were very pure. Figure 6.3 shows the currents in H_2. Here low-order self-sustaining currents of 10^{-11} amp are indicated by the dotted lines. At increased potential, at the vertical line with the arrow pointing up, the current abruptly jumped to $\approx 100\ \mu A$ and was limited only by the external resistance in the power supply line, as Miller found in N_2. Reduction in potential after the increase in current then showed the initial overshoot in potential and a hysteresis loop. The offset of the heavy current thus set in at a lower potential. This appeared notably as with Miller, only at the higher pressure. The large current jump was indicated visually by the appearance of one or more luminous spots that one by one went out as potential decreased to offset. The initial discharge was probably by photoelectric γ_p, but the jump came at an appropriate potential when the wire clean-up yielded a γ_i at some spot.

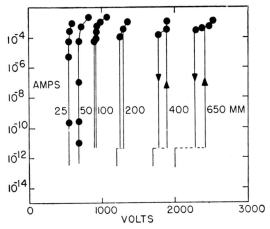

FIG. 6.3. Lauer's current-potential curves in pure H_2 with negative wire in coaxial cylindrical geometry.

The negative wire data are given for A in table 6.2. At A pressures

TABLE 6.2

ARGON NEGATIVE WIRE DATA

Pressure, mm	Gap, volts	Remarks
25	286	Onset uniform glow, 10 μamp, followed upon waiting by spark
50	392	Onset uniform glow, 1 μamp
	424	Spark
100	529	Onset uniform glow, 1 μamp
	530	Spark
200	900	Spark
400	1000	Spark

greater than 200 mm no current was observed with increase in potential until a spark appeared. This is what Weissler observed with the negative point in A. At lower pressures corona started as a blue glow spread uniformly over the wire. This contrasts to the negative wire in H_2. Unfortunately, not too much detail was obtained in this study.

A differs markedly from N_2 and H_2 in its properties. It would appear that the H_2 used by Lauer was relatively not as pure as his A or the N_2 of Miller, since the H_2 showed presence of a surface gas coating that was removed on bombardment. This was probably H_2 on Pt or a hydride layer, as this gas is readily adsorbed by Pt. The H_2 showed no negative ion space charge resistance, since the current was limited only by external resistance. Probably pure N_2 and certainly A are chemically inert relative to the Pt filament. The H_2 coating on Pt is photosensitive, as Hale showed, but rather insensitive to γ_i for H^+ or H_2^+ ions. On bombardment it cleans up, and, when the value of X/p at the cathode is adequate, the γ_i discharge takes over as in N_2. The γ_p pre-discharge of 10^{-11} amp current was verified by Lauer with the positive wire.

In A the situation is quite complicated. First, the potentials involved in A are on the order of one-half those in H_2 and one-fourth those in N_2. Thus X/p at the cathode is smaller. Above 200 mm A does not yield much of a γ_i. Breakdown is largely due to a γ_p, as noted by Colli and Facchini.[3.71] This is created by a three-body impact of A^m metastables with two A atoms to give excited A_2 molecules emitting a delayed photon of 10 ev after a 3.6 μsec delay. These are all created very close to the wire. As it is likely that the A^m are destroyed by electron collisions and lost by diffusion out of the region of r_0, it is possible that there is no relatively low potential breakdown *at the cathode*. The spark breakdown implies that the breakdown occurred through an anode streamer mechanism, as Lauer's A contained traces of N_2 and Kr. It had been shown by the author, Westberg, and Huang[3.72] that streamer sparks occur in positive point to plane in this same A down to 100 mm. Below 100 mm Lauer and Colli and Facchini[3.71] observed a γ_i from clean Ni in A. This probably accounts for the glow phenomena at 100 mm and below. It appeared unnecessary to have a clean-up of the clean Pt wire used in A. It is surprising that a streamer breakdown from the anode cylinder should be initiated at a potential as low as 1000 volts at 400 mm. Apparently it does. With the impurities present, the A probably corresponds to a Penning gas with 10^{-5} to $10^{-3}\%$ of Kr and N_2 present. This in the limited high field region may account for lack of sufficient A^m to yield a γ_p corona. Further speculation is futile.

It is to be noted that in general all of these pure free electron gases act very much in the same fashion with highly stressed cathode.

Sparking probably follows from a cathode streamer that initiates a positive streamer from the anode cylinder at higher pressures.

B. HIGHLY STRESSED CATHODE IN ELECTRON ATTACHING GASES

1. MILLER'S DATA FOR N_2 WITH 1% O_2

Addition of 1% O_2 to pure N_2 enhanced the conditions created by slight soiling of N_2 by the discharge. Figure 6.4 shows the current-potential curves under these conditions for various pressures. Here the lower current data are defective in consequence of malfunction of the electrometer. The current rose steeply from 10^{-13} amp, indicating a Townsend discharge below threshold, presumably from a γ_p. The threshold was complex. If the tube was not used for 30 or more min and the potential was then raised, a discharge appeared that was much greater than for pure N_2 and nearly equal to that to be reported for air. The discharge appeared as a stationary spot glow with a steady current, as revealed by the oscilloscope. The current was fairly high. At 715 mm threshold appeared at 5150 volts with a rise from 10^{-10} amp or more to 100 μA. The discharge pattern is shown in figure 6.5 for 20 μA.

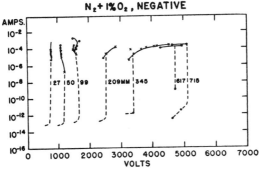

FIG. 6.4. Miller's current-potential curves for pure N_2 with 1% O_2 with negative wire. Note the very steep rise of current from low values below threshold. Note the heavy overshoot at higher pressures and the steep curves at lower voltages when limiting resistors were needed to prevent spark-over.

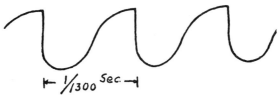

FIG. 6.5. Oscilloscope pattern with a visual corona at 20 μA in N_2 + 1% O_2.

After this discharge had run for half a minute, or after the potential had been reduced or the tube restarted after the discharge extinguished, a new phenomenon appeared. Onset of steady corona with visible spot *was preceded* by an oscilloscope pattern consisting of single pulses of duration of 7×10^{-4} sec randomly spaced. Radioactive triggering increased the number but did not alter the size. These pulses occurred over a range of 200 volts leading to onset of steady corona. Increasing potential led to a visual corona at 20 μA. The oscilloscope then showed a pattern of regular pulses with a frequency shown by the schedule of table 6.3. Further

TABLE 6.3

PULSE FREQUENCIES AT 715 MM IN N_2 AND 1% O_2 AFTER STABILIZATION

At 10 μA	500 pulses per second
20	1300
40	2500
80	6400
90	9400

increase in potential caused the pulses to cease. The frequency depended on potential and/or current, but was independent of external circuit constants. Steady running of the discharge for six minutes reduced the frequency at a given current. At 9000 cycles and 100 μA the pulses merged, giving a steady corona. Once the current in either pulse or steady mode is operating, reduction in potential by 500 to 1000 volts continues the discharge at a fairly high current before offset. This overshoot was observed down to 200 mm and showed vestiges of such behavior down to 90 mm.

In this borderline region of oxygen concentration, the situation is complex indeed. There is not enough oxygen to cause negative ion formation close to the cathode, and thus cause the regular Trichel pulses near the cathode which clear in a few microseconds. Photoelectric ionization of O_2 by N_2 photons is certainly present. The range in distance of this ionization could be fairly great; however, the high field region is limited to the cathode. Propagation of the discharge along the cathode should be possible. However, the confinement of the ionizing field probably makes it exceedingly slow. The initial breakdown in a tube that has stood for some time parallels that for N_2. The gas breaks down to a very low-order current Townsend γ_p discharge that with all the O_2 present cleans a spot on the cathode. Thus a local breakdown occurs with typical spot. Space charge is not adequate to choke it off locally because of the high overvolting at the start, so that it is pulseless. This is a γ_p-initiated breakdown leading to an overvolted γ_i breakdown with much overshoot.

Running the discharge for some time, say 30 sec or more, produced a change in the cathode by a preliminary partial cleansing by ion bombardment. Recovery on 30 min standing came from the re-forming of the oxide layer that had caused the overshoot needed for the initial type of breakdown. Thus restarting after an initial cleansing discharge led to a lower threshold for spot clean-up and localized γ_i corona. In consequence the space charge formation by negative ion creation choked the discharge and gave pulses at the lower starting potentials. These ceased at higher potentials.

The space charge was formed so far away from the point, because of the low O_2 concentration, that a complete cleansing of the charge from the gap was required before the next discharge could occur. This slow creation and cleansing of the negative ion space charge, taking 10^{-3} or so sec, indicates that the attaching molecule was formed slowly and over greater depth and thus was not present in high concentration. The higher the current density the more rapidly the charge built up, and at the increased potential of some 1000 volts clearing times decreased from 2×10^{-3} to 10^{-4} sec. The negative ions had a low mobility and crossed the gap in about the expected clearing times for ion crossing with space charge fields. It is noted that, once pulses began, there was a counting range of 200 volts over which triggering events caused the localized discharges to occur. The breakdown at all times was of a localized glow discharge spot type, and there is no evidence of a spread along the cathode wire, as in a Geiger counter. Counts were triggered by γ ray bursts from the anode by the radioactive source.

Once potentials were sufficient so that photons produced by neutralization of the negative ions at the anode could possibly trigger a new breakdown at the cathode, or else the last incoming positive ions from the plasma sheath near the cathode could trigger a new breakdown after clearing of the space charge at the anode, the self-counting region was reached. This, as noted, extended over a large range of current densities, for example, from 10 μA to 90 μA. Then the potentials were adequate to clear the space charges as fast as they accumulated and the steady discharge set in. It could be remarked that this sort of counter action was the type first tried by Geiger with concentric spherical geometry in 1910. It was not very successful, as noted.

Probably the most interesting other feature of this discharge is that below 200 mm pressure the slope of the curves became so steep that *the discharge could only be studied with a limiting* resistor. Hence, except near threshold, the negative ion space charge resistance is inadequate to prevent arc-over. Thus at 99 mm *without* limiting resistor, threshold *accompanied by Trichel pulses* began at 1550 volts and some tens of microamperes. The currents then rapidly rose to higher values and the Trichel pulses ceased.

With limiting resistor the current could be followed upward for more than 100 μA, with no pulses, down to 100 μA, when pulses appeared. The spot extinguished at 10–20 μA. Below 99 mm the discharge, once started, rose automatically to such magnitude as to blow a 10 mA fuse unless a limiting resistor was present. This situation indicates that the oxygen concentration needed to yield a current limiting space charge resistance lies in the region of 0.1 mm partial pressure, or roughly $10^{-2}\%$ or one part in 10^{+4} of N_2. This suggests that indeed a good deal of O_2 was liberated by the Pt filament on bombardment, and places the quantities estimated above (p. 519) at the upper limit suggested. However, it must also be pointed out that with 1% O_2 the threshold for 617 mm pressure was 4500 volts, while for pure N_2 it was below 2250 volts. Thus for the same tube the space charge resistance had to overcome not 1000 volts but 3000 volts. This would possibly reduce the amount of O_2 needed in pure N_2 to ≈ 0.02 mm or $2 \times 10^{-3}\%$ O_2. If efficiency of negative ion formation by attachment is reduced by the high fields, these data are not inconsistent with the estimates for pure N_2.

2. MILLER'S DATA IN AIR

In this gas the curves, if anything, are more confusing than with 1% O_2. The curve for 752 mm, shown in figure 6.6, was made with a fresh filament and clean air. After it had been run, the filament and/or gas was so altered that the ensuing curves at 617 mm and below represent an altered system.

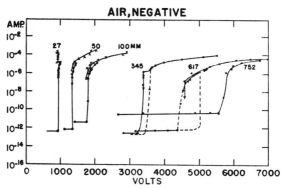

FIG. 6.6. Miller's current-potential curves for air with negative wire in coaxial cylindrical geometry. The curve at 752 mm was made with a freshly outgassed filament in air. Note the varied levels of the pre-threshold currents and sharp irregular thresholds with overshoot extending to low pressures. The curves at 617 mm and below 345 represent the altered system after those at 752 mm.

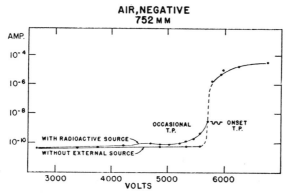

FIG. 6.7. Shows another curve at 752 mm with clean filament as in figure 6.6. Here curves were taken with and without radioactive triggering. The triggering raised the pre-threshold current and rounded the rise to the breakdown above that with no external triggering. Note the appearance of Trichel pulses.

Another curve for a new filament and clean air is shown in figure 6.7. Here curves were taken with and without external radioactive triggering. In both cases curves were nearly constant at about 10^{-10} amp from 3000 volts to near 5700 volts. The radioactive triggering increased the current by a factor of two except near threshold. At threshold the radioactive radiations appeared to initiate Trichel pulses with a current increase, below values without the radioactivity. Once the discharge had been initiated, the sequence of events was much more like that at 617 mm and below. Here the initial currents appeared to be on the order of 10^{-12} amp and led to discharges showing much hysteresis or overshoot of potential. In fact, starting conditions varied from day to day depending on the preceding treatment and how long the wire remained standing after discharge. Dashed rises indicate the absence of reproducibility. At 10^{-12} amp near 4400 volts, an increase in potential would result in abrupt transitions to currents of 0.1 to 1 μA. Transitions did not occur at once but only after waiting some time at 4400 volts.

At threshold the coronas showed continuous current with a slight ripple. The visual manifestation was a glow along the whole length of the wire, particularly if the system had been standing overnight. After some time the glow contracted to a spot, frequently with increase of current at the same potential. With appearance of the spot, Trichel pulses began. If the discharge was interrupted for a short time only, it restarted at once as a spot at the same location with Trichel pulses. After 10 to 12 hours standing it began as a glow and ultimately contracted. Once the spot and pulses formed, the potential could be reduced materially without much reduction in current, until potentials on the order of 4400 volts were reached. At this point the discharge extinguished and currents went to low values. Similar effects were observed at 345 and 200 mm pressure. If the potential was reduced during the diffused glow phase, the extinction curve for the glow paralleled the curve for the pulsed spot curve but at lower current values. The wire after the Trichel pulse phase and spots showed the characteristic pitted craters on the cathode wire observed by Hudson with Trichel pulses in point-to-plane corona.

Interpretation is, in the light of the introductory factors, fairly clear. A clean cathode Pt wire apparently yields a rather high γ_p relative to a wire that has been subjected to chemical action by O_3 and NO_2. Thus the initial field intensified current or a pre-Townsend discharge current of low order will give a high value relative to that from an oxidized Pt wire. Whether the 10^{-10} amp current vis-à-vis the 10^{-12} amp current below onset of a glow was a field intensified current or a Townsend discharge is not certain. A field intensified current could hardly remain constant from 3000 to 5000 volts. It should increase exponentially with potential. A glow discharge which starts at one point and spreads over the cathode as applied potential increases—that is, a glow discharge in the normal mode— would show a slow current increase at increasing applied potential over a large range as the discharge spreads over the surface to points of lower γ_p with potential increase. Here increase in space charge across the tube might but would probably not be a factor. Such a surface spread represents merely an increase in the number of localities capable of yielding a Townsend discharge because of the large range of variation of γ_p values over either the still slightly or, with NO_2 exposed Pt, the more heavily oxidized surface. This does not increase the over-all density of ions in transit. Again radioactive triggering would not be supposed to increase the current if it were a Townsend discharge. Here again such triggering could assist in initiating breakdown at untriggered spots of the same low γ_p at a given potential along the wire. Until the whole wire is covered with glow, radioactive triggering will assist in the spread and thus increase currents. Triggering would also, where a few spots cleared up enough to trigger Trichel pulses, ensure their earlier appearance by triggering events.

Thus doubling of currents and rise of the triggered curve near onset need not be arguments against a low-order Townsend pre-discharge that cleans the surface. In any event one must assume that currents of 10^{-10} and 10^{-12} amp which increase slowly in an almost linear fashion with potential are either saturation currents due to outside sources or else space charge controlled or slowly spreading surface discharges of the Townsend type. Since the initial currents in H_2, in A, in N_2 with 1% O_2, and probably in N_2 with clean cathode wires lie around 10^{-14} to 10^{-13} amp without external triggering and these are the "saturation" ionization currents, it is likely that the currents here observed are localized, self-sustaining, weak

Townsend pre-breakdown currents which vary in magnitude at a given potential with the number of spots of sufficiently low γ_p to give a discharge. When the potential becomes adequate and/or one has waited long enough for a spot at the surface to clean up so as to yield a γ_i, or when the positive ion space charge in front of the cathode reaches some critical value so that bombardment yields a spot that is highly overvolted, then the localized corona takes over.

The whole picture is consistent with the glow over the whole wire surface once the potential reaches an adequate value so that the positive ion space charge field augments γ_i. The pre-discharge slowly cleans the surface, but the ozone and NO_2 foul it. If the potential and current become high enough once the whole surface is covered with glow and spread is no longer possible, then a self-sustaining discharge of greater magnitude sets in. Once a spot cleans up on the fouled surface, the spot discharge takes over. Here current density is sufficient to give a localized negative ion space charge, and Trichel pulses appear. If the wire is initially relatively clean, then the initial currents are greater by a hundred-fold. The discharge still spreads over the wire, but, until an appropriate potential is reached so that currents reach about 10^9 amp, no spot can clear up enough to give the contracted glow, probably because of re-formation of the oxides.

It is uncertain whether the critical potential is determined by the energy of the incoming positive ions needed to cause sputtering, which would depend on the positive ion space charge density in the glow, or whether it is purely a matter of current density, that is, rate of sputtering and removal of oxide faster than it can form. The influence of the triggering events in figure 6.6 would point to current density at some point. The sharpness of the current jump might indicate positive ion impact energy. The lack of reproducibility of the curves and the time factor, which was not adequately studied, probably point to current density, though ion impact energy may also be a factor. The data presented do not show whether after breakdown and discharge the threshold is lower or higher than with cleaner wire. After a discharge, currents are 10^{-12} amp and clean-up is slow, so that overshoot should be great.

Once the Trichel pulse corona starts, as with point to plane, the frequency increases with potential. Unlike the case of the *point-to-plane* corona, increase in frequency at one spot to over 10^6 cycles/sec does not occur for the wire cathode. At 10^6 cycles for point to plane corona, space charges clear faster than needed to choke off the discharge, and lead to a steady discharge for the point. Instead, as soon as a spot along the wire begins to be overloaded, the extensive equipotential surface of the wire leads to creation of a new spot. With spots near overload there is also a tendency toward the extinction of one spot, particularly if the second one can carry the discharge at a lower potential. This tendency is caused by extensive crater formation and perhaps local vaporization in an old spot. With the appearance of more than one spot, the oscilloscope pattern becomes confused. Two Trichel pulse spots acting at the same time can give pulses of differing amplitude. The offset potentials of multiple spots appear *not* to be the same. Thus with multiple spots the oscilloscope shows fluctuations on a much longer time scale than those caused by the pulses. At low pressures with multiple spots the spots were randomly spaced along the surface but tended to approach an equally spaced distribution because of mutual repulsion. At higher pressures the spots appeared preferentially along one side of the wire, possibly because of some slight axial asymmetry.

In air the general glow and spots were visible at threshold; at 750 mm it was visible at 10 μA, and at 617 and 345 mm it could be seen at only a few microamperes. In general at lower pressures the glow appeared at somewhat lower currents. Overshoot at threshold was observed down to 200 mm. Here currents required for visibility were 0.1 μA. At the lower pressures spots appeared superposed on the general glow as potential increased. There was always a dark space between glow and spot because of the potential distribution about the spot. After the Trichel pulses appeared together with the glow, subsequent onsets had many bright spots superposed on the glow. It is possible that the discharge was alternating between spots and glow and that the composite picture caused by continuity of visual impressions was misleading. In some instances spots migrated up and down the wire. The glow followed the spots, always separated by a dark space. This could have indicated coexistence of glow and spots.

On occasions when the glow had not contracted to a spot and extinguished on decreasing potential, the dying glow persisted in diminishing intensity at spots showing no pulses. The uniform glow must have left the surface unevenly conditioned. This could have been surmised by the appearance of spots at isolated points.

On occasion, after standing, the newly initiated discharge appeared as a uniform glow. If potential was slightly increased, random Trichel pulses appeared, presumably emanating from all points of the wire but giving the appearance of a uniform glow. Further increase in potential caused the glow to disappear, giving way to a few spots with sequences of pulses instead of random ones. Pressure reduction generally diffused the phenomena in space and time.

3. MILLER'S DATA IN PURE O$_2$

Thresholds in pure O$_2$ were from 500 to 1000 volts higher than in air. This is not surprising, since the first Townsend coefficient is lowered by negative ion formation, though how much it is lowered in the high X/p region near the cathode is not known. Here again the presence of the O_2 materially affects the surface of the Pt cathode wire. Notable in figure 6.8 are the relatively high starting currents of 10^{-11} amp, even with the coated wire, which on the initially cleaner wire required around 5300 volts before a steep current jump from 5×10^{-10} to 10^{-5} amp. Thereafter the rise of the curve at the start was still more rapid, and a transition at pulse onset was only by a factor of two or more in current. Once established, the curves after the first discharge from a clean wire were repeatable with no notable overshoot or hysteresis. This applied to all pressures except at 27 mm, where there was hysteresis. The changes notable in O_2 are the gradual, relatively steeper rise before pulses appeared at onset, the small transition at pulse appearance, and the lack of hysteresis. Transitions at lower pressures were more marked.

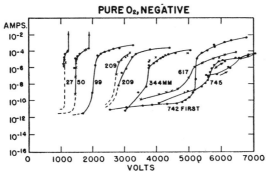

FIG. 6.8. Miller's current-potential curves in pure O_2 with negative wire in coaxial geometry. Note the unusually high pre-threshold currents in O_2 and the relatively small potential jumps at onset, and note that these jumps occur at potentials considerably greater than for the corresponding pressures in air. There was no hysteresis in O_2.

One must assume that even at lower potentials a low-order Townsend current appeared with a γ_p cathode action. It is possible, as some work of Theobald[5.3] showed, that oxidation or coating of a Pt surface with O_2 by bombardment gave a large increase of photoelectric γ_p which was, however, never very stable. It increased, in fact, by a factor of several hundred without changing the work function or energy of photoelectric electron emission. This could account for the very much increased Townsend currents. The oxide layer probably reduced γ_i to near zero. Thus conditioning pre-discharge currents were able to clean up the surface more rapidly, so that no overshoot in onset potential for burst pulses occurred, since γ_p is not materially changed and γ_i is small. It is possible that in air the NO_2 reacted with the Pt to give very different, tougher, surface layers.

The appearance of pulses was not accompanied by light emission because of the very low luminosity of the O_2 bands in the visible. Even the Trichel pulse did not exhibit visible spots. The current distribution to the guard rings and central section indicated an initial uniform discharge down the whole cathode wire. It is not known how much of the spread along the wire and cylinder was caused by diffusion and how much was caused by photoelectric ionization of the gas near and out from the high field region. At any rate, the first discharges must have created O_3 and much O which did not rapidly recombine to O_2. This two-component gas is readily ionized by its own photons, as was shown by Waidmann[2.36] in another connection. The wire length distribution held, even with the appearance of pulses. Thus it seems that the pulses near their threshold cannot maintain a clean surface against the oxidizing action of the plentiful O_2. Thus a spot and pulse would start but shortly extinguish, giving a spot at some other point along the wire. As potential and current densities increased, the spots were able to anchor at one point. The pulses were then more regular. The uniform glow and current distribution along the wire ceased. Occasionally the uniform glow might persist after a spot anchored and became stable, but usually not. This glow could have been active during off periods of the spot. With transition from random pulses to a fixed regular pulse system at one spot, the current increased by a factor of two.

It is clear from all this that in O_2 the photon γ_p-created Townsend discharge is heavier than in air. It produces its own clean-up without delay as potential increases and yields an adequate current density at all times. However, the clean-up is not sufficient to anchor a spot until current density is further increased. Possibly here a γ_i may act to anchor the spot. Then the pulse system becomes regular. Potentials for this situation are materially higher than in air.

The Trichel pulses in O_2 at 750 mm lasted from 1 to 5 μsec in contrast to those in air, which last 0.4 μsec. It is probable that these values of the duration may be valid relatively. However, the oscilloscopes of English and of Miller could not temporally resolve the Trichel pulses in point-to-plane corona in air at 760 mm. These had a real duration of 2×10^{-8} sec at 760 mm in air, as Amin showed with better oscilloscopic resolution. If Miller's comparisons of the duration of the pulses in air and in pure O_2 are correct, then choking of the pulses is more rapid in air than in O_2. This could only occur through accumulation of NO_2 in the gap and more rapid dissociative attachment of electrons to NO_2 than to O_2. This seems unlikely, as it would make Trichel pulse duration and frequency depend on the degree of NO_2 formation, which would be time-dependent. It is probable that, as Miller suggested, the instruments were not reliable in this regard. The choking is caused by dissociative attachment of electrons to O_2 molecules, giving O^-. Thus, in agreement with observations of Das[3.70] in point-to-plane geometry, the pulses in pure O_2 are shorter than in air.

Because of overloading and multiple spot formation in air, it was impossible to follow the increase in frequency of the pulses with increasing current and potential to the point where they merged. In O_2 at low pressures the pulses followed each other so rapidly that, as soon as one pulse extinguished, the next one would start. The current increase in this region was such as to blow a 10 mA fuse. The space charge choking, local both near the cathode and in the gap, no longer functioned, and the discharge went over into an incipient power arc. At higher pressures the space charge persists, and, as one spot overloads a second one appears, as in air. In general at least 10 μA of current are required for visible glow in O_2. These glows appeared at the single anchored spots at slightly lower currents.

With decreasing pressure the details of the phenomena changed somewhat. At 209 mm Trichel pulses marked the threshold at 2×10^{-9} amp and continued to 0.15 μA. These pulses were uniform in amplitude but random. A continuous ripple background of current was always present, owing to the accompanying Townsend discharge. At 0.15 μA the pulses became more frequent but were no longer uniform in amplitude. The visible glow appeared along the *whole wire* at 40 μA and was still visible at 100 μA. Here it seems as if the current density was not able because of diffusion to achieve a sufficiently clean spot to anchor the spot discharge. The regular pulses were due to a *semi*-anchored spot of uniform work function.

The more frequent irregular pulses at higher current came from many spots of different work functions along the wire. It was clear that the glow could have been caused by a myriad of temporarily active spots spread all along the wire in this region and not due to a continuous background of Townsend discharge. However, in the absence of regular spots that carried the full current load at these lower pressures, it is possible that the γ_p conditioned glow discharge persisted into the pulsed regime.

At lower pressures the evolution of the pulses, current, and spots underwent a number of confusing transitions, some without further increase in current. It was impossible to account for these with the data at hand.

C. RELATIVE THRESHOLD POTENTIALS
OF POSITIVE AND
NEGATIVE CORONAS

This discussion has been left for the last section of this chapter as irrespective of the geometry of the system; it involves the same principles and is best discussed after point-to-plane and coaxial cylinders have been considered.

The threshold for all breakdowns is governed by the generic Townsend type of condition set by (eq. 2.9) $\gamma e^{\int \alpha dx} = \gamma e^{\int \eta dV} = 1$. Here γ represents any secondary mechanism emanating from the avalanche process described by $e^{\int \alpha dx}$. γ represents the probability per avalanche ion or electron that one new electron will be created under conditions and position suitable to initiate a new avalanche, which can continue the sequence. This is given by the product $\gamma e^{\int \alpha dx} = 1$. Continuity of avalanche sequences on the average is then assured. The new electron need not be created directly by the ion generating process but can be created by non-ionizing collisions of the avalanche electrons in proportion to their number at any point in the gap. In principle these secondary electrons should be created in temporal synchronism with the avalanche. Actually, there can be some delay as long as it does not exceed an ion or sometimes an electron transit time, so that the ensuing avalanche occurs on the heels of the preceding event. When the chance γ multiplied by the number of electrons or ions in the avalanche becomes unity, one has the theoretical condition for a continuation of the sequence of avalanches constituting a self-sustaining discharge. This threshold marks the observed transition from a field intensified externally created ionization current $i_0 e^{\int \alpha dx}$ to a current sustaining itself with no outside aid.

Much has already been said to indicate that, when the generic condition based on average values of γ and α equals unity is satisfied, there is the *theoretical possibility* of such a self-sustained sequence, or the breakdown. However, statistical fluctuations of γ and of $e^{\int \alpha dx}$ make this unlikely at the theoretical threshold, so that breakdown in practice appears at somewhat above the potential satisfying the average conditions. The situation is either improved or made worse by the favorable or inhibiting action of space charges. Of course the breakdown threshold due to the relation using average values of the constants will the more nearly be approached in practice the greater the number n_0 of triggering electrons to create avalanches that are simultaneously available. For, if single electron sequences soon break off through bad luck, more than one ensures a longer continuation.

Thus the α particle triggering system of Lauer in positive wire coaxial geometry gives an ideal method of ensuring a fairly accurate measurement of the threshold potential. The basic concept implies that the quantities γ and α are constant in time in the tube used and do not depend on current density or other conditions except for potentials and the fields created by them.

Since the $\int \alpha dx$ comes in as an exponential, it is the critical term. As one also deals here with starting potentials, it is convenient to express the integrals in terms of the so-called Philips-Stoletow efficiency function η and the potential applied V. Since α/p is a function of X/p, it is possible to derive a new quantity $\alpha/X = \eta$, which is also a function of X/p. Then, since $\gamma e^{\alpha x} = 1$ sets the threshold, and since for a uniform field $X = V/x$, then $\gamma e^{\alpha x} = \gamma e^{\alpha X x} = \gamma e^{\eta V}$. If in a uniform field $\gamma e^{\eta V} = 1$, then V has arrived at its threshold value V_s. Thus threshold is expressed as $\gamma e^{\eta V_s} = 1$. Since in non-uniform fields α and X vary with x, the equation would take the form $\gamma e^{\int_0^{V_s} \eta dV} = 1$, where the upper limit of the integral represents the applied breakdown potential at the highly stressed electrode.

Now $\eta = \dfrac{\alpha}{X} = \dfrac{\alpha/p}{X/p}$ is also a function of X/p and in a single-component pure gas has the form of a sort of inverted parabola starting low at low values of X/p and rising to a maximum at some particular value of X/p. Actually α/X expresses the number of ions and electrons created per centimeter advance in the field direction per volt/centimeter field gradient. It is the efficiency of ionization expressed as ions per volt. The reciprocal represents the volts required to create an ion pair. Thus when X/p is low, electrons lose much energy to elastic and lower loss inelastic impacts and create few ions. At very high energies the electrons move so rapidly that, although they lose little energy to elastic and exciting elastic collisions, they are also inefficient ionizers. The peak of the curve therefore is a most important quantity. Since η varies over a very great range of values (10^{-4} to 10^{+2}) between $X/p \approx 1$ to $X/p \approx 1000$ for all gases, it is the critical factor in determining V_s in the relation $\gamma e^{\int_0^{V_s} \eta dV} = 1$.

The quantity γ, the nature of which we shall discuss at a later point, is a probability. It can vary over a considerable range of values, for example, from 10^{-5} to perhaps 10^{-1}. It depends on the processes involved for its value and rate of variation. If the process is a γ_i process by ion bombardment of the cathode, the product varies directly with the number of active ions created, and with the energy of ion impact. Thus γ_i can be a function of X/p, increasing as X/p increases. But in general it increases relatively slowly as a $f(X/p)$ owing to the presence of two different processes, kinetic and potential electron liberation. Its variation on this score will not be by much more than a factor of 10 over the observed X/p ranges in gaseous discharges. More important for all γ processes at lower X/p values on the cathode is the back diffusion loss factor ν. This can vary at least 100-fold from values of $X/p \approx 1$ volt/cm per mm to $X/p \approx 400$ volt/cm per mm. However, here again it is only where $X/p < 10$ that its variation is rapid and it is an important factor.

More important for over-all values are the surface conditions of the cathode. The secondary photoelectric emission from the cathode, γ_p, is generally lower than γ_i, but except for the ν factor it is not critically affected by X/p. Where the effective γ-like term is dependent on photo-electric ionization *of the gas* of critical value, the important factor is the relation of the photoelectrically active absorption coefficient μ in the gas, its variation with pressure, and the relation of its reciprocal, $1/\mu = \lambda$, the photoelectric ionizing free path, to the length of the ionizing zone r_0–a in the neighborhood of the anode. The γ of this type can vary over a very wide range of values as p, r_0, and the gas change.

Thus here again γ depends on X/p determining r_0 and μ or λ, as well as on p independently. Not much is known about this quantity and its values. In any study at fixed geometry and constant p, it will not vary over a great range in a breakdown study. Where p varies over a large range, it can be important in altering γ. γ varies over great ranges between different gases and gaseous mixtures. Since the effective μ has recently been shown to depend on X/p, the γ of this type will vary widely with geometry which affects the X/p range active.

What it is desired to establish here is that in any study of V_s as a function of γ and on that account of X/p, the range of possible variations at constant pressure is not more than, say, 100, while η will vary by factors of 10^4 and that in an exponent. The significance of this statement can be seen in taking the logarithm of the expression $e^{\int_a^{V_s} \eta dV} = 1/\gamma$, which reads $\int_a^{V_s} \eta dV = \log_e 1/\gamma$. Thus the righthand term varies very little relative to the lefthand term as X/p varies, so that the form of the $V_s = X/p$ curve is largely dominated by η. If one transforms the η–X/p curve so as to give a curve of V_s as a function of the product pd for a uniform field gap, where $\eta V_s = \log_e 1/\gamma$ with $1/\gamma$ essentially constant, the function will have V_s high at low values of p and d corresponding to high values of X/p and $1/\eta$ and small values of d. It will rapidly decline to a single minimum very near the value of X/p at which η has its peak, and then rise more slowly, gradually approaching, but never reaching, a line of constant slope. This is shown relative to an η/p–X/p curve in the V_s–pd plot of figure 6.9.

The latter is the characteristic form of all breakdown curves and follows what is known as Paschen's law. While the exact form may vary, because Paschen's law is not always followed, it is almost universally observed for all breakdown phenomena. Paschen's law as plotted in figure 6.9 is for a uniform field. For a non-uniform asymmetric field it becomes difficult to define a proper effective gap length. Here d would probably best be replaced by r_0–a. However, as most studies do not evaluate r_0, or cannot do so, it is usual to ignore the d value to the extent of putting in the anode-cathode distance b − a or d. Frequently with fixed conditions, measurements only relate V_s to p. The V_s–p(b − a) or V_s–p curve then is not strictly a pd curve because in such gaps, as X/p changes, r_0 will also vary. However, for experimental purposes they suffice.

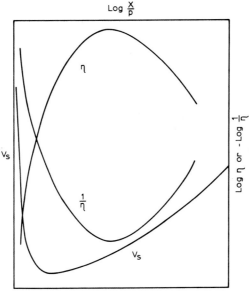

Fig. 6.9. Schematic plots of log η against log X/p and of log $1/\eta$ against log X/p. Finally by a proper transformation of variables this leads to the plot of V_s against pd, since X depends on V/d. The complex transformation is not shown.

The minimum of V_s in the V_s–pd, or V_s–p, curve is called the *minimum breakdown potential*. It occurs in uniform field gaps in the range of about 2 to 10 mm Hg $\times$ cm.

Since gas composition and η dominate the relation, the order of magnitude of the quantity V_s for any gas will be set by η and its properties as a $f(X/p)$.

It now happens, as was discovered by Penning, that if one mixes small traces of a gas of an ionizing potential which is close to but lower than some metastable or longer lived excited state of the main gas constituent, or vehicular gas, one gets an exceedingly high η at a relatively *lower* X/p. This occurs because it takes less electron energy to excite the lower states of the main gas than it does to ionize it. Once the main gas is excited, it completely ionizes the added lower ionizing potential gas by inelastic impacts of the second class if the trace gas is present in the right amount. The η–X/p curves for Penning type gases show *one maximum at low X/p* and *another maximum for the main constituent* or vehicular *gas* at high X/p. Since the V_s–pd curves are inverted and transposed relative to the η–X/p curves, there will be two *minima* in the V_s–pd curves. The first one corresponds to the maximum of the η–X/p curve for the main gas at high X/p but low pd or p, and the second one for the Penning effect at low X/p or higher pd and p.

The minimum of the second efficient Penning mixture, which may be much lower than the first, will appear at higher pd values and show a much broader minimum. Penning gases known to be effective are He–A mixtures and Ne–A mixtures, He–Hg and Ne–Hg mixtures, and A–I_2. Ne and N_2 should exhibit the phenomenon. A similar effect is also produced in A with much higher concentrations of Hg than that of the most effective Penning mixture. This has been shown by Nakaya[6.1] to be caused by metastable Hg_2 molecules. The amounts of lower ionizing impurity for most effective action in some cases need be very small. Thus the most effective Ne–A mixture is Ne + 0.002% A, or 2×10^{-5} parts of A to one of Ne. However, from 0.1% to 0.002% of A in Ne is very effective. The best quantity of Hg in Ne at 20 mm pressure is that given by the vapor pressure of Hg at a temperature of about 30° C, for A at 20 mm pressure, and again for a pressure of Hg at 40° C, and for I_2 in A at a temperature of about 0° C. In all cases the breakdown threshold with uniform fields was lowered in Ne + Hg from 400 to 150 volts, and in A + Hg from 520 to 150 volts. The minimum rises for Ne at about 0.01% Hg and for Ne with about 0.1% A.

The Penning effect is reduced if the discharge region is illuminated by Ne light on Ne, A light on A, and Hg or white light on the Nakaya high temperature A–Hg mixture. This comes from the destruction of the active metastable states of Ne by Ne light, A by A light, and $Hg_2{}^m$ molecules by white or Hg light. The $Hg_2{}^m$ molecules that lead to the second or Nakaya lowering of V_s by Hg at higher temperatures need A or a neutral gas to allow them to form, and Hg metastable atoms to create them. The metastable $Hg_2{}^m$ molecules that are ionized by A are destroyed by white light and by Hg light.

Unfortunately, the study of Penning effect mixtures has never been carried further, but there are many gases which in traces with gases having metastable atoms or molecules should give the Penning effect. One certainly notorious case is that of the influence of NO and NO_2 on a N_2 plasma. These are readily ionized by something of long life.

In any event it is clear that the order of magnitude of the threshold potential will largely be governed by the values and variations of η or α and by the ensuing integral $\int_0^{V_s} \eta \, dV$. It is also clear that, in view of this functional variation over large ranges of X/p, it will lead to a threshold curve of V_s against pressure times an effective gap length d of such form that V_s declines sharply as X/p decreases at very low values of pd, goes to a minimum value, and rises again more slowly, roughly approaching a straight line at high X/p. The value of η in some cases is very sensitive to traces of more ionizable impurities, and, where these are present and lead to the Penning effect, V_s is much decreased and the V_s–pd curve will *show two minima,* one for the vehicular gas and the other characteristic of the Penning mixture. In the less favorable percentage mixtures the two minima may merge to a broad one. If by chance the polarity of an asymmetrical field can influence the value of $\int_0^{V_s} \eta dV$, then the starting potentials in such Penning mixtures will be strongly influenced.

While for extremely different conditions the maximum variation of γ values can be very great, this will not be likely in any general V_s–pd or V_s–p curve with fixed conditions at the cathodes. γ will vary very much with the *nature* of the cathode material, X/p, at low X/p, and the state of the cathode. For example, for the element C it is very low both for γ_p and γ_i. The same is true for heavily oxidized surfaces[2.1] and liquid water.[3.68] γ_i appears to be very low for Hg, especially with Hg ions.[2.1] For some composite surfaces, such as AgO–Cs, it can be high, as well as for γ_p. H^+, H_2^+, and possibly H_3^+ ions have low values of γ_i on most surfaces until high energies of impact are achieved.[2.1] On the other hand, γ_p is fairly high for some hydrides and hydrogen exposed metal surfaces,[2.1] as well as for partially oxidized or O_2 coated metal surfaces.[3.1] γ_i if caused by a potential emission process is fairly high in vacuum (≈ 0.25), if the ionization energy of the ion is twice the value of the work function of the metal surface.[2.1] Some adequately gassy surfaces have a high γ_i.[2.1] All ions of sufficient kinetic energy, that is, in the hundreds of volts, can yield a γ_i.[2.1] Some pure hydrocarbon gases such as methane give very low γ values for metal surfaces, probably because of C deposits on the cathode. Raether[3.55,3.56] has shown that, in uniform field geometry with such gases, γ is so very low that a streamer spark by photoelectric action at very high V_s will cause breakdown with no Townsend cathode pre-discharge.

The equivalent of γ in the case of the highly stressed anode, caused by photoionization of the gas by photons from the avalanches, is in general small compared to the values ordinarily encountered for γ_i and γ_p of $\approx 10^{-4}$ to 10^{-2}. Thus a highly stressed cathode initiated discharge will in general have a very much lower V_s than is caused by a γ *in the gas.* Since in most earlier work cathodes and anodes were more or less equally highly stressed, most breakdown occurred through a cathode agency. It was only when cathode fields were so low in asymmetrical gaps with the right gas combinations and highly stressed anode that the γ phenomena caused by photoionization in the gas were observed and studied.[3.1,3.2]

The question as to the relative thresholds of V_s with positive and negative wire corona came to the fore through early studies carried out in Townsend's laboratory, which were later extended by Huxley, Boulind, and Bruce.[5.1] Curves for supposedly pure He, Ne, air, O_2, N_2, and H_2 are shown in figures 6.10 to 6.15, respectively, in the low p region at constant geometry. Here it is seen that in He the value of V_s for the negative corona lies *below* that for the positive past the minimum. It then rises *above* V_s for the positive wire. In Ne the same phenomenon occurs and the positive curve does *not* show a minimum at around p = 2 mm Hg, but at p = 30 mm. Now Penning was surprised at this crossing in those two gases, for as clean free electron gases Ne and He should have a lower V_s at all p values.

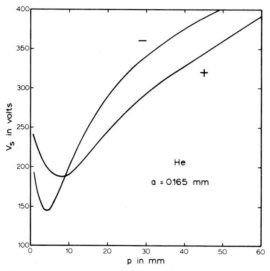

Fig. 6.10. Townsend group curves for threshold potentials V_s against p for coaxial cylindrical geometry in relatively pure He for a 0.165 mm diameter wire, showing a crossing of the two curves around 11 mm.

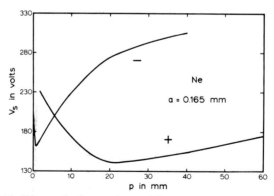

Fig. 6.11. Curves similar to those of figure 6.10 for supposedly pure Ne under the same conditions.

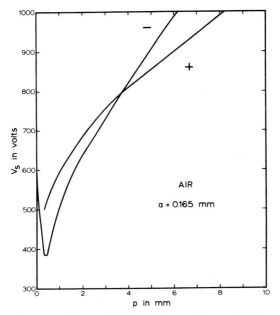

FIG. 6.12. Curves similar to those of figure 6.10 for air.

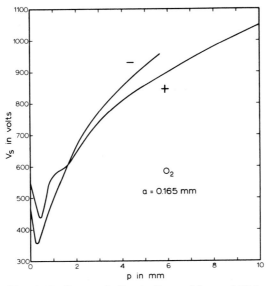

FIG. 6.13. Curves similar to those of figure 6.10 for O_2.

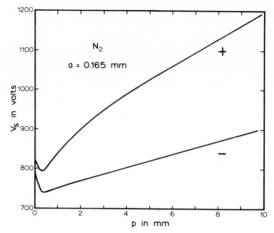

FIG. 6.14. Curves similar to those of figure 6.10 for sufficiently pure N_2.

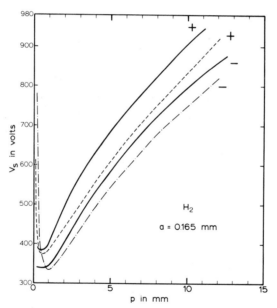

FIG. 6.15. Two sets of data from the same group, similar to the curves of figure 6.10 for relatively pure H_2. The differences probably represent the effects of varying purity, since the two sets of data are not contemporaneous and were not made by the same workers.

He repeated the measurements with truly pure He and pure Ne devoid of A and of Hg. In those gases he observed that V_s for the highly stressed anode lay considerably *above* that for the negative wire at *all pressures, as it should*. For Ne and He are free electron gases giving a high γ_p and γ_i at low p and high X/p. Thus V_s for the highly stressed cathode for equal η should with its high X/p and energetic ions lie below the positive. It is unlikely that much photoelectric ionization by their own photons occurs in these pure gases, so that the γ action for the highly stressed anode must have been perhaps some γ_p, but more or less a weaker γ_i at the cathode cylinder. These results are shown in figures 6.16 to 6.20.

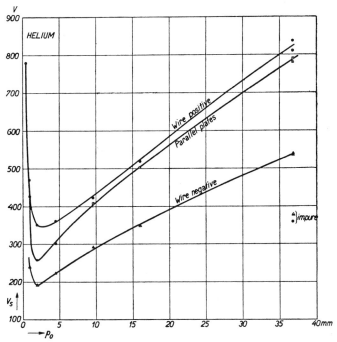

Fig. 6.16. Penning's curves for the breakdown threshold V_s of pure He with positive wire, negative wire, and parallel plates. It is strange that in He the breakdown for parallel plates should lie below that for positive wire. However, the plates require a γ_i and γ_p, while the wire requires photoionization of pure He by its own photons. Note that with reagent grade He the two points on the lower right give V_s for negative wire slightly greater than for positive wire.

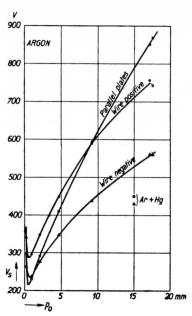

FIG. 6.17. Penning's threshold potential curves V_s against pressure for relatively pure A. The curves for positive and negative wire do not cross. Here the uniform field breakdown curve lies above both curves above 5 mm. This arises as Penning's A was probably not quite pure enough, because of Kr, so that photoionization occurred. Note the reversal of the positive and negative data and lowering when Hg is admitted to A at low pressure on the right.

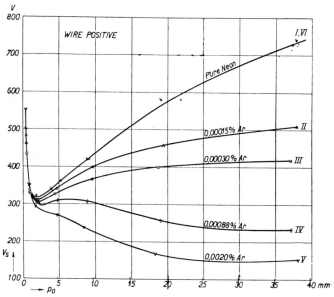

FIG. 6.18. Penning curves such as in figure 6.16 for Ne with various amounts of A added as impurity for a positive wire. Note the strong lowering for $8.8 \times 10^{-4}\%$ and for $2.0 \times 10^{-3}\%$ A, curves IV and V. The double minima due to the Penning effect are clearly noted. The effect here as well as for uniform fields is quite pronounced, since with positive wire all Ne metastables that are created have a maximum chance, even when they diffuse, to ionize outside r_0.

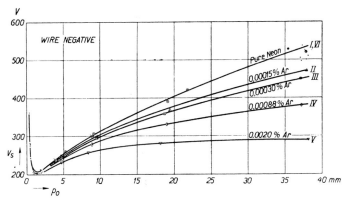

FIG. 6.19. The same as figure 6.18 for a negative wire in Ne–A mixtures. Here the Penning effect is hardly noticeable. The Ne metastables are created inside r_0, but most diffuse out of this small volume because of the high gradients and short distances before they make enough impacts to find A atoms and ionize them. These cannot create new avalanches, and the negative wire in these mixtures does not differ much from that of pure Ne or the positive wire.

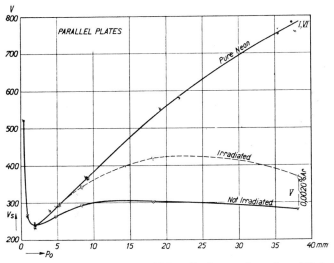

FIG. 6.20. This shows the influence of Ne radiation on the value of V_s in uniform field geometry with pure Ne and Ne with 2×10^{-3} % A. It is seen that irradiation raises V_s materially above the value without that light. If the Ne radiation could be made strong enough, the curve would approach that for pure Ne.

Penning then introduced a trace of impurity in He by heating his iron wire electrode. In A he introduced a drop of Hg. The values of V_s for the positive and negative coronas in these two gases at one point each for anode and cathode stressed are seen to be much decreased, and, what is more, the V_s for the anode lies slightly below V_s for the cathode.

This peculiar action is caused by the Penning effect and not by a γ. Now while α or η alters avalanche size and should be independent of the sign of the field, this is not so when metastable atoms produce the ions in asymmetrical gaps. The $\int_{r_0}^{a} \alpha dx$ occurs close to the wire in both cases. The metastable atoms are produced within the ionizing zone r_0 of the wire. To create new ions by the Penning effect, the metastable atoms must encounter the impurity after a number of collisions and before they are destroyed by triple impacts and what not. That is, the He^m in He or Ne^m in Ne will experience some 10^5 or more collisions before encountering an impurity atom or an A atom. They do not create their new Penning effect electrons at their place of origin but some distance away. In general they diffuse away from the confined zone between a and r_0 where they were created. With wire positive they will produce new electrons outside r_0, these will be drawn within r_0, and all will produce effective avalanches, increasing η and the $\int \eta dV$.

With highly stressed negative wire, all the ionization is produced in the positive ion zone or negative glow region within r_0 or just beyond it, but very little will be produced near the cathode surface. Here again diffusion of metastable atoms will in general be away from the high concentration where they are produced, and thus away from r_0. However, *unless* an electron is produced *at or very near the highly stressed cathode surface*, it does not multiply in the high field region and α remains the same, as it would for the clean gas. Thus with *highly stressed anode, a Penning gas benefits to the full value* of its increased η and decreased V_s, while the *highly stressed cathode* merely follows close to its pure vehicular gas curve. For this reason in Penning type mixtures, the peculiar phenomenon observed by the Townsend workers occurred for He and Ne. Their gases were contaminated by Hg from their gauges, or else were not pure. At that period only the Philips group at Eindhoven had achieved the degree of purity in inert gases to deal with such studies.

The question may be asked as to why the minimum in He was lower for the negative than for the positive wire in the Townsend group studies. At a sufficiently low pressure of He, for example, < 5 mm, the number of impurity atoms in the Townsend group's He was not enough to give much Penning effect lowering. It was only when the number of impurity atoms reached sufficient values that the Penning effect occurred. The Townsend group curve for Ne strongly points to A as the contaminant, since the minimum of V_s for the positive curve occurs at 30 mm, which is precisely the value to be found for Ne with positive wire with 0.002% A, as seen in figure 6.18 of Penning. The influence of the Penning mixture of A in pure Ne is shown in figures 6.18 and 6.19. Final proof of the Penning effect in their studies came through destruction of Ne metastables by irradiation with Ne light, as shown in figure 6.20.

It is possible that, had Penning and the others extended their studies to pressures as high as 760 mm, the diffusion of A^m and He^m would have been so far decreased that positive and negative wire curves would have had V_s about the same, or in A with V_s higher for the negative electrode.

While this explains the behavior of those gases with metastable states which can ionize impurities, it does not account for observations in O_2. In the case of N_2 and H_2, as seen in figures 6.14 and 6.15, the values of V_s for the negative wire lie below those for the positive wire at all times. For air and O_2 the results are not unexpected (figs. 6.12 and 6.13). At very low values of p, below 4 mm for air and 2 mm for O_2, V_s for the negative wire lies below the positive. Here, despite an oxygen coated cathode, γ_p and γ_i are high enough to favor the cathode. Photons created in the ionizing zone can penetrate to the cathode in air more readily than in O_2, clean-up can occur quickly, and V_s is low. For positive wire in O_2, the action at the cathode will be largely γ_p. It is weak, but X/p at the cylinder is not too low. Threshold, however, is lower for the cathode with its high X/p. Photoelectric ionization of the O_2 at these low pressures by O_2 or N_2 photons is weak and occurs too far outside the ionizing zone to be effective. However, as the partial pressure of O_2 increases, the cathode fouling becomes worse, but the photoionization in the gas increases and *a discharge sets in at the anode through photoionization in the gas* and gives an effective higher γ than the γ_p at the highly stressed cathode. At this point the two curves cross over. That is, the discharge changes from a cathode γ mechanism to a photoelectric mechanism *in the gas* which is more effective with the highly stressed anode and lowers V_s.

The question of how the discharge is influenced by the ratio of O_2 to N_2 was in part answered by Miller in his studies. Figure 6.21 shows the relative threshold potentials for the coronas at various pressures in N_2–O_2 mixtures. Data on starting potentials of the negative coronas are uncertain because of overshoot or hysteresis. This is also true of the threshold in pure N_2 for positive corona, since it depends on spot formation. Miller used the onset of steady corona or the current jump as the criterion instead of that for the pre-threshold Townsend discharge. This certainly applies to the more obvious discharges with currents in the 10^{-6} amp range. It is noted that at all pressures in pure N_2, in conformity with the Townsend group observations, threshold V_s for the highly stressed anode lies around 20% to 30% higher than for the cathode.

With as little as 1% O_2 in N_2, there is a rapid drop in the positive threshold and a steep rise in the negative threshold. At around 10% O_2 the

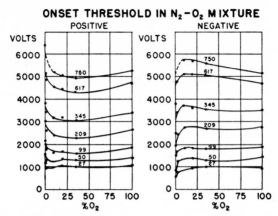

FIG. 6.21. Miller's values for onset of steady corona V_0 in N_2–O_2 mixtures at various pressures for positive and negative wires.

negative threshold lies above the positive one. This is to be expected, as Huber[2.22] showed that with 5% O_2, γ_i was reduced at least 5-fold and that positive corona threshold was set off by burst pulses rather than by a glow corona. The positive corona threshold for burst pulses seems to reach a minimum at around 30% O_2 at all pressures down to 99 mm. Thereafter there appears to be relatively little decrease in threshold with further addition of O_2. The increase in threshold above 40% O_2 is to be expected on the basis of the very intense absorption of the photons as O_2 content increases. N_2 gets too scarce to furnish enough photons to ionize O_2, and pure O_2 is not readily ionized by its own photons until much O is formed by the discharge. Again the study of streamer thresholds and propagation has recently shown that propagation is much reduced by increase in O_2 concentration. That the threshold for highly stressed anode should rise above that for highly stressed cathode, at the highest concentrations of O_2, is not too surprising. What is surprising is the decrease in the threshold for the highly stressed cathode beyond 15 or 20% O_2. This could mean that the photoelectric ionization in the gas with increasing O_2 content is being more and more closely confined to the zone between d_0 and a and that the short wave-length photons produced in O_2 are so copious and so near the cathode that γ_p is increased considerably beyond what it was at 15% O_2. At 15% O_2 many of the photons were escaping to the outer cylinder and being absorbed away from the cathode. Thus, to the extent that Miller's observations about the second reversal of threshold, making $V_s^+ > V_s^-$ in pure O_2, are accurate, it can be accounted for by high photo-ionization in the gas near the anode and by increase in burst pulse and pre-onset streamer thresholds by high photon absorption in the critical zone. If this is the case, as it appears to be, reduction in pressure of the mixture should reduce or reverse the trend.

It should be noted that Miller used the threshold potentials for the clean wire and fresh gas for the data in air and O_2 at 517 mm. This appears to have been done in some of the other values given. If there was more overshoot in the N_2–O_2 mixtures, the peaks and decline could be ascribed to uncertain readings. Certainly any decline at pressures below 345 mm is within experimental uncertainty. If the decline of V_s^- at 750, 617, and 345 mm is real, the absence of a decline at lower pressures would seem to substantiate the suggestion above concerning the cause for the decline in V_s at higher pressures with O_2 content.

In H_2, as noted by all observers, V_s for the highly stressed cathode is uniformly lower than for the anode, as it should be.

In any event, the question of the relative starting potentials has been described in sufficient detail and with enough examples, showing the causes for deviations from the generally anticipated relative values, to alert the reader to the complexity of the phenomena and the nature of some of the factors involved. Generalization is not possible, and the variation of V_s with p and/or pd in any discharge with asymmetrical electrodes must be studied with great care to ascertain all the factors involved. Perhaps a striking example of what can happen is the effect of running a discharge at high potential and low pressure between two non-uniform field but perfectly similar electrodes for some time. It will be noted that after some time the discharge will cease to pass from the highly stressed electrode. If the polarity is reversed, the discharge again runs nicely with what was the anode acting as the new cathode. Either on standing or on running at reversed polarity for some time, the original cathode recovers its high γ and will again pass the discharge acting as a cathode. Here γ is probably altered by intense bombardment or sputtering to a higher value. This may be associated with denudation of the γ_i augmenting gas film, since it recovers in time. Thus factors determining starting potentials in a tube never remain constant, certainly not under discharge conditions.

REFERENCES

6.1   T. Nakaya, J. Phys. Soc. Japan *11*, 1264 (1956).

7

Symmetrical Gaps
with Non-Uniform Fields
at Both Electrodes

A. GAP FORMS; AVOIDANCE OF FIELD DISTORTIONS

In the previous chapters attention has been focused on asymmetrical gaps with non-uniform fields about the electrodes. Because this book deals with the basic processes and the breakdown mechanisms active, attention has been focused on essentially two forms: (a) those with extreme asymmetry, best typified by the point-to-plane geometry with relatively longer gap lengths; and (b) the coaxial cylindrical geometry with relatively shorter gap lengths. These represent the minimum action of the less stressed electrode and maximum action of such electrodes in asymmetrical fields. Obviously this limitation omits a large number of variant gap forms, such as a large and a small sphere, a point and a large sphere, a flat cylinder and a plane, and so on.

This group should include plane capped cylindrical gaps, which have been investigated by Hermstein in a study of insulator breakdown. The latter is primarily an engineering field of study, and does not belong in this book. However, from the viewpoint of basic mechanisms, his excellent investigation is such that it merits reading. This book has also failed to discuss the shorter point-to-plane gaps, which would more or less fall into group b.

There is essentially not much to be gained by the analysis of such breakdown geometries. If the spheres are too much on the same order of magnitude, then the case approaches the symmetrical sphere-to-sphere gap, especially if the gap length is great. If they are more disparate, the smaller, more highly stressed sphere will break down first and will show in general the same behavior as the highly stressed point or wire, now with larger surface areas involved, inviting multiple spots above threshold. The degree of asymmetry will govern the extent of the range of breakdown phenomena which will occur between the potential at which the smaller electrode breaks down and the larger one breaks down.

Except for extreme differences in size and certain unique gas combinations discussed at the end of chapter 6, the disparity in corona breakdown and spark breakdown values will be greater for the arrangement in which the smaller electrode is the cathode and the larger one the anode. If the smaller electrode is the anode, the two breakdown thresholds will be closer together, and ultimate spark breakdown may come at a lower potential for the more highly stressed anode, say in air, though the highly stressed cathode will usually show the lower *corona breakdown* threshold.

Needless to say, at the corona breakdown threshold or near it, all asymmetrical gaps will preferentially pass greater currents from the more highly stressed cathode, and perhaps from the cathode in general, than from the anode. Thus all asymmetrical gaps and many more symmetrical ones close to and above the lowest breakdown threshold will show a strong rectifying action under alternating potentials of half periods greater than the formative time lags of the breakdown processes. Where the extreme form of asymmetry of the point to plane is carried down to short gaps, the highly stressed electrode no longer governs the breakdown, but secondary actions at the plane electrode come into play, placing it in class b above. In fact, a point very close to an infinite grounded plane may cease to be an asymmetrical gap, since the electrical image force field of the highly stressed electrode produces a high induced field at the plane surface which appears to come from a point of opposite polarity at the same potential at a distance of twice the gap length. In fact, such a situation would conceivably yield such high electrical fields at the plane that under some circumstances spark breakdown would initiate from the plane. The exact analogue cited above may be altered in static breakdown by space charges from the point. While such geometry would perhaps be of interest in applications, it is of no particular value in the analysis of basic processes.

Returning to the non-uniform field with *symmetrical* electrodes, one may consider three possible geometrical forms. These are (1) point-to-point, presumably two identical hemispherically capped cylindrical point electrodes; (2) two identical spheres; and (3) two small plane discs normal to the field direction with the common axis at their centers. The point-to-point electrodes are most commonly encountered in engineering installations and occasionally as electrodes in low pressure gaseous discharge tubes. The discs are used in glow discharge tubes at lower pressure, since they present a maximum surface to the positive column. The sphere gaps are used for calibration purposes in spark breakdown studies and in alternating potential breakdown studies at atmospheric pressure. There is today a tendency to use coaxial cylindrical cups with open ends facing each other—that is, *hollow cathodes*—in many discharge tubes at low pressure. The hollow cathode conserves photons, has its plasma sheath within itself, and gives a copious electron emission. The larger area of the cylinders also reduces current density at the electrodes and thus reduces destructive sputtering outside the electrode. Such electrodes are, however, of no interest in this basic study, although the physics of the hollow cathode is most interesting.

The two-disc arrangement is too asymmetrical about the surface to treat theoretically, owing to the high fields at its edges. This circumstance plays only a minor role in their use in a glow discharge tube, since the electrodes are used well above breakdown when the space charge sheaths have formed, and wall charges on the glass tube cause the chief current to flow normal to the disc surface after the surface is covered with a glow in the abnormal mode.[7.1] In such usage pressures are so low that the characteristic glow discharge structure is well-delineated. The cathode will have a surface glow (the cathode glow) at low pressures and higher fields, owing to excitation of the incoming positive ions and their neutralization at the surface. The sputtered surface material or metal vapor may add to this glow. Then comes the Crookes dark space, in which the secondary electrons from the cathode are accelerated to high energies. They show little ionization in the dark space of the gas, because in general they have too few encounters. The high field at the cathode falls off nearly linearly with distance from the cathode for some millimeters, and is caused by the space charge of the incoming positive ions, which have energies appropriate to the cathode potential fall region at the cathode.[7.2] As the field of the ion space charge declines, the electrons that are no longer being appreciably accelerated begin to scatter randomly through impacts at high speed. Here they create many ions and excited atoms in a region called the negative glow.

Thus the negative glow represents a nearly luminous neutral plasma of high charge density. This furnishes the positive ions that go to the Crookes dark space charge and the now slowed electrons that proceed on toward the anode to carry the current and keep the positive column conducting. The loss of slower electrons from the high density plasma of the negative glow by diffusion to the glass walls, and the decrease in energy of the electrons by their inelastic impacts in the glow makes it impossible for the electrons created in the glow to supply the positive column adequately and maintain current continuity. Thus a situation is created in which there must be an additional high field region, called the Faraday dark space, in which the electrons receive enough energy from the field to enter the positive column and keep that column conducting. The column is an equilibrium, a nearly electrically neutral conducting region. Here the potential gradient is low, but produces just enough electrons and ions to maintain its conductivity in the face of a radial ambipolar diffusion loss to the walls.

In front of the anode there must be another relatively dark space in which the electrons leaving the column get enough energy to ionize the gas adjacent to the anode surface in order to supply the flow of positive ions that maintains the column. Thus there is a new small but sharp rise of potential at the anode, with its dark space, and an anode glow at the surface. The greater proportion of the potential across the glow discharge tube is, however, taken up by the cathode fall and the Crookes dark space, with a little more in the Faraday dark space and a small but sharp rise at the anode. The potential gradient in the column is very slight. All these attributes of the two-disc electrode system belong in the realm of study allotted to glow discharges and really have little to do with coronas.[7.1] The description above, however, serves to indicate the nature and role of the disc cup and/or rods in low-pressure discharges, which will not be considered beyond this section.

Needless to say, the breakdown threshold is initiated by local discharges at the edges of the highly stressed cathode, in most cases followed by a similar corona discharge at the anode, or in rare cases initiating with an initial breakdown at the anode. As potentials increase at the low pressures, generally considered, the localized breakdown discharges push out into the tube at both ends. At a relatively small increase in potential above threshold, they break down into the continuous glow discharge along the whole tube. This occurs with an increase in current as soon as the ionic space charge clouds created by the current meet. At the low current densities near threshold, the glow will not cover the whole cathode, and one has the so-called *normal* mode of the glow. Increase in potential goes to increase the area of the discharge at the cathode surface, increasing the current in proportion. Potential across the tube, however, remains constant. This increase occurs through a momentary increase in ionization in the initial limited area of the glow spot at the cathode. As soon as this occurs, the increased space charge density of the column temporarily takes up the added potential. However, as seen in negative point coronas, the region of conducting column relative to the adjoining nonconducting region causes the radial expansion of the discharge. Furthermore, the high density positive space charges in the Crookes dark space are repulsive and at lower pressures permit the discharge to expand.

Thus with increase in current density the column and cathode fall region expands over the surface, the initial current *density* is restored, and the current as a whole has increased. Thus with greater current and decreased space charge resistance at the initial density, the potential across the discharge goes back to its initial value, while the iR drop in the external circuitry takes up the excess potential. This continues until the glow can no longer spread. Once the cathode is covered with the glow, increase in current requires an increase in potential across the dark space and column to maintain the current flow in the *abnormal* mode. The glow does *not usually cover the whole anode,* but just a spot on it, unless the pressures are very low.

The discussions to follow will largely deal with point-to-point or sphere-to-sphere gaps, except for the sections on rectification (D.5) and on the influence of illumination on the breakdown thresholds (E). Here any electrode form is adequate and disc electrodes will be considered.

One of the items that appears to have been ignored or neglected in many investigations of breakdown in non-uniform but symmetrical geometries, such as point-to-point or sphere-to-sphere gaps, has been that of making sure that the fields are truly symmetrical. Quite thoughtlessly, results have been seriously compromised by neglect of field distortion, which renders gaps asymmetrical when they are assumed symmetrical. This results in falsification of data when potentials are reversed on the supposedly symmetrical system.

Ideally a truly symmetrical gap of the point-to-plane or sphere-to-sphere type should have the electrodes infinitely removed from all other surfaces. The ideal gap is one in which surrounding space walls are symmetrically placed far distant from the electrodes relative to gap length and electrode diameter. The walls should be grounded and the electrodes at equal positive and negative potentials relative to ground. If the gap lengths of the electrode systems are in the 1 to 10 cm range, then good symmetry is assumed if the walls are, say, 50 times the electrode separation distant. Unfortunately, the usual convenient practice is to have one electrode highly insulated and isolated and at the positive or negative potential of the potential source, while the other electrodes of source and gap are grounded. Economy of space then demands that the grounded electrode be placed on a grounded base that is not too far removed from the electrode and the gap spacing. In consequence, the potential of the grounded electrode is influenced by the other parts of the grounded system on which it sits, through image forces of the isolated electrode on those surfaces. Thus lines of force diverge from the isolated high tension point, and *not all the lines end on the second point* as they should.

In fact, unless the second point, or sphere, is at least ten times the gap length distant from its grounded base, the system is more point-to-plane than symmetrical. Thus the relative initial breakdown thresholds for the grounded point will not correspond to those for the high tension point at the same polarity, as they should. The asymmetry of the field can be estimated in a measure by observing threshold potentials for breakdown at *each* of the two points, with the positive high tension first at the isolated point, the negative being grounded, and then with the negative high tension on the isolated point, the positive being grounded. The *differences* in thresholds of the two negative or two positive points will indicate the degree of asymmetry, as tip fields for threshold should be the same irrespective of polarity. The error is not only in the relative starting potentials when asymmetry is present, since the field even at the positive point is not precisely what it is at the positive point in a truly isolated bipolar symmetrical system. The errors made and erroneous deductions drawn on the basis of such arrangements are unfortunately all too common in the literature of the past. The influences of this sort will be noted in the analysis of data in table 7.1.

It should also be added that the use of insulating materials that can "see" the gap in a geometrical sense must at all costs be avoided, as these can charge up and introduce surface charges distorting the gap. This is equally true for steady and alternating potentials and for all but the first impulse potential application of a series. Insulators used as bases for grounded electrodes, to remove them from surrounding ground, do not effectively screen off the distorting base field. Insulating surfaces or metal surfaces with a floating potential anywhere near the gap are worse than useless; they are seriously detrimental. Such precautionary warnings ought to be unnecessary, but published works indicate the contrary.

B. THRESHOLDS AND THE ASYMMETRICAL CHARACTER OF THE INITIAL BREAKDOWN

1. CORONA THRESHOLDS IN POINT-TO-POINT GAPS

As potential is raised statically—since the threshold for a Townsend discharge at the highly stressed cathode is in general lower than for the anode—a local breakdown will appear at that electrode, provided that adequate triggering electrons are present to prevent overshoot and that the point is not heavily oxidized. The breakdown potential will be materially higher than that at the equivalent point-to-plane gap with the same radius of curvature of the point and the same gap length. This might appear surprising but should not be so. When a potential V is applied at the point the large proportion of the fall of potential occurs close to the point. This is the region which is responsible for the avalanche formation leading to the Trichel pulse, burst pulse, or streamer. In theory the field configuration for a point to an infinite grounded plane gap of length d is the same as that for a pair of point charges of *opposite sign* at potentials $+V$ and $-V$ separated by 2d, or a total potential difference of 2V at 2d. Thus one could expect threshold V_S for the two-point system at a distance

TABLE 7.1

CORONA AND SPARK THRESHOLDS IN KILOVOLTS FOR 2 AND 4 CM GAPS

Room air 22° C, hemispherical points 1 mm, rods 5–10 cm long. Point-to-point gaps, one point grounded (g). Point-to-point or point-to-plane, center grounded (f). Measurements by Winn (W), Waidmann (Wa), and Lam (L).

Corona Threshold

	Impulse	Static	Static	AC
4 cm Gap				
+ > < −		7.95(W)(g)	10.7(L)(f) 8.7(L)(f)	12.0(L)(f)
+ > | −	≈7.8(g)	7.11(W)(g)	7.5(L)(g) 9.0(L)(f)	7.5(L)(f)
− > | +			8.1(L)(g)	7.5(L)(f)
2 cm Gap				
+ > < −		7.8(W)(g)	10.0(L)(f) 7.0(L)(f)	10.(L)(f)
+ > | −	≈7.0(g)	6.5(W)(g)	6.7(L)(g) 7.3(L)(f)	6.9(L)(f)
− > | +			7.1(L)(g)	6.9(L)(f)

Spark Breakdown

	Impulse	Static	Static	AC
4 cm Gap				
+ > < −	52(Wa)(g)	37(Wa)(g)	37(L)(f)	48(L)(f)
+ > | −	52(Wa)(g)	52(Wa)(g)	45(L)(f)	>40(L)(f)
− > | +	>52(Wa)(g)			>40(L)(f)
2 cm Gap				
+ > < −	34.5(W)(g) 28(Wa)(g)	28(Wa)(W)(g)	13.8(L)(f) 28.2(L)(f)	24(L)(f)
+ > | −	28(W)(g)	38(Wa)(g)	26.5(L)(g) 36 (L)(f)	31(L)(f)
− > | +	42(W)(g) 31(Wa)(g)			31(L)(f)

Measurements such as shown in this table are difficult to achieve in an accurately reproducible fashion. Difficulties involve the exact duplication of configuration of the curvature for a hemispherically capped point 1 mm in diameter, the triggering of the gap with impulse and alternating potentials, especially for the negative point, where at least 1 electron per microsecond from the point surface is needed, and finally because of the field conditions in the gap owing to the influence of grounded regions about the gap. This is very important in point-to-point gaps, where there is a marked difference if one point is grounded or each point is raised to a floating potential. The data here derive from several different measurements by three observers, using two sets of power sources. Results are generally consistent except for a few unexplained glaring exceptions. They throw light on mechanisms active.

2d at a potential difference of $2V_s$ to be equal to that for a point-to-plane gap of length d at a potential V_s. In actual practice this is not observed. For example, the more reproducible burst pulse corona threshold for gaps with long 1 mm diameter point-to-point configuration with d ranging from 8 cm to 1.5 cm had a point-to-point threshold V_s of 1.25 to 1.3 V_{s+} for positive point-to-plane gaps ranging from 4 cm to 0.75 cm with the same point diameter. This is seen in table 7.1 for 2 and 4 cm gaps. The value did not change appreciably as the gap lengths were increased. The deviation here arises from the fact that the points were very long and the plane was not infinite in extent. Thus the point-to-point gap had its field considerably more axially concentrated than would be the case for a spherical charge and its image force in an infinite plane. That is, with point-to-point geometry the *axial field* at the *surface of the negative point* at 0.65 V_s in the point-to-point gap was equal to that for the shorter point-to-plane gap at V_s. With points positive and negative relative to ground (that is, with floating potentials), the ratios lay around 1.5 to 1.6 for these gaps. Here fields at the points were materially reduced by the surrounding geometry.

Attempts to get suitable threshold values for the negative point-to-plane gap were unsuccessful, as adequate triggering could not conveniently be applied.

If properly triggered in air for a point-to-point gap of length from 2 to 8 cm, the first threshold will appear as a Trichel pulse corona at the cathode at lower pressures with a 1 mm point in air. At higher pressure it will lie slightly above.[3.20] In electron free gases it will appear as the glows described under negative point-to-plane geometry. The condition of the point will influence the appearance potential somewhat, depending on the work function. Again triggering electrons must be present to avoid potential overshoot. The current at potentials *above threshold* for point-to-point gap at 2d and 1.3 V_s will be less than for the equivalent point-to-plane gap of the length d with threshold at V_s, since the low field region is now double that for the point-to-plane gap and the fields at the center are lower. The plane in point-to-plane geometry also serves as a very good ion sink, which is not so for the low field gap region of the point-to-point gap.

Depending on pressure, point size, and gas nature at potentials not much greater or less than the threshold for the negative point, the positive point will also undergo a local breakdown. In air and impure gases generally it will partake of the form of a burst pulse and pre-onset streamer breakdown. In pure H_2, N_2, and so on, it will be at significantly higher potentials and it may not show intermittence. When the breakdown of the second (positive) electrode occurs, or shortly above this potential, the current should undergo a sharp increase to more than twice the previous value, and pulse formation at both electrodes may be much reduced. This comes about as follows.

First the anode breakdown adds its share of current. Then as soon as the *ions* of opposite signs from the two points come into the neighborhoods of the electrodes (a positive ion cloud near the cathode, and negative ion cloud near the anode), the ion space charge resistance should decrease materially in the low field regions. This is especially true when the breakdown currents from anode and cathode become comparable.

As Hermstein has shown, negative ions at the anode form a sheath and terminate the streamers and burst pulses, yielding a steady negative space charge sheath glow corona. The positive ions converging on the negative ion space charge near the cathode, causing Trichel pulses, should weaken this action by reducing the quenching negative ion space charge. The space charge of negative ions in midgap is much reduced, and the positive ions reaching the quenching zone near the cathode will reduce the speed of quenching and prolong the pulses. Thus Trichel pulses should weaken or cease. What results then is a steady corona with glows at cathode and anode, initially probably modulated by a Trichel pulse ripple. Such action was observed by Miller[1.9] in coaxial cylindrical geometry with positive wire in slightly soiled N_2. Spark breakdown by streamer is temporarily prevented by the Hermstein sheath. As potentials and currents increase, the glows should lengthen axially and become more intense. If the space charges are quite well mutually neutralized with negligible ion loss to the surrounding gas and to recombination, the regime may be unstable, and at an appropriate potential a power arc will develop on a time scale of milliseconds or more without a streamer. The discharge channel with current increase will become luminous over its whole length, connecting the glows.

If the gap is very long then it is more probable that the carrier loss in midgap and the long low field region will prevent arc growth before the Hermstein sheath at the anode is broken down. At this point breakdown streamers appear and cross the gap under more favorable circumstances because of the weakening of the positive ion space charge by negative ions.

In free electron gases, the initial breakdown of the cathode in point-to-point gaps will be even more likely to lead to an instability yielding a power arc by a gradual build-up on a millisecond or greater time scale before the breakdown at the anode can occur. This is the more likely as in relatively pure gases photoionization is not very efficient and the streamer threshold may be relatively high. With free electrons the discharge at the cathode is not pulsed. It is also *not likely* that near the anode the *electron* space charge can form an effective Hermstein sheath. However, the copious electron current coming into the anode can lead to a small reduction of positive ion space charge in low field regions. What is more, the heavy electron current can start an externally triggered positive ion glow corona at the anode without the need of photoionization in the gas. Such action was observed by Kip[3.1] in his first study of streamers with adequate external electron triggering. Thus even before a normal breakdown could be expected at the anode, an electron-triggered field intensified current of considerable magnitude can start at the anode and increase as potential increases. While one might not expect an appreciable electron space charge action across the gap, with electron velocities more than 200 times those of the positive ions, this expectation may be in error. It is possible that, as in the glow discharge, the field distortion at the electrodes may yield a neutral plasma in between. If it does not, the current will be controlled by the positive ion space charge in the midgap and anode region.

Thus there will early be anode and cathode spots that project out into the gap further and further as potential increases. The main action will, however, be from the cathode. It is probable that the ultimate breakdown to a spark will come as it does in air and negative ion forming gases. Streamer action from the anode, if potentials are high enough, could readily occur, as no Hermstein glow forms and streamer sparks are not ruled out. The arc-over then will occur at relatively lower potentials for such gases, and will perhaps consist of a filamentary streamer spark from the anode on a time scale of 10^{-7} sec once the avalanches feeding into this point become large enough. In fact, if an arc does not occur through cathode instability near the cathode threshold, the transition to an arc might be near the streamer threshold for the anode in that gas, in the absence of a Hermstein glow for a point-to-point gap.

With impulse potentials of, say, a millisecond's duration and a rise time of a microsecond or two, the picture will be somewhat altered. Breakdown again will probably start at the cathode, except in air at 760 mm. The potential will depend on its past history and conditions. No time exists for a clean-up by low-order Townsend discharge. The cathode breakdown will also be delayed or made capricious as regards each potential impulse applied, in consequence of the statistical time lags. It will lie at a potential generally above the static threshold for the same cathode. When breakdown does occur, a highly overvolted Townsend discharge will materialize at the cathode. The cathode will very likely be highly unstable, and if clean-up can occur under a millisecond bombardment at the higher current, an arc might develop in pure free electron gases. Miyoshi's studies indicate that a negative glow with overvolting leads to a negative streamer that then induces a positive streamer from the anode. Here the copious electrons at the anode could lead to early streamer formation.

In electron attaching gases, however, the Trichel pulse has a rise time in air at 760 mm of 10^{-8} sec and quenches by space charge formation in 10^{-8} sec more. Unless the gap is excessively overvolted, a Trichel pulse discharge will carry on for a few cycles in each impulse duration if its repeat rate is on the order of 10^4 per second. Moore and English[3.19] in a positive point-to-plane gap got two or three streamer pulses in 2 μsec. The negative and positive ions cannot cross the gap in 1 msec. Thus the Hermstein negative ion sheath will not form at the anode in one pulse by *ions from the cathode*, as with static potentials. In gases like air at the threshold and for some range of potentials above, it will be a Trichel pulse or a pre-onset streamer corona current with more and more pulses during the milliseconds of the potential application. In air for points of about 0.5 to 1 mm diameter at 760 mm, the burst pulse threshold at the anode and Trichel pulse threshold at the cathode are not very different. The anode will break down into a pre-onset streamer corona or a mixed streamer and burst pulse corona. For points of 0.1 mm diameter, the negative threshold is about 400 volts in 2000 lower than positive.

Thus onset of the coronas in electron attaching gases such as air will occur at nearly the same potential for both electrodes. The onset discharge will also be intermittent at the anode. In 1 msec the negative ions *from the short streamers can reach the anode* and Hermstein's space charge sheath will form *somewhat above threshold*. Space charge resistance in the midgap will be high, and currents will be low. As potential is raised, spark transition will occur as a result of anode and cathode streamers, in accordance to what has gone before.

As indicated with free electron gases, the cathode is highly overvolted and its threshold is relatively low compared to the streamer threshold in the pure gas. Thus breakdown at the cathode leads to highly unstable glow discharge with current increasing during the millisecond because of clean-up. Electrons from the cathode can in 1 msec conceivably reach the anode if it is not too distant. This can lead to a breakdown at the anode. Arc-over could conceivably ensue through cathode instability without a streamer spark for a short gap. This could be detected by a formative time lag study in which breakdown extends to 10^5 sec instead of 10^{-7}. It could also occur through a streamer at sufficiently high potentials for longer gaps.

Impulse arc-over transition could require higher potentials than the static case, but should occur at definitely lower values than for the anode streamer spark in point-to-plane gaps of half the length. It could also happen, as in pure A gas for point-to-plane corona, that there is no corona, but only a spark transition to an arc at threshold.[4.16] In any event, here there should be relatively little difference between static and impulse spark breakdown except that the latter should occur at somewhat higher potentials. Generalizations of this sort are in principle dangerous, as witnessed by the changes in relative sparking potentials for point-to-plane gaps in mixtures ranging from slightly soiled N_2 to pure O_2. Here the N_2 threshold is markedly lower for the static than for impulse potential. Addition of 2 to 5% O_2 raises the static threshold far above the impulse threshold. For air depending on point diameter and humidity they are about equal, and in pure O_2 the static threshold is two-thirds of the impulse value.

No oscilloscopic study of the point-to-point breakdown under static conditions has been made in air at atmospheric pressure for different gap lengths to distinguish the arc growth by slow transition at short gaps from the more rapid streamer transition at longer gaps.

If the gap is not too long and electrode conditions are right, the power arc materializes. At higher pressure radial diffusion is reduced, since there is no wall to which the electrons can diffuse. The only factors causing carrier loss from the highly ionized plasma are recombination and diffusion cooling by axial thermal gradients.[7.3] In all cases the mass motion of gases which can distort or blow out the plasma in low field regions must be avoided. If the pressure is low, the more or less wire-like discharge channel a millimeter or less in diameter expands to the walls of the containing envelope. The theory yielding the conditions under which this expansion occurs has been indicated by R. G. Fowler.[7.3] The glow discharge then appears, with ambipolar diffusion to the walls and thus a sufficiently rapid loss of carriers, electrons, and ions from the plasma, to reduce the carrier concentration in the column and delay the appearance of the arc to much higher potentials.

2. Spark Breakdown in Point-to-Point Gaps

As stated, data of a comprehensive comparative nature for static, impulse, and alternating potential corona for point-to-plane gaps of different polarities, and point-to-point gaps for various gap lengths and a single geometry of point configuration are lacking. Meek and Craggs[7.4] in their book *Electrical Breakdown of Gases* show data of Strigel with point to 5 cm diameter sphere gaps from a fraction of a millimeter to 1.6 cm length in comparison to those between two 5 cm spheres and a point-to-point gap. Most of the sphere-to-sphere values are very high and about double the point values. For gaps of less than 2 mm the point-to-point and positive point-to-sphere values lie together and slightly above the negative point-to-point. Above 2 mm the positive point encounters the Hermstein sheath and spark breakdown nearly reaches the sphere-to-sphere value. It lies well above the negative point-to-plane values and requires a jump from 3 kv to around 6 or 7 kv to cause a streamer. At 2.5 mm the negative point-to-point sparking threshold falls well below both point-to-plane potentials. In fact, at 1 cm it occurs at 6 kv, while negative and positive point-to-plane gaps are nearly the same and lie about 15 kv. Here space charge influence is again noted, as in table 7.1. Data by Uhlmann[7.4] for point-to-plane gaps with conical tips, or cylindrical rods with hemispherical tips of 0.4 cm diameter against planes are shown. Here for the conical tips the positive point spark breakdown value lay below the negative, the negative potential being about double that for the positive point. With hemispherically capped cylinders the negative spark breakdown lay possibly slightly below but probably at the positive value out to 2 cm. From 2 cm to 9 cm the negative values lay materially above the positive but at most no more than 25%. It is clear that, with the stronger guiding field of the hemispherical points, the negative streamers were able to contribute more toward the breakdown. Impulse potentials of short duration for gaps up to 300 cm and more again showed negative point-to-plane values nearly twice the positive point-to-plane value.

As shown in table 7.1, comparative data for corona thresholds and spark breakdown thresholds have been obtained for point-to-plane and point-to-point gaps with impulse, static, and alternating potentials. The data for the corona thresholds have been presented in the section 1 just preceding. The observations on spark breakdown shown may be summarized as follows:

The ratio of sparking potentials between a 4 cm point-to-point gap and a 2 cm positive point-to-plane gap lie at 1.8 for the impulse, 1.3 for the static, and 1.5 for the alternating potential. The impulse potential of rise time $\approx 10^{-8}$ sec allows no chance for space charges to form. It is basically a streamer process, where the streamer crosses the gap. Since streamer length and vigor are proportional to point potential and the point field plays less of a role, it takes nearly twice the potential to launch an anode streamer 4 cm than 2 cm. That the ratio is somewhat less possibly indicates some action of the fields at the points but more likely some interaction of the positive and negative streamer tips on each other. In the point-to-plane gap the negative streamer does not materialize. The negative point in point-to-point geometry yields a shorter negative streamer. Positive streamers have probably crossed nearly to the negative point, even below breakdown, and negative streamers form to meet them in midgap.

For static potentials the predicted space charges of negative ions at the anode and positive ions at the point cathode obviously eventually enhance streamer formation at both electrodes, lowering sparking potentials materially by a factor of 1.4 for a 4 cm gap and 2.5 for a 2 cm gap. With alternating potentials as corona thresholds are exceeded, the growing space charge at the end of each half-cycle of sign opposite the rising potential in the next phase causes phase advance of corona threshold. This causes reduction of the space charge field as the potential rises, so that at peak the conductor field is perhaps nearly that which might exist if no space charge of the opposite sign were present. With static potentials the space charge fields at the electrodes build up and remain high. However, with alternating potential owing to diffusion, a conducting plasma of negative and positive ions builds up in midgap. Thus static sparking potentials are 0.77 and 0.58 of the alternating values for 4 and 2 cm gaps, respectively. They are reduced by factors of 1.08 and 1.43 below the impulse values for the same gaps.

The ratio of the potentials for 4 and 2 cm point-to-point gaps are 1.5 with impulse, 2.1 with static, and 2.0 with alternating potentials. In the point-to-point impulse gap the cathode was grounded, and space charges do not interfere with impulse. There is no Hermstein sheath. Both points, however, give streamers in point-to-point gaps, and again it is clear that their interaction even for a 4 cm gap is quite effective, as was indicated in comparison with the point-to-plane gap of half the length. However, the point potentials must suffice to drive strong streamers across the gap. With static and alternating potentials floating potentials were of necessity used. Here negative ions yield Hermstein sheaths. While the absolute values of the potentials are lower than for the impulse potential, it is seen to take twice the potential to break down the sheath and yield a streamer spark on doubling the gap length. Here it is not basically the launching of interacting long streamers but of dissipating the Hermstein sheaths. With floating potential the fields at the points are weaker and it took twice the potential at the point to overcome the Hermstein glow.

The ratio between the sparking potentials for 4 and 2 cm positive point-to-plane gaps with cathode grounded was 1.8, indicating again that the positive streamer was largely responsible. However, the anode point field must have contributed in the propagation as well as point potential, since there are no negative streamers and the point fields are accentuated with cathode grounded.

The sparking potentials for negative point to plane where they were observed are from 1.3 to 1.5 times the potentials for positive point to planes. This is in keeping with data from many other observers. It is to be expected from the short range and inefficiency of negative streamers, as shown by Nasser, Waidmann, and others.

As indicated in the table, the point-to-point sparking potential on impulse in a cleared gap is 1.4 and 2.5 times that for static breakdown on 4 and 2 cm gaps, respectively. This must be ascribed to the strong influence of the space charge fields at the electrodes and the neutral plasma in midgap. The influence appears to be greater the shorter the gap. This action occurs in spite of Hermstein sheath development. Possibly with the heavy negative ion current and adjacent neutral plasma structure the Hermstein glow is weakened. The actions of space charges and neutral plasma in interelectrode regions are also shown by the lowered alternating sparking potential, although here the action is materially less.

In the case of positive point to grounded plane spark breakdown, the impulse and static values are the same for the 4 cm gap. Comparison between a positive point with grounded plane and one with a floating potential for the 4 cm gap with impulse and steady potential gave a ratio of 1.15, indicating clearly the reduction of point field with floating potential. All the ratios are unity for the 2 cm point-to-plane gap. It must be noted that the equality of impulse and static sparking potential values with planes grounded in positive point-to-plane configuration applies to the case of 21% O_2 in N_2, as in air only, but not for other mixtures, as Waidmann has shown.

At thresholds the space charges are not great enough to alter offset potentials observationally.

C. SPHERE-TO-SPHERE GAPS

Such gaps may be considered in two categories. In the first the diameter of the spheres is small compared to the gap length. In the second the diameters are comparable with or larger than the gap length. Except for the fields immediately in the neighborhood of the electrode, the form of which differs from the hemispherically capped point, there is not too much difference in behavior to be expected. Where the length is large, the general behavior will closely resemble that of the point-to-point case. Where the length is short with the large spheres, the gap will range from one in which pseudo corona behavior with both electrodes participating will predominate to one in which the field approaches that of near uniform field geometry, with the discharge regions tending to be confined along the axis of the line of centers. With the larger spheres, aside from the condition of geometrical symmetry, the symmetry must be equally extended to include that of identity of electrode material and condition. For in this case the work function of the highly stressed cathode is important, since both electrodes participate.

Any dissimilarity of electrode condition and material will in principle render the gap performance definitely different in properties when the polarity on the electrodes is reversed. Possibly this should also have been pointed out in connection with point-to-point gaps. However, there the distances are relatively greater and, while there will be some differences in starting potentials for the threshold of the negative electrode, the over-all behavior will not differ too much. The effects will be most pronounced in free electron gases, with outgassed and clean metal electrodes, and at lower pressures. If not recognized, they can lead to erroneous conclusions, especially in the larger spheres.

Probably the greatest difference between the point-to-point and the sphere-to-sphere system with large gaps will be in the more radial character of the field of the spheres relative to the spherically capped cylinders, where the field is more axially directed. Thus burst pulse spread and streamer branching should be greater for the spheres than for the points. In consequence, the spread of the burst pulse glow and the Hermstein sheath glow should be greater over the spheres, and the cathodes in air should show a readier tendency to have multiple Trichel spots in air and electron attaching gases. Currents with increase in potential can be larger, and the ring discharge observed by Greenwood[4.3] should be prominent. For equal radii the thresholds may be slightly higher with the spheres than the points. Streamer development from the anode will not be as easy with spheres. Hence breakdown to a spark may be higher. However, by and large the two systems should yield similar results.

With larger spheres and more uniform gaps, some modifications can be expected. For very short gaps, as indicated, conditions will approach uniform field conditions and a Townsend breakdown will precede the streamer spark. Its breakdown will set in at lower thresholds than for the equivalent uniform field gap between planes. The space charge from the avalanches will be created at the anode, but the discharge will not be confined to the anode surface. A plasma will build up at the anode rather rapidly, and as long as α/p increases faster than linearly with X/p, avalanche size will increase and lead to a streamer spark. The observed sparking potential, as anticipated, being initiated by the Townsend pre-discharge, actually falls below that for the uniform field gap as gap lengths increase relative to sphere diameter. This arises from the earlier initiation of the Townsend discharge at the high field of the curved surface, if the separation is not too great. The direct streamer spark transition on impulse where Townsend pre-discharge cannot arise will be higher. This arises from the lower midgap field region in contrast to the uniform field in a plane parallel gap. Thus the Köhrmann[3.39] type of spark transition will be higher while the Townsend conditioned spark will be lower than for the plane parallel gap.

In free electron gases with gaps intermediate between the two cases we have just considered, another mode of breakdown may occur. Here the initial breakdown will be a Townsend breakdown, which covers the cathode high field area diffusely and then contracts to a point discharge of small diameter and high current density as the surface cleans up. This sends a localized beam of electrons to the anode, which converge on the anode in a more or less localized area of high field. A Hermstein sheath does not form and the burst pulse-like discharge does not fluctuate, but is sustained as a glow by the influx of large avalanches at the point of impact. Thus, while there may be a tendency to spread over the surface, the concentration of avalanches and conductivity along the axis and at a point leads to a localized self-contained glow. Because of the shape of the field and localization, this manifests itself as a luminous anode spot that extends outward into the gap as current increases with potential. Overloading may give rise to multiple spots on the cathode with opposite spots on the anode. Through self-repulsion they will arrange themselves in uniform patterns around the sphere surface. How the transition to a spark occurs here is not known. The distance between spheres is too great and the longer low field region too extensive for the space charge accumulation at the anode which carries the streamer spark. It is possible that at sufficient potential a breakdown streamer starts from the anode. It is not inconceivable that the arc materializes gradually above a certain potential by increase in current along the avalanche path between the anode and cathode spots which become connected by a conducting path. In some cases, because of cathode instability, there may be no corona, as Miller found in pure N_2 and Weissler found in A with point to plane, unless there is an internal space charge resistance or an external limiting condenser.

Except for the convenience of the geometry, since the fields are readily calculable along the gap, if space charge is not present, these gaps present no particular virtue for corona studies. In industry and in engineering practice they are very useful as spark gaps. Since geometry is simple and adjustments are easy, they are widely used in the measurement of breakdown potentials and as safety gaps. The point-to-point and infinite plane gaps have disadvantages. The point-to-point is hard to calibrate with uncertain fields, while the infinite plane uniform field gap takes up too much space if it is to give really uniform fields. The sphere gap is also readily illuminated with ultraviolet light to reduce the statistical time lag.

Perhaps a word might be said about space charge clouds in gaps in which each electrode has its own separate breakdown potential and breaks down more or less independently of the electrode of opposite sign, as with point-to-point and sphere-to-sphere gaps of greater length. The very weak midgap field region, especially in electron attaching gases, will permit accumulation of a considerable ionic space charge. Some of this will go to the anode, especially in free electron gases. Otherwise, a rather extensive diffuse cloud forms. This is aided also by the electrical wind. If the anode has not reached threshold, there will be considerable current loss to the surrounding space and the anode will collect only a part of the cathode liberated current. The negative space charge cloud will, however, increase the potential gradient near the anode and on the anode side of midgap. When the anode breaks down, a simultaneous emission of ionic space charge clouds moving in opposite direction from anode and cathode takes place. If negative ions form, the mobilities will be about the same. Near positive threshold the negative ion current will exceed the positive and the clouds will be disparate. In a large measure what happens depends on field arrangements.

If the gap is truly symmetrical with the surrounding walls of the gap at ground and the anode and cathode at equal potentials above and below ground, the clouds will attenuate by radial flow to the walls. The components that approach each other from anode and cathode will interdiffuse both by field action and mutual attraction, making a diffusely conducting plasma of low ion density in midgap which will increase anode and cathode falls. Loss of carriers in this situation will be radial by diffusion and by ion-ion recombination. Thus the current through the gap will not be equal to the current from either electrode measured individually, as there will be a wall current and loss by recombination. In principle this sort of action does not play an important role in the breakdown sequence or phenomena with static potentials. It can and does become a very important feature in alternating potential flashover at high potentials.

D. NON-UNIFORM BUT SYMMETRICAL FIELDS, AND ASYMMETRICAL FIELDS WITH ALTERNATING POTENTIALS

1. INTRODUCTION

The case of alternating potentials introduces several new considerations into the discussion of the breakdown. These involve the influence of frequency, the rectifying action in regions near thresholds for the two electrodes, and the influence of space charge clouds. In speaking of alternating potentials, discussion will be limited to relatively low frequencies, thus leaving out of consideration the microwave region, where frequencies are so high that electrons traverse distances on the order of and less than a free path. It will also leave out of consideration frequencies in which the half-cycle is short compared to the formative time lag for breakdown at an electrode. It must, however, consider frequencies sufficiently high so that carriers of one sign or both signs fail to cross the gap in one half-cycle. Under these conditions, space charge clouds accumulate which may influence breakdown. Thus even for moderately long gaps, 60 cycle AC would come under this heading.

2. THE INFLUENCE OF FREQUENCY

If the frequency is very low or the gap is quite short, so that electrons and/or negative ions can cross the gap from cathode to anode in the time of one half-cycle, and the positive ions can do the same, the situation is much like that of static breakdown. There will be more uniformity in the gap, since any residual asymmetry of positive and negative space charges in the low field regions will be wiped out. Loss of carriers to the walls will be less, and recombination loss perhaps increased. The initial breakdown potentials for the electrodes will not differ materially from those for steady potentials. What happens subsequently is complicated by the fact that, once threshold has been reached for Trichel pulse or streamer and burst pulse breakdown, the clouds of negative or positive ions left at the end of each positive or negative half-cycle are drawn to the conductor or point during the early part of the ensuing opposite half-cycle. If the breakdown currents are high the positive space charge during the negative half-cycle and the negative space charge during the positive half-cycle increase the fields at the surface of the conductor. This means that breakdown occurs before the peak of potential, and with considerable phase advance over the peak of the potential. Thus, while the peak alternating potential V must exceed V_{s+} or V_{s-} to start breakdown once space charges begin to be of importance, breakdown will be sustained at a lower value of V than, say, V_{s+} and V_{s-}. Since at V_{s+} and V_{s-} currents are weak the effect will be so small as to elude detection. If the AC peak potential is raised above V_{s-} and V_{s+}, then the phase advance will follow and currents will be large. The spark breakdown potentials are thus altered by space charge action. Once the frequency becomes so high for a given gap length that electrons, negative or positive ions, or carriers of both signs cannot cross the gap in one half-cycle, the situation alters. Space charge clouds of opposite polarity push out into the gap but do not cross. Before they cross, potential reverses, and the space charge clouds of opposite polarity are pulled back toward the electrodes. Thus the first space charge clouds move back into the electrodes of origin, but are not completely collected because of diffusion and self-repulsion, so that more and more space charge of both signs works its way out into the gap by diffusion and self-repulsion and forms a conducting plasma, reduced by recombinations in midgap.

The situation can become very confused and complicated where crossing times are on the order of but longer than the half-period, since disparate clouds near threshold and small differences of mobilities can lead to various complicated situations which, except near sparking threshold, are not too significant. At higher frequencies the alternate breakdowns will project alternately charged space charge clouds into the midgap region. Thus there will accumulate in time, with internal repulsion, diffusion, and mutual attractive force between clouds, a conducting plasma of continually increasing density. If dissipative actions to the walls and recombination do not take too large a toll, the space charge can quite strongly alter the fields at both electrodes. Thus both anode and perhaps even cathode streamers will develop at relatively lower potentials than with static potentials. This action is complicated by the action of more local space charges of opposite sign at the conductor surface on reversal of polarity. In consequence, spark breakdown by streamer action will occur at lower potentials. This has been observed in shorter sphere gaps in air at atmospheric pressure, where diffusive loss to the walls is not great.[7.5] Quite generally in fact, spark breakdown studies in both uniform and symmetrical non-uniform gaps have shown sharp changes in threshold when the carriers of one sign or the other, or of both signs, cease to cross the gap in a half-cycle.[7.5]

In exceedingly long gaps on the order of meters in length, with relatively small electrodes at 60 cycles, the breakdown is rather interesting. As spark-over potential is approached, which is well below the static threshold, the fields at the instantaneous anode and cathode become exceedingly high, because of the oppositely charged space charge streamers projected into the gap ahead of them in the preceding half-cycle. Thus in each succeeding half-cycle, the streamer processes from negative and positive points become more vigorous and extend further out into the gap. Eventually the stream-ers advance so far on each side toward midgap that their highly charged space charge clouds create a field in midgap between them sufficient to launch a midgap streamer process such as Raether's in overvolted gaps. A streamer spark develops in midgap which rapidly propagates in both directions to illuminate the channels left by the streamers that created them and thus yield an arc.

3. The AC Behavior in Asymmetrical Gaps,
   Corona Phenomena, and Radio
   Interference

As indicated, not much has been done in a basic study of the AC corona in terms of more recent knowledge. The problems inherent in the increased radio interference extending from radio communications through television and into radio astronomy, as a result of the increase in potential of high power transmission lines has called forth a number of isolated, good investi-gations of various phases. These have so far not been integrated into a consistent scheme as related to basic phenomena. Involved are transmission lines of conductors of various form ranging from 3 to 6 cm in diameter with potentials ranging from about 70 kv to 400 kv. The potential is in general 60 cycle, but where conductor systems of three conductors appropriately spaced carry three phase 60 cycle, peak stresses *between* conductors are more involved. One need only consider the corona and the radio inter-ference, RI, produced from the single conductor relative to ground at 60 cycles, which may be on the order of 20 m distant. In what follows the author collects and interprets the findings in terms of corona processes. These yield a good introduction to the physics of the AC corona.

Radio interference comes from the shock excitation of conductors as a result of rapid changes of current. Here the excitation results from the cutting of the conductor by changes in magnetic flux created by changes in the current. Since dB/dt is proportional to di/dt, then aside from di/dt the circuit conditions linking the flux between conductors through the geometry of the circuit determine how great the influence will be. Impulsive dis-charges such as the Trichel pulse corona and still more the heavy breakdown streamers have di/dt on the order of 10^6 to 10^7 amp/sec. Small sparks where instantaneous currents in the hundreds of amperes arise in 10^{-6} sec can lead to still higher values. These induce surges of current in any conductor cut by the flux. They lead to damped electrical oscillations in the conductor lasting for some time after the triggering source has ceased. Such oscillations are further transmitted for some distances along the line and are also radiated to yield a broadcast. The current pulses, being of complicated form, give rise to many harmonics covering the usual broadcast range and invading the high frequency field.

The sources of such radio interference are essentially three: (1) Electrical breakdown of insulators inside the generating and distributing complex. These involve corona and sparks in the insulating oils of circuit breakers, switches, and transformers. (2) Breakdown over the surface of insulator strings and at regions of high electrical stress where supports of various types are embedded in cement. (3) Corona discharges from the conductors themselves. The type 1 interference does not concern us. The second type at the insulators has been extensively studied and improvements in design that are not too costly have, and in the future can, reduce this source to a minor one. It is true that insulator strings have the insulator surfaces coated with dusts, sea salts from fogs, fume condensates from industrial contamination ranging from sulfuric acid and hygroscopic sulfates to carbon and oily deposits. The dusts will be coated with salts and hygroscopic materials so that they become conducting. They are most troublesome when the humidity is such as to give a conducting coating, leaving small dry isolated patches over which small sparks can occur. On the other hand, the discharges from metals embedded in cement are less serious when the humidity is high, since these surfaces become sufficiently conducting so that sparks and discharges from small sharp points no longer occur, as these are smoothed over and gaps are bridged.

The real concern is the corona from the long high-tension conductors. To analyze this one must consider the nature of the breakdown from such an isolated cylindrical conductor at some distance from ground subjected to a 60 cycle alternating potential that is being gradually raised. While the pressure, temperature, and relative humidity as well as air currents in nature vary, the influence of the changes under normal conditions is relatively trivial. The same probably applies equally to changes in conductivity of the free atmosphere. The influence of dusts and particulate matter including precipitation will be discussed at a later point.

If the conductor has a clean, smooth metal surface it is probable that the negative Trichel pulse threshold V_{s-} still lies slightly above that for pre-onset streamers and burst pulses on the positive half-cycle. If the surface is coated with oxide films, less conducting dust layers or carbon, V_{s-} will be still higher. In any event breakdown will start with Trichel pulses on the negative half-cycle, and burst pulses and pre-onset streamers on the positive half-cycle. These will at first occur near the peak and will perhaps be a bit noisy, although the Trichel pulses and streamers will not be too vigorous or too frequent. Here pulses will occur near the peak of potential for each half-cycle. Very slightly above this threshold, which in actual observations may be passed over without notice unless it is carefully looked for, a more regular regime will set in.

An analysis by Cobine[7.6] (1941) before too much was known about corona mechanisms indicates that, once corona currents of an appreciable magnitude appear, there is a change. Starting with the first positive phase when the AC was applied at t = 0, the breakdown will start somewhat below potential peak at a phase angle $\phi_1 < 90$ deg, and terminate at about the peak or a bit before because of the positive space charge in the gap near the conductor. The breakdown consists of pre-onset streamers and burst pulses scattered here and there along the conductor where electrons trigger avalanches. They will cluster on portions of the conductor under the highest stress. These send a cloud of positive ions out during the last half of the positive half-cycle. As the potential passes through zero there will be a positive ion space charge surrounding the wire which moves toward the conductor as negative potential increases. At some point of phase angle ϕ_2 relative to the rising negative potential, the rising negative potential and the field of the positive ion space charge will initiate a Trichel pulse corona at points along the wire, with ϕ_2 less than ϕ_1. That is, there will be a phase advance for the negative half-cycle over that advance for the initial positive one caused by the fact that V_{s+} was below the applied peak potential V_P. If V_{s-}, the Trichel pulse threshold, is slightly above V_{s+} then ϕ_2 will be behind ϕ_1 by a very small amount. During the rise of the negative potential the positive space charge will in part be drawn into the conductor and in part be neutralized by the electrons and negative ions from the Trichel pulse discharge. The Trichel pulse discharge spots will appear near where the space charge of the streamers and burst pulses was created. At the peak of the negative potential phase, or possibly even before, the negative current will terminate by space charge choking and the charge will move out from the conductor. As the potential again passes through zero and goes positive, the space charge of negative ions (all electrons having attached by then) will again move into the conductor, causing a new corona starting either at ϕ_2 or possibly a bit later than ϕ_2 at ϕ_3, with $\phi_1 < \phi_2 < \phi_3$. This would occur if $V_{s-} > V_{s+}$. From here on there will be some changes. The positive space charge began a bit earlier because $V_{s+} < V_{s-}$. Again many electrons attached dissociatively to form O^- ions near the conductor. Those that did not moved rapidly away to form O_2^- ions at somewhat greater distance than the other negative ions. Thus the negative ion cloud is more diffuse than the positive ion cloud. Differences in ion mobilities could also favor this action. Thus the negative ion space charge field may not be as great as that of the positive ions. This could make ϕ_3 about equal to ϕ_2 or perhaps slightly smaller. The difference is probably immaterial, and what will be observed is merely the phase advance ϕ_2 in each half-cycle as regards positive and negative current relative to the applied potential. A discussion of the consequences of the greater dispersion of the negative space charge will be given later. It will also be noted that the current is confined largely to the rising quarter of each cycle, since the space charges cut off current at, or possibly even slightly before, each potential peak near the corona threshold. As potential increases, the currents on each half-cycle increase, space charges and their fields increase, and ϕ_2, the time from potential zero to corona current rise, decreases. For high potentials space charges may initiate breakdown at current zero or even a bit before. It also will appear that space charges on the rising potential may choke off the corona before peak. If the potential rise above this point to the peak is great enough, a new breakdown potential may occur. There is nothing to prevent new corona at undisturbed points during and after the rise. It is thus not unexpected that well above threshold there may be perhaps two or more corona peaks during the rising quarter-cycle and extending somewhat beyond it. Usually there is no current beyond 120° after zero.

The changes noted are, however, not of more than passing interest. Of much greater significance is the fact that the plentiful negative ions following the first negative peak are drawn to the surface of the conductor shortly before the breakdown on the positive phase begins. The negative ion concentration is increased because diffusion of the ions during the times involved broadens the boundaries of the space charges caused by ion movement in the field. Hence there will be enough negative ions to yield a negative Hermstein ion sheath at the conductor surface at the time preonset streamers should start. In consequence there will be *no streamers* and the positive discharge will be a burst pulse and Hermstein glow for which di/dt is relatively low. Burst pulse currents are of the order of 10^{-1} to 10^{-2} or even less than for Trichel pulses. Hence the positive discharge is a uniform glow corona spread over the whole conductor surface, which is practically devoid of RI noise.

On the other hand, on the negative half-cycle the positive ion space charge enhances the field, giving the Trichel pulse corona. It also makes quenching by negative ion space charge more difficult. In consequence Trichel pulses are increased in current and lengthened in duration. They thus give a rather brilliant spectacle seen in the dark. The current increase as potential increases is due to proliferation of Trichel pulse spots over the surface. One spot operates within the high field duration, giving a regular series of n Trichel pulses per second. Here the value of n lies between 10^3 and 10^6 pulses per second, depending on the changing potential, yielding from about 1 to 1000 pulses during each quarter-cycle of discharge time. The discharges are located all over the conductor, usually at fairly clean metal spots. The spots do not remain anchored on clean surfaces but usually move from time to time as the metal surface is altered by the discharge. Weathered or aged conductors look as if succeeding discharges on both cycles were anchored at bare spots on the coated surface.

In consequence, oscillograms published in various studies by different observers, perhaps most exhaustively by Denholm,[7.7] indicate the quiescent Hermstein glow during the positive quarter-cycle with its phase advance. It shows virtually no RI. The corresponding Trichel pulse corona on the negative quarter-cycle with its phase advance yields considerable RI and even an audible sound if currents are high. Thus RI appears only at the negative quarter-cycle of potential rise at an angle $\phi_2 < 90$ deg after the zero phase. The RI may be modulated by the n Trichel pulses during each quarter of negative 60 cycle. As potential varies over the half-cycle, n also varies.

Denholm has made an excellent study of the Trichel pulse corona phase on clean conductors in the laboratory and has analyzed the RI noise spectrum at different potentials. It should be noted that the RI noise appears only during the rising negative quarter-phase extending relatively little beyond 90 deg. With a small 0.05 in. wire the pulses per millisecond were observed to have two or perhaps more unresolved peaks in the range from 20 kv to 50 kv. The value of V_{s-} was 16.5 kv, and at this potential there was one peak at nearly 90 deg. At 20 kv there were two peaks, the first at 60 deg, the second at 80 deg. This represents the influence of the rising potential and perhaps clearing of the longer range negative ion space charge cloud caused by the first burst of pulses from the neighborhood of the wire. The pulse frequencies here ranged around 50,000 cycles/sec at 17.5 kv and reached 5×10^5 at 50 kv, with the peak at about 30 deg and a second higher one at 80 deg. There were some pulses with decreasing frequency down to 120 deg. This was with a small wire. The RI voltage generated had a narrow, sharp peak at ≈ 0.5 mc and weaker peaks at 2 and 4 mc, leveling to a plateau that declined sharply by a factor of ten at 10 mc with a smaller peak at 20 mc. These data apply to the 43 to 48.5 kv range. The radio interference voltage, RIV, microvolt peak per kilocycle band width in general fell monotonically by a power of 10 as the noise meter frequency rose from 0.2 mc to 30 mc. The curves were parallel, and the peaks for 18 kv were about one-ninth those at 48.5 kv. As wire diameter increased, the RIV quasi-peak at 1 mc rose to about 10^3 μv at 24, 43, and 61 kv for 0.05 in., 0.125 in., and 0.25 in. diameter wires, respectively. The RIV quasi-peak at 1 mc in microvolts rose according to the empirical law to about RIV = 109 D(E − 39) μv, where D is wire diameter and E is the field at the wire surface in kilovolts per centimeter. Duration of the 60 per second noise at the RIV quasi-peaks at 1 mc rose roughly linearly with RIV intensity from around 20 μv to above 300 with intervals of a duration of 2×10^{-6} to 10^{-3} sec, the latter being nearly continuous over the quarter-cycle. The rise time of the Trichel pulses was found to be near 10^{-9} sec with a peak amplitude of 8 mA and a pulse duration of 15 nannosec. These pulses are even sharper than those reported for point-to-plane geometry. It is also clear that the 1 mc frequency at peak potential reflected the repeat rate of the Trichel pulses from one spot. However, in this case many spots were operating, kept nearly simultaneous by the phase relations with the 60 cycle AC.

A further study by T. W. Liao, W. A. Keen, Jr., and D. R. Powell[7,8] adds some more information from laboratory study. Corona with 60 cycle AC using metal points produced a discharge on the negative half-cycle in the form of Trichel pulses. A $\frac{1}{32}$ in. diameter point gave pulses of amplitude 10 times that of a needle point. The larger diameter point had a lower repeat rate of pulses by a factor of 5, as would be expected. Two needle points gave pulses of differing amplitude and extended to longer times during the negative cycle. Four needle points gave four levels of current pulses, one being twice as high in amplitude as for one point, the weakest being the same as for one point. These extended over a still larger portion of the quarter-period potential rise. The pulses were shorter at the beginning and end of the period. A fiber phonograph needle that had been wetted to make it conducting gave *only positive streamer pulses* that were about ten times the amplitude of the $\frac{1}{32}$ in. diameter metal point and 100 times those of the Trichel pulses from the steel needle point. Increasing the number of fiber points to three perhaps trebled the number of pulses. As is usually the case with streamers, these were not of uniform amplitude at any time. The low conductivity of the fiber points did not permit a Hermstein glow to develop, as the corona was too weak and there were no negative ions in the absence of a Trichel pulse corona. It gave only streamers. Putting one fiber needle point and two steel needle points onto a plane surface gave a few weaker streamers early in the rise of the positive phase, and two very heavy streamers, three times the amplitude for those from a single fiber needle, coming after a delay of 4×10^{-3} sec. These were followed in the negative quarter-cycle by a very prolonged series of much weaker Trichel pulses. These had 30 times the amplitude of the same pulses from the two needles alone. Here one sees clearly the influence of space charges from the previous half-cycle initiating breakdown at each rising potential quarter-wave with much phase advance but increasing the amplitude of both streamer and Trichel pulse breakdown.

With metal points only, when potential was increased to several times V_s, streamer pulses began to appear on the positive half-cycle. Trichel pulses for a single needle point began at 2 kv and required 14 kv before streamer pulses appeared on that half-cycle. This gave a large increase of RI. The number of pulses on the negative half-cycle decreased with positive streamer pulses. When isolated points were used, and points protruding through planes or spheres, the character of the RI changed, indicating that this does not depend on the point alone but on the point in relation to the rest of the conductor. The effect of the plane or sphere was to reduce the noise lying below 200 μv in the region from 2 to 13 kv. Once the streamers began, the rise at higher voltages to 1000 μv and above was steeper and increased to much greater levels. The sphere caused the jump in RI and streamers at about the same 12 kv as had the needle, but it took 18 kv to create the noise with the plane. This no doubt is related to the influence of the surfaces on space charge and the suppression of streamers and/or Trichel pulses by the charge. It was observed that the heat of the corona discharge current slowly dried out the fiber points, at which time the streamer corona grew weaker and eventually ceased, as did the RI.

Next a metal point was placed on each side of a wetted fiber point with a plane near them, and the oscilloscopic patterns as well as the visual effects were noted as potential increased from 10 to 16 kv. At 10 kv only the two needles gave a Trichel pulse corona of two different amplitudes. The characteristic negative corona was observed with multiple spots on the point. At 14 kv the negative corona had extended to cover most of the negative *half*-cycle. Amplitudes of pulses were constant as before. Positive streamer corona appeared at the beginning, at the peak, and near the end of the positive *half*-cycle. Those at the beginning and end were much weaker than those at peak. It is clear that the initial pulses come from the negative space charge. Then the positive space charge created stopped the streamers for a clearing time of about 3×10^{-3} sec after the first group and after the group of peak streamers before the streamers could resume. At 15 kv the positive streamers extended to the plane electrode and return strokes were occurring with bright trunks developing. Streamers were then continuous throughout the whole half-cycle with intensities proportional to the potential. The negative streamers associated with Trichel pulses appeared unchanged. The Trichel pulses had by this time also developed long axial spikes (incipient negative streamers) which extended about half as far into the gap as the positive streamers. It is clear that above 15 kv the space charges do little more than stimulate early breakdown on both phases. The streamers are no longer suppressed by their own space charges, and the gap approaches a breakdown by streamer spark. It is to be noted, however, that the later observations deal with a short point-to-plane gap and not with an isolated electrode. This renders phenomena different from those noted by others on more isolated transmission lines where flashover to ground does not occur.

Another study of interest presented photographs of a conductor corona on a wire. Here there were many corona spots. The whole conductor appeared to glow with the Hermstein glow corona along its length. The Trichel pulses appeared at various isolated points along its length. There were not very many. Had the potential been raised to high values, both Trichel pulse and streamer corona would have been observed at spots all along the wire on a still photograph. With stroboscope the two would have appeared separately. The conductor, which had been treated in a salt fog cabinet for a week, was then tried at the same potential and showed many more bright spots with great increase in RI. These were positive streamers that were no longer suppressed by the Hermstein glow arising from the sharp semi-conducting points of the NaCl no doubt coated with moisture films.

Tests run by Pakala, Fahrnkopf, Lippert, and Bartlett[7.9] on the lines at the Tidd test site showed that a dry *new* conductor at 288 kv had marked RI largely during the first half of the negative phase. There was no sign of noise on the positive rising quarter-cycle. On a dry weathered conductor the RI started at 178 kv and showed a feeble RI at 90 deg in the negative phase. The increased RI on the weathered conductor at 228 kv was far less than that on the new clean conductor at 228 kv. It is seen that, if potential is high enough and the conductor is rough, the Hermstein sheath condition begins to fail and streamers can appear. However, a weathered conductor reduces RI due to the Trichel pulse corona over that on the clean conductor.

One more observation can be ascribed to the action of the diffusion of the space charges: the differences in V_{s-} and V_{s+}, the differences of positive and negative ion mobilities and the slower attachment of electrons to air molecules to give negative ions. All the charges created during the corona are not collected on each half-cycle. There is thus created outside the conductor a plasma, which is not very dense, of positive and negative ions, *with a slight excess of negative ions*. This increases the conductivity of the air and yields a slight negative current to ground. In a slight wind these charges are carried away.

Actually the RI from the negative Trichel pulse corona is not very serious, and one could perhaps get along with it. The real trouble comes with streamer corona on the positive phase. This produces very serious noise and is only inferior to that caused by small surface sparks. It was noted that it does not occur for a clean metal corona at reasonably low potentials. It appears on a weathered conductor but is very unpredictable. At some times at a given locality it is small; at others it is serious. When it does occur it occurs at various points along the wire. It appears on very new conductors, but after a short time of operation (a day or two) it wears off and ceases. This is particularly true of clean conductors in the laboratory. Streamer corona then does not reappear until the voltage is raised appreciably. At such a time it greatly exceeds the RI from the Trichel pulse corona. Its appearance is clearly associated with longer elongated particles ($\approx\frac{1}{32}$ in.) on the conductor surface. Here the corona streamers extend several inches out into the gap, forming luminous plumes. These are indeed electrically noisy and yield a sharp crackling audible noise, less than but akin to that of a small spark. The streamer corona appears especially strongly in rain. Rain nearly suppresses the RI from the Trichel pulses but enhances the RI from streamers.

With these observations it pays to refer to some excellent work done by Miller.[7.10] A laboratory test was run on smooth conductors 25 ft long mounted horizontally 11 ft above the floor level, which acted as the ground plane. The conductors consisted of single 1 in. and 2 in. diameter aluminum tubes and a horizontal 2-conductor bundle of $\frac{3}{4}$ in. diameter tubes spaced 16 in. apart. Various materials were dropped onto the central portion of the energized conductors by a man on a platform 12 ft above it. They were dropped so as to maintain a more or less continuous stream of particles falling onto and past the conductor. Very fine particles were dusted onto the conductor, using a garden-type dusting gun. This gave a slowly falling cloud of particles which covered the conductor for a length of several feet. Water was applied in several ways. In some cases the conductor was wetted before the test. In other cases precipitation as in rain continued during the RI test. The wetted section ranged from 1 ft lengths to 25 ft.

Apparently enough of the particles were collected so that remarkable corona displays were observed. Nine kinds of test particles were used; water, iron shot, fly ash from the base of a factory chimney, SiC grains, dry white sand (SiO$_2$), dry sawdust, powdered aluminum (Al$_2$O$_3$), powdered sulfur, and smoke from burning kerosene. The particles ranged in sizes of 10, 20, 35, 100, 120, 240, and 300 mesh, with mesh openings of 0.79, 0.0331, 0.0197, 0.0059, 0.0049, 0.0029, and 0.0024 inches. Measured were RIV in microvolts at 100 kc/sec, ranging from 10 to 23,000, for a range of line-to-ground potentials of from 60 to 280 kv. The curves were more or less parallel for different particle sizes of fly ash, the noise beginning at lower voltages for the larger particles but crossing at higher RIV microvolts in some cases. The finest particles still started noise at around 110 kv, while the clean conductor started noise only at about 195 kv. In all these cases the noisiest corona was the streamer corona where plumes extended as far as several inches. There were also numerous Trichel pulses with their characteristic structure, some of them extending out at least an inch, as might be expected for such large conductors and high fields with space charges.

The allowable level of RIV was on the order between 10 and 1000 μv. A diagram could be plotted for different sizes of particles of the various substances in a sort of histogram. Ordinates represent the ratio of voltage needed to yield a certain RIV relative to the voltage needed for a clean conductor, ranging from 0 to 1.0. The values of the ratio for 10, 100, and 1000 μv RIV levels were indicated by crosshatching. Lowest and thus noisiest was water with the 100 μv level at 0.30 with largest drops; a pre-wet conductor at 25 and 1 ft was next. Then came iron shot at 0.04 to 0.05 in. diameter around 0.48; greater than 20 mesh fly ash came next at 0.40; next came SiC with 60 mesh at 0.55; and next dry sand of 10 to 20 mesh, at around 0.80. The finer particles in all cases had higher thresholds. Less than 100 mesh dry sand was at 0.90. Dry sawdust was variable but lay around 0.80. Its shape factor, which could not be controlled, and its variable moisture made the results inconsistent. Powdered S was around 0.90. Smoke from kerosene with the source removed was 0.90 but was less while burning. Powdered Al_2O_3 *surpassed* the clean metal at the RIV level 100. This meant that it was a corona depressant.

It is noted that, aside from H_2O, the effectiveness in producing RIV fell off with increasing electrical resistance and size. One other interesting histogram showed the maximum surface potential gradient that gave 1000 μv. Here water came in at 30 to 50 kv; fly ash from 35 to 52 kv; SiC from 40 to 60 kv; shot, 45 to 50 kv; smoke, 55 to 65 kv; sawdust, 55 to 67 kv; dry sand (SiO_2), 58 to 68 kv. Powdered S around 68, a clean conductor at 70 kv/cm, and Al_2O_3 at 75 kv/cm.

It is seen that particles impinging on the surface lead to RIV noise. The larger particles used were the more effective. The more conducting the particle, the more effective it was. The voltage gradient is the controlling factor, rather than diameter or configuration of the conductor, as one might expect. Water, despite its being less conducting than some particles, was the most effective in creating RI.

The explanations of these phenomena are clear. Solid conducting particles that pass by the conductor have sharp corners that are set into corona by induction and on the positive induced polarity create noisy streamers. This occurs despite negative ions because stresses are so high and time of passage so short as to prevent or disrupt the Hermstein sheath leading to breakdown type streamers. The same applies to conducting particles captured.[1] Conducting particles in general probably do not remain pinned to the conductor surface for long, as they are repelled on contact charging. It could happen that the corona discharge from the sharp points would be adequate, owing to local restricted conductivity, to leave sufficient residual charge of the opposite sign to hold them. Good conductors like iron should not stick for long. Semi-conductors such as SiC can remain pinned by discharge and induction. Water, as can be guessed from the work of Macky

---

[1] The study could not differentiate between noise from inductively charged falling particles close to the conductor and those actually making contact. Larger particles separating from the surface give heavy corona and possibly even small sparks.

and English, is particularly bad, as its droplets deform in the electrical field. These are drawn out in filamentary points and droplets, giving vigorous streamer corona. Water suppresses negative corona, since its secondary γ_p and γ_i coefficients are very low. It does cause disruption and noise by charged disrupted particles. Departing heavily charged larger droplets are in themselves noisy. It appears that the positive streamer corona from water drops is the greatest source of RI in open air.

The fly ash, collected near the base of a chimney with much C as well, apparently was of a glassy nature and doubtless conducting because of H_2SO_4 and other hygroscopic substances on its surface. It will tend to be pinned to the surface. It may also have had elongated shapes which in the surface fields always align at right angles to the surface. This gives a heavy positive corona. Such ash can also give a *feeble* negative Trichel pulse corona as well. Any fine projecting particles, especially with poorer conductivity than the conductor, will not develop the Hermstein sheath, but will penetrate the sheath and give streamers. The finely powdered *nonconductors* in many cases will remain pinned to the surface for long periods. They will charge up in falling through ion clouds, and being light will on reversal of field have a very good chance of being collected. They remain pinned because the ions from the air charge them on the outside oppositely to the charge on the conductor. The influence is two-fold. First, they are too small in size and too resistant to give large streamers, and if they have a low second coefficient they may suppress Trichel pulse corona. Studies using fine Al_2O_3 and SiO_2 dusts to trigger Trichel pulses from a clean conductor where insufficient electrons were present to sustain continuity of pulses have indicated that, even with steady potentials, these particles were pinned for hours.[1.10] Some of them also yielded very weak corona currents that were not sufficiently heavy and intermittent to cause any appreciable noise, probably through a field emission process.

Carbon deposits on conductors are created by the decomposition of carbon-containing gases such as CO_2, CO, CH_4, and other hydrocarbons. These are deposited in thin films of ultrafine particles. Certain tarry residues are also produced from organics and are captured. These, together with fine SiO_2 and Al_2O_3, and so on, build up the coatings on weathered or aged metal conductors in the open. It is stated that some of the coatings were not oxides of the conductor metal. This need not always be the case, but fine nonconducting particle films are most probable. As long as larger, more conducting elongated particles are not captured and included in the weathered surface, the coating will act as follows. It may yield a low-order silent discharge current, perhaps below threshold for the bare metal, depending on the surface conductivity of the grains as caused by adsorbed salts and humidity. The coating suppresses Trichel pulse corona because of its low secondary coefficients. It will not encourage, and may discourage, streamers by covering sharp points that would otherwise give trouble, especially under high humidity. These films, being by nature in the open and consisting of a mass of fine bound particles, will not develop cracks and flake off on changes of temperature and chemical reactions, as would industrially applied coatings. Any voids or breaks will be filled up by newly deposited particles.[2] Such films constitute the best surface coating corona-suppressing agency so far discovered. Unfortunately they only relatively slightly reduce the serious RI.

One of the most annoying and frequent causes of streamer corona and RI comes from insect wings and parts, especially those that are elongated and adhere to the conductor at one end. These give very bad RI with large plumes.

It appears that fog or mist, unless it condenses to large liquid drops on the surface, is beneficial rather than deleterious. It renders surface discontinuities superficially conducting and effectively smoothes the surfaces with a poor corona-generating material. Usually the fact that the conductors are slightly warmer than the air prevents serious condensation, producing water drops and noise. Where liquid water collects to form drops, RI is prevalent and bad.

Snow appears to cause relatively little RI unless the flakes are very large and disrupt in the high fields. Then there is noise. Actually, fine ice crystals can collect on the surface and, owing to their low conductivity, can give both streamer corona and Trichel pulse corona. The currents, being orders of magnitude smaller than their metal counterparts, yield little noise.

The audible sound produced by the Trichel pulse and large plumed streamer discharges is the direct consequence of imparting considerable momentum to the gas surrounding the conductor at each phase through the drag of moving of ions on the air molecules. The motions of the ions to and fro over large areas of conductor create atmospheric oscillation that yields sound waves. The frequency of the sound should be around 120 cycles. However, the pulses and discharges are produced irregularly, multiple Trichel pulse spots go on and off, and so on, so that there are random motion and amplitude fluctuations superposed on the regular motion. When the positive streamer burst extending out 4 in. from the conductor suddenly is projected out in some 10^{-6} sec, a considerable number of ions are set into motion and bursts of sound of high intensity and short duration are created in the trunks of streamer channels akin to those in sparks.

The normal color of the coronas in air is a bright blue, since the most

---

[2] Recent microscopic observation of such weathered conductors indicates small, clean metal spots apparently blasted through the granular black coating by Trichel pulse corona.

frequently excited radiation is from the second positive band group of highly excited N_2 molecules. Both the negative glow of the Trichel pulse close to the conductor and the Hermstein glow sheath at the surface of the wire should show a more bluish white color owing to higher states of excitation. Yellow flashes are frequently reported as having been seen, especially in the streamer corona. The most likely cause is the contamination of the glow by Na atoms from the omnipresent NaCl on the surface of all matter exposed to the open air for some time. If minute carbon flakes are ejected and rendered incandescent in corona bursts, these could give red or yellow streaks. No doubt certain salts such as Ca and Cu salts, if the conductor is copper, can be injected into the corona discharge, be vaporized, and thus mask the usual blue color of the discharge. In the lower voltage regions of plumes where fields are low, recombination of ions and chemical reactions may occur and the purple glow associated with oxides of nitrogen and low voltage excitations may appear.

4. Threshold for Symmetrical and Asymmetrical Corona Gaps with Alternating Potential

Table 7.1 indicates the values for corona and spark thresholds for 2 cm and 4 cm point-to-plane and point-to-point gaps with alternating potential for comparison with the same gaps using steady and impulse potentials. As these indicate, the static and AC corona thresholds for *point-to-plane* gaps of 2 and 4 cm differ relatively little, being 7.5 kv for a 4 cm gap compared to 7.1 kv, and 6.9 kv relative to 6.5 kv for a 2 cm gap. These differences could be in the statistical time lag factor for the short duration of the AC peak. With floating potentials the thresholds are slightly higher as fields at the points are reduced by the surroundings. For floating point-to-point gaps the difference is more pronounced, with 12 kv for the AC against 10.7 for the static at 4 cm, and 10 kv against 10 kv for the 2 cm gaps. There was a difference in the measurements of the AC point-to-point values in that two transformers were used giving peak potentials of $+V$ and $-V$ relative to ground, thus ensuring gap symmetry. The points were thus acting more nearly as separate point systems. In consequence the ratio for corona thresholds for the 4 cm *point to point* and 2 cm *point to plane* was 1.5 to 1.6 instead of the value 1.3 observed with the steady potential where one point was at high potential and the other was at earth potential. The midgap fields in the longer gaps were thus more disparate on AC, and the point fields were not as axial as with steady potentials. The ratio of the static starting potentials for point-to-plane gaps of 4 cm and 2 cm was 1.13, while that with AC was about 1.09, that is, not very different. For the point-to-point gaps 4 cm and 2 cm it was 1.07 for the static and 1.2 for AC, again indicating a greater relative potential needed with the longer gap and more isolated electrodes on AC.

The measurements of peak potential sparking thresholds were again made with the symmetrically arranged point potentials. With this arrangement the 4 cm point-to-plane AC sparking threshold could not be observed. It was certainly greater than 40 kv. For the 2 cm point-to-plane gap it was 31 kv as against 28 kv for the same gap with steady and impulse potentials, both floating and with plane grounded. This is not surprising. The sparks are streamer sparks. The negative potential pushes the negative ions and streamers out into the gap, so that a very effective Hermstein sheath forms on the positive potential phase. Thus it may take higher potential at the positive peak of short duration to overcome the Hermstein glow and yield streamer sparks. The weak plasma of diffusion escaped ions apparently plays little role in short gaps.

The point-to-point gaps required 48 kv for 4 cm on AC, and 24 kv for 2 cm gaps on AC. Here again the proportionality of streamer length and gap potential is manifest in the 2 to 1 ratio of potentials for the two gaps. The potential for the 4 cm *point-to-point* gap was higher than that for the 2 cm *point-to-plane* gap, in the ratio of 1.55. Here again the influence of the greater point isolation is noted but is not as extreme as for the corona thresholds, obviously because of some space charge effects. It is more interesting to note that the AC *point-to-point* threshold at 48 kv is lower than that for an impulse potential where streamers alone form in a clean gap but is *markedly higher* than *37* kv for the *point-to-point* steady potential breakdown. It is clear that with steady potentials of opposite sign the ionic space charges interpenetrate in such a fashion as to facilitate spark breakdown by space charge fields at both electrodes. With AC the alternately positive electrodes have negative ions present to make Hermstein sheaths at *both* electrodes. Thus breakdown potentials are higher than where the negative streamers at the steady cathode are stimulated by positive space charge and negative ions are so prolific at the anode as to alter the nature of the Hermstein glow. However, with AC, the ion clouds projected into the space on alternate half-cycles eventually create enough of a plasma to break down the Hermstein glow and yield streamer sparks even below impulse thresholds. The fact that point-to-point AC breakdown occurs at lower potentials than point-to-plane breakdown definitely indicates the ameliorating action of diffusion conditioned space charge plasmas in midgap relative to the rigid Hermstein sheath situation in point-to-plane gaps with efficient ion removal by the plane.

In general terms these results corroborate the inferences drawn from the static corona and spark breakdown procedures.

5. THE RECTIFYING ACTION

It is clear that any difference between the two electrodes causing the threshold at one side to differ from that on the other will, with an alternating potential in the region of potential between the anode and cathode thresholds, cause the current to pass in one way only. Thus if one has an asymmetrical gap and applies an alternating potential, the side with the lower threshold will break down first on the proper potential phase and the current will pass in only one sense. Assume that $V_{s^-} < V_{s^+}$. Then whenever the smaller electrode is negative, it will pass the current of electrons or negative ions toward the positive larger electrode. When potential reverses, the small positive electrode will not break down. And while some of the charge in midgap will be drawn back, some will remain in the gap or even reach the other electrode. Thus current pulses starting near or above V_{s^-} will always send a negative ion group across to the larger electrode every other half-cycle. This then leads to a pulsed unidirectional current from cathode to anode. The pulsed current can be stored in a condenser if desired. When $V_{AC} > V_{s^+} > V_{s^-}$, some rectifying action will also occur, but the net negative current will be less. It depends also on whether one or both currents can cross the gap in one half-cycle. This current will persist in a decreasing measure to higher potentials until the current from the small electrode as anode equals that from the small electrode as cathode. Thus, geometrically, asymmetrical gaps will always act as either complete half-wave rectifiers, if $V_{s^-} < V_{AC} < V_{s^+}$, or partial rectifiers, if $V_{AC} > V_{s^+} > V_{s^-}$.

Since generally the cathode has a lower threshold than the anode, the rectifying action will be negative with the smaller electrode. In air at atmospheric pressure, where $V_{s^-} \approx V_{s^+}$, rectifying action will be slight and may pass unnoticed. In free electron gases and at appropriate pressures it can be marked.

A similar result is achieved if a symmetrical gap has materials of different work function as identical electrodes. Here the value of V_{s^-} will be lower than V_{s^+} at one polarity and $V_{s^-} > V_{s^+}$ at the other polarity. As long as $V_{s^+} > V_{AC} > V_{s^-}$, half-wave rectification is achieved, and, until the current at a given $V_{AC} > V_{s^+} > V_{s^-}$ is about equal for the two electrodes, partial rectification is achieved. This fact was and is used in electrolytic rectifiers in solution. Before the three-electrode tube gave continuous wave oscillations, such unidirectional action was used to achieve oscillations in communications. Thus Cu and Al electrodes in H_2 have different thresholds. Below a certain potential this yields favored breakdown at one polarity. The Chaffee arc imposed a DC potential at relatively low value in the sense favoring the electrode of low V_{s^-} as cathode. As the condenser charged through a circuit of RC time constant, say at 10^{-1} of the desired frequency, it surpassed V_{s^-} and a current surge flowed through an inductance coil of low impedance. Closely coupled to this was a slightly damped circuit of time constant such that its frequency was 10 times the discharge rate frequency of the existing circuit. This was set into oscillation by the primary current surges, and every tenth cycle was reinforced by the new surge from the rectifier. This came each time the capacitor discharged by the breakdown recharged over a time RC. In this way with proper tuning, a train of quite useful, slightly damped oscillations could be obtained.

If one has a completely similar pair of electrodes and a symmetrical field with $V_{s^-} > V_{s^+}$, then as the alternating potential is applied, with $V_{s^+} > V_{AC} > V_{s^-}$ or $V_{AC} > V_{s^+} > V_{s^-}$, there should in principle be no rectifying action. Alternating current will pass through the gap. The breakdown will take place at each instantaneous cathode in turn first and will deliver all the current or more current from the instantaneous cathode as V_{AC} exceeds V_{s^+}. No rectification is to be expected, since the fields and electrodes are symmetrical. What will occur when $V_{s^+} > V_{AC} > V_{s^-}$ is that the portion of negative current that does not cross the gap will cause the negative charge to gradually build up in midgap. This process will aid in the ultimate spark breakdown by a positive streamer, unless negative ions, as in air, build up a Hermstein sheath.

E. INFLUENCE OF ILLUMINATION OF
THE GAS ON BREAKDOWN THRESHOLDS;
THE JOSHI AND ALLIED EFFECTS

1. INTRODUCTION

In 1857 Werner von Siemens invented his ozonizer.[7.11] In the problem of ozone generation, ozone, O_3, which is very unstable, is readily decomposed by metal surfaces. Thus von Siemens used a double-walled cylindrical chamber with O_2 between the inner and outer glass cylinders. The inner cylinder was made conducting by filling with Hg or even a conducting aqueous solution. The outer cylinder was surrounded by a conducting layer of metal, usually in the form of a metal copper spiral wound about it. One end of the coaxial cylinders was sealed off, and the O_2 at low pressure was circulated through the system by inlet and outflow tubes. Low-frequency alternating potentials of some 50 to 60 cycles/sec were applied between inner and outer walls, causing an alternating current breakdown of the O_2 in opposite directions each half-cycle across the gap when the applied potential reached high enough values during the peak periods. The method is still in use in industry. The physics of this discharge was never properly investigated, nor could it have been until the advent of fast oscilloscopes. The breakdown was complicated by the fact that the glass walls were not good conductors, so that wall charges played an important part in the phenomenology.

S. S. Joshi,[7.12] a chemist, and his students, working with these discharges and measuring the breakdown current between the surfaces, noted that near threshold the current, or better still its rectified component in one direction, was influenced by illumination of the glass chamber with the light of an incandescent lamp as well as by more intense light. In some cases the current was decreased by the light; in others it was increased by the light. This was designated by them as the $\pm\Delta i$ effect. It was called, in honor of its discoverer, the Joshi effect. It was noted that the effect was greater or more impressive if the walls rather than the gas were illuminated. Gases studied were air or O_2, usually with some humidity, and gases containing the halogens or the halogens themselves. None of these gases was pure. Elaborate and controversial theories were developed, usually without adequate experimental investigation of the basic processes involved. Oscilloscopic study was eventually made, and the current was found to consist of a series of short pulses on the ascending quarter portion of the potential half-wave in each phase.

2. BACKGROUND PHYSICS OF THE BREAKDOWN

The first basic investigation of this breakdown was undertaken by Harries and von Engel,[7.13] first on Cl_2, with some work for which data were not given in air. A later study used quite, but not absolutely, pure Ne. Oscillographic study showed that the breakdown pulses had a rise time of hundreds of microseconds downward for Cl_2, and from microseconds downward for Ne as potentials increased above threshold. At threshold just one pulse occurred near peak on each half-cycle. The multiple pulses extended from the peak of potential wave to earlier and earlier times on the ascending quarter of the wave as potential increased. The analysis indicated that the breakdown in Cl_2 and presumably in air was caused by a Townsend discharge with a γ_i due to the bombardment of the glass with positive ions. In the Ne it was a similar breakdown but with a γ_p on the glass. The discharge built up exponentially, in time reached a peak, and extinguished itself by the space charge of the ions and electrons from the plasma on the poorly conducting glass. This produced a counter potential that decreased the field across the gap, so that discharge ceased.

The charges liberated per pulse were roughly estimated to be adequate to create such annihilating fields. In Ne the value of the threshold potential was taken from the sustaining peak AC when the gap was broken down. The Ne was assumed to be pure, whereas it was contaminated by A to yield a first-stage Penning gas. The sustaining AC potential has since that study been shown to be lower than the static breakdown potential V_s, which Harries and von Engel did not measure.[7.14] Using the sustaining potential and Townsend coefficients for pure Ne, they underestimated the value of the Townsend integral $e^{\alpha d}$. This overestimated the γ_p required by *Ne photons on glass*, which appeared to have a value on the order of unity. This is an unprecedentedly high value for any γ. They also studied the influence of light on the $\pm \Delta i$ with Cl_2, and found that the changes produced appeared to reside on the walls, as Joshi's workers had stated.

The next step came from a doctoral dissertation of Bhatawdekar, who used relatively pure Ne and A in a typical Joshi experiment.[7.15] He was unable to observe any $\pm \Delta i$ effect in these gases, in conformity with the observations of Harries and von Engel in Ne, and as is to be expected from basic theory. However, on one occasion his gas was contaminated with Hg vapor, whereupon he observed a very strong $-\Delta i$ effect on illumination. This prompted him to study the effect of Hg vapor at various temperatures. In Ne and A the $-\Delta i$ effect (reduction in current by light near threshold) was strong not only at very low Hg concentrations, such as would be used in a Penning gas mixture, but at some very much higher temperatures and partial pressures of Hg as well. In addition he found that Ne light was not more effective than incandescent lamp light and that light from an Hg arc was much more effective. As comparable intensities were not used, the observation was of questionable significance. However, a later study by Nakaya[7.16] revealed that the observation would have been similar had comparable intensities been used. In part, the action that caused the $-\Delta i$ effect was unquestionably of the Penning type.

More work was obviously required. This came from two directions. The first were studies in the author's laboratory,[7.14] and the second was a most remarkable and exhaustive independent study of the influence of light on the breakdown potentials of A–Hg mixtures by Nakaya.[7.16]

The first study was a direct measurement by Rohatgi[7.17] of the quantum efficiency of the 2537 Å Hg light on Pyrex glass, such as was used by Harries and von Engel. The resulting value was used to determine the same quantity for radiations in a Ne discharge as was used by Harries and von Engel in their breakdown studies. The values of γ_p observed were 10^{-4} for $\lambda 2537$ Å of Hg on Pyrex and Na glass, and 10^{-3} for the Ne radiations. It had been expected that glass, as a relatively poor conductor, would make such studies difficult. Rohatgi observed that the ultraviolet light, in liberating electrons from the surface of the glass, increased the conduction through the glass in the measure that the photoelectric current density increased. It was later proven by Muray[7.18] in the author's laboratory that the ultraviolet light liberated electrons from states of O^- ions in the glass surface, thus freeing bound Na^+ ions to migrate in the glass.

Since the current in glass is carried by the Na^+, or alkali ions, this liberation accounted for the increased conductivity. Glass yields a photoelectric quantum efficiency in proportion to its alkali atom content in the outermost 10 Å units of the surface. When a potential is applied to the glass, the current resulting, if the influence of the time constants of the measuring circuit is deducted, at first declines in a linear fashion on a time scale of tens of minutes, leading to a constant value after sufficient time.[7,18] The linear decline is caused by a polarization current charging the glass surface through transport of alkali ions to the electrode surface in which they are bound. This current is reversed if the glass cell is shorted and that charge is recovered. After polarization is achieved, the steady conduction current flows with electrons from the cathode, neutralizing the space charge of alkali ions next to it and electrons going to the anode from the glass through the action of the high field across the boundary.[7,18] Thus part of the initial current flow through the glass observed in studies of breakdown of the cell is a slow polarization current and must be differentiated from the true conduction current involved in the action of the cell.

The breakdown of these cells was investigated in detail by El-Bakkal[7,14] and the author, using spectroscopically pure A gas with Alpert vacuum techniques. The cells were Pyrex glass cylinders with suitable guard rings to reduce surface leakage. The electrodes, 1.2 cm apart, were 3 mm plane plate 7440 Pyrex glass silvered on the outside, closing the two ends of the cylinder. Measurements were made on the A at 7, 15, and 35 mm pressure, using both steady and alternating potentials, as well as timed impulse potentials applied through an RC circuit. Oscillograms of the breakdown both on fast sweep for steady potentials and on fast and slow sweep for 60 cycle alternating potentials were made. By means of a larger capacity series condenser, accurate electrometer and recorder measurements of the charges developed could be made. In addition, the potentials placed on the glass electrodes by the discharge could be estimated and losses by leakage plus glass polarization could also be determined. The results may be summarized as follows:

In describing the results it is at first essential to distinguish several thresholds. Since, as Harries and von Engel correctly inferred, the breakdown in the inert gas system is a diffuse Townsend discharge initiated by a γ_p photoelectric emission from the glass, there must be a Townsend threshold potential V_T set by (eq. 2.7) $\gamma_p e^{\alpha d} = \gamma_p e^{\eta V_T} = 1$. In practice, since critical traces of gas impurity and uncertainty in γ_p values precluded accurate information, this threshold V_T was never accurately established. It is the lowest of the thresholds encountered. There is an *observed* but more *nebulous* breakdown threshold V_s, which lies above V_T. It is a *prac-*

tical threshold at which a discharge can maintain itself in a sufficiently large sequence of avalanches to build up a back potential ΔV_{Gd} across the gap, because of electronic charges placed on the glass at the anode by the sequence of avalanches. The applied potential must be great enough to ensure the sequence of avalanches which allow the significant charge accumulation measured to be generated.

From what has been stated earlier in this text concerning corona and Geiger counter pulses, it should not be surprising to discover that *this threshold is fuzzy* and depends on the number of electrons involved in the triggering event. That is, in the small glass cells used and at the necessarily low gas pressures, the breakdown is usually triggered by bursts of ionization from cosmic ray or other externally produced events. Use of ultraviolet light as a trigger was unsatisfactory. At the potentials used, ultraviolet radiation of the glass increased the background field intensified current, producing spurious effects, so that it could not be used. Thus, depending on the magnitude of the triggering event, the value of V_s observed for each pulse varied. A mean value called V_s was chosen at which the gap usually broke down within a few seconds of applying the potential. This fixed the V_s for a size of triggering pulse which was conveniently frequent. If the applied potential was designated as V_a, then from V_s upward the charge was proportional to $V_a - V_s$. In consequence, the inhibiting potential on the glass which interrupted the discharge $-\Delta V_{Gd}$ was equal to $V_a - V_s$. That is, the current once launched above V_s flowed until the electronic charge on the anode glass was such as to lower the field within the cell to V_s or slightly below.

The oscillograms of the fast electronic component of the breakdown gave curves which rose from zero exponentially, reached a peak, and then declined more slowly. The duration of the rise was roughly one-third the duration of the pulse. At threshold the pulses lasted on the order of 10 μsec at 15 mm pressure. As potential increased the duration of the pulses rapidly decreased, but the amplitude increased. Over a range of values of V_a from V_s to $3V_s$, the duration of the pulse decreased by a factor of 20 and the amplitude increased by a factor of 200.

At about $1.2V_s$ the mechanism of the discharge altered so that q was proportional to V_a and *not* to $V_a - V_s$, while $-\Delta V_{Gd}$ was equal to V_a. Thus q increased linearly with V_a. This regime continued up to $3V_s$. At that value the charge q was still proportional to V_a, but the current rose on a time scale of 3×10^{-8} sec instead of 2×10^{-7} sec. Instead of declining smoothly, the current declined in steps. The transition from a $\Delta V_{Gd} = V_a - V_s$ to $\Delta V_{Gd} = V_a$ was also marked by a backward discharge inside the cell on shorting out the applied potential. This followed, since if ΔV_{Gd}, the field between the glass electrodes, exceeds V_s because of accumulated surface charges, the surfaces are partially able to discharge the cell through the gas by a back discharge once the binding charges giving V_a outside the glass are removed.

The explanation of this action is relatively simple. When q is proportional to $(V_a - V_s)$, the plasma formed by the succession of avalanches is not very dense. Free diffusion of the electrons to the anode glass wall quickly creates a back field that wipes out the field due to V_a and reduces it to V_s. At about $1.2V_s$ the avalanche succession creates a sufficiently dense plasma adjacent to the anode so that the electrons are not rapidly removed. That is, at this potential the anode plasma reaches a density yielding ambipolar instead of free diffusion. This has the same effect as with a uniform field gap, in which the space charge plasma of the Townsend discharge increases the avalanche size. Thus in this region, a sort of "run away" discharge occurs that continues until the quantity of q liberated is equal to the capacity of the glass C_g gas cell at the applied potential, that is, $q = V_a C_g$. At this point it terminates.

The peculiar transition which occurs when $V_a \approx 3V_s$ has all the earmarks of a streamer-type breakdown at low pressures. Here the decline is step-like, since the discharge, instead of covering most of the glass electrode, now covers just a small fraction of it. The step-wise decline comes from readjustment discharges that equalize the surface charge coverage of the electrode. Little is known about such breakdowns, but a recent theory indicates a possible mechanism.[7.19]

The discovery of the back discharges raised an interesting question. In all the 60 cycle-type oscillograms, the separate pulses which were observed occurred only on the rising quarter of the 60 cycle potential wave. There were no apparent back discharges on the declining quarter-cycle of the potential wave corresponding to the shorting of the external electrodes. To ascertain if such events occur, the potential increase and decrease were applied by a charging capacitor through a resistance on a 2 sec and 6 sec time scale. The resulting events were displayed on the recordograms so that the charges and back potential charges produced on both rise and decline of the potential could be evaluated. Above $1.2V_s$ generally a forward discharge would be followed by a back discharge on the declining potential. It was expected that perhaps one forward or normal breakdown would take place for each whole multiple n of V_s applied, that is, when $V_a = nV_s$. This would have been observed with 7 mm pressure A in the cell, had it been investigated. At 15 mm pressure n was greater than expected, and at 35 mm n was several times as great as anticipated.

That is, above $1.2V_s$ and especially above $3V_s$ for pressures of 15 and especially 35 mm, V_a was not just 1, 2, or $3V_s$, but n was more, and at 35 mm it went as high as 5 below $V_a = 3V_s$. When the n breakdown discharges occurred, there were always $n - 1$ back discharges on declining potentials. The sum of the charges, or ΔV_{Gd}, put on the cell on successive forward discharges on the rise in the n breakdowns, was equal to the charges recovered in the n − 1 discharges if these were added to the leakage loss through the glass and the residual charge on the glass of the cell after back discharge had ceased.

This phenomenon of n > V_a/V_s was explained once the discharge was seen in the dark on alternating potential. At 7 mm the discharge glow covered the glass electrodes in a single breakdown path. At 15 mm there were about 1½ breakdown paths over the surface, and at 35 mm there were 3 to 4 breakdown paths. Thus at 35 mm each breakdown path covered about 25 to 30% of the glass electrode. This indicated that whereas the first breakdown, at V_a > 1.2V_s at 35 mm, covered just one-third to one-quarter of the electrode charged to ΔV_{Gd}, a new breakdown could occur at a slightly higher potential as this charge on the glass spread over the surface. Thus several smaller breakdowns occurred on the rising potential until the whole surface of the cell had been charged to ΔV_{Gd} at the peak. This could lead to n breakdowns where n exceeded V_a/V_s by a factor on the order of the number of separate breakdown paths. Even at 7 mm above 3V_s, the streamer-like breakdown was so patchy that more than one breakdown could occur. This study with the condenser circuit indicated, first, that given time, the forward discharge would be followed by a back discharge on declining potential, and such discharges would lead to n − 1 back discharges. It further indicated that where multiple breakdown paths at higher pressures occurred, there would be a succession of n discharges on rising potential where n might be several times V_a/V_s. Such a situation was also suggested by Harries and von Engel to explain the multiple peaks observed on the rising potential wave with a 50 cycle potential. In that case, many more peaks were observed than were expected from the ratio of V_{AC}/V_s at the high pressures which they used.

The techniques used in this study were then applied to 60 cycle breakdown of the cell at 7, 15, and 35 mm. Oscillograms at fast and slow sweep of potential and current enabled the charge q per pulse, and the phase of the current pips relative to the potential wave, to be measured. The potential and current through the cell were observed on separate oscilloscopes. Fortunately, the current drain in each pulse put a small nick on the potential wave trace. Thus phase relations were quantitatively portrayed.

It must next be indicated that in addition to the Townsend threshold V_T and the breakdown threshold V_s mentioned, disregarding the transition potentials for the three discharge forms, one more potential must be considered. It appeared in the course of measurements that alternating potential peak during an observed breakdown is not V_s. It was found on study that in general an alternating peak potential of at least V_s, or above V_s, is needed to *start* the AC breakdown. Once the breakdown has begun, the applied peak potential V_{AC} is in general below V_s. This at first appears strange. This very significant circumstance seems to have been missed by Harries and von Engel, as they studied only alternating potential breakdown and were not aware of V_s, the static value. The explanation of this peculiarity is not difficult. After the first few discharges, the quantity q leading to a negative ΔV_{Gd} on the anode glass from the preceding negative half-cycle of discharge is left on what (with the change of polarity) will be the glass cathode of the newly rising potential wave. Thus at any point on the potential curve the cathode will not have a potential $V_{AC} \sin \theta$ but $\Delta V_{Gd} + V_{AC} \sin \theta$ acting across the cell.

Hence, if at 7 mm, where only one channel of discharge occurs, $\Delta V_{Gd} + V_{AC} \sin \theta > V_s$, the first breakdown takes place. Furthermore, if $V_{AC} = V_s$, the discharge would not occur just at the potential peak, with $\sin \theta = 1$, but in advance of it at the point where $\Delta V_{Gd} + V_{AC} \sin \theta \geq V_s$. Thus even near threshold, once the gap has broken down there will be one peak on the rising quarter of each half-cycle starting from zero on both positive or negative phases. This pip or current pulse will appear *before* the peak. The higher the value of V_{AC}, the larger ΔV_{Gd}, and thus the earlier on the potential rise the current pip will appear. As ΔV_{Gd} is soon proportional to V_{AC} with $V_{AC} > V_s$, it is clear that the advance in phase of the pip can be very large. In consequence, it was not surprising to note that at a relatively low value of V_{AC}, the first current pip appears at $\sin \theta = 0$, that is, at the rise of the wave. At 7 mm only one pip was observed, appearing earlier and earlier on the rise, and only at $V_{AC} \approx 3V_s$ did a second smaller pip appear.

At 15 mm and more effectively at 35 mm with V_{AC} near V_s, one pip appeared near peak potential, but with some phase advance. At a slightly higher peak potential, a second pip appeared, the first pip being advanced in phase, then successively by the time $V_{AC} \approx 1.3V_s$, there were four pips, the first one being now advanced some 40 deg. Evaluation of the charges q showed that, as at 7 mm, where just one pip per half-cycle appeared, q was about the same or slightly less than the q statically observed for the same potential. The value of q then increased in proportion to V_{AC}. However, it was clear that, with the pip appearing with phase advance on a rising potential wave, the breakdown current flow continued at low value until the peak. It, however, did so at a lower rate than during the rise to peak value. At 35 mm with multiple pips, the value of q for each pip was limited to a fixed small value on the order of $\frac{1}{4}$ q static, but the sum of the q_s for all the pips on the rising half-cycle was equal to the q observed statically at about the value of the peak V_{AC}.

No back discharges were observed on the *declining quarter-cycle* of each half-cycle. This followed because at 60 cycle frequency the whole remaining half-cycle after the breakdown pips was taken up with collecting the charges created in the gas on the glass. It was these charges carried over to the reversed polarity in the ensuing half-cycle that led to phase advance. Thus slow charge segregation on the 60 cycle time scale and the phase advance replaced the back discharge which had been observed on the half-cycle per second frequency observed with a longer RC.

Having accounted for the discharge and the mechanisms active in the Siemens' ozonizer type of discharge, one is in a better position to discuss the nature of the Joshi effect. If one notes that the study of this effect generally measured the rectified current on the AC breakdown, it is clear that the current i, which will be influenced by light, is composed of the sum of the q's of the separate breakdown pips for each half-cycle, divided by the time of the half-cycle. However, the Joshi effect is not manifested for values of V_{AC} very much above V_s. In fact, there is no change in current by illumination of the cell at high values of V_{AC}. The Joshi effect and the Δi changes in current are only observed for V_{AC} close to the sustaining threshold for the discharge, that is, to V_{AC} or V_s. As will later be shown, in a physical system open to study the Δi effect depends critically on changes in V_s produced by illumination of the gap. Why this is so will follow directly from the change of a q near V_s and from the influence of light on V_s. It is thus essential at this juncture to follow Nakaya's beautiful study of the influence of light on V_s in an A–Hg cell.

3. BREAKDOWN STUDIES OF NAKAYA WITH A–HG MIXTURES

Nakaya[7.16] employed low-pressure glow discharge tubes, of both the hot and cold cathode types. The tubes were made of varying materials, from quartz to different glasses, and in a wide variety of shapes. The starting potential of a DC or AC glow discharge was measured. The tube was kept in the dark in an oven at a controlled temperature to vary the vapor pressure of mercury vapor. The vehicular gas was A at pressures from 0.7 mm to 20 mm. Figure 7.1 shows the starting potential in volts of the discharge as a function of the wall temperature or, what is more significant, the corresponding density of Hg atoms/cm³. The values V_{sn} are in the dark, and V_{sr} and V_{sr} sat. are in the presence of light. The solid curve with black dots and circles with a minimum at 3×10^{14} atoms/cm³ represents a lowering of V_{sn} that corresponds to the optimum Penning mixture. The black dots represent the potential without illumination, V_{sn}, and the white circles represent V_{sr} under illumination of light from a 100 watt high pressure Hg arc in a glass globe. The illuminated curve rises at 7×10^{14} atoms/cm³ and continues its upward trend. The other dots on this curve represent various illumination effects with other light sources, giving V_{sr} sat. The second minimum for V_{sn} with the solid line is broader and extends from 2×10^{15} to 3×10^{16} atoms/cm³. This occurs without illumination.

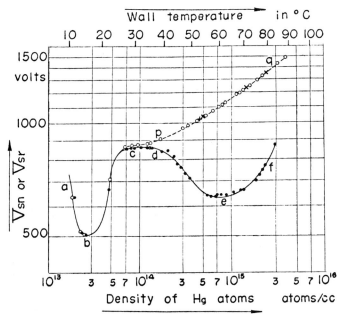

FIG. 7.1. Nakaya's curves for sparking potential V_s plotted against wall temperature, or the equivalent number of Hg atoms per cubic centimeter, in A gas at 15 mm pressure. Solid line and black dots give the value of V_{sn} in the dark, not illuminated. The dashed line and open circles and crosses show the same tube illuminated with Hg arc light and a 100 watt incandescent lamp at 100 cm with addition sign and 200 cm crosses. The first minimum is the Penning effect, which is sensitive only to strong A radiation; the second broad minimum, on an "aged" tube, is the Nakaya effect also noted by Bhatawdekar in his thesis.

The lowering by Penning effect and by the new higher density effect is strikingly large and amounts to 300 to 500 volts. Thus there is a new density Hg vapor effect *which lowers V_s, but the lowering is quenched by almost any light,* and especially effectively by Hg light. The use of light from an A lamp would have wiped out the Penning mixture minimum at lower Hg pressures. It was not tried in that region, as the consequences are well known.

Hg or white light does not affect the Penning lowering. This indicates that A metastables cause the Penning effect, but some Hg states must cause the lowering at the higher Hg pressures. This new $-\Delta i$ effect has all the earmarks of the sort of lowering which the Joshi effect workers have noted: appearance in impure gases, effects of aging, effects of localization of the most effective illumination, and so on. It was observed that if the tube was fresh, the lowering was confined to a narrow region at 5×10^{15} molecules/cm³. If the tube had run for some time, the minimum was very broad. If illumination was weak, the log of V_{sr} under radiation was proportional to the log of the intensity. Beyond a certain light intensity, V_{sr} showed a saturated value. The saturation curve occurred for strong illumination from whatever the source.

If the increase of V_s by light is to be observed, the A *pressure must exceed some minimum value.* The higher the A pressure, the more sensitive the tube becomes to visible long wave-length light. At 7 mm and lower pressures, the curve without radiation, V_{sn}, is monotonic and lies *above* V_{sr}, the one that is irradiated. This is the equivalent of a $+\Delta i$ effect. Somewhat above 7.5 mm Hg the illuminated curve, V_{sr}, lies above V_{sn}, the unilluminated one, which is much lower and shows a broad minimum. This is equivalent to a negative Δi effect. At 15 mm the increase of V_{sr} above V_{sn} is most prominent.

It was found that the sensitive portion of the tube for the effect of illumination was in the neighborhood of the anode. When illumination is adequate to give the saturation curve, there exists a definite point in the critical screening distance which can be screened from the cathode side toward the anode without seriously impairing the quenching effect of light raising V_s. Through the use of the critical screening distance it was established that the Hg light from a high pressure Hg arc was by far the most efficient. Screening the anode side produces no effect.

As the wall temperature and thus the concentration of the Hg vapor decreases, the critical screening distance extends to within 1 cm of the cathode. It was shown that the saturation voltage under radiation may also be achieved by light of Ne, A, and an incandescent lamp. In other words, while the Hg arc light is most effective, any source of visible and near ultraviolet light of sufficient intensity will raise V_s.

If the A pressure is high, the effect of $\lambda2537$ is to raise the starting potential. The $\lambda2537$ light of Hg, if not pressure broadened, *lowers* the starting potential. This effect of $\lambda2537$ in *lowering* the starting potential V_{sr} is greatest when little or no A is present. Under these conditions the effect increases as temperature is raised as well. The anode side is also sensitive to $\lambda2537$.

The effect of aging the tube, by running the discharge for long periods, on the shape of the unilluminated V_{sn} curve made it desirable to study the effect of impurities that affect the metastable Hg atoms. Addition of small amounts of H_2 gas suppress the effects of illumination by suppressing the lowering in the dark. In other words, H_2 gas destroys the V_{sn} curve and even at low concentrations gives V_{sr} sat. This is shown in figure 7.2. Before adding H_2 the curves (1) indicate the illuminated (upper) and unilluminated curves (lowest). Curves (2) indicate the effects of a trace of H_2. The lower solid curve near A' is without illumination, and the curve (2) under illumination coincides with the saturation–illuminated curve with no H_2. Further addition of H_2 yields increased potentials with no effect of illumination.

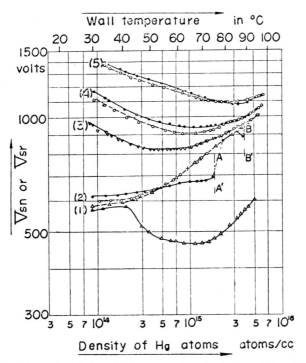

FIG. 7.2. Effect of addition of H_2 gas on changing V_{sn} and V_{sr}. (1) ▲V_{sn}, △V_{sr}: before adding H_2 gas. V_{sr} in this case gives the value of V_{sr} sat. (2) ●V_{sn}, ○V_{sr}: after adding a small quantity of H_2 gas. Measurements were taken while raising the wall temperature. At A–A′ a small quantity of H_2 gas was added again, and at B–B′ some more H_2 gas was added. (3) ▼V_{sn}, ▽V_{sr}: after the measurement of (2) had been finished. (4) ●V_{sn}, ○V_{sr}: after adding more H_2 gas. (5) ●V_{sn}, ○V_{sr}: after adding still more H_2 gas.

That is, H_2–A mixtures have higher V_s because Townsend's α is decreased by addition of H_2.

One peculiar effect was noted. If illumination suppresses V_{sn} but current flows for some time at V_{sn}, then if the illumination is discontinued, the potential does not at once fall to V_{sn} again. At times with DC measurements only, the low, dark breakdown potential V_{sn} was not restored for minutes or occasionally hours after the light was cut off, if the current had been run at V_{sr} sat. for 30 sec or more. This effect of illumination was most prominent in the cold cathode tube with a long gap. It was also characteristic of lower temperatures and thus lower Hg pressures.

If alternating potential was used and the voltage was raised slowly, an intermittent but feeble series of discharges were visible to the naked eye over a small range of potential. The significance of such a threshold with AC will not come as a surprise to the student of coronas. Illumination with weak ultraviolet light at once stops the intermittent discharge. Here the effect lasts only as long as the illumination, since there is no division of the discharge into permanent anode and cathode regions with alternating potential as with the steady potential. Once the potential reaches a point where the stable glow discharge operates, even intense radiation cannot stop the breakdown. Nakaya suggests that at threshold the two electrodes are not in the same state. Thus the feeble intermittent current is a rectified unidirectional discharge mentioned earlier in this chapter as resulting from electrodes in different states. This was confirmed by oscilloscopic study. When the current was weak, illumination anywhere stopped the discharge. At higher currents only the region of the electrode (assumedly the rectified anode) was effective.

The explanation of these phenomena follow. It is clear that the high temperature, higher Hg pressure minimum V_{sn} is not suppressed by light, even white light from an incandescent light by absorption and destruction of metastable *atoms* of *Hg* or *A*. The effect is not the Penning effect in A, which occurs below 30° C, or 5×10^{14} atoms/cm³ of Hg. The effect can only be ascribed to some *metastable states, but these must be broad, not narrow, atomic states.* That is, they must be metastable molecular states capable of absorbing and being destroyed by all wave lengths. Nakaya then reasons as follows: Assume that the density of Hg atoms is quite high and the A pressure is sufficiently high to prevent undue loss of atomic Hg metastables and molecular Hg_2 metastables to the walls. Conditions are then favorable for the creation of large numbers of metastable Hg molecules of the $6^1S_0 + 6^3P_{2,6}$ states. These are created through the three-body impacts according to the relation

$$Hg^m + Hg + A \rightarrow Hg_2{}^m + A.$$

The $Hg_2{}^m$ serve as the principal source of $Hg_2{}^+$ and electrons by stepwise ionization at low-energy electron impacts.

The $(6^1HgS_0 + 6^1HgS_0)$ molecule normally dissociates at 0.07 ev. Thus in a discharge at 100° C, kT $\approx$ 0.03 ev, these molecules cannot exist. Excited Hg atoms $6^3HgS_{1,0}$, 6^1P_1, 7^3S_1, 7^1S_0, and so on, do exist and, combined with normal 6^1HgS_0 atoms in triple impact, can form stable excited molecules. The energy level diagram of the Hg_2 molecule in figure 7.3, taken from Mrozowski, which needs some modification in the shape of the $^3\Sigma^-ou$ state, gives the information needed. The metastable atom 6^3P_2 forms the $^3\pi_{iu}$ and $^3\pi^-ou$ molecules, the latter being unstable. Transition from the $Hg^3\Sigma^-ou\ 6^1S_0 + 6^3P_0$ to the ground state $^1\Sigma^+g$ is not entirely impossible, and its life is 1.8 msec, according to Holstein, Alpert, and McCoubrey.[7.20] Formation of molecules by binary collisions between the metastable atom and the neutral molecule is very rare.

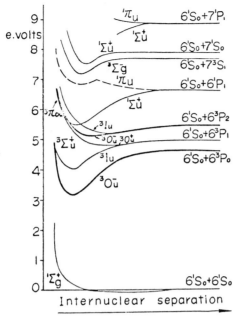

FIG. 7.3. The energy level diagram of Hg_2 states as given by Mrozowski.

The formation of $Hg_2^m(^3\Sigma^-ou)$ by ternary collisions in Hg at 200° C has been studied by McCoubrey,[7.20] in which $Hg^m(6^3P_0) + 2\ Hg(6^1S_0) \rightarrow Hg^m(^3\Sigma^-ou) + Hg(6^1S_0)$. The conversion frequency was estimated at $100 \times 10^{-32}\ (nHg)^2$. Here nHg is the density of Hg atoms. In the present study, A is the vehicular gas and temperatures can be correspondingly lower, for example, 60° C and not 200° C. In McCoubrey's case, $Hg^m + Hg + Hg^{200}$ cause the reaction. In the case of A the reaction is caused by $Hg^m + Hg + A^{40}$. The collision diameter of Hg is 3.01×10^{-8} cm, and A is 2.95×10^{-8} cm. Choosing 100° C as the tube temperature, and correcting for velocity changes as well as the relative masses of the third body, Nakaya calculates that his reaction should be 1.77 times as fast as McCoubrey's. However, since A and Hg are not too similar, as energy remover the factor 1.77 may have to be reduced by a factor of 2. Thus Nakaya sets the reaction rate as on the order of $10^{-30}\ (nHg \times N_A)$ in Hg–A mixtures such as gave low V_{sn}.

Calculations and considerations by Nakaya lead to the conclusion that A acts in two ways. First, it prevents rapid diffusion to the walls, which destroys both the metastable atoms and molecules. This is shown by the fact that at 7 mm A there is *no* lowering of the breakdown potential V_s in the dark, while at 15 mm the effect is pronounced, even though both breakdown potentials are increased at higher pressures. The second important function of the A atoms is to yield convenient atoms at the lower temperature kinetically to remove the heat of formation of the metastable molecule. In Hg at 200° C in McCoubrey's work, the Hg atoms carried out both functions. Reactions of A which could interfere with this molecular metastable state formation are not likely. A does not readily deactivate Hg^m. Again, while 2 Hg^m can react to give Hg^+ or Hg_2^+ in impacts removing Hg^m, this reaction requires much higher concentrations of Hg^m than are here involved.

Nakaya then made spectroscopic studies of the discharges as a function of the current in his A–Hg tubes at 100° C. The green continuum gives a measure of the metastable molecular level. This increases with current from 1 to 14 mA. The Hg triplet intensity emitted from the positive column of the tube increased according to the second power of the current, showing that while Hg^m, the metastable atoms, increased proportionally to the current, the green continuum did not increase as rapidly. This is due to the fact that the spectral distribution of the green continuum varies with A pressure, density of Hg atoms, and current. All these factors were altered by current increase. The intensity of the green continuum was measured as a function of the density of Hg atoms by changing the wall temperature across the tube at a constant current of 5 mA. The intensity increased approximately as the second power of the Hg concentration, as required. If the current at 137° changed over a wide range of values from 1 to 100 mA, the intensity of the green continuum increased to a maximum at about 15 mA and then declined.

Nakaya then lists all the ionization processes in Hg–A, which fall into four classes. These involve first the Penning effect by A metastables on Hg, which occur at lower Hg pressures but occur only feebly in the pressure range here used. They involve secondly ionization of A^m, Hg^m, and Hg_2^m states by electron impact. They involve thirdly the interaction of metastable atoms of Hg and A leading to atomic or molecular ions. Fourthly, they involve the direct ionization of normal atoms.

Discussion rules out the Penning effect as active here. Stepwise ionization is then considered. It is clear that stepwise ionization accomplished by lower energy electrons will be highly efficient. Once conditions are such that sufficient A^m, Hg^m, or Hg_2^m is present, the sparking potential goes down. However, to begin with, if this occurs for A, the metastables of A require 11 volts and the Hg metastables require 4 volts. Creating A metastables depopulates the high-energy electron states quickly. The Hg^m requires between 4 and 5 volts, and the Hg_2^m takes but 4.66 and 5.43 volts. Thus if Hg^m is created and can accumulate and form Hg_2^m, a system exists, that is excited, created, and ionized below 6 volts for each step, which is not the case with any A states. The cross section and possibilities of creation of the various states involved are presented and discussed by Nakaya

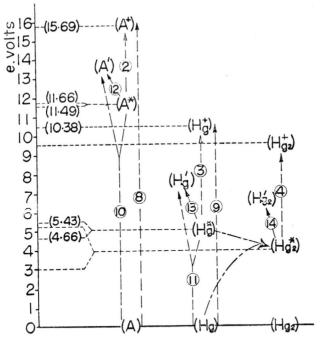

FIG. 7.4. Ionization processes and their energy levels in the Hg–A mixture.

in terms of Druyvesteyn and Penning's 1940 article.[7.21] The various ionization processes are indicated in figure 7.4 in terms of energies in electron
volts.

Call α the density of Hg atoms relative to A atoms. The terms involving
Hg_2 molecules are proportional to α^2, and those involving Hg atoms vary
directly as α. In the region of low-order currents, the densities of Hg^m are
proportional to the currents. As the ratio of X/p increases from low values
in a gas, the electron energy is given up to elastic impacts with atoms.
As X/p further increases, the fraction going to inelastic losses increases,
and a fraction is consumed in raising the energy of the newly formed electrons. Three ranges of values of α density of interest are shown in figure 7.1,
as noted: region a-b with an α of $5 \times 10^{-5} - 1 \times 10^{-4}$; region b-c with an
α of $1 \times 10^{-4} - 3 \times 10^{-4}$; and region d-ef with an α of $6 \times 10^{-4} - 7 \times$
10^{-2}. When current density is not large, the density of radiating atoms is
small. As the 6^3P_1 state at $\lambda 2537$ is heavily absorbed by imprisonment of
resonance radiation, this state accumulates, as do the Hg^m states. In fact,
many of these $\lambda 2537$ states by collision end up as Hg^m. Now regard the
changes with α. For a very small α excitation of A^m, A and ionization of A
take place. Here the stepwise ionization of A^m influences V_s.

In the region a-b, besides the A interactions, only a slight ionization of Hg and excitation of Hg to Hg^m and the resonance line occur. Thus the A^m will produce Hg^+ by inelastic impacts of the second class, and one has the Penning effect. In the region b-c, the Penning effect still dominates, but inelastic impacts to yield Hg^m, Hg, and the singlet metastable Hg^m increase. This wastes energy, as it does not lead to ionization. Thus V_s increases. In the region d-ef the creation of metastable Hg^m is large. Now, however, Hg atoms are sufficiently dense to permit triple impacts between Hg^m, Hg, and A atoms, leading to Hg_2^m molecules. These molecules accumulate and are ionized as rapidly as Hg^m are being created by new impacts. Thus now the Hg^m, by a two-step process involving ternary collisions, are converted into ions at low average electron energies. Hence V_s falls. If for some reason the Hg_2^m cannot form, then the curve for V_{sr} sat. is the result of the direct ionization of normal Hg atoms. Thus all actions that quench or destroy the Hg^m or the Hg_2^m will lead to the V_{sr} saturated type of ionization curve.

The direct interpretation of observations follows at once:

According to the energy level diagram for Hg_2 molecules, the Hg_2 molecules $^3\pi^-$ou and $^3\Sigma^-$ou can be photo-excited to higher radiating states by *absorbing light of any wave length from ultraviolet to visible*. Thus the Hg_2^m molecules are destroyed and the low-energy two-step ionization at 5-6 ev ceases. Under these conditions only direct ionization of Hg at 10.38 ev can occur, and one observes V_{sr} saturated.

If the A pressure is too low, then Hg_2^m cannot form. Losses by diffusion and rate of triple impact generation are too low. Thus only direct impact ionization of Hg occurs; again one has V_{sr}. The increase of all potentials with or without light with A pressure, seen in figure 7.5, is due to the usual increased loss of electron energy to elastic impacts with A atoms.

The *lowering* of V_s by λ2537, or in pure Hg at low A pressure, probably arises as follows. At 40° to 90° C this radiation still penetrates well into the cell. This yields added 6^3P_1 states. These 6^3P_1 states, being resonance-absorbed, have a very good chance of going to 6^3P_2 by absorbing λ4385, or by λ4047 after having gone from 6^3P_1 to 6^3P_0 by impacts. If pressure is low, then 6^3P_2 and 6^3P_0 are ionized by electron impact. The influence on V_s is slight in any event.

At higher A pressures λ2537 is one of the most effective radiations at *raising the starting potential*, changing V_{sn} to V_{sr} sat. However, this light comes from a hot Hg arc. In this there is a narrow unreversed portion of the λ2537 line which can be absorbed by Hg atoms and yield Hg^m as above. However, most of the light in λ2537 is from the pressure-broadened Hg line. This is very intense in the arc. But the transition from Hg_2 $(6^3P_0 + 6^1S_0)$ to Hg_2 $(7^3S_1 + 6^1S_0)$ is very close to λ2537. Thus Hg_2^m is readily destroyed by the broadened λ2537 from the arc. This raises V_{sr} very effectively, as seen in figure 7.1.

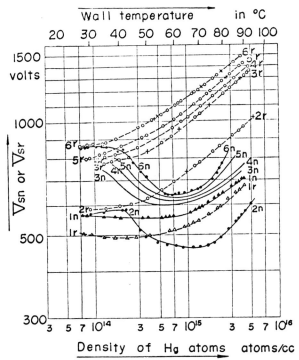

FIG. 7.5. Relation of V_{sn} and V_{sr} to A pressure. Suffixes n and r mean V_{sn} and V_{sr}, respectively. ○ or △: illuminated by a 100 watt high-pressure Hg arc lamp enveloped in a glass globe placed at a distance of 100 cm. +: illuminated by a 100 watt incandescent lamp placed at a distance of 100 cm. ×: illuminated by a 100 watt incandescent lamp placed at a distance of 200 cm. As illustrated in the figure, V_{sr} in these cases gives the values of V_{sr} sat. (1) A pressure of 0.7 mm Hg; (2) 2.6 mm Hg; (3) 7.5 mm Hg; (4) 8.5 mm Hg; (5) 10.0 mm Hg; (6) 15.0 mm Hg. In (1), the A pressure is too low for the effect of illumination on raising the starting potential to take place, and, contrary to this, the effect of illumination with visible light on lowering the starting potential is noticed.

The increased range and lowering of the potential in the dark produced by aging are readily explained. Hg vapor when fresh is rarely pure. It has H_2, N_2, O_2, CO_2, and CO. All these are very active in destroying both Hg^m and Hg_2^m. Kenty[7.22] has shown that prolonged running of a discharge gradually removes these impurities to form HgH, Hg_3N_2, and HgO, which go to the walls as solids. Thus as the Hg cleans up, the lifetime and reaction rate of Hg^m and Hg_2^m in the discharge increase.

Addition of H_2 gas acts at once to remove or destroy Hg^m and Hg_2^m by being dissociated by it via the formation of an excited HgH molecule that dissociates on losing its activation energy.

The reason for the localization of the sensitive region for destruction near the anode is relatively simply explained. In the glow discharge tube the two important regions for ion production and maintenance of the discharge are at the anode and cathode. At the cathode, owing to the positive ion space charge, the fall of potential is high and the direct ionization of both A and Hg atoms takes place there. In fact, most of the fall of potential across a glow discharge lies across the Crookes dark space. As shown in this discussion, for point-to-point systems the breakdown of the whole system to a glow or an arc depends on a breakdown at the anode as well. At the anode of a glow there is no high potential gradient. The nature of the small anode spot in such glows means that, in order to yield the electrons to create the connecting positive column, a different process must occur in the anode fall. This depends on the current density and on a low field ionization process. In consequence, one would expect the coefficient of ionization at the anode at low X/p afforded by the Hg_2^m process to be most important if it can occur. Thus at onset the destruction at the anode of the Hg^m and Hg_2^m by light will be critical. Since the process at the cathode, which breaks down at lower potentials and first builds up to its space charge and high field, does not require the low X/p processes, that region will be little affected. But before the glow can develop, the stepwise efficient ionization dependent on *current density must be achieved at the anode.*

The one most difficult aspect of the observations, in Nakaya's view, was the explanation of the effect of duration of illumination in the anode region, which produced a V_{sr} sat. that lingered for minutes or hours after the light was cut off, depending on the form of the tube and the duration of the illumination. It appeared only for the cold cathode discharge and was greatest for the long, narrow tubes and low Hg pressures. It was absent with alternating potential.

Some study of the phenomenon was made by investigating the influence of the potential on the tube during and after illumination, on the recovery of the tube at different temperatures, 48° C, 65° C, and 76° C. If the potential applied during illumination was low, the value of V_{sa} after illumination was the same as V_{sn}. That is, for low potentials during illumination there was no increase in potential to V_{sa} on cutting off the illumination and measuring the starting potential. The illumination in all cases lasted 30 sec. It was also noted that the potential required to cause a change toward V_{sr} during illumination became higher the higher the temperature. It is unfortunate that currents during illumination were not measured instead of potentials, since the resulting currents during illumination and not potentials are the important factors needed for a proper interpretation. It was, however, noted that, as the potential during illumination increased, the restarting potential increased in proportion.

The explanation to follow is that of the author, reported here for the first time. It has been accepted by Nakaya in personal correspondence. It is clear that the starting and/or restarting potential is equal to V_{sn}, the dark value, and is low if in the critical anode region the Hg concentration is such that Hg^m and Hg_2^m are created so as to give a stepwise ionized gas of high ionizing (low X/p) efficiency. It is also well known that Hg^+ and other more ionizable impurities in inert gases like A are transported from anode to cathode in the phenomenon known as gas cataphoresis. In it a minority of more ionizable substances, for example, Ne and A in He, and Hg in Ne and A, are selectively removed by transportation to the cathode. This phenomenon is somewhat complex. It is favored by long, relatively small diameter discharge tubes at pressures on the order of 5 to 20 mm of the vehicular inert gas and by the ionizability and concentration of the impurity. The rate of separation depends on current density, and increases with current density up to certain limits. The method is currently being used to produce very pure inert gases, and removal of the impurities is very rapid and effective.

If now in the gases studied here the potential applied during illumination is high and yields a high current near the anode, the anode region will have lost Hg atoms faster than diffusion and the wall counter current can replace them. During illumination V_{sn} is replaced by V_{sr} sat. This means that ions are being supplied by direct ionization of Hg at a high potential and with a high current. It also means that if current density at the anode is high, Hg^+ ions are being removed very rapidly. Thus it is most probable that the region for a few millimeters from and about the anode is cleared of Hg atoms. Hence the lowered concentration of Hg atoms will render it impossible to restore V_{sr} sat. to V_{sn} immediately and for some time after the current has ceased. It is to be noted that at 76° C where the Hg concentration was much higher, the potential applied during illumination had to be higher (higher current density) to raise the restarting potential above V_{sn}. Large diameter tubes and alternating potential do not cause sufficient removal of Hg atoms from the anode region. Recovery time to reduce the restarting potential to V_{sn} is the diffusion time of Hg atoms to restore the anode concentration to normal. The hot cathode undoubtedly sets up convection currents to prevent the cataphoretic action from removing too much Hg. It also furnishes many electrons to start and sustain this discharge.

The author's tentative explanation of the only serious difficulty encountered by Nakaya's excellent investigation of the nature of the change of starting potential of an A–Hg discharge at larger concentrations of Hg leaves little doubt about its nature and the validity of Nakaya's interpretations. As a consequence of this interpretation of one recognized Joshi $\pm \Delta i$ effect observed in an inert gas–Hg mixture, it is now possible to indicate the direction which future investigations of the Joshi effect must take in order to achieve a correct and sound physical interpretation of the phenomena. To understand fully the significance of Nakaya's study of V_s, together with the knowledge gleaned from El-Bakkal's study, the interpretation must be applied to the explanation of the Joshi-type of effect as observed by Bhatawdekar.

Joshi effect studies are in principle concerned with the rectified currents obtained under a 50-60 cycle alternating potential through effectively plane parallel glass electrodes with external electrodes. These are so contrived in shape that the gap can be illuminated with light from various sources. The effect of the light is to increase, $+\Delta i$, or decrease, $-\Delta i$, the rectified current, through the cell. It appears in more or less soiled gases, that is, mixtures of gases as with A–Hg. The effect is observed above or very near the threshold for the AC discharge. As noted in El-Bakkal's study, it represents the integrated multiple breakdown current loops on the rising potential quarter of each alternate half-cycle. There will be one or more peaks advanced in phase, depending on how high above V_s the combination $V_{AC} \sin \theta$ and ΔV_{Gd} lie, as well as on how many parallel breakdown channels are required to cover the electrode surface at the existing pressure. The important feature to note is that the current i in this near threshold region is

$$i = \frac{\Sigma \Delta q}{\tau} = C\Sigma \frac{(V - V_s)}{\tau},$$ where V is related very closely to V_{AC} peak +

V_{Gd}, and τ is the time of a half-cycle. If now by action of light V_s is raised in this region, then i will be decreased. If V_s is lowered by light action, i will be increased. One will have a $-\Delta i$ or $+\Delta i$ effect. The $-\Delta i$ effect is in fact the most common type of Joshi effect. It is the one observed in the controllable mixture of Ne or A and Hg. It will also be noted that the Δi effect, through raising of V_s from V_{sn} to V_{sr} sat., was not the Penning effect, though at lower Hg pressures the Penning effect did occur. Instead it was a new and very intricate effect through destruction of a stepwise ionization mechanism created by triple impacts by the light. The destruction is also created by nondescript incandescent light, though more effectively by pressure-broadened $\lambda 2537$.

It will also be noted that the effect is most actively suppressed by illumination of the anode region, as the anode in this type of discharge breaks down through a stepwise process requiring low voltages. This gives an example of the situation in nearly uniform field or plane parallel geometry where the glass walls appear to be the active regions. For the glass walls are the alternating anode and cathode in Joshi's tubes. In Nakaya's case the phenomenon is not caused by the influence of light on adsorbed gases on the glass wall, but through the destruction of Hg^m at the anode surface.

While it is as unforgivable on the author's part to speculate as to the exact mechanisms active in the Joshi effect cells with the halogens and oxygen as it was for past investigators to speculate on insufficiently controlled studies, it may not be amiss to make a suggestion. The $\pm \Delta i$ effect is produced when at constant V_{AC} near the breakdown threshold of the cells, the value of V_s for the cell statically observed is *raised* ($-\Delta i$ effect) or *lowered* ($+\Delta$i effect). How this comes about in any one case requires investigation. Since light can produce states susceptible of stepwise ionization in certain gases as well as destroy them, the effects are capable of interpretations. What is needed now are a series of basic, carefully controlled studies of V_s on *static breakdown* in cells with electrodes, or if needed, glass cells with external electrodes, of each of the Joshi effect gaseous mixtures. These must be carried out with the same searching quality and excellence which characterized the work of Nakaya. Only in that way can a reasonable and correct interpretation of the Joshi effect be achieved.

REFERENCES

7.1 L. B. Loeb, *Fundamental Processes of Electrical Discharge in Gases* (John Wiley, New York, 1939), pp. 560 ff. *Encyclopedia of Physics* (Springer, 1956), vol. XXII, "The Glow Discharge at Low Pressure," by Gordon Francis, pp. 53 ff.

7.2 R. W. Warren, Phys. Rev. *98*, 1650, 1658 (1955).

7.3 R. G. Fowler, Proc. Phys. Soc. (London), *B68*, 130 (1955).

7.4 J. M. Meek and J. D. Craggs, *Electrical Breakdown of Gases* (Clarendon Press, Oxford, 1953), chap. VII, pp. 321 ff.

7.5 L. E. Reukema, Trans. AIEE, *47*, 38 (1938). J. Thomson, Phil. Mag. *10*, 280 (1930); *18*, 696 (1934); *23*, 1 (1937). *Encyclopedia of Physics*, vol. XXII, "Breakdown in Gases at Alternating and High Frequency Fields," by S. C. Brown, pp. 531 ff.

7.6 J. D. Cobine, *Gaseous Conductors* (McGraw-Hill, New York, 1941), sec. 8.19, pp. 265 ff.

7.7 A. S. Denholm, Trans. AIEE, Atlantic City, June 19–24, 1960, paper 60-755.

7.8 T. W. Liao, W. H. Keen, Jr., and D. R. Powell, Trans. AIEE, New York, Jan. 21–25, 1957, paper 57-167.

7.9 W. E. Pakala, C. D. Fahrnkopf, G. D. Lippert, and C. C. Bartlett, Westinghouse Engineer (Nov., 1951), p. 190; also R. L. Tremaine, A. R. Jones, and O. Naef, Westinghouse Engineer (Sept., 1951), p. 150.

7.10 C. I. Miller, Trans. AIEE, Chicago, Oct. 1–5, 1956, paper 56-1046.

7.11 W. v. Siemens, Pogq. Ann. *102*, 66 (1857).

7.12 S. S. Joshi and P. G. Deo, Nature *151*, 561 (1943); *153*, 434 (1944). P. G. Deo, Proc. Indian Acad. Sci. A*19*, 117 (1944). S. S. Joshi, Nature *154*, 147 (1944). V. L. Talekar, J. Electronics, Studies of Rectification in a Gas Discharge Between Coaxial Cylindrical Electrodes, Parts I, II, and III, Nov., 1956, Jan., 1957.

7.13 W. L. Harries and A. von Engel, Proc. Phys. Soc. (London), *B64*, 951 (1951); A*222*, 490 (1954).

7.14 J. El-Bakkal and L. B. Loeb, J. Appl. Phys. *33*, 1567 (1962).

7.15 M. G. Bhatwadekar, doctoral dissertation: Studies on the Joshi Effect in Especially Pure Rare Gases, University of Rajputana, Jaipur, Rajasthan, India, 1957.

7.16 T. Nakaya, J. Phys. Soc. Japan *11*, 1264 (1956).

7.17 V. K. Rohatgi, J. Appl. Phys. *28*, 951 (1957).

7.18 J. J. Muray, J. Appl. Phys. *33*, 1517, 1525 (1962).

7.19 J. Lucas, Fifth International Conference on Gaseous Electronics, Munich, Aug. 28–Sept. 1, 1961, paper 2, sec. EIII.

7.20 T. Holstein, D. Alpert, and A. O. McCoubrey, Phys. Rev. *76*, 1257 (1949); *85*, 985 (1952).

7.21 M. J. Druyvesteyn and F. M. Penning, Rev. Mod. Phys. *12*, 87 (1940).

7.22 C. Kenty and D. A. Larson, Phys. Rev. *89*, 180 (1953). C. Kenty, Phys. Rev. *80*, 95 (1950). C. Kenty and J. R. Cooper, Trans. Am. Electrochem. Soc. *87*, 397 (1945).

Appendixes

I

Recent Advances Concerning the Mechanisms of the Streamer Spark in Air

After the text was completed, the Sixth International Conference on Ionization Phenomena in Gases in Paris in July, 1963, brought together workers who reported important advances, many of which had not been and as yet are not published. The author has had access to reports and has carried on correspondence with many of these workers, and the summary of their advances is here presented. This report covers the work of five significant groups. These are as follows:

1. The very significant work of Tholl,[1] at Hamburg, using pulsed over-voltages, a uniform field gap, and numbers of simultaneously triggered avalanches from a small cathode spot with photomultipliers and photographic observation to evaluate velocities of avalanches and streamers and to determine the effect of avalanche space charge fields on the effective value of α. He photographically recorded the development of midgap streamers to an arc.

2. The application of an intensified image converter to the growth of single avalanches in a uniform field with pulsed voltages yielding the velocity of avalanches over a range of X/p values up to the critical field for streamers, as well as electron energies, by Wagner[2] at Hamburg.

3. The studies with a super-image converter and fast oscilloscope of the growth of streamer sparks to an arc in a point-to-plane gap of small dimensions in air at atmospheric pressure by a research team under Goldman[3] at Ivry, France. A series of still photographs as well as current-

potential and potential-frequency plots covering the regions studied were also provided. Here time-resolved photographs of single streamer branches were obtained with the image converter for the first time.

4. Photomultiplier and oscilloscope studies, as well as still photographs, were made of long point-to-plane gaps using impulse potentials with both positive and negative points. These photographs show return strokes and corroborate the Lichtenberg figure observations in many respects. Mixtures of O_2 and N_2 were also studied. The work was begun at the University of Witwatersrand in South Africa as a doctoral dissertation by Kritzinger[4] and continued at Central Electricity Research Laboratories, Leatherhead, England.

5. Phillips and Allen,[5] at Associated Electrical Industries in Manchester, England, carried out extensive studies with impulse potentials in uniform field gaps in air. They used improved cloud track techniques, measured breakdown thresholds, and got still photographic and spectroscopic measurements of the pulsed discharges. They made time lag studies over a range of pressures in which they corroborated and extended Köhrmann's study of the transition overvoltage from Townsend to streamer development of the spark, showing that it depended on a critical avalanche size. Gases used were air, N_2, O_2, A, and CO_2, none of which was pure. Finally, attempts were made to analyze the growth by a streak camera with rotating mirror optics. Spectra of all the discharge forms were observed.

A few general remarks are in order before the results are presented.

Because of the branching of streamers, their rapid tip motion, and the fact that the high potential tips alone are luminous, these cannot well be seen by the eye or recorded directly by still camera. Superposition of some hundreds of branched streamers in a continuous exposure with the best optics makes them visible as a sort of haze. Strong axial streamers of few branches can be detected by photomultiplier. It now seems that super-image converters can photograph single avalanches and yield traces of time-resolved primary streamers. Streamer branches illuminated by return strokes can be photographed. The streamer tips streaking in a gas less than 0.1 mm from the surface of a photographic film will record their autograph on the sensitive photographic film.

Avalanche speeds lie in the range of 10^7 cm/sec, streamers range in speed from 10^7 cm/sec to 5×10^8 cm/sec, and return strokes move on the order of 10^9 cm/sec. Accordingly, these can only be resolved in their motion and have their velocities measured by oscilloscopes having a recording speed in excess of 10^{-8} sec and for return strokes in the nanosecond range. The phenomena can also be studied by square wave-form pulses of short rise time and of length appropriate to the phenomena it is desired to observe. Unfortunately, it is hard to get high potentials in pulses of short rise time. These are required for longer gaps where the time resolution of existing available equipment is adequate. It becomes very hard to observe faster events with lower potentials in the shorter gaps. Uniform field gaps are difficult to study and have short formative times. Point-to-plane gaps having long ranges of potential between threshold and spark have some advantages and are not hampered by Townsend discharges that go over to sparks. Here the points must be properly contoured in order to assure one high field point of origin. Flat-ended cylinders are very poor. Complications follow if the spark forms while potential is still rising, and too much series resistance with small point capacity can cause complications by repeated discharges down channels that remain conducting. All these make observations difficult to interpret.

1. Tholl's work made use of the light as detected by photomultiplier from a group of avalanches triggered simultaneously from the cathode and allowed to advance for a certain time in order to measure the velocity and growth of the avalanche in N_2, CH_4, and $N_2 + CH_4$. Previous work had revealed an approximate ratio between the number of ions and the light emitted in the spectral range used. Here it was possible to show that as the number of ions exceeded 10^6, the space charges of the positive ion and electron clouds reduced the field so that an effective α less than the conventional value α_0 appropriate to X/p could be measured. Around 3×10^8 ions reached a value of about one-half α_0. Below 10^6 ions α equaled the established values. The data permitted calculation of the avalanche diameters and thermal energies. These lead to values of 2 to 2.5 ev for N_2, and from 1 to 2.0 for CH_4 and $N_2 + CH_4$. Above about 2×10^7 ions the avalanche diameter increases. Above 5×10^8 ions the apparent α increases suddenly because of photoionization and streamer growth. The velocity of the avalanche tips in this region is not reported. One interesting item on velocities where α is still normal shows that for N_2 as 1.4×10^7 cm/sec, for CH_4 as 0.67×10^7, and for 390 mm N_2 and 10 mm CH_4 (that is, at about 2.5%) as 1.6×10^7, at p = 400 mm and a gap of 3 cm, over-voltages are 20.7% in N_2, 11.3% in CH_4, and 18.1% in N_2–CH_4 mixtures. Here clearly is seen the effect of 2.5% CH_4 in increasing the drift velocity in a pure gas such as N_2 through more inelastic impacts.

One very significant picture is given by Tholl in the *Zeitschrift für Naturforschung* paper. Here the development of the streamer from the avalanche was observed quantitatively by photomultiplier and followed photographically. The electron swarm must have been liberated from a very small area of the cathode. A sequence of photographs is presented. The first, taken after 105 nannosec, shows the tip of the avalanche at 1.9 cm from the cathode with a luminosity extending back to 1.7 cm from the cathode. The avalanche had in that time advanced 1.7 cm at a speed of

1.6×10^7 cm/sec. Thus the added 2 mm of track indicated that the avalanche had started to launch a fast negative anode-directed streamer that jumped to 1.9 cm. At 3.8 nannosec later the negative streamer had advanced to within 6 mm of the anode, and at 6.4 nannosec it was within 2 mm. At 6.4 nannosec a positive streamer had started from the cathode end and advanced perhaps 3 mm toward the cathode. The anode-directed streamer in the N_2–CH_4 mixture of 390 mm N_2 and 10 mm CH_4, with an overvoltage of 20.7%, with $X/p = 49.02$ and $\alpha = 13.5$ cm^{-1}, had a velocity of $1. \pm 0.2 \times 10^8$ cm/sec. After 9.2 nannosec the cathode streamer tip had reached the anode, and the high field distortion there produced a marked increase in luminosity at the anode, which can be seen in the next trace at 11.2 nannosec to have caused the whole channel to become more luminous. This is a result of an ionizing space wave of potential that moves from the anode toward midgap along the negative streamer channel, increasing luminosity as it moves but attenuating near midgap. At this time the cathode-directed positive streamer is within 6 mm of the cathode. It reaches the cathode at 13 nannosec and has a speed of $3.1 \pm 0.3 \times 10^8$ cm/sec. Its arrival sends a new ionizing space wave of potential, making a very luminous channel at 17 nannosec. This ionizing space wave of potential has traversed 2.8 cm of gap in less than 4 nannosec with a speed of between 7.0×10^8 and 10^9 cm/sec. Similar changes in luminosity had been noted by Loeb in Hudson's crossplots.[6] The ionization in the tracks of the positive and negative streamers before this return stroke or ionizing space wave of potential was weakest at 7 mm from the cathode, since the wave from the anode had not reached it and the field in the positive streamers was stronger as it approached the cathode. Thus the return stroke from the cathode left a pinch or faint region there. The channel, being completed across the gap after 28 nannosec, then draws enough current to become highly luminous, and the arc is completed in 33 nannosec instead of 90 nannosec after the streamer mechanism began. This first detailed record of the avalanche streamer sequence for midgap breakdown of a highly overvolted uniform field gap is most significant, as it shows the ionizing space waves of potential. The gas used was chosen for its luminous brilliance, making photography possible with pulsed avalanches.

2. Wagner's study with the image converter gives electron drift velocities, as well as energies of the electrons in avalanches in N_2, A, and Xe, carrying data to the point where streamers form. The velocities were linear in X/p for N_2 but not so for A and Xe; X/p varied from 50 to 200 in N_2, 25 to 50 in A, and 45 to 100 in Xe. Electron energies in N_2 lay between 3.5 and 5 ev and varied little with X/p in a given range of X/p. This value is in satisfactory agreement with Loeb's theory[7] for N_2. The low values in the presence of N_2 with CH_4 come from inelastic impacts.

The value does not agree with Tholl's value, derived from theory, connected with the reduction ion α at 10^6 electrons.

3. The study carried out by Goldman's group used conventional point-to-plane gaps in room air largely with point diameters of 0.34 mm and gaps of 1.5 cm and 3.1 cm length. Details of the arrangement could unfortunately not be given in their short paper. The point was separated from the potential source by a 35 megohm resistor which exerted an important influence on observations. Besides their use of an image converter which was amplified 10^3-fold by a device of their own and which gave records with a fast time sweep, the merit of their study lay in the coördinated observations of different phases of the breakdown. They recorded potential-current plots to a semi-log scale as well as observations of oscilloscopically determined pulse frequencies relative to current. Pulsations were observed over the whole range of currents from onset to the arc phase for the large diameter point and over most of the range except in the Hermstein[8] glow regime for the small point. At the higher currents the pulses, aside from the breakdown streamer regime, derived from discharge of the capacity of the point, and recharging was through the 35 megohm resistor. This complicated observations at the higher currents. Accompanying the above-mentioned plots were a series of still photographs of the discharge at various states of current. This was the first time a whole sequence had been associated with currents. It placed the breakdown streamer photographs such as Trichel's and Kip's[9] in a meaningful relation to other events. In the breakdown and arc regions single time swept image converter photographs were taken with simultaneous recordings of the current growth during breakdown. Here results were somewhat obscured by lack of knowledge as to whether the breakdown photographed was an initial one or one of a second breakdown along a pre-ionized arc channel.

The association of frequency, current, and still photographs for the small point as potential increases beautifully shows pre-onset streamers, the pulseless Hermstein glow region, and resumption of breakdown streamers. As potential and current increase above the reappearance of streamers, these lengthen but the main stems are seen largely without branches. When streamer branch tips reach the cathode, the light flashes can just barely be seen. Then, as potential gradually increases, the haze caused by return strokes up the branches, which at first attenuate before reaching the conducting stem, is seen. As potential gradually increases, the return strokes are seen eventually to reach the more conducting stems of the vigorous unbranched streamer channels near the anode point. Here the luminosity is enhanced all the way across the gap. At this point or just above it at 10^{-4} amp the point is discharging and recharging, since the arc quenches because of the current drain, so that after the first break-down the succession of breakdowns occurs along the pre-ionized path of the initial arc channel and reaches a frequency of breakdown of about 5500 times per second. Thus the channel narrows down as photographed. Here some trace photographs in the arc phase show the peculiarities noted by the earlier investigators.[9] The channel seems to oscillate; it shows evidence of lateral branches, and in a final form, called by them the "eye of the needle," is straight except for a dark loop near the anode. In this region the potential falls with current increase after reaching a peak potential of 29 kv and 10.4 amp. This accurately plotted fall of potential in the unstable negative resistance region can only be explained if one assumes that per second the breakdown potential of 22 kv of the 5499 succeeding discharges is enough to cause a re-ignition of the first arc channel. This mystery can only be clarified by actual oscilloscopic analysis of the potential on the point over a sequence of breakdowns.

As regards the peculiar forms of the arc channel photographed, Waidmann[10] reports that visual observation of a series of pre-breakdown streamers in a point-to-plane corona made while he was photographing spectra indicated that they appeared to be slowly weaving over the point surface. This sort of action in a sequence of streamers could result from the action of space charges left behind by antecedent arcs which deflected and interfered with the progress in one direction. The heating along arc channels could also cause them to be disturbed by convection currents, so that the later breakdown paths were no longer straight. Finally, if in a sequence of many such breakdowns the arc channel was occasionally dissipated so that a new spark channel had to form the path, this could well deviate from the initial one. In any event, lack of control of the successive events precludes any certain explanation.

The image converter photographs covered only the breakdown phases of one of the breakdowns, in the sequence of arc figures. In two at lower voltages are seen the bright stem of the streamer system near the anode advancing for a third of the gap and then two obviously primary streamer tips crossing the gap. The more axial tip has a constant slope and a velocity of 1.7×10^7 cm/sec; the second, more radial, one has an initially slower speed but increases its speed as it approaches the cathode. There is no indication of a return stroke. The velocities are in agreement with Hudson's[6] observations on primary streamers. At a somewhat higher potential the second tip shows four branches, the more axial one of constant speed, the others of progressively slower speeds. There is no visible evidence of a nearly vertical return stroke. It could be too fast and faint to be photographed. However, in both cases current oscillograms show that the arrival of the streamer tips in the photographs at the cathode causes a very sharp rapid increase in current for a very short time. This can only occur if the streamer tip on arrival at the cathode gives rise to a very sudden burst of current, and if the completion of the bright spark is prevented by fall in potential with high resistance. This would be recorded on the oscillograph as applied to their circuit. Why the return stroke was not visible across the film on the inverter photograph is not clear. The succeeding time photographs, which are farther along in the arc phase, the last one applying to the eye-of-the-needle form, show the following. The third frame, taking only nannoseconds, shows one primary streamer crossing which increases in velocity and very much in luminosity as it crosses the gap. It leads to a very sharp and somewhat prolonged increase in current on arrival. A strong luminosity of long duration develops at the anode and extends farther into the gap. In the last trace the streamer crosses very rapidly and is very bright near the cathode, and a small spur hints at the return stroke which is nearly contemporaneous with the development of the arc phase of long duration that now crosses two-thirds of the gap from the anode. This anode luminosity is identifiable with Kritzinger's leaders.

Through the kindness and coöperation of R. L. Hetrick of STL Products, Dawson and Winn were able to apply the Model 1D STL Image Converter Camera to the study of the breakdown in gaps using the impulse generator of Waidmann and Winn. A 6 cm gap at 17 kv was studied, using N_2 with 1% O_2 at about 150 mm to get luminous intensity. Using a time sweep transverse to the point-to-plane axis, they photographed on film three to six superposed single streamer pulses. The pulses were triggered at intervals of about ½ sec apart. Photographs, therefore, had superposed the traces of three to six streamers. This procedure paralleled the technique of the Goldmans except that the Goldmans with their intensifier observed only a single streamer sweep. Their sweep was perhaps the nth in a sequence of streamers being repeated 5500 times a second, with light therefore following the streamer channels of previous breakdowns. This was not the case in the study reported on. The traces were similar to those of the Goldmans except that there were more streamer branches crossing and the traces were somewhat fainter. The primary streamers crossing the gap followed clearly delineated paths with velocities increasing as the cathode was approached and bright flashes seen at impact on the cathode. By drawing tangents to the trajectories, Dawson and Winn could estimate velocities at various points in the gap. The streamer traces were continuous with no beading or intermittence in a horizontal direction, as reported by the Goldmans. Where the streamer tips reached the cathode, the secondary or leader luminosities of Kritzinger could be observed near the anode. The inclination of these secondary traces permitted rough estimates of the velocities which were of the order of one-tenth those of the primary streamers. There were no vertical traces from the bright spots on the cathode to the appearance of the secondary leader stroke near the anode. The return strokes were either too fast or too faint to record. However, several photographs in which primary streamers crossed the gap giving bright flashes indicated a marked increase and lengthening of the secondary leader luminosity almost vertically above the impact points of the primaries.

More interesting was the observation of later new *primary types* of streamers originating in the trace of the secondary leaders and progressing to the cathode. These definitely confirm Kritzinger's observations of repeated strokes between the secondary leader tip and the cathode following the return strokes. These later streamers from the secondary luminosities were clearly marked by their bright impact points on the cathode.

Velocities for primary streamers for various points were estimated. These velocity values were confirmed in another fashion through the use of the converter camera. By triggering on the electrical rise of the streamer, three separate images of the gap could be observed. The rise of these image exposures occurred at 50 nannosec intervals following the triggering on current rise. The duration of the exposure of each of these images was 20 nannosec. Thus the first image rose at 50 nannosec after the breakdown began and lasted 20 nannosec, the next image appeared 30 nannosec after the decline of the previous one and lasted 20 nannosec, and was followed by the third image 30 nannosec after its decline. These images were sufficiently bright with few repetitions of pulses to give excellent photographs of the range of advance of luminosity across the gap at the times noted. By delay line the beginning of the sweeps relative to the rise of the current could be varied. In this way more accurate determinations of the velocities of the streamers across the gap were possible. The values agreed with those for the velocities of the primary streamers measured with the transverse sweep. The values of the velocities observed were in excellent agreement with those of Waidmann, Hudson, and Dawson taken by different techniques in equivalent gaps. In none of these observations was the intermittence noted by the Goldmans observed.

One phase of the Goldman observations does not appear to belong to the breakdown mechanism. The paths of some of the primary streamer traces appear stippled along the track. These interruptions, or reductions in luminosity, are on time scales too short to be associated with the basic processes active. They could be caused by some ultrahigh frequency ringing in the point circuit or they may be inherent in the electronics of the complicated inverter system.

Some work was done on erosion of the points. Here radioactive tracers were used to measure sputtering. The positive point obviously gave no sputtering. The negative point gave heavy sputtering. Strangely, two peaks were observed as a function of potential for both Pt and gold, one between 5 and 10 kv and another at 17 kv. From earlier studies of their own, Goldman's group estimated ion energies of 40 ev for the first peak in air. This is very probably an overestimate, as Weissler[11] observed energies in the tens of electron volts using a calorimetric method. Spectra of the Trichel pulse discharge showed metal lines. These come from sputtering.

4. The study of J. J. Kritzinger was carried out in a doctoral dissertation in electrical engineering at the University of Witwatersrand, South Africa. Work was continued at Central Electrical Laboratories, Leatherhead, Surrey. A brief note in *Nature*, a report before the Sixth International Conference on Ionization Phenomena in Gases, Paris, 1963, and his doctoral dissertation are all that are available to the author.

Despite the fact that Kritzinger, as an electrical engineer, worked in some isolation without any knowledge of the considerable later physical literature on the subject, his contribution to the knowledge of streamers is monumental. He had good facilities in the form of adequate high-voltage sources of fairly rapid rise time which enabled him to use the photomultipliers and oscilloscopes of the more conventional types to resolve phenomena on gaps ranging from 8 to 120 cm to obtain meaningful data. Beyond this the seven years devoted to this study permitted his remarkable experimental versatility to manifest itself in well-directed, meaningful investigations. His inferences from the results, despite his ignorance of recent progress, were remarkably good and permit correlation of his results with other data. In principle his work was a continuation of the fruitful work of Meek and Saxe[19] but with superior facilities. He used point-to-plane gaps with a $\frac{1}{8}$ in. rod having a 30 deg. conically tapered point or else $\frac{1}{2}$, 2, and 5 in. spheres to a large grounded plane in room air. Potentials varied from 150 to 600 kv, and rise times of his potentials ranged from 0.07 μsec for the lower potentials to 0.25 and 3.0 μsec for higher potentials and special investigations. The potential pulses could be chopped at convenient times for photographic purposes. Observed were light pulses arriving at a given plane or off-axis point in the gap as a function of time; the currents as a function of time; and the luminosity of the light pulses as a function of time. Under his conditions luminosities were considerable compared to those in smaller gaps. Currents in the streamers ranged from about 1 and 2.5 amp up to 6 amp. The sweeps of the two oscilloscopic recorders registering current and luminosity as a function of time were triggered by electrical impulse from the firing of the impulse generators. They were carefully synchronized and were photographed to be displayed together in a given observation. Thus current pulses and luminosity could be correlated. The relative luminosity of the pulses was fairly accurately measured. Positive and negative point coronas were studied at potentials below and above breakdown, the pulses being chopped in the latter case short of spark-over, or at convenient times when the streamers had advanced a certain distance.

It soon became clear to him that in using a single long slit transverse to the gap axis he was observing not the passage of a single pulse of light across his slit but several, at times nearly simultaneous. Use of a short slit on the axis and off the axis revealed that he was seeing a group of discrete light pulses projected in different directions from the point. The short slit off-axis did not show a pulse for every impulse, but only when a pulse was projected in that direction. These light pulses, which he called "balls of light" and later "globules," left a dark interval behind them once they passed in any direction. From the still photographs of arrested breakdowns, Kritzinger recognized these pulses as representing the passage along what later were streamer channels. He was also aware of the fact that they usually did not represent a group of simultaneous streamers from the point surface but were branches of one or perhaps several streamers. He noted, as was well known from the time of Kip's studies,[9] that, once a streamer had propagated, the space charges for a time prohibited further streamers from the point even on a rising potential. Thus Kritzinger was observing the tips of streamers that flashed by his slits, as had Amin,[12] Hudson,[6] and later Nasser,[13,14,15] as indicated by his Lichtenberg figures. Loeb has indicated that the channels behind streamer tips are relatively dark. Hereafter we will call Kritzinger's "globules" or "balls of light" *streamer tips* that flash by, leaving the channels of the streamers relatively dark. By varying his slit width and observing the duration of constant light intensity, Kritzinger established that the length of illuminated parts in the primary streamer tips was about 1 cm in his long gaps. In this connection the studies of Amin[12] on pre-onset streamers is of interest. The relative darkness behind the tip is discussed by Loeb.[16] Hudson[6] could not do this, as his time resolution was not adequate in the short gaps. While Hudson had recognized the heavy branching of streamers, this phase of streamer advance in point-to-plane gaps was not clearly discerned until Nasser's[13] work, of which Kritzinger was ignorant.

Kritzinger first observed the time-distance curves of streamers for the different gaps under different potentials. The velocities of his tips ranged from 1.5×10^7 cm/sec to 5×10^8 cm/sec. The tips which were more axial had the higher velocities. In agreement with Nasser's[13] findings the streamers slowed down and decreased in luminosity as they crossed for lower potentials. He also noted, as had Nasser,[13,14] that streamers crossed the gap without causing breakdown. As potential increased so that the streamers crossing could cause a spark, the streamers were in the critical range. For these, in conformity with Hudson's[6] findings, the velocities were high and nearly constant across the gap. Above the critical potentials the velocity might decrease slightly near the anode but usually did not; it increased strongly as the streamers approached the cathode, in conformity with Waidmann's recent observations.[17] The luminosity for critical and super-critical streamers increased in conformity with Hudson's observation for his primary breakdown streamers. The velocities are in accord with the primary streamer velocities observed by Hudson[6] and by Waidmann.[17] The velocities of noncritical streamers decreased across the gap in the same fashion as had Hudson's[6] secondary streamers, thus probably establishing the tentative explanation by the author of Hudson's secondary streamers as representing the light from the more radial streamer tips that were slowing as they advanced. Many graphs of distance and luminosity of the tips across the gap are given for the various gap lengths, potentials, and electrodes used by Kritzinger.

The use of slow rise potential times showed, as Kip[9] had shown, that a second streamer will succeed an earlier one if the positive ion space charge has time to clear sufficiently.

Above critical voltage for a given gap the observations become interesting indeed. As the streamer crossed to the cathode it gave rise to a light pulse moving from cathode to anode at about the same speed, usually somewhat faster than the streamer tip pulse. This could be ascertained by observing at a plane *near the cathode*. With the known velocity of the streamer tip from the anode, its arrival there could be quite well determined. Within a short time thereafter the pulse reappeared at the plane. The farther the viewing plane was removed from the cathode, the longer the return pulse took to reach the plane. The light pulse due to this return stroke, or ionizing space wave of potential (Kritzinger's globule), going from the cathode after the arrival of the streamer tip at the cathode did not always travel all the way up to the anode. If the potential was at or slightly below the critical value, the return stroke would fade before it reached the anode. This is also seen in Goldman's still photographs. In the breakdown region the return stroke would reach the anode, or the anode glow, and cause a prolongation of the anode luminosity. However, in Kritzinger's long gaps the single return stroke or returning globule did *not complete the arc channel*. It was followed by a second anode streamer of about the same speed as the first one, and then by a new return stroke. The return strokes were always indicated by a current pulse so that current pulse and return stroke could be associated. This justifies the author's identification of a current pulse in Goldman's oscillograms as marking a return stroke which is not seen in the inverter photographs. The many back and forth pulses needed to establish Kritzinger's arc channel produced such a multiplicity of pulses and light pips that it was not possible to follow these for more than three sequences. In the case of the negative streamers, where greater regularity occurred, many more (up to five or more) could be followed. As the still photographs showed, when the spark was interrupted at various stages of advance, the phenomena could quite clearly be seen. The still photographs, interrupted at various stages of development from the time the primary streamers crossed the gap, were very revealing. As soon as the single anode streamer and its myriad of branches reached the cathode, the first return strokes made them bright enough to observe. These photographs show at the anode the primary streamer branches extending from the stem of bright luminosity, which may also have had several branches, to the cathode. As each successive streamer and its return stroke reach the anode, the bright stem of light from the anode advances farther across the gap. Its branches proliferate, although there is always a main stem which is brightest and will be the final arc channel. It is most revealing to see the paths of the many initial primary streamer branches now photographable and ending to feed the brighter branches of the main trunk. Some photographs of Kritzinger show this bright process proceeding two-thirds of the way across the gap. He calls this bright phase the "leader," presumably in analogy to lightning.

At this point Kritzinger's study reveals a good deal more. The bright stem, which was noted in Hudson's work and also appears in Goldman's stills, appears to start and advance a short way across the gap *before the return strokes prolong it*. This is undoubtedly the luminosity produced by the very heavy streamer near the anode before it attenuates by branching. It is unquestionably identical with the beginning of Hudson's secondary streamer, but in none of Hudson's crossplots does it advance *all the way across the gap at breakdown*, as Hudson's secondary streamers do. Thus Hudson's secondary streamers probably are not the same as this luminosity or "leader" all the way across the gap. His secondaries seen near the cathode are the delayed, more radial branch streamer tips that arrive at higher speeds than Kritzinger's bright leader luminosity, which is built up and extended by a succession of streamers and return strokes.

In his long gaps with adequate power to build up an arc that will continue to burn, the conductivity of the channel must be built up to a higher level. Thus the ultraluminous stem that is photographed by Kritzinger as advancing across the gap as a leader represents an ionization that is building up the channel. Oscillograms at reduced sweep speed indicate that this is not a continuously luminous channel but that the bright light is built up by the streamer or dart leader pulses and return strokes to the leader pulse and moves across the gap as a leader luminosity with a speed of some 1×10^6 cm/sec to 1×10^7 cm/sec. From its characteristics it is identified by Kritzinger with Allibone and Meek's[18] and Saxe and Meek's[19] streamers. It may possibly be an attribute of long gaps. The luminous length as a light pulse determined by slit widths is about 5 cm. From the length of 1 cm of the primary streamer tip light pulses and their speed of 10^8 cm/sec, the duration of excitation at the peak in primary streamer tips can be estimated as $\approx 10^{-8}$ sec. The bright luminosity or leader stroke, as Kritzinger chooses to call it, will have a duration of 10^{-6} sec. This identification of the leader brightness seen in still photographs as a continuous channel thus proves to be a light pulse of 1 μsec duration. That it does not form the arc arises from collapse of anode potential by current drain. This identification helps to account for the appearance of the trunk of the branched streamer near the anode. From Nasser's[13] observations, the trunk is the channel for the currents from the many branches. It thus is initially very bright as a heavy streamer tip and continues to build up its intensity for a microsecond before it decays. By this time it is joined by the first return stroke, and the bright luminosity advances as a pulse out into the gap. In a still photograph, as with Nasser's streamer paths, it appears as a continuous channel, writing its autographs on the film as it goes by. Actually in Nasser's Lichtenberg figures it appears as the bright trunk that fades across the gap on branching, as there is no return stroke.

Kritzinger did not understand the nature of the return strokes and repeated streamers, since he was not aware of the author's article on ionizing space waves of potential.[20] The clue, of course, lies in the high potential at the streamer tips[7,16,21] and the tremendous fields built up when such a high potential approaches the cathode. The burst of photoelectrons from the ultraviolet light under the influence of these fields yields the ionizing space waves of potential.[20] These are akin to return strokes in lightning, as Kritzinger notes.

Unfortunately Kritzinger quoted as the equation for the velocity of propagation of such pulses an equation given by Schonland, which was derived by him from a theory prepared by Cravath and the author in 1934.[22] That such pulses give rise to new streamers from the anode results from the intense field distortion and excess ionization produced by the return stroke in the more heavily ionized leader tip which it encounters. Nasser could not observe the return strokes as his film interfered with the progress of the streamer tips to the anode.

Kritzinger next investigated the streamers from a negative point. He observed negative streamers which moved out into the gap. At 0.07 μsec rise time the negative corona streamer started at about 50 kv. The corona speed was almost independent of gap length, with speeds from 6×10^7 to 1.5×10^8 cm/sec. The intensity decreased very rapidly as it moved along, and the negative streamers only reached the plate for a critical gap length of 10 cm. The light intensity decreased rapidly with distance. The currents were initially high and around 7 amp but decayed rapidly. At higher potentials the negative streamers moved farther. For 300 kv with 0.25 μsec rise time the speed was 1×10^8 for a 30 cm gap, and 3.5×10^8 for a 20 cm gap.

Photographic recordings show that the negative streamers were few and very little branched. They followed the field lines more closely than did the positive streamers. With the ½ in. sphere at 300 kv they formed a hemispherical volume extending halfway across the gap. At gaps just above this critical length some managed to cross the gap. When this happened there was a peculiar circumstance that Kritzinger failed to note— that when they reach the anode the streamer tips in the section within one-third of the distance above the anode plane are all straight, normal to the plane, and have *upward branching* where they meet the negative streamers; in fact, they are much brighter than the negative downward descending branches. Some of them even showed that a return stroke met them from below. Sometimes there were upward vertical off-axis streamer paths that did not directly match up with any negative streamer branches. With slowly rising potentials, successive pulses occurred along the same channel. The breakdowns here showed considerable delay from the start up to the final phase.

There is a general observation by Kritzinger that the negative streamers find it difficult to cross the gap compared with their positive counterparts. This is in excellent accord with Nasser's findings and agrees with Loeb and Meek's 1941 predictions.[23] On the other hand, once the negative streamer appears to cross, the return stroke is much more easily achieved than that of the positive. The up and down movements of streamers and return strokes keep much more in step for the negative point. Four or five pulses follow the crossing of the first streamer. The speeds for movement of the streamer and return strokes range from 1.5×10^8 cm/sec to 5×10^8 cm/sec. The still photographs show the streamers moving downward to give a brush crossing one-half to two-thirds of the gap. The streamer tips are met by upward flaring branches moving from *straight stems of high luminosity normal to the anode plane.*

What Kritzinger here failed to notice was that the *negative streamers did not really cross the gap.* Instead they were met at midgap by positive streamers moving upward from the plane. In all cases of breakdown this appears to have been the case. This could have been expected from Allibone and Meek's[18] 1938 photographs with moving film. What in reality happens is that, as the negative streamers advance out across the gap, their image force fields in the plane anode are capable of calling forth positive streamers to meet them. Since positive streamers are more vigorous and appear at lower thresholds, this is quite possible if the gaps are short. However, in Kritzinger's case the gaps were long, and the uniform length and variable course of the positive streamers indicate that it is not the tip fields of the negative streamers that cause them to move but a negative ion space charge field at the anode. Positive streamers are then launched upward to meet the downward branching negative streamers. This space charge blanket, being of considerable area, brings out positive streamers off-axis even where there are no negative streamers meeting them. The fact that branching is upward indicates the sense of motion to and from the anode. Branching is always in the direction of motion. The space charge build-up explains the delay often observed in the course of the breakdown.

That this is the mechanism is also borne out by the conclusions drawn by the author on the basis of Miyoshi's[23] data on negative point breakdown. Once negative and positive streamers effect a junction, ionizing space waves of potential will travel from the junction point to the anode and cathode, respectively, being then reflected or returning as anode and cathode streamers from midgap. The two waves moving in opposite directions will appear to pass through each other at the junction point. Since they have only half the distance to travel each time and interact mutually, they will be much more regular. Attenuation will be less, and so the regularity and the sequence will be less interrupted.

It is clear from this brief résumé that Kritzinger has succeeded by the use of long gaps in corroborating much that has gone before. However, above this he has unquestionably established the sequence of streamers and return strokes needed to build up the channel to arc proportions. This, together with Tholl's observation in uniform fields, has once and for all established the return strokes and ionizing space waves of potential as an important aspect of the spark breakdown. He has further established the nature of the process, which he called the leader, and related it to earlier observations of poor light-gathering power. His data establish the rather ineffective nature of the negative streamers, as previously anticipated, and show that the ultimate breakdown with the negative point is in part due to creation of positive streamers by image force in the anode or to space charge in the gap which meets the negative streamer in midgap.

Kritzinger's report on the work done at Leatherhead appears in a later report.[25] The work was done on a point-to-plane gap with a potential source in which rise time of the pulse was reduced to 50 nannosec, with potential decay to half value in 100 μsec. Oscillations at the top of the pulse were $\pm 5\%$ and lasted less than 0.2 μsec. The plane was negative and on top. The point was a 0.5 cm diameter brass rod with a flat end. This undesirable shape was essential to provide triggering electrons by field emission from the sharp edges of the cylinder needed with the short rise time, since in general ultraviolet illumination was not practical. Comparative tests were made with a hemispherically capped cylinder with ultraviolet light, and any data and results that could be compared showed no significant difference. The gap was enclosed in a Pyrex glass cylinder

75 cm long and 45 cm in diameter. It could be evacuated to a degree such that the residual contamination was no more than that normally existing in the gaseous mixtures used. A double beam 555 Tektronix recorded the pulses of current and photomultiplier or of two photomultipliers. The sweep could be triggered by a photomultiplier at the anode or by the rise of the electrical pulse. Then the one PM placed at various regions of the gap or the two PM's separated by suitable distances could be used to give the pulse velocities in that region. The potential rise and fall could be oscilloscopically observed. An attempt was made to determine the electron density in the streamer channel near the anode by microwave absorption. The microwave beam covered some volume in the gap and did not really scan just one streamer branch. So that the potential of the streamer tip could be studied as it approached the plane, the plane had a probe at its surface. Potential arrangements required the plane and probe to be at zero potential, so that in this instance only negative point streamers were observed. Probe voltage and currents could simultaneously be observed. Still photographs of the discharge were also made.

Some few observations extended work already reported. The Pyrex cylinder let out light above 3500 Å, so that the PM saw only light of wave lengths beyond this. This light intensity comprised 70% of the total light output.

Much of the work was done with mixtures of N_2 and O_2 and Xe, N_2 and A, and N_2 and CO_2, as well as A and O_2. Some significant differences were observed with A. However, N_2 in high concentrations gave the brightest streamers, so that N_2 mixtures were largely studied from an estimated 0.01% of active gas in N_2 to high percentages. The light in this gap again followed the pattern of streamer tip pulses or, as Kritzinger called them, globules, the same term of Kritzinger's applying to the return strokes. The data for 1% Xe in N_2 clearly show the primary streamer tips and the return strokes. Clearly all data given are for the positive point.

With N_2 containing from 0.001% Xe to 1% Xe the velocity of the streamers ranged from 10^8 cm/sec near the anode to a minimum of 4×10^7 cm/sec at 10 cm and rose again to 5×10^7 at 25 cm. The velocities decreased some 60% at minimum in going from 1% Xe to 0.001%. The curves for each mixture were roughly parallel. These measurements are in agreement with streamer behavior below breakdown observed by Waidmann and by Dawson, to be reported later. Generally speaking, N_2–Xe mixtures are typical of other mixtures. Light intensity of the streamer tips was observed at different points in the gap as a function of the percentage of Xe in N_2. The variation was least near the anode and was greatest in midgap. There was still some variation near the cathode. All curves peaked at between 0.1 and 1% Xe. The time of arrival of the primary streamer tips at various points from 5 to 25 cm from the anode as a function of the percentage of Xe in N_2 was given, and the velocities at 5 and 17 cm from the anode derived therefrom were given as a function of the percentage of Xe in N_2. The velocities at 5 cm ranged from twice that at 17 cm at the lowest Xe concentrations and were 30% higher at 2% Xe. The velocities at 17 cm lay along a curve slightly concave upward and ranged from 2.0×10^7 cm/sec to 7×10^7 cm/sec between 0.001% and 2%.

The curves for light intensity in N_2–O_2 mixtures were more highly peaked at all points except at the anode than for Xe and were a maximum at 1% O_2 in N_2. The time of passage of streamer tip to a given point in O_2 plotted against percentage of O_2 decreased in time from that at 0.01% O_2 and increased again above this. This is different from the curves in N_2–Xe mixtures, where the time decreased but did not rise at higher percentages. The velocities at 5 and 17 cm from the anode peaked at between 10 and 3% O_2, respectively. The change in velocity from 0.01% O_2 at 5 cm was from 2.5×10^7 cm/sec to a maximum of 5×10^7 cm/sec. The velocity of the streamers at 17 cm rose from around 1.2×10^7 cm/sec to a peak of 3.5×10^7 at 3% O_2, falling to 2×10^7 at 20% O_2. This change in velocity was not noted by Waidmann. On the other hand, no streamer range measurements as a function of O_2 content were made by Kritzinger. He thus failed to note the critical role of the absorption coefficient and attributed the phenomena observed to ion formation. With A in N_2, light amplitudes remained very weak at all points except at the anode. They were constant up to 10% A. Thereafter they rose sharply. Velocities were not determined, as the light was too faint.

Kritzinger reported measurements on the conductivity of the channel with time for streamer channels for varying percentages of O_2. It is difficult to see how he determined this and how significant are the values he was able to derive relative to the true conductivity. Conductivity varies with position along the streamer and in time, as will later be noted. The decay time reported was nearly constant at around 20 μsec from 0.01 to 0.5% O_2. This was ascribed to negative ion formation. It could also have been in part due to recombination. The curves for impulse spark breakdown voltage in various O_2–N_2 mixtures were very similar to those published by Waidmann.[25]

Kritzinger estimated what he believed to be the charge in the streamer tip by current increase when the streamer tip struck the anode. He arrived at 3×10^{10} electrons. Dawson, as will be seen, has interpreted the current peak in this fashion. Here Kritzinger was measuring the charge on a negative streamer tip striking the probe. If, as he assumes, the negative streamer tip was 1 cm long, then the value 3×10^{10} electrons/cm of streamer length is commensurate with and about three times those observed by Kip, English, and Winn. However, Kritzinger's currents were much larger and his streamers were broader and may have contained more charge. There unquestionably was some multiplication of electrons in the field between streamer tip and anode, as indicated by the light flashes observed at the cathode.

In keeping with Dawson, Kritzinger observed that streamer tips did not always strike the probe; they hit in 30% of the corona pulses.

The microwave measurements did indicate a definite attenuation, of the order of 5%, near the anode because of streamers. This indicated that electron densities as high as 10^{13}/cm³ existed near the point. Despite the difficulty in interpretation this value is not at all out of line with what would be expected in the trunk of a streamer channel near the anode. O_2 concentration reduced the attenuation, and it disappeared for air with 20% O_2. Kritzinger ascribes this loss of electrons to attachment. It could also have been due to the attenuation of streamers in range and vigor as the O_2 concentration increases, as Waidmann[25] reported.

With negative point, as has been repeatedly indicated, spark breakdown required about twice the potential of the positive point. How far this reflects lack of adequate triggering is not known. The luminous intensity rose to a peak at around 0.3% O_2. The negative streamer tip velocity increased rather sharply from around 5×10^7 to 2.5×10^8 cm/sec at 2.4 cm from the anode in going from 0.005% O_2 in N_2 to 20%. At 7.5 cm the velocity rose from around 1×10^7 cm/sec at 0.005% O_2 to 2×10^7 cm/sec at 20% O_2. Here the potential was 100 kv and the gap only 10 cm. The same lower actual velocities for the positive streamers were observed at 100 kv on a 25 cm gap and 5 and 17 cm from the anode, so that the difference of speeds is owing to different field conditions and may not be significant. Waidmann, in work to be reported for air, found negative streamer velocities to be lower. Nasser[15] working at nearer breakdown reported higher negative streamer tip velocities.

5. Phillips and Allen[5] have carried on a series of studies on streamers using various techniques, including spectroscopic as well as cloud track photographs of pulsed overvolted gaps with a pulse duration short of crossing the gap, and time lag studies of breakdown in air and other gases. The work was written in a series of reports of which portions have been published. Most of the work was done with methods involving time resolutions inadequate to delineate events in their short uniform field gaps. They did confirm and improve on many of the pre-World War II studies. Report 3[5] is an excellent piece of work using time lag studies of breakdown in air and other gases over an extended range of pressures and gap lengths. The technique used was essentially that of Fisher and Bederson. The particular merits of the studies lie in finding the percentage of overvoltage needed for the transition from a Townsend discharge that leads to a spark through space charge distortion at and above threshold and the direct initiation of a fast streamer breakdown. In this respect they corroborate

Köhrmann,[3.39] whose excellent work had largely been ignored, and go further. They covered a large range of pressures and gaps and indicated that the percentage of overvoltage in air required for the transition declines as the product of pressure times gap length, pd, increases. This can be related to a constant avalanche size for streamer advance. It deviates at larger values of pd.

In all gases at appropriate pressures, because of impurity, streamer sparks occur at sufficiently great overvoltages if pressures are not too low. If pd is too small for avalanche growth to streamer-forming proportions, the discharge is a Townsend discharge. In the time intervals covered in these studies there is no evidence of the shrinkage of a glow along the axis to yield an arc channel, as Ward postulates from computer studies. In some cases where the potential was in the right range for the glow to lead to space charge distortion yielding a streamer spark, photographs show the glow with the streamer spark channel superposed, just as Bandel's oscillograms and those of the Hamburg group would lead one to expect. In cases where the Townsend glow did not lead to a spark, the luminosity is brightest near the anode. In this study excellent spectra of the various discharge forms are shown as indicated above. From the continuous presence of Zn and Cu lines in most of these spectra, it appears possible that trace amounts of these elements as oxides may have been sputtered in past breakdowns and suspended in the gas as dust, so that they appear to have diffused across the gap as ions and sputtered atoms even in the very short times, when in reality they are omnipresent. It should be indicated that such elements are very readily detected spectroscopically because of low ionization potentials, so that their lines dominate in any discharge.

REFERENCES

1. H. Tholl, Z. Physik *172*, 536 (1936) and especially Z. Naturforsch, *18a*, 587 (1963). K. Richter and H. Tholl, Sixth International Conference on Ionization Phenomena in Gases, Paris, 1963, paper Vb, 13.
2. K. H. Wagner, Sixth International Conference on Ionization Phenomena in Gases, Paris, 1963, paper Vb, 14.
3. G. Buchet, G. Hartmann, and A. Goldman, Sixth International Conference on Ionization Phenomena in Gases, Paris, 1963, paper Vb, 12.
4. J. J. Kritzinger, Nature *197*, 1165 (1963); doctoral dissertation in electrical engineering, University of Witwatersrand, Johannesburg, 1962; and Sixth International Conference on Ionization Phenomena in Gases, Paris, 1963, paper Vb, 11.
5. K. R. Allen and K. Phillips, Associated Electrical Industries, Ltd., Research Laboratories, Manchester, England, reports 1, 2, 3, and 4.

6. G. G. Hudson and L. B. Loeb, Phys. Rev. *123*, 29 (1961).
7. L. B. Loeb, Spark Breakdown in Uniform Fields, Office of Naval Research Tech. Report (Naval Research Laboratory Press, Washington, D.C., 1954, actually published in 1955).
8. W. Hermstein, Arch. Elektrotech. *45*, 209 (1960); *45*, 279 (1960).
9. A. F. Kip, Phys. Rev. *54*, 139 (1938); *55*, 549 (1939). G. W. Trichel, Phys. Rev. *55*, 382 (1939).
10. G. Waidmann, personal communication, Oct. 10, 1963.
11. G. L. Weissler and M. Schindler, J. Appl. Phys. *23*, 844 (1952).
12. M. R. Amin, J. Appl. Phys. *25*, 358 (1954).
13. E. Nasser, Arch. Elektrotech. *44*, 157 (1959); *44*, 168 (1959).
14. E. Nasser, Z. Physik *172*, 405 (1963); also Office of Naval Research Tech. Report (Naval Research Laboratory Press, Washington, D.C., Nov., 1961).
15. E. Nasser, Dielectrics *1*, 110 (1963); E. Nasser and L. B. Loeb, J. Appl. Phys. *34*, 3340 (1963).
16. L. B. Loeb, Phys. Rev. *94*, 227 (1954).
17. G. Waidmann and L. B. Loeb, Sixteenth Annual Electronics Conference, Pittsburgh, Pa., Oct. 16–18, 1963, paper 3A. *Z. Physik 179*, 102 (1964).
18. T. E. Allibone and J. M. Meek, Proc. Roy. Soc. *A166*, 97 (1938); *A169*, 246 (1938).
19. R. F. Saxe and J. M. Meek, Inst. of Elec. Eng. Monograph No. 124M (April 1955); also Allied Brit. Ind. and Research Assoc. Report L/T, *183* (Fall, 1948); Nature *162*, 263 (1948).
20. L. B. Loeb, The Role of Ionizing Potential Space Waves in Spark Breakdown, Third International Conference on Ionization Phenomena in Gases, Venice, 1957, invited paper, published by Italian Phys. Soc., p. 646.
21. L. B. Loeb, J. Appl. Phys. *19*, 896 (1948).
22. A. M. Cravath and L. B. Loeb, Phys. *6*, 125 (1935) [now J. Appl. Phys.].
23. L. B. Loeb and J. M. Meek, Mechanism of the Electrical Spark (Stanford University Press, Stanford, Calif., 1941).
24. Y. Miyoshi, J. Inst. Elec. Eng. Japan *78*, 1413 (1958).
25. J. J. Kritzinger, Central Electricity Research Laboratories, Leatherhead, England, Laboratory Report RO/L/R, 1197 (Aug. 6, 1963).
26. G. Waidmann, Dielectrics *1*, 81 (1963).

II

Recent Advances in
the Author's Laboratory

While the text was being edited, the author's group of research men was augmented by a postdoctoral fellow, Acting Assistant Professor of Physics at the University, Dr. G. A. Dawson, on leave from the University of Keele, England, and a graduate student, William Winn, in addition to Dr. G. Waidmann from the University of Bonn. These men, stimulated by Nasser's techniques, by the work of Kritzinger, and by the technique of the Goldman group at Ivry, France, made remarkable progress in the observation, interpretation, and theory of the streamer spark mechanisms. This book would be incomplete without inclusion in abbreviated form of these important advances. (See chapter 3 for an account of earlier research.)

1. G. Waidmann[1] completed his study of the velocity of positive and negative streamers in crossing the gap. He used impulse potentials from 5 to 40 nannosec, from a 1 mm diameter point and for 2 and 4 cm gaps. Potentials were such that streamers crossed to various points in the gap up to the cathode during a pulse interval. Thus streamers longer than the gap, including the breakdown, could not be studied. In general, velocities were high near the point and declined across the gap for streamers that were shorter than the gap length. The higher potentials and the shorter gaps gave streamers, the velocity of which increased again near the cathode. For the closest approach to breakdown conditions open to study in the gap, the streamer tip velocities were uniform across the gap. This applied equally to positive and negative streamers. The speed of the negative streamers was somewhat less than the speed of the positive streamers for the same conditions. The negative streamers were, however, materially shorter under similar conditions. The tip velocities for the streamers nearest breakdown were thus constant across the gap and had velocities comparable to Hudson's primary streamers under similar conditions. This

631

completes the identification of the tips of the Lichtenberg figure agencies with the primary streamers of Hudson. It still leaves undecided the exact status of Hudson's secondary streamers and their possible relation to the late arrival of more lateral branches, to Kritzinger's so-called leader strokes, or to the photographic bright trunks photographed by various observers.

2. W. Winn took over a modified form of Waidmann's apparatus for observing Lichtenberg figures with better control of conditions for the study of streamers in N_2-O_2 mixtures of lower O_2 content. He also improved the long duration pulse. With this a number of exploratory studies on streamers were made.

A. He made a more careful study of the range of streamers in room air from 15 kv to 50 kv as a function of applied potential for various gap lengths and point diameters. He observed the following:

(1) The length of the streamers is proportional to the applied potential when the gap length is sufficiently greater than the streamer length. It is about the same whether the film is 1 mm from the point or against the point.

(2) The maximum axial length under these conditions is, within limits of error, the same as the sum of the distance x of the photographic film transverse to the field axis from the anode and the length l of the longest residual Lichtenberg figure traces on the film at that point. That is, the deflection of the streamer by the film does not seriously alter the streamer length, and the residual guiding field along the axis under these conditions does not materially affect the streamer length. This has very useful theoretical implications, as will be seen.

(3) The streamer length for the same gap under the conditions above is the same whether the point used is a needle point or a 2 mm hemispherically capped cylinder. It is the total potential near the anode and not the field distribution near the point that fixes streamer length.

(4) Where the length of the gap is more nearly commensurate with the streamer length, there is a marked influence of a field between streamer tip and cathode and the axial streamer has its range extended by this field. Thus such axial streamers will be about 10% longer than the radial streamers.

(5) When the gap becomes relatively short, that is, in the 0.5 to 1 cm range, there is some influence of point diameter on streamer length. That is, in general, if the gap geometry becomes such that the field across the gap has appreciable values near the cathode—or, in other words, the field becomes more uniform—the streamer length is influenced by the field distribution and by the effect of the cathode on the streamer tip.

B. Winn succeeded in observing Lichtenberg figures with photographic film in the plane of, and parallel to, the anode, its plane thus being normal to the cathode surface. Here the Lichtenberg figures were displayed in a direction parallel to the axis. It was clear from these figures as well as from photographs of streamers illuminated by weak return strokes in N_2 with a few percentages of O_2, where streamers were exceptionally luminous, that there were several streamers starting nearly simultaneously from the point and radiating in some cases at 90 deg with the point axis. There were also some streamers along the shank of the point. In Nasser's studies with the film plane normal to the point at the point tip, or even a millimeter or two from it, there appeared to be only one main streamer normal to the point at the tip, which branched very little until it struck the film. This observation indicates that because of the strong induced field of the point in the semi-conducting film, the field at the point, with film normal to the axis, was so much along the axis that streamers were induced to move along the axis. The multiple streamers at the point surface, in the absence of the film normal to the point, were replaced by a single streamer at the point tip when the film was present.

It is probable that the film, even at greater distances, in addition to increasing sparking potential by diverting and lengthening the streamers over its surface, may have altered the branching pattern in another fashion: once it has arrived at the film the streamer tip can no longer advance along the axis, branching as it moves in three dimensions, a branch at a time. Instead, its heavy charge concentration is now constrained to progress on the plane in two dimensions only by the induced fields in the film. Since branches repel each other the plane at any point x intercepting a particular streamer branch causes an increase of radial branching on the plane at x, relative to what it would have done at that same plane in free space advance. Thus at each x the branching pattern on the transverse plane is distorted from what it would have been in free space. Whether the total number of branches created by any streamer branch intercepting the plane is altered by this circumstance has so far not been determined by statistical analysis. If the streamer is intercepted at smaller x values, branching might be increased. That is, assuming a normal rate of branching along the film surface and in free space, if multiplication is increased several-fold at a low value of x, the total number of branches resulting may be increased by that factor. If interception occurs at larger x values, the theory of Dawson and Winn indicates that it should not increase. However, the statistics of branching obtained by Nasser could well have been falsified. In any event the traces of Winn parallel to the field show less radial branching and many more short branches, as well as longer distances x along the axis with no branching, than Nasser's statistics suggest. This circumstance may also lead to a slower decline of the number of branches with x than his curves indicate.

One of the most valuable features of this new aspect of the figures is the opportunity it gives of observing the nature and diameter of the streamer tip near the end of its path where halation is weak and branching ceases. This gives a radius of 0.0035 cm with a micrometer and microscope. This is far smaller than the 0.05 cm assumed by Raether and his group for the diameter of the space charge at the anode in a uniform field when the breakdown avalanche has advanced 1 cm. In Raether's case, however, there is a strong guiding field.

It was also noted that branching into two *equal* branches appeared less frequent than on planes. While the branches diverged initially because of mutual repulsion, it was observed that the branches showed a tendency to curve in the field direction when streamers were fairly well across the gap. This indicates that the field still exerts some influence. With dense branching at higher fields, crowding of branches brings tips of slower, weaker branches near the negative stems of vigorous branches. The weak branches often merge with a small spark.

C. With a 1 mm diameter anode in air and an impulse potential of 25 kv, the following sequence of streamer length and properties are noted as gap length d is decreased.

d = 14 cm:	streamers stop 5 cm from the anode
d = 8 cm:	streamers cross the gap by Lichtenberg figure observations
d = 6 cm:	streamers can be seen by the unaided eye to cross the gap
d = 3 cm:	bright spots appear at the cathode where streamers impinge; this probably also occurs at somewhat longer gaps where streamers cross but are not seen
d = 2.8 cm:	bright trunks extend out 0.5 cm from the anode

D. It was observed that in N_2 gas with 1% O_2 the luminosity of the streamers was much greater than for air. The sequence of events under part C is recapitulated, but the gap length required for each new event is much greater than for air. In air the bright trunks at the anode observed in the still photographs of Hudson, Kritzinger, and the Goldman group are faint and diffuse, while in this mixture they are ribbon-like and much brighter. They can readily be photographed. This gave an opportunity to determine whether the bright trunks or *leaders*, as Kritzinger calls them, are an inherent property of the streamer channel near the anode—as might be inferred from Nasser's Lichtenberg figures—or whether they were invoked by return strokes, as Kritzinger claimed.

It was observed that the bright trunks began to appear as soon as the streamers were observed to cross the gap and give bright flashes at the cathode. In fact it appeared only then that individual streamers became bright enough to photograph by the fastest polaroid film. The higher the impulse potential after the streamers were seen to cross the gap, the farther the bright trunks penetrated into the gap.

At the potential where the streamers can be photographed and bright flashes first appear at the cathode, giving the beginning of the anode trunks, a film is placed on or near the cathode. Then the bright flashes do not occur and the trunks are not seen. This would appear to bear out Kritzinger's interpretation derived from long sparks. If the potential was raised to a value of twice that which would cause a spark with no film on the cathode, the bright trunks reappeared despite the film. Under these conditions there is evidence of return strokes. These bright trunks were accompanied by bright segments of light occurring along the surface of the film, which was placed 2 mm above the cathode. These segments appeared to arise from some sort of streamer-like discharges over the film, radiating from the impact point of the streamers. They appeared to call forth a bluish discharge between the lower film surface and the cathode. These direct photographic observations are in conformity with what Nasser observed under similar circumstances with Lichtenberg figures.

These observations would then indicate that the anode trunks or branches are associated with Kritzinger's leaders and with return strokes. From this evidence they appear as not identical with Amin and Hudson's secondary streamers, which occur long before streamers can cross the gap and cause secondaries. Hudson's secondaries, also observed by Dawson, at this point appear to be the delayed integrated arrival of the more radial streamer tips at a given plane.

E. Winn has roughly studied streamer length in N_2–O_2 mixtures from about 1% O_2 up to 80% O_2 in N_2, using wider films than did Waidmann. He has verified the law for change of streamer length S as a function of the concentration of O_2 and has shown that it follows the exponential law deduced by Waidmann, that $S = S_0 e^{-cp_0}$. Here p_0 is the concentration of oxygen and c is a constant. The slope was sensibly the same as that of Waidmann.

The streamers reach their maximum length at 5% O_2. Thereafter the length of the streamers appeared to decrease as the percentage of O_2 gas decreased down to 1%. From 5% of O_2 on down, while the active photons produced in N_2 are still very plentiful, the concentration of O_2 molecules became such as to reduce the photoelectron production in the critical zone to values that are ineffective. At 5% and 1% O_2, respectively, the absorption coefficients for N_2 radiations producing photoelectrons from O_2 become 1.25 cm^{-1} and 0.25 cm^{-1}, so that the resulting photoionizing free paths L are 0.8 cm and 4 cm, respectively. Now the rate of production of photoelectrons is the quantity $dn/dr = (dr/L)(e^{-r/L})$. The very critical region of photoelectron production for streamer advance at a distance r_1 from the streamer tip, to be defined in the theory to follow, lies between at most 0.1, and more usually 0.02, cm. If the appropriate convenient value of 0.03 cm is used, the exponential term has a value between 0.96 and 0.90, but dr/L is markedly less. The width of the critical zone at r_1 must be of the order of 5×10^{-4} cm, as the theory will show. The solid angle available will be less than 1 radian at r_1. The fraction of the $2 \times 10^8 \times 2 \times 10^{-3}$ photoelectrons stated to be produced by Przybylski's studies of light from sparks that are absorbed in the volume $r_1 dr$ of space will be

$$(1/4\pi)(dr/L)e^{-r_1/L}, \quad \text{or} \quad 4.4 \times 10^{-5} \text{ at } 5\% \text{ } O_2 \quad \text{and}$$
$$9.5 \times 10^{-6} \text{ at } 1\% \text{ } O_2.$$

Thus at 5% there will be 9.6 photoelectrons, and at 1% there will be 0.38 photoelectrons produced in the zone. Thus streamer advance, if one considers statistical fluctuations, will become rather difficult, altering streamer propagation requirements.

Kritzinger observed similar effects of reduced O_2 concentration in N_2 but did not apply it to the absorption coefficients, as did Waidmann and Winn.

3. As a result of the work of the Goldman group on streamers in which a high series resistance was used with the gap in breakdown studies with pulsing potentials to limit the heavy spark current which obscures observation, G. A. Dawson proceeded to use the same device with a steady potential. To this end he used the Van de Graaf static generator of Hudson, charging a small capacity gap of 10 pF or less through a 30 kilomegohm series resistor. The time constant for charging was probably less than the estimated RC value of 3×10^{-3} sec because of the inadequacy of these resistors at the high potentials used. As the capacity charged up, since no pre-onset streamers, burst pulses, or Hermstein glow fouling the gap appeared to precede the spark, the time of charging must have been less than the statistical time lag for spark breakdown. Thus streamers were propagated into a relatively clean gap, and all that was observed were the sparks as potential was raised. Potential at breakdown was never just threshold, but slight overvoltages of differing value occurred at each breakdown. The anode, which was at high potential, consisted of a point and more often a 0.5 cm diameter steel sphere facing a plane. The plane was a small disc of 3 cm diameter grounded through a resistor to measure current pulses. Gaps from 1 cm to 5 cm long or more were used. At times an auxiliary condenser was placed across the large resistor to increase the capacity to about 10 pF in order to get sufficient light to study. The charge stored under these conditions was too small to lead to the very brilliant main stroke of Hudson, or to the normal transient arc. Changing the capacity altered the streamers in other ways than in luminosity. Increase in capacity increased speeds as well as luminosity under comparable conditions. Observations were recorded by a 519 Tektronix oscilloscope, the

traces being photographed by the highest speed Polaroid camera. This gave time resolution on a nannosecond time scale. One or two 1P21 photo-multiplier tubes, PM, scanned the gap at one or two points with suitable slits of the order 0.5 mm aperture and from 2 cm to 0.5 mm length with proper optics. Either one PM could be used in conjunction with current rise, as did Kritzinger, or the two photomultipliers could be used in succes-sion, scanning the gap at the same or different points, as did Hudson.

The current curves were of considerable interest, potential and gap lengths being such that the streamers crossed the gap. In one series, for example, the currents rose sharply at onset and then more gradually to reach a nearly constant value. With a 3.8 cm gap length the rise time was on the order of 10 nannosec. The plateau lasted about 60 nannosec. There-upon there was first a gradual rise in current in 5 nannosec, and then a very steep rise in about 0.5 nannosec to a second plateau of superposed oscillations or pips. This plateau was followed by a decline showing several pips over some 60 more nannoseconds. The curves were remarkably repro-ducible except that the reproduction of the oscillations at peak and there-after were never accurately duplicated on successive breakdowns. With a somewhat longer gap the initial rise was followed by a dip in current before the sharper rise at the approach to the second peak.

The interpretation of these current curves is that they represent the development of the positive space charge streamer tip at the anode (the initial rise) and its motion across the gap. For shorter gaps the speed is nearly constant across the gap, giving the horizontal plateau. The second rise of current represents the approach and arrival of the streamer tip at the cathode. The peak is occasioned by the burst of current corresponding to the flashes of light and return strokes on arrival of the streamer tip at the cathode. The sharp pips on the plateau peak might represent the arrival of several different streamer branch tips or else several streamers at the cathode within 5 nannosec of each other. The later pips possibly represent return strokes and subsequent streamers going up and down the several streamer channels. They would correspond to Kritzinger's more completely resolved oscillograms and his "dancing" globules of light. In longer gaps the streamers noticeably slow down as they leave the high anode field. This causes the current decline in the first plateau region. They speed up again as the high field between the streamer tip and the cathode acts on the space charge of the tip. Similar changes in speed and luminosity were observed for longer gaps by Kritzinger. Dawson recorded average velocities of streamers crossing the gap for a number of gap lengths, together with photomultiplier velocity data at a later point.

To confirm this interpretation the same gap was viewed by photomulti-plier at the cathode. Using a 5.5 cm gap he marked the rise and decline of the speed as shown by the current; the time from current rise to arrival of the peak was 200 nannosec. Small pips followed for about another 100 nannosec. The photomultiplier showed the arrival of the light flash at the cathode in just 200 nannosec and the appearance of light flashes which occurred during the next 100 nannosec, with a later flash at 200 nannosec after the arrival at the cathode. These light flashes corresponded very well with the oscillations of the current.

Techniques were then improved so that by a delay line the photomultiplier trace could be recorded on the same sweep with the current. This permitted correlations in shape of the current pips with photomultiplier pips after arrival at the cathode. Since there was no background current the photomultiplier pips were more clearly resolved than the current pips. But it was clear that current surges and luminosity increases, or bursts, were closely correlated at the cathode.

By cutting off the beginning and end of the current signal the current peaks and the light signals on arrival at the cathode and later could be more closely compared. Correlation was still good. Placing the PM at various distances above the cathode in the hope of observing the return strokes gave negative results, as the correlation of PM output and current pulses rapidly decreased with increasing PM distance from the cathode. The difficulties here must be ascribed to the complications inherent in the arrival of branch streamer tips and possibly tips of several nearly simultaneous separate streamers at the cathode. Furthermore, return strokes up the individual streamers were unquestionably occurring, but because of a few nanoseconds' difference in arrival they did not give a single integrated return stroke, but each one had its own. The PM with long transverse slit observed the sum of the advancing streamer tip and their branches. If the return strokes were along individual branches and were spread over 5 nannosec, each one giving a relatively weak signal, it became very difficult to identify streamers and return strokes. The pips as much as 60 nannosec after arrival of the first tips could only have been new anode streamers along these channels called forth by return strokes, such as Kritzinger observed. Even with Kritzinger's long gaps it was difficult to identify more than the first one or two sequences of return strokes and streamers.

In an attempt to improve resolution Dawson used two photomultipliers. Here the trigger PM scanned the cathode and noted arrival of the multiple tip branches. The second PM starting at the cathode observed the signals at varying distances from the cathode. Again at the cathode both photomultipliers recorded the same sequence of events with excellent correspondence, although they viewed the gap at right angles to each other. As the signal PM was moved successively to 1, 1.5, 2.0, 2.5, 3.0, 4.0, and 5.0 mm from the cathode the *rise* of the first peak in this PM came progressively ahead of that in the trigger PM at the cathode. The gap length was 3 cm.

From this the velocity of the streamer tip in the last 1 mm, the next four 0.5 mm, and the subsequent 1 mm and 1 mm could be determined. These were in succession

0–1	5×10^7 cm/sec
1–1.5	1.7×10^7
1.5–2	5×10^7
2–2.5	2.5×10^7
2.5–3	2.5×10^7
3–4	2.5×10^7
4–5	5×10^7

The displacements were obtained for different breakdowns in succession. With what has been said about lack of constancy of potential on successive breakdowns, no accurate conclusions about speed at various points near the cathode can be drawn, but the speed appeared constant disregarding fluctuations. It is clear that in this gap the speed averaged 3×10^7 cm/sec in that region. Considering the weak character of the charge, the value is in keeping with the data of Hudson, Nasser, and others.

Another set of data was taken at d = 5 cm and measurements were made at various distances from the anode ranging from 5 cm to 2 cm. Here the primary streamer tip velocity was constant at 1.7×10^7 cm/sec. It is noted that for the longer gap the velocity is less. These data were taken from the rise of the current pulse to the arrival at the cathode peak, the velocity being determined over the stretch of 0.5 cm at 2 cm from the anode to 2.5 cm from the anode, from 2.5 cm to 3.0 cm, and so on. These data are in good agreement with Waidmann's findings for streamers that cross the gap.

Again by using the current signals the average velocities across the gap were determined as a function of gap length. In all these cases the capacity was sensibly the same and the same spheres were used.

Gap Length, cm	Velocity, cm/sec
1.9	6.3×10^7
2.8	4.7×10^7
3.1	4×10^7
3.5	3.4×10^7
3.8	2.9×10^7
4.5	2.1×10^7
4.9	1.63×10^7

The velocity in the last increment of distance before the cathode can be had from the succeeding traces. These data are subject to some fluctuation. The values are

2.8–3.1	1.87×10^7
3.1–3.5	1.67×10^7
3.5–3.8	1×10^7
3.8–4.5	7.8×10^6
4.5–4.9	5×10^6

These values are in good agreement with Nasser's findings from the Lichten-berg figures. As the gap lengthens for approximately constant potential, the average velocity of the streamers decreases and the streamers are very much reduced in velocity near the end of their paths at the cathode. For more vigorous streamers and shorter gaps the velocity is more nearly constant across the gap. These results also agree well with Hudson's observations on primary streamers.

These agreements lead one to the conclusion that in all these varied techniques the same phenomenon, called the primary streamer, is being observed. What are the secondary streamers of Hudson and the leaders of Kritzinger? In this direction Dawson applied his two-photomultiplier tech-nique with results of considerable importance. Before discussing these, we will consider what Dawson observed in still photographs of streamers with different gap lengths. In many ways they resembled those of Winn. A number of different series were taken, the first set beginning at long gaps apparently with smaller capacity observing single breakdowns. Here at the longest gap there was a very short, straight, bright trunk at the anode with a few very faint diffuse streamers extending into the gap for some distance. As the gap was decreased the streamers became slightly clearer but the bright, straight trunk lengthened progressively to about double its initial length, although there was no evidence that the streamers reached the cathode. As soon as streamers crossed the gap a very few of the longest streamers showed bright flashes at the cathode and became distinctly more visible. The brighter ones could be seen to prolong the trunk. With still shorter gaps more and more streamers and/or branches reached the cathode with more bright flashes. Some of these flashes appeared to extend their intensity part way up their respective branches. At this point the bright trunk had advanced a third of the way across the gap and several branches had formed a brightly branched trunk system connected to the brightest branches. It was clear that here return strokes had been active in making filaments visible and in extending the leader trunks of Kritzinger.

Dawson observed two other series with measured gap length. The one at a maximum gap of 3.4 cm showed a 3 mm long straight bright trunk. Here streamers did not cross the gap but extended faintly out across only one-half of the gap. The upper ends of the streamers were brighter near the anode but rapidly faded. At 2.8 cm five streamer branches clearly crossed and the bright flashes could be seen. Here the trunk had fissioned into two bright branches and advanced slightly into the gap. At 2.2 cm about eight streamer branches had reached the cathode. There were three bright trunk branches now 5 mm long. The flashes at the cathode increased in luminosity as the gap shortened. At 1.3 cm there was one very bright streamer straight across the gap. The trunk extended halfway across the gap. Here the cathode luminosity extended in a strong fashion for 2 mm up the streamer from the cathode. Note that these represent a series of separate photographs at each gap length and that no two streamer systems photographed are identical but they show similar behavior. With gaps from 5.4 to 1 cm in length, the sequence was similar. Here streamers did not cross until 2.2 cm was reached. Individual streamers varied widely irrespective of distance. With the largest capacity used and repeated exposures to a number of repeated streamers at different gap lengths extending from 10 to about 2 cm, the many superposed tracks made the photographs more legible. Streamers crossed at about 4 cm. At this point it was clear that the cathode had exerted considerable influence on the streamer tips. Here they all had paths normal to the cathode surface for at least 1 cm. The same held for shorter gaps. The central trunk at d = 2.6 cm extended halfway across the gap. At 1.9 cm the bright trunks crossed the whole gap. For a 6 cm gap the bright streamers extended out to 1.3 cm, but the streamer branches no longer reached the cathode and faded out so that they were no longer discernible beyond 3 cm. In the 10 cm gap the bright trunks extended out about 1 cm, and the streamers faded out by 1.5 cm. Here one notes the strong influence of the cathode and of the guiding gap field as well as of the tip potential on the progress and range of the streamers.

Using a 3 cm gap with a PM viewing various points out in the gap to perhaps 0.5 cm, Dawson was able to observe a sequence of what resembled Hudson's primary and secondary streamers. At the anode the two processes were not separated, but at 0.2, 0.3, 0.4, and 0.5 mm the primary pulse readily separated out from a slower pulse. The primary pulse was of fairly uniform height and in the earlier experiments appeared to have a duration of 12 nannosec, representing a transit time across the slit of about 6 nannosec. It was followed by a marked dip in luminosity, to nearly zero. Dawson states that in his many other observations there was virtually no recorded luminosity in the region between the primary and the subsequent gradual rise to a secondary light peak, which at 0.5 mm from the anode rose to a peak in 50 nannosec and had a total duration of 125 nannosec. Its amplitude initially was equal to or greater than the primary, but at 0.5 mm it was definitely declining. In none of Dawson's gaps did this secondary reach the cathode. In the shortest gaps it may have crossed two-thirds of the way across the gap. In all cases it had a much longer duration than the primary streamer. Its speed was difficult to estimate, but if one took its rise point the speed was 10^7 cm/sec or less as against a speed five or six times as great for the primary near the anode. According to Dawson, with improved techniques the effective length of the primary streamer tip pulse, which was not corrected for geometry of the optical system, was 0.5 mm, and that of the secondary was about 12.5 mm. These can be compared to Kritzinger's values with his heavy currents and long gaps of 1 cm and 10 cm.

Dawson carried on some further crucial measurements. These were taken under great technical difficulties, where, for example, in long gaps the region viewed was not much more than 1 mm, which begins to be comparable with the size of the region viewed. He cut his PM slit down to 0.5 mm in length so that he viewed an area 0.5 mm by 0.5 mm square at various points along the axis and off-axis as well. In all cases off-axis he observed an occasional light pulse due to a primary streamer *followed invariably by a secondary of the type described above.* The streamers were not as clearly defined because of the weak luminosities involved. Not every streamer that started was visible. It was only when a chance streamer branch passed the region viewed that both primary and secondary were seen. Many times no branches passed the field of view. When the aperture was placed along the axis, insofar as the streamers were axial farther out in the gap, streamers would pass, followed by an attenuated secondary, much more frequently than off-axis. However, these secondaries were never observed to reach the cathode with sufficient intensity to be recognizable.

Finally the primary and secondary sequence was always noted close to the anode, whether the primaries crossed the gap or not, if the gap were viewed close enough to the anode. When the primaries crossed the gap and return strokes up the branches occurred, the secondaries appeared more prominently and extended farther into the gap.

From all these observations certain conclusions may be drawn:

(1) The secondary streamers observed by Dawson with PM and photographed as bright trunks by Hudson, Kritzinger, the Goldman group, Winn, and Dawson under differing conditions are one and the same phenomenon. They are essentially what Kritzinger called leaders.

(2) Despite the evidence offered by Kritzinger and by Winn's photographs, these bright trunks do not first appear when streamers cross the gap and yield return strokes. They actually *appear before this* and are an inherent property of streamer advance.

(3) In the Goldman photographs as well as those of Dawson they appear before streamers cross, and they begin to develop as the streamers extend farther and farther out into the gap. The identification with secondary streamers is established by the PM studies of Dawson. The fact that they occur off-axis occasionally and axially more frequently, being preceded by the primary streamer which delineates the path, is a most convincing argument.

(4) These secondary streamers are therefore not to be associated with the delayed arrival of radial streamer branches, but are part of the streamer system themselves.

(5) The fact that in Dawson's observations the bright trunks did not cross and the secondary streamers also did not cross adds further proof of their identity. Had Dawson had more energy in his capacity to give a

more powerful arc, or main stroke, as did both Hudson and Kritzinger, his secondary streamers would also have crossed.

(6) Before the streamer crosses the gap, if it is sufficiently vigorous to advance into the gap and give many branches, the current in the trunk near the anode will increase continually. It probably increases in bursts of current from excess electrons poured into the streamer channels at each branching process. The conductivity near the anode will thus increase as current increases. When this current and conductivity reach an adequate level, the field at the anode causes further ionization. Thus the anode end or stem of the branched streamer system in time becomes sufficiently ionized to be luminous. This luminosity fades as the energy is drained from the capacity and the potential falls.

(7) At the point where streamers cross, the succession of return strokes up the branches still further increases the conductivity of the most highly ionized upper region of the streamer trunk, and the luminosity rapidly extends farther into the gap.

(8) At the point where the most conducting branch of this streamer system reaches the cathode, it is probable that the spark channel is sufficiently ionized to lead to a power arc if the energy available is adequate.

(9) This interpretation of the secondary streamer is compatible with Nasser's Lichtenberg figure traces. Here it is noted there are not and cannot be any return strokes. Streamers become very vigorous and the current and luminosity continually increase along the streamer channel as the current from the branches augments. Thus the broadening of Nasser's Lichtenberg figure traces as the anode is approached is due to halation from the ever-increasing conducting channel as the anode is approached. Since the nature of the tracks created in the first place by bright streamer tips that diminish in energy as they progress outward, followed by growing luminosity up the channel, causes sufficient darkening of the film, the apparent abrupt termination of the bright stems seen in direct photographs is not seen in the Lichtenberg traces. The photomultiplier also does not evidence any abrupt termination of the secondaries as they approach the cathode. The signals are decreasingly weak nearer the cathode, which is more or less in agreement with Lichtenberg figures.

It is believed, therefore, that these findings in a large measure clarify the difficulties encountered in the past.

4. All these new data together with some original thinking have led Dawson and Winn to achieve a first step in the theory of propagation of the streamer across the gap. This great advance came through Dawson's own observations, aided by Kritzinger's data, together with the concept that the primary streamer continues to advance in low field regions, where the influence of the applied field is virtually zero. Until it approaches the

cathode and senses the cathode-to-tip-potential created field, it progresses as an ion creating mechanism, carrying with it a small mass of excited and ionized gas and leaving behind a virtually nonluminous conducting channel of plasma. The question then is, by virtue of what properties does this streamer advance?

Dawson and Winn assume that the advancing streamer tip consists of a small sphere of positive space charge of radius r_0 backed up by a non-luminous neutral plasma of electrons and ions created in its advance. It is shown that by a suitable choice of the charge Q_0 and radius r_0 of the space charge sphere the sphere will propagate itself by virtue of its space charge field and adequately placed photoionization. If one assumes such a sphere of n_0 positive ions with its center at a given point, then it will have ahead of itself an electrical field owing to the space charge. This field will attract photoelectrons created in a critical small, shell-like region ahead of itself, and for simplicity only one of these theoretically will advance in the field to create an avalanche of electrons and ions. The field of the space charge attenuates rapidly, so that only at some critical field strength and distance will photoelectrons created by photons during the growth of that space charge be able to start an avalanche that will propagate the charge. The critical distance is designated as r_1, and this was chosen as a point where the field of the space charge reaches a value at which dissociative electron attachment is just equal to the first Townsend coefficient for ionization by collision. At distances less than r_1 the value of α exceeds loss to attachment and the avalanche begins to grow. Geballe and Harrison[3.58] fix this ratio of needed field strength to pressure X/p as about 30 kv/cm $\times$ mm of Hg. The avalanche will continue to grow in magnitude as $e^{\int \alpha dx}$ in the increasing space charge field. To be accurate, classical values of α can be used until the avalanche reaches 10^6 electrons. From there on Tholl's value α' must be used until streamer forming proportions of around 3×10^8 ions are reached. For simplicity Dawson did not make this correction, as it is in fact not too critical. In a more accurate study it should be used. However, in advancing from r_1 toward r_0 (the radius of the assumed spherical space charge), the avalanche will expand in radius because of electron diffusion. The value of r_0 can be calculated from the relation $r_0 = \sqrt{6Dt} = \sqrt{6 \int (D/v) dr}$. The limits of this integration are to be from r_1 to a distance Δ from the center of the space charge at which the integral $e^{\int_r^\Delta \alpha dx} = n_0$, the number of positive ions in the sphere to be propagated. This is necessary in order that three basic conditions for propagation are fulfilled. These are that (1) the space charge must be maintained at its value n_0 positive ions; (2) the diffusive radius must not become larger than r_0, for if it does even if n_0 is the same, the field in advance will not be maintained; (3) the amplification must reach n_0 before the electrons of the new avalanche have penetrated seriously into the initial sphere of space charge, that is, $2r_0 \leq \Delta$.

The charge on the sphere Q_0 is merely the number of ions n_0 times the electronic charge. The value of the potential at the surface of the sphere $V = Q_0/r_0$; the field strength at the surface is Q_0/r_0^2. The energy of the charge inside r_0 is $Q_0^2/2r_0$. The electron velocity can be had from the data of Wagner, or extrapolated values with $v = k\sqrt{X/p}$ can be used. Calculations yield the table given here.

n_0	10^6	10^7	10^8	2×10^8	10^9	assumed
r_1 in cm	1/400	1/125	1/40	1/28	1/12.3	calc.
Δ in cm	10^{-4}	1.5×10^{-3}	6×10^{-3}	10^{-2}	2.9×10^{-2}	calc.
r_0 in cm	1×10^{-3}	1.6×10^{-3}	2.85×10^{-3}	3.4×10^{-3}	—	calc.

Here various possible values of n_0 are chosen and the values of r_1, Δ, and r_0 are calculated. It is at once apparent that for $r_0 = 10^6$ and 10^7 $2r_0 < \Delta$. This is not permitted. Hence the sphere must contain more than 10^7 ions to propagate itself. For $n_0 \geq 10^8$ all results of the calculations are self-consistent and reasonable in the light of recent data.

(1) The absorption coefficients for the most active radiations for photoionization by Przybylski[3.52] taken from sparks are about 8 cm^{-1} and 5 cm^{-1} for O_2 and N_2 radiations acting on O_2 in *air*. The photoionizing efficiencies of these are about 5×10^{-5} and 1×10^{-4}. The number of photoelectrons created within 0.001 cm of r_1 (about $\frac{1}{6}$ of Δ) at r_1 in a solid angle of 1 radian will then be about 50 for N_2 radiations acting on O_2. This appears rather large in view of the small branching ratios. However not much is known about the photoionizing efficiency in an avalanche compared to the sparks used by Przybylski, and even less about the number of photoelectrons that are adequate for propagation. However, the number must be such as to give the chance of branching observed, as discussed further in Appendix IV.

(2) The radius r_0 of the sphere is here about 2.85×10^{-3} cm, while the value of the radius of the streamer from Winn's tracks is 3.5×10^{-3} cm, which is also in reasonably good agreement.

(3) Since Δ, the length of the ionizing advance between space charge centers for each new avalanche, is 6×10^{-3} cm, one can calculate the number of excess ions per centimeter path of streamer advance as $10^8/6 \times 10^{-3}$ or 1.7×10^{10} ions per centimeter of streamer. Values calculated from observed lengths of pre-onset streamers by camera, visually, and by photomultiplier by Kip, English, and Hudson range from 5×10^9 to 1.5×10^{10} per centimeter. These values represent the only measured number of ions per streamer which can be compared to these here treated. These excess ions imply that perhaps from 2 to 5 times that number of ions and electrons are created at each step, as all 10^8 electrons cannot be separated from the ionic plasma created.

(4) The value n_0 of 10^8 ions is fortuitously in agreement with the value for streamer advance set by Raether's group and established for

uniform field studies by the calculations of the author and Meek as early as 1939.

(5) There is, however, one discrepancy that needs clarification in the light of this theory. Measurements by English and others for the threshold of pre-onset streamers lead to a value of $e^{\int \alpha dr}$ of the order of 10^5 or so ions. This is incompatible with the 10^8 here indicated. It should be recalled that the formative time lag of the pre-onset streamers has been shown by both Menes and Miyoshi to be around 2×10^{-7} sec. This is much longer than the time required for a single avalanche of electrons advancing at most 0.02 cm, which is around 3×10^{-9} sec. Dawson has indicated what must happen. The value of V_S for the 1 mm diameter point is close to 8 kv, or perhaps a little less. If Tholl's α· is used, the value of r_1 for avalanche growth is ≈ 0.07 cm, and $e^{\int \alpha dx} = e^{13.6} = 8 \times 10^5$ ions. Here r_0 would be very roughly 1×10^{-2} cm. Hence the value of the space charge potential will be 12 volts at r_0, and the field at r_0 will be 1200 volts/cm. This will add to the field at the point and augment the next avalanche. At first the rate of increase will be small. But this increase takes place in 3×10^{-9} sec, so that one hundred sequential acts can take place in the 3×10^{-7} sec of formative lag. The rate of growth indicated above would probably be too slow. However, the zone for the creation of avalanches is not as critical with the point of radius 0.05 cm and its f_1 in excess of 0.07 cm. Thus several simultaneous photoelectrons are probably produced and these converge to yield perhaps 10^7 electrons and ions at the point in 0.01 cm radius. In this event $n_0 = 10^7$, and r_0 would still be ≈ 0.01 cm. The value of V would be increased to 120 volts, and X at r_0 to perhaps 12000 kv/cm, or 10% of the field at the anode surface. Under these conditions with convergence of more electrons, the needed space charge of 10^8 in an r_0 of 3×10^{-3} cm could be built up in 3×10^{-7} sec.

One must now turn to the energetics of the streamer ball in order to consider the range of its self-propagation at zero external field. Two approaches are to be employed, with somewhat different implications.

The energy to create the density of charge of 10^8 electrons in 0.00285 cm is given as $Q_0^2/2r_0$ and is 0.41 ergs. Now the energy is dissipated in creating new ions and electrons plus the accompanying photons. For as each Δ step is accomplished, 10^8 ions, electrons, and photons are created. The 10^8 electrons ideally go to neutralize the previous ball of 10^8 ions and leave the new positive sphere with center having an advance in the amount of Δ. Thus the energy at each Δ step goes to create 10^8 ions and 10^8 photons of roughly 15.5 ev each. Thus $2 \times 10^8 \times 1.6 \times 10^{-12} = 0.46 \times 10^{-2}$ ergs of energy are expended per step. That is, at each step Δ there is dissipated roughly $0.46 \times 10^{-2}/0.41$, or 0.01 of the energy in the sphere of n_0 of ions that propagates itself. Call this fraction ϕ. Then after each step $(1 - \phi)$ of the energy remains. After g discrete steps of Δ there will be a reduction of

$(1 - \phi)^g$ of the energy. Call U_i the initial energy of the sphere and U_f the final energy at which advance ceases. Thus $(1 - \phi)^g = U_f/U_i$. Here U_f/U_i is proportional to $(n_{0f}/n_{0i})^2$. Whence $(1 - \phi)^g = (n_{0f}/n_{0i})^2$. The range of a streamer going from U_{0i} to U_{0f} in g generations or steps will be $g\Delta = S$. If we use $n_{0i} = 10^8$ and $n_{0f} = 10^7$ as indicated by the table $(n_{0f}/n_{0i})^2 = \frac{1}{100}$. This leads to a g of 420. If $\Delta = 0.006$ (actually the value of Δ declines from 6×10^{-3} to 1.5×10^{-3} in this range), then $g\Delta = S = 2.5$ cm, which is a maximum value. The value of S is sensitive to the value of Δ but insensitive to the values of U_f/U_i because of the logarithmic ratio. If $U_f/U_i = 1/10^4$, as would be the case for change from 10^9 to 10^7 ions, the value of S at $\Delta = 0.006$ would be 4 cm. However, it is clear that Δ is not constant in this region and declines from 2.9×10^{-2} to 1.5×10^{-3}. The actual energy decline will be rapid at first but will decrease as energy decreases. Hence the shorter values of Δ will predominate and the path luminosity will decline gradually toward the end, as it is seen to do. Hence the range of such a streamer with a starting $n_0 = 10^9$ would best take $\Delta = 3 \times 10^{-3}$. S would then be 8 cm.

There are no data for comparison, and the question of S for a zero field streamer is in doubt. All this assumes a single streamer channel with no branching. It is possible using pulsed square wave potentials of duration t sec, for example, 40 nannosec, to record with photomultiplier the rise and decline of the pulse and to observe the time and distance S_a of actual advance of the streamer. Waidmann only observed S_a and assumed that it occurred in t sec. Thus in this technique one can observe on midgap ending streamers the actual time t_a of arrival of the tip at its farthest photo-multiplier-detected S_a if the pulse signals record as pips because of the rise and decline of noise signals on the photomultiplier record, marking the beginning and end of t. Then the advance at zero field is obtained by deducting vt from S_a. Here v is the velocity of the streamer to S_a given by $v = S_a/t_a$. Whence S calculated above can roughly be compared to $S_a - S_{at}/ta$. Since between t and t_a the speed may fall, there may be some error, but this is probably not too great. Waidmann using the Lichtenberg figure technique has estimated v from S_a/t with S_a given by darkening on the film. His v therefore is too high in the ratio $S_a/(S_a - S)$. The measurement is exceedingly difficult because of electrical pulse noise levels compared to intensity of the PM signal.

A preliminary measurement by Dawson, using Waidmann's 40 nannosec pulse with 40 kv on a 2 mm diameter point and a 5 cm gap, indicated the following: S_a was close to 4.5 cm and perhaps longer, and $t_a = 60$ nannosec. This gives v_a as 7.5×10^7 cm/sec. In the 40 nannosec of pulse time the streamer tip advanced 3.5 cm. Thus the tip advanced 1.5 cm after the anode potential fell to zero. However, since the high field region during t extends only 1.5 cm from the anode, the advance in low and zero field was more nearly $\approx$3 cm. This rather well confirms the values of advance of 2.15 cm in zero field given by the theory. Waidmann's velocity as computed from S_a/t was about 1.12×10^8 cm/sec instead of 7.5×10^7, if his photographic end point of the streamer coincides with the photomultiplier end.

A new phenomenon appeared during this study. When the PM was anywhere between the anode surface and a point within about 1 cm of the anode tip, a *second light* pulse was observed. This moved from the anode toward the cathode. This light pulse started at exactly 40 nannosec, that is, at the end of the potential pulse t long and when the anode potential fell to zero. This pulse progressed to about 1 cm from the anode before falling to zero in intensity, having started with an intensity just less than that of the initial streamer. The time taken to traverse the 1 cm from the anode was not much more than 2 nannosec. This new pulse had a speed of at least 5×10^8 cm/sec.

The interpretation of this pulse is simple. The ionized trunk of the streamer system that in 40 nannosec had progressed 3 cm into the gap was somewhat conducting and at a potential gradient paralleling that of the strong field decline out from the anode. When the anode potential went to zero the field between channel and anode reversed. This projected an ionizing space wave of potential down the channel for the extent of what had been the high field region of the point. If one assumes that the high created field extended over about $r_0 = 0.003$ cm and the wave progressed with a speed of 5×10^8 cm/sec and the electron velocity was 1.5×10^7 cm/sec, then from the relation $v_w = x_0 v e n^{1/3}$ the value of n turns out to be 1.3×10^{12} electrons/cm^3. This value of ion density is perhaps low when one considers that the small tip created somewhere around 10^{10} ions/cm length of channel, that is, initially the ion density n is 10^{15} ions/cm^3 in the sphere of radius r_0 of space charge. However, it is probable that n in the channel 30 nannosec after the tip had passed n had decreased materially. If the coefficient of recombination of $O^- - O_2^+$ or $O^- - N_2^+$ and the dissociative recombination of $N_2^+ + e$ is as observed under normal conditions $\approx 10^{-6}$, then by recombination n will have been reduced by a factor of 0.025. If in addition the channel by diffusion and ion self-repulsion expanded from r_0 to a value $3r_0$ in 40 nannosec, the density computed would be reasonable. That a channel with 10^{12} electrons and ions per cubic centimeter should be relatively dark is not unexpected. The radiation and emission due to excitation died away in 10 nannosec, so that behind the bright tip of 0.5 mm length there was no more strong radiation. As the recombination processes lead to little or no light, there will be no emission unless the field gradient in the dark channel exceeds $X/p \approx 20$. Such gradients cannot exist in these channels, as was long ago shown by the author.

The question of streamer range may be extended to include the whole travel of the streamer for a given tip potential in the presence of the field. It was noted that, if the cathode influence is avoided by using a large d, S is proportional to V_0 at the anode both in axial and total range of a single channel. The total input of energy at the anode on the theory proposed for zero field cannot be more than $\int_{V_0}^{0} Q_0 dV$, and, if Q_0 is constant, as required for a given sphere of n_0 charges, $Q_0 V_0$ is the energy input. Again, if the ionization energy is assumed to be ϵ_a and is doubled for excitation if the energy loss step is Δ, disregarding branching temporarily, the loss is $2\epsilon_a n_0 (S_a/\Delta)$. If now $Q_0 V_0$ is expressed in electron volts to equate with loss, $Q_0 V_0$ becomes $n_0 V_0$, so that $n_0 V_0 = 2\epsilon_a n_0 (S_a/\Delta)$ or $S_a = V_0 \Delta / 2\epsilon_a$ in centimeters. If one chooses $\Delta = 0.006$ and V_0 is 25 kv, the value of S computed is 5 cm. The observed S_a for 25 kv is 5 cm. This is a fortuitously good agreement. There are a number of flaws in this oversimplified calculation but not in the principle involved. It will be noted that n_0 *appears* to drop out of the equation. It is, however, implicitly contained in the value of Δ. Again Δ is a variable but does not change too greatly with n_0. It could alter the value S_a computed by a factor of perhaps two or three.

The real error lies in the fact that a streamer from the anode whose tips reach a maximum range of S_a will branch and each branch will contribute to the dissipation of energy in proportion to its length because of its creation of new ions. Thus the quantity S in the equation should be S_b, which represents the *total* length of the streamer and all its branches. No accurate data on S_b/S_a are at hand, but a value of 10 is reasonably approximate. Thus the value 5 chosen for S_a is in error and $S_b = V_0 \Delta / 2\epsilon_a$. This equation as applied above is wrong in order of magnitude, despite the fact that as applied to the unbranched streamer it yields S_a correctly.

If one neglects any influence of the tip field of the streamer caused by the anode potential, of which more will be said later, then one must ascribe the difference between the S_a and the S_b to the fact that for full-length branching streamers Q_0 and perhaps r_0 are not constant along the path, being greater at the anode. Not much amelioration can be had by altering r_0, and the energy content through r_0, in view of the size of the streamer tips. Doubtless it does vary. The only significant change can come in assuming that n_0 is initially 10^9, as it may well be near the anode, and that over its major range of advance it drops to 10^8. This would be in keeping with the more rapid initial dissipation in the region where branching is greatest, and the last of the advance where branching is less or zero. This situation appears from Winn's Lichtenberg figures along the axis. If this is done $V_0 n_{0i} = (2\epsilon_a n_{0f} S_b)/\Delta$, with $S_b = V_0 (n_{0i}/n_{0f})(1/2\epsilon_a)$. If Δ is used as 0.006, as before, on making $n_{0i}/n_{0f} = 10^9/10^8$, S_b at 25 kv becomes 50 cm instead of 5. This is probably reasonable, but statistical analysis of S_b/S_a must be made to check the theory.

One is still faced with the apparent paradox that V_0 is observably proportional to S_a if one keeps far away from the cathode. This relation is borne out by the fact that the streamer tip potential fall in longer gaps is sensibly that given by the relation between V_0 and S_a if the cathode region deviations are ignored. The only solution to the paradox can be had if in some fashion branching so varies with tip potential that there *is a constant ratio* between S_b and S_a, that is, that S_b/S_a is constant. A recent check indicates that S_b/S_a is not constant. Thus one cannot neglect the influence of anode potential.

So far the streamer has been discussed only where it moves without influence of the cathode and of any guiding field. However, this is an idealized situation. Even for the cases considered, the tip potential–x curves across the gap depart from the ideal linear fall indicated by the theory outlined. At lower potentials the sharper decline near the anode comes from a rapid decrease in n_0 and loss of energy owing to branching. The later departures indicate that the tip potential of the streamer is something more than just that of the sphere of charge n_0 having a radius r_0. This potential, though about 5 kv for $n_0 = 10^8$ at r_0, falls off rapidly and is ineffective beyond r_1. On the other hand, as the breakdown streamer intensity is approached, the streamers carry a large fraction of the potential of the anode out into the gap at their tip. This influence, aside from the relation between S_a and V_0, or S_x and V_x, is substantiated at any point x by a large number of observations. These are:

(1) Winn's and Nasser's observations that S_a increases when the tip approaches the cathode.

(2) The production of cathode or negative streamers that register on a double emulsion film when the heavy streamer tips strike the film, even if the film is at an x 1 or 2 cm short of the cathode. The field at the cathode is never that high *until a streamer approaches it.*

(3) The brilliant flashes of light from the photoelectrons emitted by the cathode as the streamer tips reach it, and the presence of return strokes in consequence.

(4) The small electron tracks in a double emulsion on the cathode side when a heavy positive streamer strikes the anode side. The same electron tracks in the emulsion are recorded in a red-sensitive plate facing the anode. This shows virtually no blue light positive streamers, but each point of streamer impact has the minute dot-like electron tracks.

(5) By the use of a single emulsion facing the anode, there is further proof that emulsion tracks are produced by high fields. If the anode side has black paper on it the dots appear at the point of streamer impact, though no light penetrates the paper. If the paper is replaced by a thin extensive piece of aluminum foil *no traces are observed*, as the field does not penetrate. If a cellophane tape is placed on the anode side over the metal foil the dots are produced in profusion, as the metal by induction projects

the high field of the impacting streamer tip on the tape through to the emulsion.

(6) The observations of Kritzinger, of Dawson, and of Waidmann that, as the near breakdown primary streamer tips approach the cathode, they increase their speed. Kritzinger and Hudson observed that they increased their luminosity as well.

Since the field of the isolated ball of charge is incapable of achieving such effects, the anode field must in some fashion be projected out to the tip.

Now the tip of the breakdown type of streamer is in fact not isolated from the anode. There is a continuous conducting channel of plasma that feeds to the anode the current of electrons generated by the tip and permits continuity of advance. The fact that a current flows means that there is a potential gradient along this channel. Call the resistance of a section of the channel 1 cm long w_0 and let the current i flow in it, assuming w_0 independent of x the distance of the tip from the anode. Then for a single unbranched channel, if one designates V_{r0} as the surface potential of the sphere or charge of radius r_0, one can write that V_x, the tip potential at any point x in the gap, is of the form $V_0 - iw_0x + V_{r0} = V_x$. If the potential of the anode is suddenly reduced to zero, then V_0 and i fall to zero. If there are potential gradients in the channel caused by space charges, i will fall to zero a bit more slowly. It does so by an ionizing space wave of potential recently observed by Dawson. However, if V_0 is approximately 0, then $V_{r0} = V_x$, and the zero field advance calculated above and postulated and observed by Dawson occurs. If V_{r0} is known and V_x is estimated from the value of S calibrated in terms of potential, the value of V_0 gives iw_0x, and if i is known w_0 can be determined. In practice this is difficult, since the streamer branches. Thus as one proceeds from the tip at x up the channel, each branch adds its own quota of i to the channel. Accordingly i increases by steps along the channel, depending on the i of each branch as the anode is approached. It is also probable that, as i increases, w_0 will decrease. Thus, as the streamer tip advances, each branch as it develops injects a new quota of electrons into the channel. This must send a pulse of potential and added ionization up the channel for some distance, reducing w_0. Thereafter there is a steady current at that point from the branch as long as it advances, unless the branch itself branches. Thus as a heavily branched streamer moves out across the gap the current in the trunk near the anode is continually increasing with current surges as each branch develops. The values of w_0 and i are not constant but increase in time at any one x along the channel, while at the same time w_0 progressively decreases and i progressively increases along the channel at any time t as the anode is approached. This action and the surges of potential on branching are doubtless what keep the channel conducting during streamer advance. It is quite possible that toward the end of the advance, even before any return stroke from cathode impact, the trunk has become luminous at the anode. This in fact is what Dawson photographed, as is proven by his observations with a very small, square PM slit. The luminosity could be such as to produce the broadening of Nasser's Lichtenberg figures as the anode is approached. However, that broadening could as well in part be produced by the streamer tips of 10^9 ions as they flash over the surface, owing to halation from intense luminosity.

In order to estimate the average streamer current in any one streamer branch, the total anode current toward the end of the streamer advance should be divided by the number of branches. Perhaps a better value is the current per centimeter length, using the product of number of branches and their respective lengths. Dawson's study of the current as the streamer system crosses the gap indicates that this is also not constant. For a long gap the current decreases as the streamer slows down in midgap, but increases again as it moves faster near the cathode. The measurements of the current should be made by a resistor in series with the anode point, the high potential being applied to the plane. More study of current as a function of time could well be made.

In conclusion, for the breakdown type of streamers it is clear that if the anode potential is projected out into the gap, as Nasser's observations indicate, iw_0 must be decreasing as i increases. This increase in current and conductivity of the channel does not come from the tip except by branching. Most of it comes from work done by the tip anode potential difference on the excess electronic carriers falling down the potential gradient from the branches. That is, additional energy is being fed into the channel beyond the initial V_0q_0 that launched the streamer. Since there are difficulties in maintaining a constant ionization gradient of adequate value along the channel, the new work in ionization must come from the surges of electron current acted on by the anode–negative potential created by excess electrons back of the tip produced by branching. This is now indicated by attraction of weak positive streamer tips by stronger branching streamers and junction sparks when high potentials cause crowding of streamers.

Since this report was written in January, 1964, two significant points have been raised. Dawson observed that the time midgap streamers advanced after the potential fell at the end of the pulse was exceedingly short if streamers ended in midgap. If the streamers nearly crossed the gap so that they could "sense" the cathode (for example, if they came within 1 or 2 cm of the cathode), they advanced for a considerable time (tens of nannoseconds) and some distance. The data given in the report were within about 1 cm of the cathode. This fact and the observation of the advance of a light pulse down the streamer for a millimeter after cutoff at the end of the pulse may place a different interpretation on the results. It is possible that, although in midgap the external guiding field is low—so that advance of the streamer is seemingly in zero field—the potential fall along the streamer is higher than had been expected by Dawson. Thus advance depends on a conduction conditioned tip potential contribution from the anode. If this is the case the travel of the tip after *anode* potential has fallen may depend on the time taken for the collapse of the potential at the anode to travel as a space wave down the streamer channel. For midgap streamer termination due to the falling off of channel conductivity with branching, the space wave on field collapse at the anode is about 1 nannosec. The speed of the space wave will be perhaps 2×10^8 cm/sec, and the 2 cm traversed down the channel could take 5 to 10 nannosec, as observed. When the streamer reaches within 1 cm of the anode the time for the collapse of the field is somewhat longer so that travel of the tip can continue for a longer time, and the field between tip and cathode increases the speed, as has been indicated earlier. In addition, the observed variation of the advance time intervals near the cathode after cutoff indicates a considerable statistical fluctuation between streamers. The shortest advance times probably correspond to the less axial streamer branches.

Further investigations of Dawson, which are contained in three papers submitted for publication, one under the names of Dawson and Winn and two more under Dawson's name alone, lead to conclusions differing and extending what has gone before. The tip field of the *primary* streamer is only that depending on its charge n_1 or n_0 and radius r_0. The channel just back of it has such a high resistance that *it does not carry some of the anode point potential with it*. Further up the channel near the anode the merging currents of the branches may make the channel sufficiently conducting so that the tip of the secondary streamer which is just developing may have some of the tip potential. The primary streamer follows the zero field theory and receives little energy from the low field near the cathode or in midgap but tends to follow the lines of force. The energy of the tip sphere of charge is proportional to the point potential V, since in the high field region the initial charge n_0 increases to n_1 and later dissipates itself in the low field region as it progresses. Hence its axial range and also its total range is proportional to V. The ratio of total range, S_T, to axial range, S_a, is nearly constant and about 6 for larger points (≈ 1 mm diameter) with extended high field regions. For fine points it decreases from 6 to 1.4 as V decreases, and the number of branches decreases as n_0 is created too far from the point in such short-range fields.

The discussion above concerning the high fields at the tips of near breakdown streamers is correct but *applies to the secondary streamer* with its highly conducting stem channel. This is particularly marked after the primary crosses the gap and augments the secondary streamer by return strokes.

With Nasser's technique, using pulses with a long decay time for Lichtenberg figures, the figures do not differentiate between primary and secondary traces because the secondaries have fronts that are not steep and because influx of current from the branches causes masking. With Dawson's technique of using square wave impulses of about 1 nannosec rise and decline times and durations of 40 or more nannosec, the secondary tips are fairly sharply delineated and both primary and secondary portions can clearly be seen in the figures. Here the conducting portion of the secondary is illuminated by the ionizing space wave produced by collapse of V at the anode in about 1 nannosec. For further details, see Appendix IV.

REFERENCES

1. G. Waidmann, Z. Physik *179*, 102 (1964).

III

Further Developments on Water Drop Corona

Boulet and Jakubczyk[1] have made an extensive study of raindrop corona in connection with radio interference from high tension lines. A coaxial cylindrical system was set up with the high tension conductor in the center and the outer conductor 2 m in diameter so arranged that drops of water or mist and fog could be introduced from the outer cylinder. The central section was connected to ground through a set of connections for current measurement and oscillographic study. The ends of the conductor were shielded against corona and surrounded by guard rings. The central conductor was stranded ACSR 1,590,000 circular mils suspended by polyethylene strings. A 545A Tektronix oscilloscope was used. Two different radio noise meters were installed. Power losses could be measured by Lissajou figures. A high-speed movie camera photographed the drop behavior.

Very valuable quantitative relations were given for the following phenomena: (1) the local increase in field created by a water drop when it enters an electrical field; (2) the elongation of the drop in external fields exceeding a certain value; and (3) the electrostatic forces on drops in a field.

1. The field distortion in space surrounding a spherical drop of radius a in a uniform strength X_0 gives the intensity at any point distant r at an angle with the field as

$$X_a = -a_r X_0 \cos \phi \left[1 + \frac{2(\epsilon_2 - \epsilon_1)}{\epsilon_2 + 2\epsilon_1} \left(\frac{a}{r}\right)^3 \right]$$
$$+ a X_0 \sin \phi \left[1 - \frac{(\epsilon_2 - \epsilon_1)}{\epsilon_2 + 2\epsilon_1} \left(\frac{a}{r}\right)^3 \right].$$

Here a is the radius and r is the distance from the center. The maximum field near the surface at $\phi = 0$ and $a = r$ is then

$$X_m = X_0\left[1 + \frac{2(\epsilon_2 - \epsilon_1)}{\epsilon_2 + 2\epsilon_1}\right].$$

If $\epsilon_2 = 80$ for water and $\epsilon_1 = 1$ for air, thus

$$X_m = 2.92X_0.$$

The maximum field induced is independent of radius a but depends on the drop shape. The field for a spherical drop with a charge q is q/r^2, and its potential is q/r, both of which depend on its radius. This is not given by the authors but, since drops on the conductor and in the neighborhood may have a charge q, the field normal to the surface adds to that of the induced charge.

2. In an electrostatic field the drop elongates in the field direction and is compressed when its surface is parallel to the field lines. The distribution of the electrical surface pressure p_X is given by

$$p_X = 4.5(\epsilon_2 - \epsilon_1)^2[\tfrac{2}{3}X_n^2 - \tfrac{1}{3}X_t^2]10^{-4} \text{ gm/cm}^2.$$

Here X_n and X_t are the normal and tangential components of the field in kilovolts per centimeter. If S is the surface tension in dynes per centimeter on the surface of the spherical drop of radius r, the pressure under the surface is $2S/r$. If, for a deformed drop of radius of curvature, $p_X \geq 2S/r$ the drop becomes unstable and elongates to disruption. The maximum potential gradient X at the drop surface relative to a field X_0 acting on a drop deformed is an ellipsoid of revolution with axis b parallel to the field and axis c normal to it, as seen in the table.

c/b	X/X₀
0	1
0.5	1.6
1.0	2.9
1.5	4.2
2.0	5.3
2.5	6.5
3.0	7.3
3.5	8.2

Even for a large drop the change in shape starts at 5 kv/cm. Once this begins, the drop rapidly distorts to a point, droplets are ejected outward, and a corona starts.

3. The critical field strength gradient when instability occurs is directly proportional to the square root of the ratio of surface tension S to the radius of the drop a. That is, $X_c = k\sqrt{S/a}$, with X_c in kilovolts per centimeter, k = 0.477, with S in dynes per centimeter, and a in centimeters.

The value of S varies from 75.6 dynes/cm at 0° C to 58.8 dynes/cm at 100° C for pure water. It is radically changed by impurities, generally being reduced.

4. If the dimensions of a spherical drop are not too great, the force acting to move a sphere of charge e in non-uniform field of gradient dX/dx in gaussian units is given by

$$F = eX + \frac{1}{2}\left(\frac{\epsilon_2 - \epsilon_1}{\epsilon_2 + 2\epsilon_1}\right)a^3\left(\frac{dX}{dx}\right)^2.$$

Here e is the charge, X the local field, and a the drop radius, with ϵ_2 and ϵ_1 the dielectric constants of water and air. This force acts to attract or repel a particle, or water, droplet to the conductor, depending on the sign of e relative to X. Under this force the particle, according to Stokes' law, would move in a viscous medium such as air according to $F = 6\pi\eta av$. Here η is the coefficient of viscosity of air in CGS units, a is the radius in centimeters, and v is the velocity in centimeters per second.

This force increases the rate of water or fine dust collection by a conductor when a is sufficiently small as in fog, but its effect on the rapidly falling larger raindrops is negligible.

5. It is stated that the condition for a complete spark discharge of the drop as a result of elongation in the field is given by Schumann's integral:

$$\int_0^b \frac{1}{X^2} e^{-6300/X^2}\, dx = 1.86 \times 10^{-4}.$$

Here X is in kilovolts per centimeter and b is the electrode spacing.

If the influence of drop size on spatial distortion is considered, it appears that a spark discharge of considerable length can occur only in rain. Fog drops are too small to produce any corona noise by their discharge or to cause sparks by disruption. They can be collected by the field and condense to produce drops on the conductor surface, which can cause noise, and do.

From the authors' experimental findings it appears that, depending on the sign of the potential of the conductor, as a raindrop passes by, the breakdown forms either a streamer (during the positive half-cycle, according to the authors) or a faint negative pulse. The question of the potential phase will be discussed later. A 3 mm drop passing the stranded conductor at a distance of 1.5 cm, with a surface field of 18.4 kv/cm at the conductor, gave a series of streamer pulses producing up to 150 mA across a 33 ohm resistor (RC = 1.2 × 10⁻⁸ sec). There were two unresolved peaks, a minor one at 0.1 and the major peak at 0.2 μsec. Two or three later peaks occurred at 0.4, 0.7, and 1.0 μsec on the positive phase. On the negative phase there was one negative pulse peak of −50 mA at about 0.1 μsec followed by peaks at 0.4 and 0.7 μsec. The wave fronts observed appeared steeper than they were in reality, as they contained components flowing through the

shunt capacity in their measuring circuit. The onset potential of the negative pulses was little lower than for the positive pulses. The positive streamers appeared much more powerful than the negative and showed much
stronger luminosity, accompanied by an audible spark-like crack. These
are particularly notable when the first heavy drops fall down past a high
voltage line at the beginning of a thunderstorm.

It is not clear from the text whether the authors actually observed the
potential phase at the time of the induced positive streamer current by a
water drop falling past the conductor at 1.5 cm. It is suspected that they
assumed that phase was positive since the current pulse was positive. A
positive pulse would have resulted from a streamer *induced* by the falling
pointed and disrupting drop with *induced* positive polarity during the
negative phase of the conductor. The subsequent pulses came from droplet
portions of the initial disrupted larger drop, probably on the ascending
phase of the negative pulse, as they were smaller. The induced streamers on
the positive potential phase would create only negatively charged droplets
moving toward the conductor from a disrupted falling drop. These would
emit a weak streamer corona from the spindle *drawn out on the drop away
from the conductor*, yielding to a capture of negative charge and a negative
current. As the drop disrupted, the smaller droplets also could have given
weaker streamers in a direction away from the conductor. These pulses
would be expected to be weaker. As English showed, water does *not* yield
a *negative Trichel pulse corona*. Where the drops are *on a conductor* the
streamers occur on the positive potential phase, while weak induced streamers
occur on the negative potential phase from the disrupting and ejected
negative droplets.

The reason for the very heavy "pops" and discharges from the large
drops in a thunderstorm is that these drops are very large in comparison
to normal raindrops. They carry a heavy positive charge. Thus on the
negative phase they would yield very heavy corona discharge with many
noisy droplets. They might even yield small sparks.

According to the authors, both main discharges were preceded by small
pre-discharge current increases lasting a few tenths of a nannosecond. These
may have been the Hermstein glow discharges before potential reached the
disruptive stages, either inductive or direct for adhering drops.

The whole mechanism of discharges, especially for the positive streamers,
was random, and the superposition of separate streamers from components
of a single drop varied from one drop explosion to the next. The streamers,
which were single or multiple, played the dominant role in noise formation.
Multiple streamers from a water drop at higher potentials often consist of
a series of individual discharges separated in time by some tens of microseconds, and may continue over some 200 μsec. An example of ten such
discharges beginning with a very strong streamer was followed by nine

others of half and smaller declining amplitudes. These could be produced in numerous ways from the component droplets, or a chain of droplets could be ejected by the one drop in the direction of the conductor. As each smaller drop in the succession reached a disruption producing distance, it yielded its own streamer. It is unfortunate that flash photographs of the drops were not obtained in correlation with the electrical signal. The authors apparently failed to recognize the nature of the rupture such as noted by both English and Macky. With the ACSR 1,272,000 cm conductor at 20 kv rms/cm the discharges extended to 200 μsec. A high-speed movie film records these as pulses with very long unresolved luminosity. The higher the gradient, the longer the conductor; and the larger the drop, the greater was the discharge current, as would be expected.

The properties of adhering water were next studied. It collects on the upper surface as small drops, with larger ones on the lower surface. Water flows down between conductor strands supplying the lower drops, which fall off from time to time under gravity as they become large enough.

Contact angle of water and surface depends on the state of the conductor, becoming lower for oxidized aged conductors, as is to be expected. New conductors are inclined to be greasy from handling. If time is adequate the surface becomes completely wet. At this point the upper droplets disappear and only the lower droplets appear. As with the falling drops, the field causes distortion of these adhering drops, as English noted. The distortion is the same as for ellipsoids in free fall if the half replacing the diameter adhering to the conductor is cut off. However, the drops on upper and lower surfaces have the superposed gravitational forces which flatten the upper ones and elongate the lower ones. The lower drops thus have the higher fields and lead to the discharges. Peek has given the critical gradient as

$$X_W = 6.4[1 + (0.815/\sqrt{r})]$$

with X in kilovolts per centimeter and r the radius of the conductor in centimeters. The pulses from the disruption of droplets during positive potential phase are the same as for falling droplets on the negative phase. Negative pulses form by induction from the end of the shattered droplets nearest the conductor on the negative potential phase. These are naturally smaller in amplitude and much less frequent, as could be expected from their nature.

In general there is not much difference between the phenomena from passing and contact breaking droplets except that they occur on opposite potential phases and the falling droplets if of appropriate size may yield pulses much larger in amplitude and duration. The droplets separated by falling are, because of the concurrence of gravity and static forces, generally restricted in size while depending on the rain; passing drops can be materially larger.

The freely falling or passing drop discharges are random. Those falling from the conductor can form pulses which repeat at each positive peak of the conductor in calm weather. Such concerted action over a long stretch of suspended conductor line can by its electrostatic changes in field between conductor and earth at a regular frequency set up mechanical oscillations of the conductor. This produces the "dancing" observed at times even in windless, foggy weather where condensation leads to a steady creation of drops.

The drip will often continue for 15 min after rain has ceased to fall, especially if air conditions are quiet, so that radio interference will increase for a while after the rain is over. In falling rain there is usually a wind that rips the droplets off the conductor before they become sufficiently large to give the big pulses that come from regularly separating large drops.

Attempts were made by the authors to reduce the corona by two devices. They thought that the noiseless Hermstein sheath observed on finer wire conductors, which was not recognized as such, could be stimulated by winding fine wires about the conductor surface, thus suppressing the noise. Under fair weather conditions such a procedure had been *reported* effective, decreasing noise by a factor of 0.7. It, however, increased corona current loss. It turned out that a 1 mm diameter wire wound about the conductor with a 3 cm pitch not only increased corona current loss in rain but was even more noisy. The other device was to coat the conductor with a thin water-repelling coating of paraffin. Falling drop discharges were avoided by using a drizzle of drops of less than 0.2 mm diameter. The radio interference did decrease slightly, but the corona current losses increased. The drops hanging below the conductor disappeared. However, the water collected as small spheroidal droplets standing upright on the upper side of the wire, each yielding a streamer corona. Though drops were smaller, there were many more of them. There was thus still considerable noise and distinctly more current loss. After the rain ceased the drops continued to remain for longer periods and the radio interference persisted for a longer time compared to the usual situation, when gravity drained off the drops from below.

REFERENCES

1. L. Boulet and B. J. Jakubczyk, Alternating Current Corona in Foul Weather, Part I. Above Freezing Point, Presented at IEEE Winter Power Meeting, N.Y., Feb. 2–7, 1964.

IV

Inclusion of Photoionization into the Dawson–Winn Zero Field Theory and Other Recent Advances

The development of the low field streamer theory by Dawson and Winn (see Appendix II) led to further experimental investigations, namely in studies on the range variation with O_2 concentration to 1% O_2 content and to a further study of the pressure variation of streamer range in air by Winn. The Waidmann[1] logarithmic range against percentage of oxygen relation was confirmed with a needle point-to-plane gap down to 10% O_2. Beyond this concentration the range reached a peak at 7% and had declined by about 20% at 1%. Lower concentration studies were not profitable owing to inadequate control of H_2O vapor and contaminants. As pressure in air decreased from 760 mm, the streamer range increased following the law Sp = constant down to about 250 mm, except that there was a superposed curvature resulting from the change in partial pressure of O_2 corresponding closely to Waidmann's relation. Below this pressure the range fell very sharply. The streamers became broad and diffuse, and branching decreased. Below 100 mm the streamers were several millimeters in diameter. They disappeared, to be replaced by a glow discharge extending from anode to cathode between 50 and 30 mm pressure depending on point diameter. Analysis in terms of the low field theory of Dawson and Winn using two different approaches indicated that, when the pressure is varied such that $p = 760/h$, if n_0 varies in proportion to h, then r_1, r_2, and r_0 also vary proportional to h. Thus it follows that S is proportional to h and thus that Sp = constant. This, as stated, was observed, except

661

for the influence of the reduction of O_2 content which alters photoionization independent of changes in n_0, r_1, r_2, and r_0 derived from the physics of the avalanche theory. This discovery points up the urgent need for a theory combining photoionization and the electrostatic criteria of streamer advance. As indicated in Appendix II, such a theory first demands a continuity relation guaranteeing that one photoelectron be present in a certain critical volume to ensure the advance of the streamer, other conditions being favorable. It happens that the continuity equation must involve the n_0 and r_2 of Dawson and Winn's equation. By eliminating these quantities from the photoelectric continuity equation in combination with the range equation of Dawson and Winn, a complete theory of streamer advance can be arrived at.

The derivation itself must be made on simplifying assumptions which while correct are not *accurately* applicable to the conditions actually existing. The limitations will be seen as the theory is developed. These assumptions include that of a single type of photoionizing photon of absorption coefficient μ_{O_2} in O_2 and certain average values of distances that are distributed over a range of values in practice. Under these conditions the theory of continuity of photoionization is capable of fitting into the Dawson-Winn theory, yielding a complete single self-consistent theory. The absorption-range curves and the pressure-range law must be consistent with such a theory. Self-propagating conditions are set by the Dawson-Winn model.

In the initial work of Waidmann the shapes of the axial range S vs. mole fraction F_{O_2} of oxygen concentration suggested an absorption coefficient μ_{O_2} for the active photons, which agreed with Przybylski's observed values for N_2 and O_2 photons ionizing O_2. This led to the use of such photons as a starting point for calculations in connection with the continuity equation. This influenced the development of the theory for some time but also gave it its correct form.

The analysis may be started by using a zero position of a nominal sphere of charge of $n_0 = 10^8$ excess positive ions within a radius r_0 that will advance a distance $+r_2$ in its next ionizing step. The generating avalanche needed for this advance starts at r_1, where $E/p \approx 30$, or at some distance r_1^1 short of this which is such that the avalanche at r_2 will still yield n_0 ions within r_0. The generating space charge n_0 at the zero position to be advanced to r_2 emits $n_0 f_1$ photons which in a solid angle Ω between r_1^1 and r_1 will create *one* propagating photoelectron. Now the number of photoelectrons created between r_1 and r_1^1 from zero with an absorbing mean free path L will be $n_0 f_1 \dfrac{r_1 - r_1^1}{L} \exp - (\bar{r}/L)$, where $\bar{r}$ is an average between r_1^1 and r_1. However, not all these potential photoelectrons will be available. The average lifetime of the excited state that yields these photons is τ. The excited states in r_0 are created in 10^{-9} or less second by electron impacts, but radiate in time and according to exp $- t/\tau$. Here the time t is fixed by $\bar{l}/v$, where $\bar{l}$ is the distance that streamer advances at a tip velocity v from the time the particular sphere r_0 was created. In the relation above, $\bar{l}$ is at $\bar{r}$/cm distant from the zero point. The time t is thus $\bar{r}/v$ with $v \approx 10^7$ cm/sec and $\bar{r} \approx 10^{-2}$ cm, giving t as 10^{-9} sec. A reasonable guess at the value for τ might be between 10^{-7} and 10^{-8} sec, so that the fraction of the photons radiated in $\bar{r}/v$ is $[1 - \exp - (\bar{r}/v\tau)]$. This is a small fraction, of perhaps 2% for air at r_2 at atmospheric pressure. However, this sphere r_0 at zero is not the only one that sends photoelectrons to $r_1 - r_1^1$, since many spheres separated at intervals r_2 back from zero will each have created $n_0 f_1$ photons. Each of these, depending on their distance $- nr_2$ from zero, where n is an integer, will suffer alteration by absorption in reaching $r_1 - r_1^1$. However, being at greater distances nr_2 back of zero, they will have radiated a larger fraction of their quota; that is, $[1 - \exp - (nr_2 + \bar{r})/v\tau]$ will be larger.

The suggestive linear log S $-$ F_{O_2} curves observed by Waidmann above $F_{O_2} = 0.1$ led to an attempt by the author to evaluate the absorption coefficient μ_{O_2}. To do this the shape of the plot had to be yielded by a relation of the form S $= S_0 \exp - (\mu_{O_2}lF_{O_2})$, for any $N_2 - O_2$ mixture with l some fixed length. Since the only length that was identifiable with the streamer process at the time of Waidmann's work was that of the ionizing zone in front of the anode, r_{1_a}, this quantity was calculated by Waidmann from field conditions for 0.1 and 1.0 mm diameter points, as 0.06 and 0.1 cm, respectively. These led to values of μ_{O_2} of 39 and 19 cm^{-1}, which are fortuitously close to Przybylski's values of 38.5 and 25 cm^{-1}. These photons were accepted as being the active ones since, conveniently, the approximate values of f_1 of 1×10^{-3} and 2×10^{-3} were also given by Przybylski. Since the more complete Winn curve was to be used in deriving the theory, the slope of this, expressed as S $= S_0 \exp - 1.8 F_{O_2}$, was adopted. The N_2 photon in O_2 was chosen as the convenient one to use with $\mu_{O_2} = 25$ cm^{-1}, $1/\mu_{O_2} = L_{O_2}$ of 4×10^{-2}, making the l chosen for this curve 7.2×10^{-2} instead of 0.6×10^{-2}. Since Winn had used a needle point, this value was recognized as not justified and the l $= 7.2 \times 10^{-2}$ now had to be considered as an arbitrary scale factor, about which more will be said later. It will be noted that with this factor the value of $L_{O_2} = 1/\mu_{O_2} = 4 \times 10^{-2}$ cm of Przybylski is to be used in the considerations to follow and is made consistent with the Waidmann–Winn curves.

As noted, the photons accompanying the 10^8 ions in the sphere at zero will be reduced by absorption by exp $- r_1/L$ at the time they have reached the origin of the future avalanche. Here L $= L_{O_2}/F_{O_2}$. However, the spheres at $-nr_2$, from zero, where n is an integer, also contribute and their photons are reduced by a factor exp $- (nr_2 + r_1)/L$ at r_1. The temporal term in emission of the photons reduces the number of photons at r_1 by a factor $[1 - \exp(n + 1)r_2/v\tau]$, sensibly $[1 - \exp - (nr_2/v\tau)]$ for larger values of n. The product of the two factors, with the absorption term, is at its maximum $\exp - r_1/L$ at zero and decreases as n increases, while the temporal factor increases as n increases, leading to a curve which rises steeply from a low value at zero to a peak and declines for values of $-nr_2$ which exceed L appreciably. A study of these curves showed that this peak occurs at about $-\bar{l} = 0.6\ L$ for the values of F_{O_2} of 0.1, 0.2, and 0.6. This was also roughly substantiated by summing the contributions of the various values of n in the region between r_1 and r_1^1, using the product of attenuation by time and distance curve as a scale factor for each contribution. One may thus for the purposes of simplification consider that all the n spheres of 10^8 ions effectively emitting photons active in advancing the avalanche can be given by $n = \bar{l}/r_2$. These nn_0 ions emit nn_0f_1 photons that act as if they came from a point $-\bar{l} = 0.6\ L$ distant from r_1 as far as absorption and temporal loss factors are concerned.

The assumption of a single point of emission also identifies the solid angle Ω once a target area $\pi(ar_0)^2$ is fixed defining the zone from which the propagating avalanches must come. Here a is a numerical factor which will be determined by the statistics of branching and the diameter of the tracks. For convenience it can be set as unity but may be larger. Thus $\Omega = \dfrac{\pi(ar_0)^2}{4\pi\bar{l}^2}$. It is seen that this yields the continuity relation for photon production as

$$n_0f_1 \frac{(ar_0)^2}{4\bar{l}^2} \frac{\bar{l}}{r_2} \frac{(r_1 - r_1^1)}{L} \exp - (1.8F_{O_2})[1 - \exp - \bar{l}/v\tau] = 1.$$

Here the term $\dfrac{r_1 - r_1^1}{L} \exp - (1.8F_{O_2})$ represents the fraction of photoelectrons created between r_1 and r_1^1 as a result of absorption. If one inserts $\bar{l} = 0.6L$ and $L = L_{O_2}/F_{O_2}$, one may write

(1) $\quad n_0f_1 \dfrac{(ar_0)^2 F_{O_2}^2}{1.4L_{O_2}^2} \dfrac{0.6(r_1 - r_1^1)}{r_2} \exp - (1.8F_{O_2})\left[1 - \exp - \left(\dfrac{0.6L_{O_2}}{v\tau F_{O_2}}\right)\right] = 1.$

The axial range S from the anode tip following Dawson and Winn's theory reads

(2) $\qquad\qquad S = \dfrac{n^1eV}{2n_0eV_iS_t/S} r_2 \quad$ or $\quad r_2 = \dfrac{2n_0V_iS_t/S}{n^1V} S.$

Placing the value of r_2 into equation 1, one has

(3) $\quad S = \dfrac{n^1V}{2V_iS_t/S} f_1 \dfrac{0.6r_0^2 F_{O_2}^2}{1.4L_{O_2}^2} \dfrac{0.6(r_1 - r_1^1)}{r_2}$

$$\exp - (1.8F_{O_2})\left[1 - \exp - \left(\dfrac{0.6L_{O_2}}{v\tau F_{O_2}}\right)\right].$$

The n^1 appearing in this expression given by Dawson and Winn requires interpretation. Near the anode a streamer tip sphere of n_0 ions emerges at some small distance from the anode surface, having traversed a potential fall V_0. In the vector sum field of its own charge $\dfrac{n_0 e}{r_0^2}$ and the anode field E at the point where it finds itself, it continues to increase in charge with relatively little change in r until it emerges from the major portion of the sharp anode potential fall region having n_1 ions within a sphere of radius r not much greater than r_0. It is this sphere which propagates across the gap in essentially a zero field region. The energy of this sphere of charge is $(n_1 e)^2/2r$. This energy propagates the streamer to give a total range, including branches, of S_t and an axial range of S. The gain of energy from the rapidly changing field E and the changing charge are difficult to compute. It is clear, however, that we can regard it as depending on the energy gained by an average of n^1 ions lying in value between n_0 and n_1 which have fallen through a potential difference of $V - V_0$. Thus one can set $(n_1 e)^2/2r = (V - V_0)n^1 e$. In this case the average charge n^1 must have been calculated to include the loss of energy emitted in radiation while the tip developed from n_0. This expression relates the energy of the streamer to the potential through which it has fallen in terms of an average charge number. If V_0 is small compared to V (it is of the order of 0.1 except at low tip potentials for small points), then the range is sensibly proportional to V as observed for measured streamers. In calculating range it is perfectly correct to use $n^1 eV$ in place of the actual emergent energy to which it is equivalent. In principle this procedure evaluates n_1 as

$$ n_1 = \frac{2n^1 V}{\dfrac{n_1 e}{r}}. $$

This implies that the emerging sphere tip potential, $n_1 e/r$, is proportional to the applied anode potential, which appears to be a rational assumption. Evidence from range studies in air, neglecting the photoelectric ionization but assuming it is adequate, indicates n^1 to be about 5×10^8 ions. Current pulse measurements at the anode for streamers indicate values of the charge created $n_1 e$ to be on the order of 5×10^8 ions varying with V, while the total streamer charge is on the order of 3×10^{10} or more ions, at $V \approx 18$ kv.

The quantity $2V_i e$ is the energy in electron volts to create an ion pair plus an estimated equal amount of the energy lost as radiation. Thus $2n_0 V_i e$ is the energy expended without branching in advancing a tip r_2 cm, with e the electronic charge. The total number of ion pairs estimated above is consistent with n_1 only if the potential fallen through is less than 18 kv and if more than $V_i e$ per ion pair is expended in excitation and loss by branching.

If the photons in r_1 to $r_1{}^1$ are plentiful, as is the case with a large n_1, the rate of energy dissipation is high, not only because n_1 is large but through branching. If two simultaneous avalanches of about equal magnitude approach from different angles so as not mutually to interfere (they must be at $r_2 > 2r_0$ from each other and from the initiating charge), the streamer has two branches of nearly equal amplitude with tip charges of about $n_1/2$ each. Thereafter these progress independently as separate branches. Each of these in turn may branch again, increasing the number of branches exponentially but reducing tip fields in proportion to the avalanche sizes. As these progress and the charge is reduced, the number of simultaneous photoelectric avalanches converging decreases and thus the exponential increase in branching ceases. As the charge approaches n_0, branching ceases. Thus the last centimeter of path of the streamers is usually devoid of branches. However, actually the chance of two strong simultaneous avalanches is relatively small. For values of V on the order of 25 kv near the anode, branches appear about every twenty or more of the r_2 steps, and most of these are composed of short 1 mm branches and the main streamer. Thus branching leads to rapid dissipation in the anode region with perhaps some four to ten branches of length significantly greater than 5 mm.

A considerable study of streamer branching has been made for various potentials and point diameters. In this S_t represents the estimated total length of all the streamer branches and S represents the axial length of the longest streamers as observed by Waidmann and Winn. The 0.1 cm diameter point from 15 to 50 kv showed that S_t/S was nearly constant between 5 and 6. A needle point whose tip had an effective radius of curvature of perhaps about 0.05 mm had the ratio of S_t/S which varied from 1.4 with but few branches at 9.5 kv to about 6 with six branches at 35 kv. In all cases the value of S was proportional to V at the anode to within 10% or better. This signifies that for the fine point used, where the range of the ionizing field is small, and at low potentials, the streamer did not reach a propagating value n_0 close enough to the anode so that the greater part of the potential V could act on it. This implies a decreased average charge n^1e and thus a lower n_1. This means fewer branches but still maintains n_1 proportional to V. The same effect applies on reduction of pressure, since the geometry of the anode field is fixed while the ionizing region expands outward.

The values of the various quantities given by the Dawson and Winn theory for air are as follows: $r_0 = 3 \times 10^{-3}$, $r_2 = 6 \times 10^{-3}$, $n_0 = 10^8$, $r_1 = 2.5 \times 10^{-2}$, $r_1{}^1 = 1.25 \times 10^{-2}$, $v = 10^7$, and $L_{O_2} = 4 \times 10^{-2}$ as indicated above. The values for f_1 for N_2 and O_2 photons in O_2 have been given. The quantities above except for f_1 are not altered by varying F_{O_2} at constant pressure. However, it has been shown independently by Dawson and by Winn that when pressure at constant F_{O_2} varies as $p = 760/h$, if n_0 is made proportional to h, that is, at a given p, so that the charge $n_0 e$ becomes $n_0 h e$, it can be proven that r_0, r_1, r_1^1, r_2, and L are also multiplied by the factor h. Under these circumstances the value of S given by equation 2 makes S vary as h so that

(4) $Sp = $ constant

if n^1/n_0 is unchanged by p_0, that is, if n^1 is proportional to h. Finally if one assumes that the Przybylski photons are active the quantity f_1 varies as $2 \times 10^{-3}[(1 - F_{O_2}) + F_{O_2}/2]$, assuming that both photons act with nearly the same value of L. The variation of this quantity between $F_{O_2} = 0$ and $F_{O_2} = 1$ is relatively small, that is, from 2×10^{-3} to 1.0×10^{-3}.

A first test of the new relation (3) is to consider how it varies with p, keeping F_{O_2} (the relative $N_2 - O_2$ concentration) constant. The solid angle expression is unchanged, f_1 is unchanged, the $(r_1 - r_1^1)/r_2$ term is unchanged, and n^1 has been assumed proportional to h so that, except for the exponential terms, S is proportional to h, or $Sp = $ constant. The term $\exp = 1.8 F_{O_2}$ is actually $\exp - \left(\dfrac{r_{1_a}}{L_{O_2}} F_{O_2} \right)$. Thus as L_{O_2} is changed by pressure and is proportional to h, at constant F_{O_2}, this term varies as $\exp - 1.8 F_{O_2}/h$. The exponent decreases as h increases, so that the exponential term increases as p decreases. If the r_{1_a} indicated in this expression is not a constant but turns out more nearly correctly to be a quantity like r_1, this exponential term will be independent of p. The expression $\left[1 - \exp - \left(\dfrac{0.6 L_{O_2} h}{v \tau F_{O_2}} \right) \right]$ has the exponent increase in proportion to h. Since, as will later be seen, the exponent is so small that the whole factor varies nearly directly as the exponent, this term will increase as h increases. Thus whether or not the absorptive exponential increases there will be an increase in S with h due to the time factor. This, superposed on the law $Sp = $ constant for air from 760 mm down to less than about 300 mm, will result in a slow increase owing to the exponential factors. Beyond $F_{O_2} = 0.07$, if equation 3 applies, then, as observed experimentally from changes of F_{O_2} and increase in h, S decreases rapidly.

These considerations then raise the obvious question as to whether equations 1 and 3 are obeyed by using the values cited above. Numerically equation 1 appears to be within an order of magnitude agreement of yielding a unit value. However, the calculations for F_{O_2} of 0.1, 0.2, and 0.6 indicate that the expression 3 will not yield the log $S - F_{O_2}$ plot observed. While Winn's observations show a peak at $F_{O_2} = 0.07$ and a linear decline to $S_0/6$ at $F_{O_2} = 0.9$, the computed values show a peak at F_{O_2} around 0.6 with a steeper decline at higher F_{O_2} values and a slow decline at lower F_{O_2} values.

Further intensive studies of the properties of this relation indicate that the prime difficulty stems from the attempt to fit Przybylski's values of μ_{O_2} into the value of L used, and in the choice of the constant $l/L = 1.8$ to fit these curves. This choice, as indicated, came from the author's early use of the length of the anode ionizing zone for streamer advance, before the zero field theory, as applied to Waidmann's data. To evaluate L_{O_2} the appropriate length to use with the new zero field theory of Dawson and Winn is to replace $l = r_{1_a}$ by $(r_1 + r_1{}^1)/2 = 1.9 \times 10^{-2}$ cm. This is $1/3.8$ of the value 7.2×10^{-2} cm used to give the constant 1.8 for the Waidmann fit. Thus the value of μ_{O_2} active in this theory is $1.9 \times 10^{-2} \times \mu_{O_2} = 1.8$ *if slopes of the Waidmann–Winn curves are to be used.* Thus $\mu_{O_2} = 0.95 \times 10^2$ or 95 cm^{-1} instead of 25 cm^{-1}, and $L_{O_2} = 1.05 \times 10^{-2}$ instead of 4×10^{-2}. These are not the Przybylski photons.

The justification for using $\mu_{O_2} = 100$ cm^{-1} instead of 25 cm^{-1} is in line with Tamura's work[2] on the threshold for streamer corona with points of small radius. Since there are many photons in air below 1000 Å which ionize air, those effective in any given field range factor will vary with the scale used. Thus as F_{O_2} or pressures reach such values that, owing to distances involved, a particular photon ceases to be effective, a more appropriate one may take over the role.

Using μ_{O_2} about 300 cm^{-1} and $\tau = 10^{-8}$ sec, the curves for equations 1 and 3 peak at between $F_{O_2} = 0.01$ and $F_{O_2} = 0.1$. They, however, decline very steeply so that at $F_{O_2} = 0.6$ the range is $0.02S_0$ instead of $0.35S_0$ as observed. The S_0 is the maximum range yielded by the Waidmann type of equation $S = S_0 \exp - 1.8F_{O_2}$. It is clear in the light of what has been stated that Przybylski's photons are always present. They are not effective at low F_{O_2}, below 0.1. But when F_{O_2} gets much higher than 0.1 these add their quota. Thus using equations 1 and 3 far closer agreement would be obtained if the photons of μ_{O_2} 300, 39, and 25 were each used in equations 1 and 3 with appropriate f_1 factors and lifetimes τ and the resulting values as functions of F_{O_2} were added together.

Under such a procedure it would obviously be possible by trial and error to fit a group of f_1 values and τ values for these photons in such relative measure as to yield the Waidmann–Winn curves. Such procedure with arbitrarily chosen values would prove little except that the curve could be simulated by appropriately chosen constants. What are now needed for a test of the theory are more data on the f_1 values and μ_{O_2} values for the photons active.

It is believed, however, that while the theory is rough it is in principle correct and allows one to understand what occurs. With better basic data an *accurate* duplication of the curves would require a computer calculation. Such a calculation would not, however, permit a single equation combining the electrical streamer criterion and photoionization and thus permit a formulation of what is occurring.

As indicated at the end of Appendix II, the high tip fields carried out into the gap for streamers of breakdown proportions lie at the ends of the secondary streamers of Hudson, Kritzinger, Dawson, and the Goldman group. Nasser's Lichtenberg figures of slow decline square wave impulses do not reveal these because of the masking effect of various agencies, including branching. With square wave impulses of about 1 nannosec decline, the ionizing space wave from the collapse of potential V at the anode makes the secondary sufficiently luminous to leave a clear record. The wave attenuates when the conductivity of the secondary channel becomes too low where it merges into the primary channel.

As noted, however, although secondary streamers are an established reality and a feature of the breakdown, the photomultiplier studies are still confused by the late arrival of radial branch tips if long slits are used.

The Dawson–Winn theory received strong confirmation during the summer of 1964. Dawson used a point-to-plane gap with a circular hole cut in the center of the cathode plane. This was large enough to transmit the streamer and its most axial branches. Equidistant and beyond the cathode was a second plane plate parallel to the cathode. The circuitry was such that a square impulse given the point anode would launch a streamer that could just pass through the plane hole in the cathode. By proper delay lines tuned so that the tip had passed the cathode plane, a pulse was applied making the cathode plane positive to three plates beyond. This gave the attenuating streamer tip new energy. In this way the axial range of the streamer and its branches was multiplied threefold. Because of branching and the diffusion of arrival times, the timing was critical and only a few of the branch tips were able to advance. Because of the timing difficulties and branch attenuation, there is not much hope of extending the range further. The observation, however, gives strong evidence for the reality of the zero field streamer concept.

During the same period Winn fairly conclusively settled the question as to whether in uniform fields streamers show the same degree of branching as do those in point-to-plane geometry. The linear character of sparks in uniform field gaps indicated the absence of much branching. Winn used an anode plate with a hole in the axis through which a point anode could be projected various distances. The plane cathode parallel to the anode plane could be displaced at will to lengthen the gap. The streamer pulses were observed. Large gaps with the anode point projecting well into the gap gave branching as for a point-to-plane gap. As the point was retracted and the gap made more uniform by decreasing the gap length, branching decreased materially so that, for a point only a few millimeters beyond the plane with a short gap, usually only one streamer appeared, though occasionally there were one or two branches. This observation is in no way in contradiction to the zero field streamer theory. As the field becomes more homogeneous the whole field is weaker but extends further out into the gap. Avalanches lead to larger positive charges, r_0 will increase by diffusion, and when n_0 approaches 10^8 ions the streamer can advance. But its tip field will now, because of large r_0, be commensurate with the gap field, and only axial avalanches in the field direction can cause it to advance except for an occasional more radial one. That is, gradually one goes from a self-propagating zero field streamer in a high field to one more and more dependent on the gap field. This in the limit yields the Tholl-like streamers in place of the Dawson–Winn streamers. Radial branching will be much reduced.

While this book was going into production a very important article came to the author's notice on an aspect of corona study not included in earlier studies. It was entitled "Characteristics of Positive Corona for Electrical Precipitation at High Temperatures and Pressures," by C. C. Shale, W. S. Bowie, J. H. Holden, and G. R. Strimbeck and appeared as RI Bureau of Mines Report of Investigations 6397, of 1964. Its value is that it deals with the properties and characteristics of this corona under conditions not previously reported. It is a scholarly and competent piece of work despite its applied implications. It also has references which may be of value.

In this connection mention should be made of the study of corona phenomena in connection with electrical precipitation. This is presented in the only complete monograph on the subject, by Harry J. White, entitled *Industrial Electrostatic Precipitation* (Addison Wesley, Reading, Mass., 1963). White has directed the research laboratory of the Research Corporation, New York, and did much work in a scholarly study of coronas in connection with precipitation problems. The book also has valuable references.

REFERENCES

1. G. Waidmann, Dielectrics *1*, 81 (1963).
2. T. Tamura, J. Phys. Soc. Japan *17*, 1434 (1962); Jap. J. Appl. Phys. *2*, 492 (1963).

Indexes

AUTHOR INDEX

Italic numbers indicate pages on which full bibliographic citations are given.

673

SUBJECT INDEX

677